The Biochemistry of the Grape Berry

Edited By

Hernâni Gerós

Universidade do Minho
Portugal

M. Manuela Chaves

Universidade Nova de Lisboa
Portugal

Serge Delrot

University of Bordeaux
France

CONTENTS

CHAPTERS

FOREWORD

Good wines can only be produced from healthy and high quality grape berries, harvested at the ripening state with the optimal composition. Fortunately, in the last decades, wine production has rapidly incorporated scientific knowledge generated by the profound advances in scientific and technological developments experienced by different disciplines, particularly biology. Within this context "The Biochemistry of the Grape Berry" represents a timely publication updating the current knowledge not only on grape biochemistry, but also on the physiology and molecular biology of berry ripening. Edited by three specialists in the area and including contributions from recognized experts in the different covered topics, the eBook describes in depth all major constituents of the grape berry as well as their related physiological processes. New information derived from the available grapevine genome sequence and related high throughput technologies (*e.g.* transcriptomics and metabolomics) provide insights into the molecular basis of berry physiology and molecular biology. In addition, the eBook addresses the variation in the biochemistry and physiology of the berries resulting from genetic as well as environmental effects. The interaction between cultivar diversity and vineyard environmental conditions is fully relevant to explain the diversity of the world wines. This eBook will become a classic in viticulture and enology courses and laboratories, but it can also be of major interest to viticulture engineers and winemakers interested in understanding the biochemical and molecular basis of grape berry ripening.

José M. Martínez Zapater
Instituto de Ciencias de la Vid y del Vino (CSIC, UR, Gobierno de La Rioja)
Logroño
Spain

PREFACE

Grape berries are sophisticated biochemical factories of major economical importance. They import and accumulate water, minerals and sugar, and synthesise amino acids, organic acids, as well as flavour and aroma compounds. The most dramatic changes in the composition of grape berries occur after veraison, during the ripening phase. Berries switch from a status where they are small, hard and acidic, with little sugar, to a status where they are larger, softer, much sweeter, less acidic and strongly flavoured and coloured. The flavour that builds up in grapes is mostly the result of the acid/sugar balance, and the synthesis of flavour and aromatic compounds, or precursors taking place during ripening. Substantial scientific progress has been achieved in understanding the physiological, biochemical and molecular aspects of grape berry maturation. Some knowledge has led to the improvement of wine quality through better grape-growing practices, but this area of basic research is still wide open due to the complexity of the biochemical and molecular mechanisms involved in grape development and ripening, and their response to the environment.

Chapter 1 by Tyerman *et al.* is dedicated to the topic of water relations in the grape berry. This is a hot research topic as the water is the most important constituent of the fruit and, thereby, of wine. Grape berries normally contain 75–85% water, which is the main solvent of solutes including sugars, acids and phenolic compounds. Thus, quality, *via* the concentration of sugars and flavour compounds, and yield of the vintage are directly affected by berry water content at harvest. The paradigm of water transport across the biological membrane has recently changed, when the molecular basis (aquaporins) was first identified by the team of Peter Agre. The role of aquaporins in the regulation of berry water relations is also discussed in this chapter, with a particular emphasis on the varietal differences observed in aquaporin functioning. This is an interesting area that is still open to further research. Finally, the impact of water status of the parent vine on berry ripening is also discussed.

Chapter 2 by Martins *et al.* deals with the mineral compounds in the grape berry. Together with water, minerals are imported to grape berries, which is why mineral soil composition has a pivotal influence on grape quality and on the organoleptic properties of wine. Grape berries are extremely rich in K, which is involved in several physiological processes: enzyme activation, cellular membrane transport processes and translocation of assimilates, anion neutralisation, which is essential to the maintenance of membrane potential, and osmotic potential regulation, which controls the plant water relations, turgor maintenance and growth. Thus, the knowledge of the mechanisms mediating the transport of K^+ to grape berries and their regulation is of the utmost importance to develop strategies optimising the levels of this cation in berries and improving fruit quality, while maintaining the pH of the wine. In this chapter, the mechanisms of K^+ transport in plant cells, and in the grape tissues in particular, are reviewed and discussed. Besides K, the present chapter provides an overview of the dynamics of other important mineral compounds in the grapevine, especially in the berry, namely N, Ca, P, Mg and S, and of the contribution of each element to berry quality and yield.

The grapevine is a good example of a crop where sugar accumulation in the fruit has an important economic role. Massive sugar transport and compartmentation into the grape berry mesocarp cells start at veraison and continue until harvest. This topic is addressed in **Chapter 3**. How does the root system (rootstock) and the wood of the canes influence the sugar status of the fruit? How does the management of the aerial system affect the berry sugar level? How is sugar imported to the fruit, stored and metabolised? Our understanding of the mechanisms of carbon partitioning and accumulation in grapevines is improving due to the use of molecular techniques. In particular, the cloning of key genes and their expression in both homologous and heterologous systems have proven to be powerful tools, and the sequencing of the grapevine genome and the development of the high throughput techniques, such as microarrays and deep sequencing for gene expression analysis, will enable rapid progress on these subjects in the future. This chapter provides an updated review of the molecular biology of sugar transport and accumulation in the grape. A complex set of transporter proteins, invertases, and finely hormonal regulated mechanisms culminate in the fructose and glucose storage in the vacuole. At the end of the chapter, a model describing the possible routes for sugar import and accumulation during berry ripening is proposed.

Chapter 4 is devoted to the biochemistry of organic acids in the grape. The author of this chapter pioneered the elucidation of the key steps involved in the metabolism of malic and tartaric acid in the grape berry. Both acids may account for over 90% of the total acidity in the berry, and they are also the largest contributors to the pH of juice, must and wine. Despite the close structural similarity of tartaric acid and L-malic acid, it was demonstrated in 1958 that different metabolic pathways led to the formation of these two acids in grapevines. What is the precursor of tartaric acid? How is the precursor metabolised? Besides the L-idonate dehydrogenase enzyme, the identification and analysis of tartaric acid synthesis genes and their encoded enzymes await further discoveries, remaining important research topics. L-malic acid is the second major organic acid that accumulates during grape berry development. In contrast to tartaric acid, malic acid formed in the berry at pre-veraison is broken down during a brief period around veraison. This process, which is regulated by temperature and by developmental cues, results in marked changes in the acid composition of grapes at harvest. As explained in this chapter, the pathways of malic acid synthesis and breakdown, and the details of its participation in a wide range of metabolic processes have been determined for the grape berry, but their dependence on environmental conditions, such as heat and water availability, is not yet fully understood. In other words, the regulation of the well-characterised switch between synthesis and breakdown of the malate at the time of veraison remains largely uncharacterised.

Chapter 5 addresses the subject of grape berry phenolics. It is well known that berry phenolics largely contribute to wine quality and have beneficial effects on many aspects of the human health. In the past decade, significant advances have been made towards a better understanding of the genetics, biochemistry and physiology governing the synthesis of this class of secondary metabolites. Three main classes of phenolic compounds are synthesised in the grape berry: phenolic acids, stilbenes and flavonoids, but phenolic composition is highly diversified among different varieties and the environments where they are grown. This chapter reviews the chemical composition and the pattern of accumulation of phenolics in the grape berry, the diversity in phenolic composition of the grapevine germplasms, and the molecular basis of their synthesis, including the effect of environmental factors. It is well known that anthocyanins are responsible for the red, purple and blue pigmentation of the grape berries. But how are they synthesised and where are they accumulated? Also, how are they transported across the biological membranes?

Chapter 6 deals with the aromas of the grape berry and wine. The subtleness, complexity and uniqueness of aromas are a source of great pleasure for wine lovers. Bouquet and flavour are obviously related to the expertise of the winemaker and the techniques used, but primarily they reflect grape composition, especially varietal character – and its particular expression in a given terroir (climate, soil, and viticultural practices). How can the chemistry of flavours explain what our senses perceive? The study of grape and wine aromas is laborious because the most important grape aroma compounds are often present in very low concentrations. The first studies of grape flavours date back to the early 1950s when gas chromatography made it possible to separate volatile components in the vapour phase. Since then, many volatile compounds have been identified, especially through the coupling of gas chromatography with olfactometry and mass spectrometry. This chapter explores the role and diversity of methoxypyrazines, terpenic compounds, C13-norisoprenoids derivatives and thiols, together with the metabolic pathways involved in their synthesis. Still, many odourless and non-volatile compounds in grapes are the source of odouriferous compounds in wine. For instance, fermentation significantly alters the monoterpene composition of grapes through chemical and microbiological processes. The most major transformation concerns the degradation of nerol and geraniol by the yeast *Saccharomyces cerevisiae via* an enzymatic reduction.

Polyamines (PAs) (**Chapter 7**) are the most important biogenic amines found so far in plants, and they are associated with numerous developmental and stress-related processes. The most abundant and well-described Pas are putrescine (Put), spermidine (Spd) and spermine (Spm). As argued in this chapter, these biogenic amines may play a fundamental role in grape berry set and development, solely or in combination with the hormones. For instance, the balance between PA and ethylene synthesis may be one of the major determinantal factors regulating the fruit-set process. Also, in berry abscission, which is largely linked to metabolic and hormonal 'disorders' and repression of the nutrient supply to the inflorescence, ABA and ethylene are considered to be among the primary drivers of the procedure, but PAs may act as regulators as well. The translocation under photoperiodic flowering induction of free Spd to the inflorescence seems to be a part of the complex mechanism which occurs during the transition of vegetative buds to flowers. Thus, this chapter illustrates important correlative roles of PAs in grape berry development and summarises the current literature, providing an accurate and updated overview of the ongoing research.

Chapter 8 details the structure-function of the grape berry vacuoles and the mechanisms of solute transport across the tonoplast. This is a particularly relevant issue since vacuoles are the main reservoirs of sugars, organic acids, aromas, flavours, ions and water in grape berry tissues. At anthesis, most mesocarp cells appear to be univacuolated, but at veraison the large vacuole splits into smaller vacuoles, generating a complex internal membrane structure. This chapter provides interesting images showing a complex array of intact vacuoles occupying much of the cytosol of a mature grape berry cell. Changes in the vacuolation degree may also be involved in the maintenance of turgor pressure, or result from an increase of vacuole storage function. The berry "fattening" that occurs at the ripening stage is mostly due to the massive accumulation of sugars, water and phenolics in the vacuole. Several tonoplast proteins including pumps, carriers, ion channels and receptors were already identified in several plant models at the biochemical and molecular levels, including the most abundant ones, V-ATPase, V-PPase, and water channels (aquaporins). But the molecular mechanisms involved in the accumulation of these solutes, and how they are regulated,is still far from being fully understood in grape. In this context, approaches aiming at the purification of intact vacuoles are developed. The diversity and storage role of the vacuole of grape cells, and the molecular mechanism involved in solute transport across the tonoplast, are discussed in this chapter.

Chapter 9 by Goulao *et al.* concerns the grape berry cell wall (CW). The CW of mesocarp cells consists of approximately 90% of polysaccharides and less than 10% of a protein fraction rich in arginine and hydroxyproline residues. Cellulose and polygalacturonans are the major constituents. In the exocarp, polysaccharides account for 50% of the cell wall material. Insoluble proanthocyanidins, structural proteins and lignin are also important constituents. Modifications of the grape pulp and skin cell wall during ripening provide the flexibility for the cell to expand during fruit growth and to modulate the final texture, with important repercussions in grape characteristics, wine quality and wine-making methods. For instance, anthocyanin extraction depends directly upon the cell wall degradation in the skin. Nonetheless, the cell wall is also an important source of biologically active signalling molecules, regulating cell-to-cell interactions, and a carbohydrate storage reserve. A number of genes related to primary CW biosynthesis and modification, and to secondary CW biosynthesis were identified in the genome of *Vitis vinifera*. The public release of the *Vitis* genome and the annotation of *Vitis* genes allowed detailed studies of the genes coding for enzymes associated with cell wall biosynthesis, modification during cell growth and fruit ripening, as well as the deposition of secondary wall polymers, providing a better picture of the associated pathways. The grape berry CW transcriptome is thoroughly analysed in this chapter.

The following chapter by Böttcher and Davies (**Chapter 10**) focuses on the role of hormones during the development of the grape berry. Hormonal coordination is a crucial issue in the development of this non-climacteric fruit, as the changes that occur at both physical and biochemical levels are considerable and rapid, occurring over only a few weeks and involving a range of tissues and cell types. This chapter illustrates the role of ethylene, abscisic acid, brassinosteroids, gibberellins, auxins, salicylic acid and jasmonates in grape berry development and ripening and how they are synthesised/accumulated. These hormones are accumulated to significant levels in flowers, and early in berry development. As extensively discussed in this chapter, hormones have the ability to coordinate changes in a large number of genes in response to both developmental and external cues. Recent technological developments now permit the detection, on a genome wide scale, of the changes in the expression of genes during development, in response to hormone treatments and in hormone mutants. The authors conclude that, despite the expansion of our knowledge on the hormonal control of berry development, there are still many questions to be answered. For instance, the role of ethylene, which is essential to climacteric fruit ripening, is still controversial in grape berry ripening. The same is true for the interactions between the different signalling pathways and how these affect berry development.

The transcriptome is the complete set of RNA molecules produced by a cell, tissue or organism. It includes mRNA, rRNA, tRNA and other non-coding RNAs, although in many cases the mRNA profile is the most sought after because it corresponds to the expression of protein-encoding genes. The metabolome, defined by analogy to the transcriptome, is the complete set of metabolites (small molecules, $M_r \leq 1,000$ D) produced in a single cell, tissue or organism. Like the transcriptome, the metabolome is complex, dynamic and varies by cell type, developmental stage and in response to internal and external cues. Berry development has been investigated at the transcriptional level to study global expression profiles during formation and ripening, while a smaller number of large-scale metabolite studies have been carried out thus far. This topic is thoroughly addressed by Tornielli *et al.* in **Chapter 11**. The availability of the complete grapevine genome sequence makes it much easier to identify candidate genes governing the key processes underlying berry development, particularly those relating to quality traits. This chapter summarizes the wealth of information generated by the grapevine transcriptomic experiments in the last decade according to the principal metabolic pathways and biological processes of berry development, such as photosynthesis, sugar, organic acid and lipid metabolism, secondary metabolism, hormone biosynthesis and regulation, cell wall metabolism and the response to stress. Although the huge amount of information provided by the transcriptomic and metabolomic approaches is frequently difficult to interpret, the investigation of key molecular events underlying grape berry development can provide useful data for the improvement of mature berry quality traits.

Chapter 12 describes the microbial community of the grape berry. How is this topic related to the grape berry biochemistry? Why do grape berry microbionts have important biotechnological repercussions? While the first question is not so easily addressed, the answer to the latter is quite obvious. Important phytopathogens responsible for grapevine diseases worldwide are the oomycete *Plasmopara viticola* (downy mildew) and the ascomycete *Erysiphe necator* (powdery mildew). The causal agent of grey rot is the saprophytic mould *Botrytis cinerea.* A wide diversity of yeast species are also common contaminants of berry surfaces, but the key agent of wine fermentation, *Saccharomyces cerevisiae*, is rarely recovered from grapes. Bacterial groups include the spoiling acetic acid bacteria and lactic acid bacteria responsible for the malolactic fermentation. As detailed in this chapter, grape berries are able to exudate a variety of compounds through the cuticle and the epicuticular wax layer onto their surfaces, including phenols, sugars, lipids, malic acid, potassium and sodium. Interestingly, grape berries exude onto their surfaces both microbial inhibitory and stimulatory compounds. Also, increasing evidence suggests that the cuticle may be a source of carbon for the berry microflora, particularly the resident community. The capacity of certain pathogens to metabolise cutin and cuticular waxes, as well as pectic and hemicellulosic components of the cell wall, may explain why certain species of the fungi and aerobic yeasts, with a high metabolic diversity, are so frequently found on the sound berry surface. However, the access to the nutrient-rich inner cells through fissures or wounds on the berry skin completely changes the berry microflora. This is a stimulating topic of research and, as explained by the authors, it is somewhat surprising to find only scarce and fragmented information on grape microbiota.

The chemical composition of the grape berry directly affects wine composition and quality. **Chapter 13** addresses this topic that has been a matter of passion and controversy both in society and within the scientific community: the health benefits resulting from moderate wine consumption. In the 1990s, people's attention was increasingly drawn to the positive effects of moderate wine consumption. Indeed, the well-known "French paradox" reports that, although the French diet is relatively rich in saturated lipids compared to that of other countries, the level of mortality due to coronary heart disease is reduced as a result of daily wine consumption. It has been reported that, besides alcohol itself, which in moderate amounts helps blood flow in the body, wine polyphenols contain antioxidant properties, which are also beneficial to health. As thoroughly reviewed in this chapter, epidemiological studies conclude that moderate alcohol and/or wine consumption may protect against the incidence of many diseases of the modern society such as atherosclerosis, hypertension and diabetes. The biochemical mechanisms that may account for these therapeutic properties are also discussed.

It is our firm belief that this eBook, which is written by an international team of leading experts, is the pioneer in offering a focussed and integrated coverage of the biochemistry and molecular biology of the grape berry, emphasising the most important aspects of grape fruit development and ripening. It is a comprehensive and updated eBook for researchers, scientists and biotechnologists, and it can also be used as a reference manual for graduate and undergraduate students as it gathers useful and updated references and original data from leading laboratories worldwide.

Hernâni Gerós
Universidade do Minho
Portugal

M. Manuela Chaves
Universidade Nova de Lisboa
Portugal

Serge Delrot
University of Bordeaux
France

List of Contributors

S. Amâncio
Centro de Botânica Aplicada à Agricultura (CBAA)/ DRAT,
Instituto Superior de Agronomia, UTL, Tapada da Ajuda,
1349-017 Lisboa, Portugal

A. Aziz
Department of Biology, University of Crete, Vassilika Vouton,
Heraklion, Crete, Greece
URVVC-Stress & Environment EA 2069, PPDD, University of Reims,
F-51687 Reims cedex 2, France

F. Barrieu
University of Bordeaux, ISVV, UMR 1287 EGFV,
33882 Villenave d'Ornon, France

L. Bavaresco
Centro di Ricerca per la Viticoltura, Consiglio per la Ricerca e la Sperimentazione in Agricoltura,
Conegliano, Italy
Istituto di Frutti-Viticoltura, Università Cattolica del Sacro Cuore,
Piacenza, Italy

E. Blumwald
Dept of Plant Sciences, University of California,
One Shields Ave, Davis,
CA 95616, USA

P. K. Boss
Commonwealth Scientific and Industrial Research Organisation,
Plant Industry, P.O. Box 350, Glen Osmond,
South Australia 5064, Australia

C. Böttcher
Commonwealth Scientific and Industrial Research Organisation,
Plant Industry, P.O. Box 350, Glen Osmond,
South Australia 5064, Australia

S. D. Castellarin
Dipartimento di Scienze Agrarie e Ambientali,
University of Udine, Italy

M. M. Chaves
Instituto de Tecnologia Química e Biológica, Universidade Nova de Lisboa,
Av. da República, 2780-157 Oeiras, Portugal

A. Cunha
Departamento de Biologia, Universidade do Minho,
Campus de Gualtar, 4710-057 Braga, Portugal
Centro de Investigação e de Tecnologias Agro-Ambientais e Biológicas (CITAB),
Quinta de Prados, 5001-801 Vila Real, Portugal

P. Darriet
Université de Bordeaux, Institut des Sciences de la Vigne et du,
Vin Unité de recherche Oenologie, EA 4577, USC INRA,
210 chemin de Leysotte, CS 50008,
33882 Villenave d'Ornon cedex, France

C. Davies
Commonwealth Scientific and Industrial Research Organisation,
Plant Industry, P.O. Box 350, Glen Osmond,
South Australia 5064, Australia

M. Delledonne
Department of Biotechnology, University of Verona,
37134 Verona, Italy

S. Delrot
UMR 1287 Ecophysiology and Grape Functional Genomics, University of Bordeaux,
INRA, Institut des Sciences de la Vigne et du Vin, Domaine de la Grande Ferrade,
210 chemin de Leysotte, 33883 Villenave d'Ornon, France

D. Dubourdieu
Université de Bordeaux, Institut des Sciences de la Vigne et du,
Vin Unité de recherche Oenologie, EA 4577, USC INRA,
210 chemin de Leysotte, CS 50008,
33882 Villenave d'Ornon cedex, France

L. Falginella
Dipartimento di Scienze Agrarie e Ambientali,
University of Udine, Italy

J. C. Fernandes
Centro de Botânica Aplicada à Agricultura (CBAA)/ DRAT,
Instituto Superior de Agronomia, UTL, Tapada da Ajuda,
1349-017 Lisboa, Portugal

M. M. Ferreira
Centro de Botânica Aplicada à Agricultura, Instituto Superior de Agronomia,
Technical University of Lisbon, 1349-017 Lisbon, Portugal

R. B. Ferreira
Centro de Botânica Aplicada à Agricultura, Instituto Superior de Agronomia,
Technical University of Lisbon, 1349-017 Lisbon, Portugal
Instituto de Tecnologia Química e Biológica, New University of Lisbon,
2781-901 Oeiras, Portugal

N. Fontes
Centro de Investigação e de Tecnologias Agro-Ambientais e Biológicas (CITAB),
Quinta de Prados, 5001-801 Vila Real, Portugal
Departamento de Biologia, Universidade do Minho,
Campus de Gualtar, 4710-057 Braga, Portugal

C. M. Ford
School of Agriculture, Food and Wine, The University of Adelaide,
5005 Australia

G. Di Gaspero
Dipartimento di Scienze Agrarie e Ambientali,
University of Udine, Italy

H. Gerós
Departamento de Biologia, Universidade do Minho,
Campus de Gualtar, 4710-057 Braga, Portugal
Centro de Investigação e de Tecnologias Agro-Ambientais e Biológicas (CITAB),
Quinta de Prados, 5001-801 Vila Real, Portugal

M. I. V. Z. Gonçalves
Istituto di Frutti-Viticoltura, Università Cattolica del Sacro Cuore,
Piacenza, Italy

L. F. Goulao
Centro de Ecofisiologia, Bioquímica e Biotecnologia Vegetal,
Instituto de Investigação Científica Tropical, IP, Quinta do Marquês,
Av. da República 2784-505 Oeiras, Portugal

M. Hanana
Center of Biotechnology of Borj Cédria, BP 901,
Hammam-Lif 2050, Tunisia

R.-A. A. Kalliopi
Department of Biology, University of Crete, Vassilika Vouton,
Heraklion, Crete, Greece
URVVC-Stress & Environment EA 2069, PPDD, University of Reims,
F-51687 Reims cedex 2, France

F. Lecourieux
UMR 1287 Ecophysiologie et Génomique Fonctionnelle de la Vigne (EGFV),
Université de Bordeaux, INRA, Institut des Sciences de la Vigne et du Vin (ISVV),
210 Chemin de Leysotte, CS 50008, 33882 Villenave d'Ornon, France

P. Lopes
Centro de Botânica Aplicada à Agricultura (CBAA)/ DRAT,
Instituto Superior de Agronomia, UTL, Tapada da Ajuda,
1349-017 Lisboa, Portugal

V. Loureiro
Centro de Botânica Aplicada à Agricultura, Instituto Superior de Agronomia,
Technical University of Lisbon, 1349-017 Lisbon, Portugal

V. Martins
Departamento de Biologia, Universidade do Minho,
Campus de Gualtar, 4710-057 Braga, Portugal
Centro de Investigação e de Tecnologias Agro-Ambientais e Biológicas (CITAB),
Quinta de Prados, 5001-801 Vila Real, Portugal

S. Monteiro
Centro de Botânica Aplicada à Agricultura, Instituto Superior de Agronomia,
Technical University of Lisbon, 1349-017 Lisbon, Portugal

M. N. Panagiotis
Department of Biology, University of Crete, Vassilika Vouton,
Heraklion, Crete, Greece
URVVC-Stress & Environment EA 2069, PPDD, University of Reims,
F-51687 Reims cedex 2, France

M. Pezzotti
Department of Biotechnology, University of Verona,
37134 Verona, Italy

P.-L. Teissedre
Institut des Sciences de la Vigne et du Vin, Université Bordeaux II,
UMR 1219 INRA Œnologie, Faculté d'œnologie,
Laboratoire de Chimie Appliquée ; 210, chemin de Leysotte, CS 50008,
33882 Villenave d'Ornon Cedex , France

C. Thibon
Université de Bordeaux, Institut des Sciences de la Vigne et du
Vin Unité de recherche Oenologie, EA 4577, USC INRA,
210 chemin de Leysotte, CS 50008,
33882 Villenave d'Ornon cedex, France

G. B. Tornielli
Department of Biotechnology, University of Verona,
37134 Verona, Italy

S. D. Tyerman
School of Agriculture Food and Wine, The University of Adelaide, PMB 1,
Glen Osmond, South Australia, 5064, Australia

A. Zamboni
Department of Biotechnology, University of Verona,
37134 Verona, Italy

S. Zenoni
Department of Biotechnology, University of Verona,
37134 Verona, Italy

ACKNOWLEDGEMENTS

The editors of this eBook are deeply grateful to the following entities, whose contribution was of great relevance to the accomplishment of this project:

- The Biology Department of the School of Sciences - University of Minho
- The Centre for the Research and Technology of Agro-Environmental and Biological Sciences (CITAB)
- The Portuguese Foundation for Science and Technology (FCT) - research project no. PTDC/AGR-ALI/100636/2008
- BabeliUM, Language Centre of the University of Minho
- Helena Martins, Communication Designer
- José Jorge Soares, Architect and Designer

CHAPTER 1

Water Relations of the Grape Berry and Aquaporins

S. D. Tyerman[1,*], M. M. Chaves[2] and F. Barrieu[3]

[1]School of Agriculture Food and Wine, The University of Adelaide, PMB 1, Glen Osmond, South Australia, 5064, Australia; [2]Instituto de Tecnologia Química e Biológica, Universidade Nova de Lisboa, Av. da República, 2780-157 Oeiras, Portugal and [3]University of Bordeaux, ISVV, UMR 1287 EGFV, 33882 Villenave d'Ornon, France

Abstract: Berry water content during development and at harvest is a crucial parameter of the vintage quality by directly impacting the concentration of sugars and flavour compounds. In the last decade, there has been considerable progress in understanding how berry water relations are involved in the ripening process as a result of exploring and integrating developmental changes in berry hydraulic conductivity, mineral nutrient loading, phloem unloading, xylem development, functionality of xylem vessels, cell vitality, cell water relations, gene expression and the molecular biology of aquaporins (water channels). In this chapter, we review the most recent advances in our understanding of the berry water influx and efflux during development and their consequences on cell turgor and apoplast solute concentrations, taking account of the changes in cell vitality observed late in ripening in some varieties. The role of aquaporins in the regulation of berry water relations is also presented, with a particular emphasis on the varietal differences observed in aquaporin functioning. Finally, the impact of water status of the parent vine on berry ripening is discussed. From this review, it appears that the varietal differences observed at different regulatory levels of berry water status represent a great opportunity to gain a better understanding of the ripening process.

Keywords: Abcisic acid, Aquaporins, Cell turgor, Major intrinsic proteins, Transpiration, Turgor pressure, Vine water status, Water deficit, Water relations, Xylem hydraulic resistance, Xylem vessels.

INTRODUCTION

Grape berries normally contain 75–85% water, which is the main solvent of solutes including sugars, acids and phenolic compounds [1]. Thus, quality, *via* concentration of sugars and flavour compounds, and yield of the vintage are crucial quality parameters that will be directly affected by berry water content at harvest.

Like many other fleshy fruits, the development of grape berries follows a double sigmoid curve [2] that can be described as three distinct developmental stages: an initial period of rapid cell division and expansion after flowering in green berries (Stage I), a short transitory phase of very little growth (Stage II), and finally a second active growth phase involving mainly cell expansion (Stage III). In grape, the onset of ripening is named veraison and is accompanied by several abrupt physiological changes such as fruit softening and anthocyanins production in red varieties.

During development, the net water flow into a berry equals inflow through xylem and phloem minus loss by transpiration and backflow *via* the xylem, and is consequently directly linked to berry volume change. Water can be imported into the berry *via* both xylem and phloem. However, it is now well recognised that the relative contribution of xylem and phloem to water inflow depends on berry varieties and developmental stages, as does water loss by transpiration and/or water backflow from the berry to the parent wine.

PATHWAYS FOR WATER INFLUX TO THE BERRY AND HOW THEY CHANGE DURING DEVELOPMENT

Changes Through Development in Apparent Hydraulic Connection Between Berry and the Vine

Before veraison, berry water is provided by the xylem and phloem, with the xylem providing the majority

***Address correspondence to S. D. Tyerman:** School of Agriculture Food and Wine, The University of Adelaide, PMB 1, Glen Osmond, South Australia, 5064, Australia; E-mail: steve.tyerman@adelaide.edu.au

Hernâni Gerós, M. Manuela Chaves and Serge Delrot (Eds)

of the water to the berry. Around veraison, sugars accumulation in berry mesocarp cells is accompanied by a strong shift in the proportion of xylem and phloem water transport [3, 4]. By examining the swelling and shrinking patterns of berries of potted and field vines, Greenspan *et al.* [3, 5] demonstrated that the role of the xylem in berry volume changes was much reduced after veraison while the contribution of the phloem increased substantially. Thus, in the final growth stage of the berries, the phloem can provide up to 80% of the berry water requirements. In addition, berry water status also becomes apparently uncoupled from plant water status after veraison. In fact, diurnal changes in berry diameters appear correlated to stem water potential of the parent vine before veraison but are greatly reduced after veraison, and not linked subsequently to stem water potential [6]. In the same way, turgor of mesocarp cells was found to be insensitive to changes in vine water status after veraison [7]. Taken together, these findings indicate that post-veraison berries appear hydraulically isolated from the parent vine, allowing berry protection against the environmental stresses experienced by the parent vine.

Evidence from Potassium and Calcium Content of Berries for a Change in Predominate Water Supply Pathway After Veraison

Some studies have examined the changes in accumulation of potassium (K) and calcium (Ca) during ripening in the context of changes in the relative contributions of xylem and phloem. This is based on K being phloem mobile while Ca is not, thus changes in the ratio of K/Ca accumulation may reflect changes in proportion of flow *via* phloem and xylem. Calcium generally tends to mirror transpiration in its pattern of accumulation in plants [8]. By examining the K and Ca accumulation per berry of Shiraz vines in the field, Rogiers *et al.* [9] concluded that xylem conduits may remain functional post-veraison since Ca accumulation continued. However, there was about a 3- fold increase in the K/Ca ratio indicating that the relative contribution from phloem increased post-veraison. A second study on field-grown Shiraz vines showed a greater reduction in Ca accumulation after veraison but also greater variability between seasons and replicates, probably reflecting variability in transpirational demand [10]. They also demonstrated *via* girdling experiments that the basic premise regarding sources of K (largely phloem) and Ca (largely xylem) were correct. From this study it was concluded from the pattern of K accumulation that phloem inflow was reduced during the berry shrinkage phase. Loss of Ca from berries after veraison can also occur (Schaller *et al.* 1992 cited in Rogiers *et al.* [10]) implicating backflow to the vine. Since Ca has a structural role in cell walls and cell membranes as well as a signalling role, it is possible that variation in Ca in the berry linked to water flow may also influence other water relations phenomena such as cell wall strength [11], osmotic competence of cell membranes and even aquaporin gating [12, 13].

Measurements of Xylem Hydraulic Conductance/Resistance and Xylem-Mobile Dyes Demonstrate Continued Xylem Functionality in Post-Veraison Berries

Changes in xylem functionality leading to the hydraulic isolation of the berries was first investigated by observations of water soluble dye uptake into the berry through the xylem, in both excised and attached berries [14-16]. Results from these studies indicated that dyes can be taken up into peripheral and axial xylem vessels of entire pre-veraison berries, whereas dye uptake in post-veraison berries appeared limited to the brush region.

By taking account of the appearance of stretched and ruptured xylem conduits in post-veraison berries, it was proposed that the expansion of the berries occurring after veraison during the second active growth phase led to a physical disruption of the xylem vessels responsible for the shift of water supply and the apparent hydraulic isolation of the berries [15]. However, these studies were based on the use of charged dye molecules of high molecular weight, and the possibility that water molecules were still able to enter or leave the berry *via* the xylem cannot be totally ruled out [16]. Indeed, several recent studies using different techniques indicate that xylem appears potentially functional and connected between the vine and the berries after veraison, with significant variations depending on the varieties studied [10, 16-21]. These findings are supported by direct observations of the xylem network in the berry, showing that the vast majority of the xylem tracheary elements remains intact after veraison despite the growth of the berry, and even suggest that xylem development continues well into the second growth phase [22, 23].

The functionality of the post-veraison xylem vessels has been demonstrated by measurements of hydraulic conductance *via* the xylem pathway to the berries [21] and the observation of dye movement between the parent plant and the berries and vice-versa, in the presence of an appropriate pressure gradient [17, 18, 22]. Thus, it was proposed that the shift of phloem unloading from the symplastic to apoplastic pathway occurring at veraison [24, 25], coupled with solute accumulation in the berry apoplast, may be responsible for the decline in xylem water influx in ripening berries because of the lack of an appropriate driving force [18], accounting for the earlier dye studies.

More recently, the hypothesis that cessation of water transport through the xylem may result from disruption or occlusion of pedicel and berry xylem vessels was tested [26]. Measurements of xylem hydraulic resistance (R_h) during berry development indicated a significant increase of R_h occurring only in the latter stage of ripening, as already reported in previous studies [21]. Choat *et al.* [26] measured R_h for a flow into the berry at one pressure (0.1 MPa) and their values are somewhat lower (2-fold) than those observed under equivalent pressure for the same variety (Chardonnay) by Tilbrook *et al.* [20]. However, Tilbrook *et al.* [20] also found that the resistance for flow out of the berry in Chardonnay was about 6–fold higher than that for flow in to the berry. Considering the sensitive nature of these measurements, the agreement between two independent sets of measurements is remarkable, and it is possible that growth conditions and viticultural differences between two continents may affect the hydraulic resistances. With whole bunches it is also possible to estimate berry hydraulic resistance using sensitive flow meters [20, 21]. Here the hydraulic resistance is proportional to the number of berries on the bunch making it possible to calculate a resistance on a single berry basis, and these agree with single berry measurements.

Are Post-Veraison Berries Hydraulically Isolated or Hydraulically Buffered?

By using measured R_h and typical diurnal fluctuations in stem water potential, Choat *et al.* [26] conclude that a relatively large amount of water may flow back to the vine, perhaps as high as 400 μL day^{-1} $berry^{-1}$. Thus rather than to consider berries as becoming hydraulically isolated they suggest that they are hydraulically buffered by water delivered from the phloem [26]. However, the 6 to 10-folds higher resistances for flow out of the berry as measured by Tilbrook *et al.* [20] may diminish this degree of buffering. In terms of the position in the berry where R_h increases, this was examined by both Tyerman *et al.* [21] and Choat *et al.* [26] by progressively cutting the berry attached to the pressure probe from the distal to proximal end. In Chardonnay there was a larger increase in R_h through the proximal (brush) region and distal part of the berry compared to Shiraz, where the increase in resistance was about half of that observed in Chardonnay. Choat *et al.* [26] found an increase in both the receptacle and fruit R_h. Tyerman *et al.* [21] could not detect changes in R_h in the pedicel or receptacle region of the berry but both studies found that the pedicel R_h did not change during development.

In summary, there is clearly a shift in the dominating pathway of water flow into the berry after veraison with xylem inflow reducing compared to phloem inflow. Findings from the past few years indicate that loss of xylem physical integrity or ability to conduct water are not responsible for this change, although there is clearly an increase in R_h. Observations of high solute content or gel content in xylem tracheary elements in post veraison berries may account for increased R_h due to an increase in viscosity [26], but so far we cannot exclude the possibility that cell membrane delimited pathways external to the xylem in the berry are also contributing to the increased R_h. The next challenge for a better understanding of berry water balance during ripening appears to be the precise quantification of the water flow occurring *via* the phloem after veraison, together with the recycling of excess phloem water through the xylem (this chapter).

PATHWAYS FOR WATER LOSS FROM THE BERRY AND HOW THEY CHANGE DURING DEVELOPMENT

Some varieties show extensive berry weight loss and shrivelling during the later stage of ripening, although the phenomena of weight loss may occur in most varieties to varying degrees [27]. The onset of weight loss after maximum weight is attained can be referred to as a final phase of berry development [28], also coincidental with the onset of loss of cell vitality in the pericarp [19]. As an example, Shiraz berries can lose over 20% of the maximum berry weight during this phase of late ripening [29-31]. This weight loss at full sugar, as opposed to other weight loss phenomena [32], has been thought to result from a combination of reduced phloem inflow and continued transpiration, but water backflow from the berry to the vine *via* the xylem has also been suggested [18, 21, 33, 34]. Indeed, water backflow has been demonstrated by dye movement from the stylar end of post-veraison berries to the xylem of the vine [18]. More recently, using dye loading and quantitative measurements of xylem hydraulic conductivity and water potential gradients, water backflow from the berry to the parent vine was found to occur in a genotype-dependent manner, indicating varietal differences in hydraulic connections of the berry to the vine [20].

Transpiration

Transpiration from grape berries not only plays an important role in water balance, but may also function to drive sugar accumulation [35-37]. Berry transpiration is dependent on climatic conditions and changes during berry development [31, 36]. Various studies indicate that berry transpiration decreases throughout development [3, 5, 31, 38]. This decline may be linked to changes in water permeability of the skin that depends on stomata and/or wax layers [39]. The stomata density on berry skin is very low, 7 ± 2 stomata per berry in Cabernet Sauvignon with slight differences in other varieties [40]. After veraison, stomata become filled with wax and lose their function, perhaps explaining the decline observed for berry transpiration [34]. At the same time, epicuticular waxes increase strongly [40]. However, Rogiers *et al.* [34] did not find any significant correlation between amount of wax and berry transpiration rate. Thus, it is hypothesised that wax structure and composition might be more important than thickness or density in the control of berry transpiration [34, 41, 42]. According to this hypothesis, varietal differences are likely, but a global decrease of berry transpiration will probably occur during development in most varieties.

Back-Flow in the Xylem

Theoretically, xylem water backflow can only occur if: (i) the xylem is functionally connected to the berry apoplast, and; (ii) there is an appropriate gradient to drive water flow from the berries to the parent vine. The effective connection of the xylem has been demonstrated by several studies (this chapter). However, the pressure gradients that drive xylem backflow from the berry to the parent vine will depend on the water potentials of the two compartments and the hydraulic conductivity between the two compartments [26]. Interestingly, the results of Tyerman *et al.* [21] indicate that, in Shiraz and Chardonnay berries, the very low osmotic potentials in post-veraison berries were not translated to negative pressures in the xylem of the pedicel or adjacent vine stem, and that in fact a gradual increase in xylem pressure toward zero and even slightly positive pressures was observed during ripening.

The hydraulic conductance between the berry and the vine has been shown to change during berry development with varietal differences [20, 21]. Thus, Shiraz and Chardonnay berries both showed a significant reduction in hydraulic conductance during ripening but Shiraz berry maintains higher hydraulic conductance back to the vine later in development compared with Chardonnay berries. In the same way, a clear late ripening increase of xylem hydraulic resistance (decrease in hydraulic conductance) was also observed in Chardonnay berries by Choat *et al.* [26]. The differences observed between Shiraz and Chardonnay berries suggest that water backflow to the vine could occur later in development and thus be more important in Shiraz berries in the presence of xylem tensions developed by the transpiring leaves of the parent vine. However, one may assume that the large negative osmotic potential of the berry sap (-4 to -5 MPa) can be effective enough to counteract the xylem and apoplast tensions transferred from the parent vine.

This last point raises the question of the membrane semipermeability of the pericarp cells that has to be maintained to effectively counter the xylem tensions generated by the vine. Analyses of cell vitality indicate that Shiraz and Chardonnay berries began to show loss of cell vitality in mesocarp and endocarp when berries reach their maximum weight, whereas cells of Thompson seedless berries show near 100% vitality even past harvest date [19]. In the same study, negative xylem pressures were measured in Thompson seedless, in contrast to slightly positive pressures for Shiraz and Chardonnay (summarised in Fig. **1**).

These findings, apart from underlying again the differences observed between varieties, may reveal different strategies. Backflow is likely to occur late in ripening in varieties where cell death takes place, at least up to the appearance of a significant reduction of xylem hydraulic conductance, either by modulation of aquaporin activity or by deposition of gels or solutes in xylem conduits [19-21, 26]. In berries where cells remain viable up to full maturity and beyond, no hydraulic isolation will be needed due to the large negative osmotic potential of the berry sap.

Berry Splitting

Apart from berry weight loss, the phenomena discussed above have been only recently considered in respect to berry splitting [43]. Here the reduction in susceptibility of Shiraz berries to splitting later in ripening was attributed to a decrease in osmotic competence and hence turgor-generating capacity within the berry due to the loss of pericarp cell vitality. Those varieties that maintain a higher hydraulic conductivity to the vine combined with maintenance of cell vitality would be predicted to show greater propensity to berry splitting under wet conditions. One such variety is Thompson Seedless (Sultana) which is more susceptible to splitting and at apparently lower berry turgor pressures than other varieties [44].

In summary, recent findings indicate that backflow from the berry to the parent vine can occur but is very much variety-dependent and may be determined by two important parameters: the membrane integrity of berry cells late in ripening and the hydraulic conductance *via* the xylem to the vine. Berry splitting under wet conditions may also be linked to these processes [43].

CHANGES IN CELL TURGOR AND APOPLAST SOLUTE CONCENTRATION LINKED TO VERAISON

There are clearly profound changes in the water relations of the berry during veraison and associated with the very large increase in sugar accumulation and berry growth over a short period of time (1-2 days) [6]. The onset of berry ripening is signalled by berry softening, accumulation of sugars and anthocyanin accumulation in the skin of red grape varieties [2]. Berry volume rapidly increases about five days after sugar accumulation and softening [2, 45]. Softening appears to be the consequence of reduced turgor pressure (P) of berry pericarp cells and subsequent changes in cell wall properties. Berry enlargement was thought to be due to a two-step process of cell wall loosening, first in the pericarp cells then in the skin tissues [45] (see the chapter by Goulao *et al.*, this edition).

Direct Measurements of Cell Turgor in the Berry Before and After Veraison

The cell turgor (P) of mesocarp cells was directly measured with the cell pressure probe before, during, and after veraison [7]. P declined to a low value of about 0.05 MPa in post-veraison berries, but not to zero. The pericarp cells also behaved as competent osmometers, indicating that the cell membrane was still intact and cells were vital. This observation is also corroborated by the measures of cell vitality and gene expression of aquaporins (membrane proteins) described below. Reduced mesocarp P was previously thought to be due to loss of mesocarp cell vitality (loss of compartmentation hypothesis) as the mechanism leading to increased phloem influx during fruit ripening [46]. The reduction in turgor maintains a pressure difference in the phloem between the source and the berry sink tissue to maximise the accumulation of sugars during ripening [25, 47]. Also, high solute concentration in the apoplast would increase the volume of water leaving the phloem (if not impeded), and contribute to the rapid accumulation of water and sugar in Stage III [48].

In a second study using the cell pressure probe Thomas *et al.* [49] examined the time course of mesocarp P for three varieties before, during and after veraison. Mesocarp P was around 0.17 MPa early in development, followed by an increase to between 0.3 to 0.35 MPa at the onset of the transitory phase, and then an immediate decline before ripening to less than 0.1 MPa. The reduction in P occurred before there was a significant increase in sugar accumulation, prompting the suggestion that mesocarp P may be a signal that triggers many other processes associated with ripening. Bulk modulus of elasticity of berries (E, units of pressure, MPa) was also measured [49], rather than deformation under a constant load. E may be more easily related to cell turgor pressure than deformation, and was found to be linearly correlated with mesocarp P analogous to the well-known relationship between cell P and cell volumetric elastic modulus that is probably related to stress hardening [50]. Another interesting outcome of this study was that the growth of cells during ripening occurs at very low turgor pressure, suggesting that cell expansion in the berry may be fundamentally different to other plant cells.

Increase in Apoplastic Solute Concentration During Veraison

Apoplastic solute concentrations and osmotic potential of the mesocarp before, during, and after veraison have been obtained using a carefully calibrated centrifugation technique [51]. This showed that the reduction in turgor was due to increased apoplastic solute concentration, mainly fructose and glucose. The estimated matric potential increased from -0.27 MPa (Stage I) to a positive value (0.12 MPa) during the transition stage (Stage II) and then declined to a slightly negative value (-0.05 MPa) post-veraison, and as a proportion of total apoplast water potential, decreased to a very small fraction. These values could explain the increase in the equilibrium hydrostatic pressure in the berry pedicel near veraison for wine grape varieties measured using a modified root pressure probe and assuming xylem tension is equilibrated with the matric potential in the mesocarp apoplast [21]. It is interesting to note that the table grape variety Thompson Seedless maintained tensions post-veraison [20] and this might imply substantial differences between varieties in the apoplast components of water potential as suggested by the differences in apoplast solute concentration recorded for post-veraison berries by Wada *et al.* [51] and Zhang *et al.* [25]. Wada *et al.* [51] concluded that because the apoplast solute potential decreased before veraison, then apoplast solutes must play an important role in changes to cell water relations (reduced P), solute transport and metabolism that instigates ripening.

A switch from symplasmic to apoplastic phloem unloading can account for the increase in apoplastic solute concentration and was demonstrated by Zhang *et al.* [25] using phloem and symplasm mobile fluorescent probes. These dyes were released from phloem strands during Stage I and Stage II, but were confined to the phloem during late Stage II. Expression and activities of cell wall acid invertase also increased at the onset of ripening as would be expected if sucrose was effluxed to the apoplast and accounting for the high concentration of fructose and glucose observed in the apoplast [25, 51].

Is Turgor a Primary Signal for Changes in Berry Physiology During Veraison?

The studies described above demonstrate that prior to substantial increase in berry sugar accumulation and growth, apoplastic solutes increase in concentration probably due to a switch from symplastic to apoplastic phloem unloading, and this results in turgor pressure declining in mesocarp cells. The decline in turgor pressure and bulk modulus of elasticity accounts for berry softening. However, they do not establish that these changes, and particularly changes in turgor, are required for other changes in cell metabolism and for growth to proceed in the ripening process, for example a signal for *de novo* synthesis of ABA may be triggered by reduced turgor. To address this it would be necessary to somehow uncouple changes in turgor from other changes in the berry. This was attempted by placing pre-veraison berries in transparent plastic boxes, thus restricting berry expansion [52]. The boxes delayed the decrease in P, where the size of the delay was reduced if some transpiration occurred (perforated versus closed boxes). The treatments also delayed the onset of ripening even though sugar accumulation seemed to proceed unimpeded. This is consistent with a role of P in the onset of ripening.

In summary, apoplastic solute accumulation, by stimulating water flow from the phloem [47], may explain the lack of negative pressure in the xylem [21] of wine grape varieties and the insensitivity of post-veraison berries to plant water deficit [3]. At the same time, the increase in apoplast solute concentration [25, 51] can explain the loss of turgor in berry mesocarp cells during ripening [7]. There are likely to be varietal differences in the degree of increase in apoplastic solute concentration and hence decrease in turgor during veraison.

CHANGES IN CELL VITALITY LATE IN RIPENING

Membrane semipermeability (*i.e.* membrane more permeable to water than solutes) is the basis of cell membranes being able to translate osmotic gradients across them into a water flow, and simultaneously to develop pressure gradients when the cell membrane is constrained by the cell wall. This property is embodied in the biophysical membrane parameter called the reflection coefficient as an important parameter that allows biophysical description of plant cell water relations [53]. Cell membranes are generally not permeable to large molecular weight molecules, charged molecules or inorganic ions, and this is often the basis for several tests of cell vitality, *i.e.* if membrane semipermeability breaks down it normally signals cell death (loss of vitality). Dyes used on grape berries that are based on the normal low permeability of living cell membranes are fluorescein (produced from esterase cleavage of fluoresceine diacetate (FDA, membrane permeable) [54], and propidium iodide [19]. Clarke *et al.* [43] used nitroblue tetrazolium which indicates oxidative metabolism. In the last few years these tests have been applied to grape berries because of the interest in the role of membrane semipermeability and cell vitality in berry water relations.

Loss of Cell Vitality Occurs Late in Ripening

To test the loss of compartmentation hypotheses [46], Krasnow *et al.* [54] examined berry cell vitality on the surface of a longitudinal cut through the centre of berries using FDA. Krasnow *et al.* [54] established that cells had intact membranes and remained vital in post-veraison berries of Chardonnay, Nebbiolo and Cabernet Sauvignon well in to ripening and some 60 days after veraison thereby supporting the conclusion of Thomas *et al.* [7] that cells remained intact after veraison and during rapid berry expansion. For each of the varieties, loss of cell vitality did not begin until maximum berry weight was obtained.

A separate study using some of the techniques developed by Krasnow *et al.* [54] examined the vitality of berry cells in Chardonnay, Shiraz and Thompson Seedless in the context of late ripening berry weight loss and shrivel [19]. These varieties were compared because Chardonnay and Thompson Seedless rarely show berry weight loss while Shiraz often shows weight loss (up to 30%) and shrivel. It should be noted that this characteristic of Shiraz occurs at full sugar accumulation and is not associated with bunch stem necrosis, sunburn or low sugar accumulation [32]. The onset of berry weight loss has been recognised as another stage in grape berry development based on the relative change in sugar and water content of the berry [28]. The onset of loss of cell vitality in Chardonnay and Shiraz occurred at about 100-110 days after flowering and corresponded to maximum berry weight [19] confirming the late onset observed in Krasnow *et al.* [54]. Thereafter the proportion of vital cells decreased linearly with time in both varieties, though Shiraz showed a steeper decline than Chardonnay. Clarke *et al.* [43] confirmed the timing for onset of cell death in Shiraz using a different stain (nitroblue tetrazolium), but in the context of reduced susceptibility to berry splitting as cells became non-vital.

Varietal Differences in Loss of Cell Vitality

In contrast to Chardonnay and Shiraz, Thompson Seedless grapes did not display loss of vitality [19] and this is interesting when compared to observations of the pedicel equilibrium pressure. Those varieties that showed cell death (Chardonnay and Shiraz) showed slight positive values, while Thompson Seedless showed negative values. These differences between Shiraz, Chardonnay and Thompson Seedless are summarised in Fig. **1**.

Many different varieties were compared for their propensity to show berry weight loss (measured geometrically as a "shrivel index") and cell death [27]. There was a general correlation across varieties between the degree of tissue vitality and shrivel of berries. Varieties with berries displaying less vital tissue also had a greater degree of shrivel. Some varieties showed this relationship within the samples taken at a single point in time; for example, Shiraz, which is known for its propensity to shrivel. Red varieties tended to show moderate to low tissue vitality and higher shrivel, while white varieties occurred at both extremes. These differences occurred independently of sugar accumulation. Table grape varieties were characterised with high vitality and a low shrivel. The presence or absence of seeds did not seem to account for differences [27]. Seeds may be implicated in cell death since it has often been observed that cell death seems to begin near the locular region. The extent of cell vitality and shrivel separated varieties into clusters that may indicate mechanistic differences in the way cell vitality in the mesocarp is linked to berry water relations.

Late Ripening Phenomena	Shiraz	Chardonnay	Thompson Seedless
Xylem Pressure	+ve	+ve	-ve
(kPa)	5	12	-24
Hydraulic Resistance	Medium	High	Low
Flow in	9	17	4
Flow out (x 10^{10} s MPa m^{-3})	29	750	nd
Berry weight loss and shrivel	High	Low	Low
	75%	75%	100%
Cell vitality	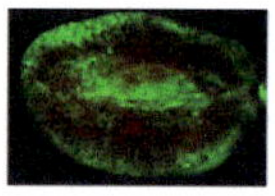	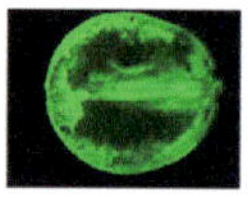	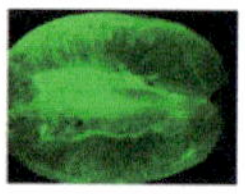

Figure 1: Summary of water relations properties of berries from three *Vitis vinifera* varieties showing contrasting features that may be related to differences in hydraulic connection to the vine and onset of loss of cell vitality. Data are summarised from [19- 21, 27].

An interesting observation from the Fuentes *et al.* [27] was a region of non-vital cells in the brush region of Shiraz, Cabernet Sauvignon, Merlot, Pinot Noir, Carignan and Grenache berries. Chardonnay berries maintain vitality of the vascular tissue within the brush region despite extensive cell death in the mesocarp and pericarp. The vitality of the brush vascular tissue may maintain some regulation of water and solute transport between the berry and the vine [26, 19, 21].

Membrane Degradation

Deterioration of cell membranes during development of other fruits is well documented, and enzymes of the lipoxygenase family and oxidative stress have been associated with ripening and flavour development in saskatoon [55], tomatoes [56] and kiwifruit [25]. Lipoxygenases (LOX) are enzymes that degrade polyunsaturated fatty acids including membrane lipids and are associated with changes in reactive oxygen species. Hydroperoxides generated by LOX are converted into oxylipins, which are biologically active, and include jasmonic acid and flavour precursors [57]. Podolyan *et al.* [57] recently characterised four LOX genes in Sauvignon Blanc berries during development. The expression pattern of VvLOXC and VvLOXD had the highest expression in the pulp but were also distributed in seeds and skin. Interestingly VvLOXC increased in expression with wounding and *Botrytis cinerea* infection, and this type of LOX has been implicated in the hypersensitivity response (programmed cell death) to pathogens in other plants [58]. An oxidative burst has been described in Pinot Noir berries with accumulation of hydrogen peroxide post-veraison (peaks 9-10 weeks post-flowering) as well as modulation of ROS scavenging enzymes [59]. The increase in hydrogen peroxide is also relevant in respect to the regulation of aquaporins (see below).

AQUAPORINS AND THEIR POSSIBLE ROLES IN BERRY WATER RELATIONS DURING RIPENING

Aquaporins (AQPs) are major intrinsic proteins (MIPs) that facilitate water flow across biomembranes [60] and are key factors in regulating cell-to-cell water flow [61-63]. They can account for the major proportion of the hydraulic conductivity of the plasma membrane and tonoplast [12, 60, 64]. The MIP superfamily in plants had been originally divided into four subfamilies: the Plasma Membrane Intrinsic Proteins (PIPs), Tonoplast Intrinsic Proteins (TIPs), Nodulin-like Intrinsic Proteins (NIPs) and the Small Basic Intrinsic

Proteins (SIPs) [65, 66]. Two new subfamilies have been identified in the moss *Physcomitrella patens*: the GlypF-like Intrinsic Proteins (GIPs) [67] and the Hybrid Intrinsic Proteins (HIPs) [68]. Another subfamily called the X Intrinsic Proteins (XIPs) has been identified, with members in a number of dicotyledonous plants including tomato [69] and grapevine [68].

A screen of PIP and TIP genes in a Cabernet Sauvignon cDNA library (young stem, leaf, tendril, petiole and roots) identified 11 full-length and two partial length AQP cDNAs. Two full length members (VvPIP2;1 and VvPIP2;5) were not represented in the published genome of a near homozygous line of *Vitis vinifera* (PN40024) [70, 71]. Of these cDNAs, five are PIP2 aquaporins, six are PIP1 and two are TIP aquaporins. Interestingly TIP2;1 isoforms had different 3′ UTRs, immediately upstream of the poly(A) tail, suggesting multiple cleavage sites for polyadenylation. Cabernet Sauvignon PIP and TIP proteins functioned as water channels with the exception of VvPIP2;5. VvPIP2;5 differed from the water conducting VvPIP2;1 by the substitution of two highly conserved amino acids probably causing blockage of the water pore [71]. The Shelden *et al.* [71] study also identified 23 full-length MIP genes from the *V. vinifera* genome sequence of the near homozygous line (PN40024) that cluster into the four main subfamilies (and subgroups within) identified in other species (*i.e.* PIP, TIP, NIP and SIP). Six partial sequences were also identified in the genome, one PIP, one TIP, three NIPs and one SIP. A different nomenclature for several of the PN40024 aquaporin genes was proposed by Shelden *et al.* [71] compared with a previous study [72]. Based on the identification of PIP2 genes in Cabernet Sauvignon that were not present in the PN40024 genome, there are likely to be more than 23 MIP genes in other heterozygous grapevine cultivars.

AQP-mediated water flow is thought to be a predominant mechanism used by plants to control their cellular and tissue flow [63], although some plants have apparently high AQP-activity at the cell level with no indication that they regulate flow across the tissue, *e.g.* roots of some plant species under certain conditions [73]. In other cases their role in regulation of flow across tissues has been indicated by correlations between AQP expression and hydraulic conductivity [74-79], and effects of inhibitors or conditions that normally inhibit activity of AQP [80-82]. Aquaporins play a dynamic role in maintaining cellular water homeostasis under conditions where modifications in water flux occur, namely in response to drought and salinity [83, 84] or to temperature and light [63]. Many studies have considered that alteration in aquaporin expression provides clues as to their role in a particular organ. This is complicated by the fact that aquaporins can be multifunctional and that they might function *in planta* to transport molecules that have not been tested [61, 83]. There are also various post-translational regulation mechanisms that affect membrane targeting and density in the membrane, as well as gating [85].

Post-Translational Regulation of Aquaporins

Two post-translational regulatory mechanisms in particular that are relevant to grape berries are regulation by reactive oxygen species [86] in relation to the peak in hydrogen peroxide observed in berries after veraison [59], and regulation by calcium [12] in relation to the variable calcium accumulation that occurs after veraison [10]. A hypothesis incorporating a feedback mechanism linking flow of calcium in the transpiration stream and aquaporins in leaves has recently been proposed that may also apply to grape berries [13].

Another interesting possibility is the role that high osmotic pressure may have on water flow through aquaporins. A study reported in Vandeleur *et al.* [84] showed that a NIP aquaporin (NOD26), which is highly water permeable in dilute solutions, had virtually no water permeability at osmotic pressures above about 2.5 MPa. Considering that grape osmotic pressures are of the order of 4 to 5 MPa it is very likely that these high osmotic pressures, which also occur in the apoplast, are also inhibitory to water permeation through berry aquaporins. It is not known if this effect is a general phenomenon of plant aquaporins, but it is well known in the alga Chara [87]. Since very high osmotic pressures occur as ripening progresses, this could result in reduced permeability of aquaporins and account for the large increase in R_h that several studies have observed after veraison.

Correlation of Expression with Function

Only one study to date has attempted to examine the role of aquaporins in berry hydraulics. Here Choat *et al.* [26] examined the expression of aquaporin genes across development in an attempt to correlate this with changes in R_h. The berry R_h was lowest in the period between 60 and 80 days, and corresponding to this was the highest expression of VvPIP1;3 and VvPIP2;1 the most predominantly expressed PIP1 and PIP2 isoforms. Fouquet *et al.* [72] also found that VvPIP1;3 and VvPIP2;1 were the most predominantly expressed PIPs and peaked in expression at 70 DAA. Both studies found that these aquaporins decreased in expression at later stages corresponding to the reduction in R_h with a larger reduction in PIP1;3 compared to PIP2;1. PIP2;3 was the only PIP that did not decrease at later stages of ripening in both studies.

In an earlier study Picaud *et al.* [88] found that PIP1 aquaporins (probably PIP1;2 and PIP1;3) increased in expression in post-veraison berries, although total PIP1s in three varieties showed slightly different patterns. One of these PIPS had very low water permeability when expressed in Xenopus and seemed to be permeable to glycerol.

It is likely that some PIP aquaporins may transport solutes in addition to water [85, 89]. In tomato the ripening associated membrane protein (TRAMP), which shows homology to PIP aquaporins, when reduced in expression, resulted in reduced accumulation of fructose and glucose but increased malate and citrate [90]. There appeared to be no obvious phenotype of fruit water relations in the TRAMP-silenced plants. The TRAMP protein was later located to the plasma membrane [91], and it remains to be determined how it influences vacuolar accumulation as was originally proposed [90].

Less is known about the changes in expression of TIP isoforms that are normally located on the tonoplast, which could potentially be important in cell expansion and transcellular flow. An ethylene induced aquaporin in grape berry has a beta TIP as its closest Arabidopsis homolog and is increased 1.8- fold by ethylene [92]. This was proposed to be involved in cell expansion since expression of various cell expansion related genes were also increased with ethylene treatment. Fouquet *et al.* (2008) suggested that decreased expression of VvTIP2-1 at veraison could increase R_h since its expression was located in xylem parenchyma early in berry development when water is provided mainly by the xylem. High tonoplast water permeability enables rapid osmotic adjustment [93] and most likely high transcellular water flow [94].

No localisation data are yet available for aquaporin gene expression in post-veraison grape berries. In this respect it would be worthwhile to examine if PIP1;3 and PIP2;1 are co-located in the same cells. Several studies have shown that members of the PIP1 and PIP2 sub-families have synergistic interactions [79, 95] when expressed in Xenopus oocytes, and this is also indicated in plant cells where PIP1-PIP2 interaction is required for PIP1 to be targeted to the plasma membrane [96]. Fetter *et al.* [95] identified amino acid residues in PIP1 that are required for this interaction, and the VvPIP1s identified by Choat *et al.* [26] in grape berries and by [79] in grape roots share 100% similarity in these residues [26]. When these residues are not conserved the synergistic interactions were not observed [97]. The simultaneous expression of PIP1 and PIP2 and their synergistic interaction may result in high berry hydraulic conductivity.

Varietal Differences in Aquaporin Expression

As reported above there are clear differences among varieties concerning xylem water backflow from the berry to the parent vine, having an impact on berry water status, with some varieties undergoing berry weight loss at earlier stages of ripening than others. These differences are related to diverse hydraulic connections between the berry and the vine and may relate to cell vitality in the mesocarp (discussed above), but there is still no clear association with differences in AQP functioning.

Differences among genotypes in AQP functioning in response to these alterations in the environment are likely to occur, reflecting the complex nature of the regulation of water transport through AQPs. This regulation may involve mechanisms of fast control affecting the constitutively expressed protein (activity, gating) or slower adaptive-responses requiring the regulation of gene expression. It is known that targeting MIPs is dependent on the presence/absence of specific sequences allowing their retention in organelles or their trafficking to the plasma membrane [98]. Indeed, re-location of AQPs in response to hormonal stimuli or environmental stresses, such as salt and osmotic stresses, has been described in plants for some TIP and PIP isoforms.

These relocations may occur *via* alterations in the phosphorylation status and thus the aperture of the channel, with important (and fast) consequences on water transport across membranes. A first observation of this phenomenon was done by [99] in *Mesembryanthemum crystallinum*, where the re-location of a TIP from the tonoplast into endosomal compartments was observed in response to hyperosmotic stress and may be part of a homeostatic process to restore cellular osmolarity under osmotic-stress conditions. The dynamics of AQP biogenesis and subcellular localisation was recognised as an important feature of their function [63] and may explain the variable modulation of activity across different genotypes and under different environmental conditions.

Differences in the regulation of AQP expression among grapevine varieties exhibiting different control of water status *via* stomata (iso- and anisohydric) were described in roots [79]. Different patterns of expression of VvPIP1;1 were observed in Chardonnay (stomatal control more of the anisohydric type) and Grenache (more of the isohydric type). While Chardonnay exhibited a significant increase in VvPIP1;1 expression under water stress, Grenache did not show any alteration. However, in roots of re-watered plants the transcript level in Grenache was higher in comparison to control roots. VvPIP1;1 may regulate water transport across roots such that transpirational demand is matched by root water transport capacity. These findings are in accordance with the difference in drought tolerance between the cultivars and suggest that Grenache may be able to better respond to rainfall and irrigation events by up-regulation of VvPIP1;1. Whether such differences among varieties are also observed in regulation of AQPs in berries and the likely impact on berry water relations is still not clear.

Differences in AQP expression throughout berry development have been described, which are partly related to the variety studied [72, 88, 100]. Picaud *et al.* [88] found that total VvPIP1 expression in berries was lowest at pre-veraison and peaked at post-veraison in Chardonnay and Ugni blanc varieties, whereas in Pinto Meunier it occurred at full maturity (harvest). Results similar to those obtained in Pinot Meunier were found in the Portuguese varieties Castelão and Moscatel de Setúbal, with increasing expression of VvPIP1 being observed from the berry green stage until harvest time (Silva, Chaves and Delrot, unpublished). Based on their functional properties Picaud S. *et al.* [88] suggest that some of the AQPs of the subfamily VvPIP1 may also be associated with transport of other solutes, in addition to water exchange accompanying berry ripening and sugar accumulation.

After working with Cabernet Sauvignon over two growing seasons, Fouquet *et al.* [72] report four distinct groups of AQP genes, whose expression ranged from no significant variation during berry development (VvPIP1;1) to expression decreasing at veraison (VvPIP1;2, VvPIP2;2, VvTIP1;2, VvTIP2;1) or after veraison (VvPIP1;3, VvTIP1;1) and finally to expression increasing at veraison (VvPIP2;1, VvPIP2;3). This seems to suggest that some AQPs are linked to specific events during berry development. Also in Cabernet Sauvignon, Schlosser *et al.* [100] found that AQP genes in berry pericarp tissues generally followed a bimodal expression profile, with high levels of expression in the two periods of rapid berry growth (Stages I and III) and low levels of expression corresponding to the slow growth period, Stage II.

In a study of Pinot Noir during three seasons, Pilati *et al.* [59] detected 13 transcripts coding for AQP isoforms. All isoforms were highly expressed in the berry in the pre-veraison phase and three of them, coding for the putative AQP (PIP2;1), were also induced during ripening. The highly dynamic pattern of expression observed in berry aquaporin genes also agrees with the findings of Deluc *et al.* [101], who provided the first functional genomic information for hundreds of genes, illustrating the complex transcriptional regulatory hierarchies occurring during grape berry development. They found out that c.a. 60% of the expressed transcripts were differentially expressed between at least two out of the seven stages of development that were considered, with 28% of the transcripts exhibiting more than 2-folds variation in mRNA expression.

VINE WATER STATUS AND BERRY RIPENING

Vine water status is known to influence berry water status, size and development [102]. Berry composition is affected by a direct effect of water availability on metabolism [103, 104] and indirect effects due to the alteration of the ratio of skin to pulp, with smaller berries occurring in vines subjected to water deficits [105]. The reported increase in skin tannin and anthocyanin that accompanies water deficits seems to result from different sensitivity of berry tissues to water deficits, with the exocarp being less affected than the inner mesocarp [104]. Proteomic and transcriptomic studies in berries from grapevines subjected to different irrigation treatments suggest that metabolic differences in response to water status occur at early stages of berry development [103] and Fig. **2**, confirming that they are partly independent of the effect on berry size.

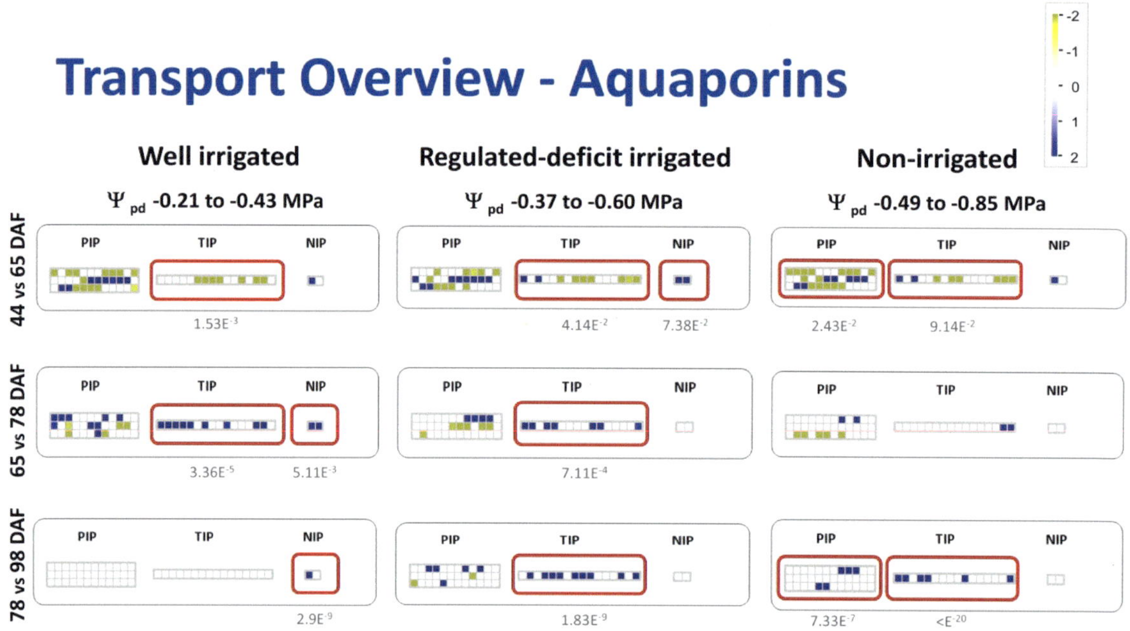

Figure 2: Mapman representation of aquaporins modulation during grape berry development and maturation 44, 65, 78 and 98 days after flowering (DAF) and under well irrigated, regulated deficit irrigation and non-irrigation conditions. Red boxes indicate the classes of aquaporins that were significantly modulated under the conditions studied (p value is indicated in grey). Francisco, Zapater and Chaves (in preparation).

Berry exposure to sunlight, with its components of light and temperature, also affect berry water relations. An excessive cluster exposure may lead to supra-optimal berry temperature or even sunburn, with detrimental effects on grape berry water status and composition [106]. This is particularly important in some warm regions, where fully exposed berries exhibit reduced or even total inhibition of anthocyanin synthesis [107], with anthocyanin metabolism responding to both light and temperature [108, 109]. While intense canopy shading down-regulates gene expression of the anthocyanin biosynthesis pathway, at photon flux densities above 100 μmol m^{-2} s^{-1} berry temperature becomes the driving variable in anthocyanin synthesis [110].

Role of ABA in Berry Ripening

A large variation from year to year in the expression profiles of the genes involved in berry development was reported by Pilati *et al.* [59], in a study performed under field conditions during three seasons. It was found that a large number of genes were modulated only during one or two seasons, with just 13% of the total 1700 transcripts being modulated in all three seasons, even though with different expression profiles. These results suggest a subtle regulation of genes in response to differences in environmental variables, likely mediated by phytohormones, such as ABA, ethylene, brassinosteroids and auxins, as well as by calcium signalling processes [101] (see the chapter by Böttcher and Davies, this edition).

The long distance chemically-mediated signalling of water-status of the root to the shoot, essentially by abscisic acid (ABA) and cytokinin, may also play a role in the alterations of berry metabolism occurring under water and salt stress [103, 111]. It is generally accepted that ABA is a signalling molecule involved in grape berry development and ripening [112, 113] and that ABA has the potential to be synthesised in the berries, since the genes involved in its synthesis were found there [113]. ABA concentration in the berry varies throughout development, with free ABA attaining a peak at veraison and declining thereafter, as observed in Cabernet Sauvignon [113] and in Aragonez (syn. Tempranillo, Zarrouk *et al.*, unpublished), or showing two peaks, at two weeks pre-veraison and at late veraison, followed by a decline in ripened berries, as reported in Merlot grapes [112]. Moreover, Wheeler *et al.* [113] reported that the expression profile of genes crucial to ABA synthesis did not correlate with ABA concentrations, suggesting that ABA accumulation in the berries is under a complex control.

Since very early on, ABA was believed to mediate drought-induced enhancement of assimilate unloading in crop grains and in fleshy fruits, including grape berry [114]. Pan *et al.* [115] reported that ABA activates both the soluble and cell-wall acid invertases during fruit development and that this was an ABA dose and a pH-dependent effect. The data suggest that ABA acts on acid invertases at all levels – mRNA transcription, translation and post-translation, the latter likely involving reversible protein phosphorylation. It has been suggested that the onset of ripening derives from the integration of multiple signals including sugars and ABA [116]. They showed that berries failed to synthesise anthocyanins when sugar import into the berry was disrupted *via* phloem girdling prior to the onset of ripening. They also found out that the onset of ripening could be induced in cultured immature berries treated with sugar and ABA. Furthermore, the expression of the grape hexose transporter *VvHT1* was shown to be regulated by both ABA and hexose [117] (see the chapter by Davies *et al.*, this edition).

ABA also regulates the expression of transcripts of tonoplast and plasma membrane AQPs [63, 89, 99] and induces transcription factors that regulate the expression of PIPs [62, 118]. Concerning the effects of exogenous ABA it seems that it affects a larger number of PIP isoforms than water deficit [119]. In addition, the increase in PIP mRNA expression is often transient and dependent on ABA concentration [120] and does not necessarily result in an increase in PIP protein content [121]. Evidence for a relationship between AQP activity and calcium was recently provided in salinity-stressed plants [122], corroborating earlier data showing that environmental stresses, such as water and temperature stresses, trigger calcium signalling cascades (see review by Maurel [63]).

Impact of Water Deficits

The change in AQP mRNA and protein contents in response to water deficit is still under debate given that different type of responses (up- or down regulation) have been reported depending on the AQP gene, the species and the organs (for a review see [89, 123]). Furthermore, the effects of AQP over-expression on drought tolerance are also not clear since there are reports of beneficial impacts, as observed by Yua *et al.* [124] with BnPIP1 from *Brassica napus* in tobacco, or negative ones, as those reported in the work of Katsuhara *et al.* [125], where over-expression of a barley AQP led to increased salt sensitivity in transgenic rice plants.

As for the effects of vine water status on grape berry AQPs, work in progress in the laboratory of MM Chaves with the variety Aragonez (syn. Tempranillo) shows that the transcript profile of AQPs throughout berry development is modulated by the irrigation regime, although there is a common tendency to all treatments - an increasing expression until veraison, followed by a decline at harvest time (Fig. **2**). In the three water regimes TIPs tend to be more affected during berry development than PIPs, except in non-irrigated vines. TIPs reached their peak of expression earlier in berry development in well-irrigated vines (with ψ_{pd} ranging from – 0.21 to – 0.43 MPa) than in vines under water deficits (both in vines under regulated deficit irrigation, with ψ_{pd} ranging from – 0.37 to – 0.60 MPa and in non-irrigated vines, with ψ_{pd} ranging from – 0.49 to – 0.85 MPa). In the latter vines, the increase in TIPs expression accentuates at maturity and PIPs are significantly up-regulated in the early stages of berry development as well as, later on, at maturity. Expression of NIPs in the berries is only significantly altered in well-irrigated vines attaining a peak at veraison. Obviously, these results have to be analysed taking into account that regulation of AQPs in response to water stress can take place at different levels - transcriptional, post- transcriptional and by post-translational modifications [126].

This type of information, together with the findings that water deficits advance the onset of colour development in berries, with flavonoid and anthocyanin specific genes reflecting that advancement [116], suggests that management of plant and berry water status (for example, *via* deficit irrigation) may be used to manipulate the timing of the onset of berry ripening under field conditions [103].

CONCLUSIONS

Loss of xylem physical integrity or ability to conduct water cannot explain the increase in hydraulic resistance (R_h) of berries occurring after veraison. The increase in R_h also does not completely explain the apparent hydraulic isolation of berries after veraison [26]. We cannot exclude the possibility that cell membranes external to the xylem in the berry are contributing to increased R_h. There is correlative evidence that changes in AQP expression may be linked to changes in berry hydraulics, but further *in situ* functional analysis is required, as well as the specific location of the likely AQP candidates in the berry. It is possible to test post-translational regulation of AQPs using inhibitors and pressure probe techniques. Comparison between varieties that show differences in R_h may also provide clues to which AQP candidates are involved. We also require more precise quantification of water flow *via* the phloem after veraison, together with quantification of recycling of excess phloem water through xylem. Transpiration from berries is also emerging as an important variable in determining the rate of sugar loading.

Backflow from the berry to the parent vine *via* the xylem can occur but is variety-dependent and probably determined by two important parameters: the membrane integrity of berry cells late in ripening and the hydraulic conductance *via* the xylem to the vine. Berry splitting under wet conditions may also be linked to these processes [43]. Determining the link between berry cell death and berry water relations and quality may unlock practices to better manage development of berry flavour. There appears to be a link emerging between the degradation of cell membranes and lipids and berry water relations and flavour development.

The rapid change in turgor that occurs at the early stages of veraison may trigger other changes in metabolism and transport that is required for ripening to proceed [49]. More precise kinetics of the change in turgor of the berry are required, and in this respect the recently developed patch-clamp turgor probe developed by Zimmermann and colleagues provides great promise [127].

Vine water status will impact on berry development and this occurs *via* a variety of mechanisms. Given that ABA is implicated in berry ripening, it is likely that this is also linked to changes in berry water relations and water flow. Given the large differences that are being observed between isohydric and anisohydric varieties in both leaf and root regulation of water flow, it is likely that this will also be observed in different patterns of water regulation in berries.

Differences observed between varieties have to be considered as a great opportunity to obtain further insights into the regulatory mechanisms of berry water balance during ripening.

ABBREVIATIONS

ABA = Abscisic acid

AQP = Aquaporin

GIP = GlypF-like intrinsic protein

HIP = Hybrid intrinsic protein

LOX = Lypoxygenases

MIP = Major intrinsic protein

NIP = Nodulin-like intrinsic protein

P = Turgor pressure

PIP = Plasma membrane intrinsic protein

Rh = Hydraulic resistance (reciprocal of hydraulic conductance)

SIP = Small basic intrinsic protein

TIP = Tonoplast intrinsic protein

XIP = X intrinsic protein

ACKNOWLEDGEMENTS

SD Tyerman acknowledges grants from the Grape and Wine Research and Development Corporation and the Australian Research Council and M M Chaves grants from the Portuguese Foundation for Science and Technology (PTDC/AGR-ALI/100636 and 099154/2008).

REFERENCES

[1] Ribéreau-Gayon P, Dubourdieu D, Donèche B, Lonvaud A. Handbook of Enology, The Microbiology of Wine and Vinification. John Wiley & Sons. West Sussex, England 2006.

[2] Coombe BG. Research on development and ripening of the grape berry. Am J Enol Viticult 1992; 43: 101-10.

[3] Greenspan MD, Shackel KD, Matthews MA. Developmental changes in the diurnal water budget of the grape berry exposed to water deficits. Plant Cell Environ 1994; 17: 811-820.

[4] Lang A, Thorpe MR. Xylem, phloem and transpiration flows in a grape:application of a technique for measuring the volume of attached fruits to high resolution using Archimedes' Principle. J Exp Bot 1989; 219: 1069-1078.

[5] Greenspan MD, Schultz HR, Matthews MA. Field evaluation of water transport in grape berries during water deficits. Physiol Plantarum 1996; 97: 55-62.

[6] Matthews MA, Shackel KA. Growth and water transport in fleshy fruit. In: Holbroo NM, Zwieniecki MA, Eds. Vascular Transport in Plants. Academic Press, Burlington, California 2005; pp. 189-97.

[7] Thomas TR, Matthews MA, Shackel KA. Direct *in situ* measurement of cell turgor in grape (*Vitis vinifera* L.) berries during development and in response to plant water deficits. Plant Cell Environ 2006; 29: 993-1001.

[8] Dayod M, Tyerman SD, Leigh RA, Gilliham M. Calcium storage in plants and the implications for calcium biofortification. Protoplasma 2010; 247(3-4): 215-31.

[9] Rogiers S, Keller M, Holzapfel BP, Virgona JM. Accumulation of potassium and calcium by ripening berries on field vines of Vitis vinifera (L) cv. Shiraz. Aust J Grape Wine R 2000; 6(3): 240-3.

[10] Rogiers S, Greer D, Hatfield J, Orchard B, Keller M. Solute transport into Shiraz berries during development and late-ripening shrinkage. Am J Enol Viticult 2006; 57: 73-80.

[11] Choi HM, Son IC, Kim D. Effects of calcium concentrations of coating bag on pericarp structure and berry cracking in 'Kyoho' grape (*Vitis* sp.). Korean J Hortic Sci 2010; 28(4): 561-66.

[12] Alleva K, Niemietz CM, Maurel C, Parisi M, Tyerman SD, Amodeo G. Plasma membrane of Beta vulgaris storage root shows high water channel activity regulated by cytoplasmic pH and a dual range of calcium concentrations. J Exp Bot 2006; 57(3): 609-21.

[13] Gilliham M, Dayod M, Hocking B, *et al.* Exploring the link between water movement and calcium storage in plant leaves. J Exp Bot 2011(In press).

[14] Creasy G, Price SF, Lombard PB. Evidence for xylem discontinuity in pinot noir and merlot grapes: Dye uptake and mineral composition during berry maturation. Am J Enol Viticult 1993; 44: 187-92.

[15] Findlay N, Oliver KJ, Nii N, Coombe BG. Solute accumulation by grape pericarp cells .4. perfusion of pericarp apoplast *via* the pedicel and evidence for xylem malfunction in ripening berries. J Exp Bot 1987; 38(189): 668-79.

[16] Rogiers S, Smith JA, White R, Keller M, Holzapfel BP, Virgona JM. Vascular function in berries of *Vitis vinifera* (L) cv. Shiraz. Aust J Grape Wine R 2001; 7: 46-51.

[17] Bondada BR, Matthews MA, Shackel KA. Functional xylem in the post-veraison grape berry. J Exp Bot 2005; 56: 2949-57.

[18] Keller M, Smith JP, Bondada BR. Ripening grape berries remain hydraulically connected to the shoot. J Exp Bot 2006; 57: 2577-87.

[19] Tilbrook J, Tyerman SD. Cell death in grape berries: varietal differences linked to xylem pressure and berry weight loss. Funct Plant Biol 2008; 35(3): 173-84.

[20] Tilbrook J, Tyerman SD. Hydraulic connection of grape berries to the vine: varietal differences in water conductance into and out of berries, and potential for backflow. Funct Plant Biol 2009; 36(6): 541-50.

[21] Tyerman SD, Tilbrook J, Pardo C, Kotula L, Sullivan W, Steudle E. Direct measurement of hydraulic properties in developing berries of *Vitis vinifera* L. cv Shiraz and Chardonnay. Aust J Grape Wine R 2004; 10: 170-81.

[22] Chatelet DS, Rost TL, Matthews MA, Shackel KA. The peripheral xylem of grapevine (*Vitis vinifera*) berries. 2. Anatomy and development. J Exp Bot 2008; 59(8): 1997-2007.

[23] Chatelet DS, Rost TL, Shackel KA, Matthews MA. The peripheral xylem of grapevine (*Vitis vinifera*). 1. Structural integrity in post-veraison berries. J Exp Bot 2008; 59(8): 1987-96.

[24] Xia GH, Zhang DP. Intercellular symplastic connection and isolation of the unloading zone in flesh of the developing grape berry. Acta Bot Sin 2000;42(9): 898-904.

[25] Zhang B, Chen KS, Bowen J, *et al.* Differential expression within the LOX gene family in ripening kiwifruit. J Exp Bot 2006; 57(14): 3825-36.

[26] Choat B, Gambetta GA, Shackel KA, Matthews MA. Vascular function in grape berries across development and its relevance to apparent hydraulic isolation. Plant Physiol 2009; 151(3): 1677-87.

[27] Fuentes S, Sullivan W, Tilbrook J, Tyerman S. A novel analysis of grapevine berry tissue demonstrates a variety-dependent correlation between tissue vitality and berry shrivel. Aust J Grape Wine R 2010; 16(2): 327-36.

[28] Sadras V, McCarthy MG. Qantifying the dynamics of sugar concentration in berries of "*Vitis vinifera*" cv. Shiraz: a novel approach based on allometric analysis. Aust J Grape Wine R 2007; 13: 66-71.

[29] McCarthy MG. The effect of transient water deficit on berry development of cv. Shiraz (*Vitis vinifera* L.). Aust J Grape Wine R 1997; 3: 102-8.

[30] Rogiers S, Hatfield J, Keller M. Irrigation, nitrogen, and rootstock effects on volume loss of berries from potted Shiraz vines. Vitis 2004; 43(1): 3-6.

[31] Rogiers SY, Hatfield JM, Jaudzems VG, White RG, Keller M. Grape berry cv. shiraz epicuticular wax and transpiration during ripening and preharvest weight loss. Am J Enol Viticult 2004; 55(2): 121-7.

[32] Krasnow MN, Matthews MA, Smith RJ, Benz J, Weber E, Shackel KA. Distinctive symptoms differentiate four common types of berry shrivel disorder in grape. Calif Agric 2010; 64(3): 155-9.

[33] McCarthy MG, Coombe BG. Berry weight loss and phloem transport. Aust J Grape Wine R 1999; 5: 17-21.

[34] Rogiers S, Hatfield J, Gunta Jaudzems V, White R, Keller M. Grape berry cv. Shiraz epicuticular wax and transpiration during ripening and preharvest weight loss. Am J Enol Viticult 2004; 55: 121-7.

[35] Dreier LP, Stoll GS, Ruffner HP. Berry ripening and evapotranspiration in *Vitis vinifera* L. Am J Enol Viticult 2000; 51: 340-46.

[36] Rebucci B, Poni S, Intrieri C, Magnanini E, Lakso AN. Effects of manipulated grape berry transpiration on post-veraison sugar accumulation. Aust J Grape Wine R 1997; 3(2): 57-65.

[37] Sadras V, Petrie P. Quantifying the onset, rate and duration of sugar accumulation in berries from commercial vineyards in contrasting climates of Australia. Aust J Grape Wine R 2011 (In Press).

[38] Greer DH, Rogiers SY. Water flux of *Vitis vinifera* L. cv. shiraz bunches throughout development and in relation to late-season weight loss. Am J Enol Viticult 2009; 60(2): 155-63.

[39] Ollat N, Diakou-Verdin P, Carde JP, Barrieu F, Gaudillere JP, Moing A. Grape berry development: A review J Int Sci Vigne Vin 2002;36(3): 109-31.

[40] Palliotti A, Cartechini A. Developmental changes in gas exchange activity in flowers, berries, and tendrils of field-grown Cabernet Sauvignon. Am J Enol Viticult 2001; 52(4): 317-23.

[41] Casado CG, Heredia A. Ultrastructure of the cuticle during growth of the grape berry (*Vitis vinifera*). Physiol Plantarum 2001; 111(2): 220-24.

[42] Leide J, Hildebrandt U, Reussing K, Riederer M, Vogg G. The developmental pattern of tomato fruit wax accumulation and its impact on cuticular transpiration barrier properties: Effects of a deficiency in a beta-ketoacyl-coenzyme A synthase (LeCER6). Plant Physiol 2007; 144(3): 1667-79.

[43] Clarke SJ, Hardie WJ, Rogiers SY. Changes in susceptibility of grape berries to splitting are related to impaired osmotic water uptake associated with losses in cell vitality. Aust J Grape Wine R 2010; 16(3): 469-76.

[44] Considine JA, Kriedemann PE. Fruit splitting in grapes - determination of critical turgor pressure. Aust J Agr Res 1972; 23(1): 17-24.

[45] Huang X-M, Huang H-B. Early post-veraison growth in grapes:evidence for a two-step mode of berry enlargement. Aust J Grape Wine R 2001; 7: 132-36.

[46] Lang A, Düring H. Partitioning control by water potential gradient:evidence for compartmentation breakdown in grape berries. J Exp Bot 1991; 42(9): 1117-22.

[47] Patrick JW. Phloem unloading:sieve element unloading and post-seive element transport. Annu Rev Plant Phys 1997; 48: 191-222.

[48] Lang A, Thorpe MR. Water potential, translocation and assimilate partitioning. J Exp Bot 1986; 37(177): 495-503.

[49] Thomas TR, Shackel KA, Matthews MA. Mesocarp cell turgor in *Vitis vinifera* L. berries throughout development and its relation to firmness, growth, and the onset of ripening. Planta 2008; 228(6): 1067-76.

[50] Steudle E, Zimmermann U, Luttge U. Effect of turgor pressure and cell-size on wall elasticity of plant-cells . Plant Physiol 1977; 59(2): 285-9.

[51] Wada H, Shackel KA, Matthews MA. Fruit ripening in Vitis vinifera: apoplastic solute accumulation accounts for pre-veraison turgor loss in berries. Planta 2008; 227(6): 1351-61.

[52] Matthews MA, Thomas TR, Shackel KA. Fruit ripening in Vitis vinifera L.: possible relation of veraison to turgor and berry softening. Aust J Grape Wine R 2009; 15(3): 278-83.

[53] Tyerman SD, Steudle E. Comparison between osmotic and hydrostatic water flows in a higher-plant cell - determination of hydraulic conductivities and reflection coefficients in isolated epidermis of *Tradescantia virginiana.* Aust J Plant Physiol 1982; 9(4): 461-79.

[54] Krasnow M, Matthews M, Shackel K. Evidence for substantial maintenance of membrane integrity and cell *via*bility in normally developing grape (*Vitis vinifera* L.) berries throughout development. J Exp Bot 2008; 59(4): 849-59.

[55] Rogiers SY, Kumar GNM, Knowles NR. Maturation and ripening of fruit of Amelanchier alnifolia Nutt. are accompanied by increasing oxidative stress. Ann Bot 1998; 81(2): 203-11.

[56] Chen GP, Hackett R, Walker D, Taylor A, Lin ZF, Grierson D. Identification of a specific isoform of tomato lipoxygenase (TomloxC) involved in the generation of fatty acid-derived flavor compounds. Plant Physiol 2004; 136(1): 2641-51.

[57] Podolyan A, White J, Jordan B, Winefield C. Identification of the lipoxygenase gene family from Vitis vinifera and biochemical characterisation of two 13-lipoxygenases expressed in grape berries of Sauvignon Blanc. Funct Plant Biol 2010; 37(8): 767-84.

[58] Gobel C, Feussner I, Rosahl S. Lipid peroxidation during the hypersensitive response in potato in the absence of 9-lipoxygenases. J Biol Chem 2003; 278(52): 52834-40.

[59] Pilati S, Perazzolli M, Malossini A, *et al.* Genome-wide transcriptional analysis of grapevine berry ripening reveals a set of genes similarly modulated during three seasons and the occurrence of an oxidative burst at veraison. BMC Genomics 2007; 8:428.

[60] Maurel C, Tacnet F, Guclu J, Guern J, Ripoche P. Purified vesicles of tobacco cell vacuolar and plasma membranes exhibit dramatically different water permeability and water channel activity. P Natl Acad Sci USA 1997; 94(13): 7103-7108.

[61] Hachez C, Zelazny E, Chaumont F. Modulating the expression of aquaporin genes in planta: A key to understand their physiological functions? BBA-Biomembranes 2006; 1758(8): 1142-56.

[62] Kaldenhoff R, Kolling A, Richter G. Regulation of the *Arabidopsis thaliana* aquaporin gene AthH2 (PIP1b). J Photochem Photobiol B 1996; 36(3): 351-54.

[63] Maurel C. Plant aquaporins: novel functions and regulation properties. FEBS Lett 2007; 581(12): 2227-36.

[64] Sutka M, Alleva K, Parisi M, Amodeo G. Tonoplast vesicles of *Beta vulgaris* storage root show functional aquaporins regulated by protons. Biol Cell 2005; 97(11): 837-46.

[65] Forrest KL, Bhave M. Major intrinsic proteins (MIPs) in plants: a complex gene family with major impacts on plant phenotype. Funct Integr Genom 2007; 7(4): 263-89.

[66] Johanson U, Karlsson M, Johansson I, *et al.* The complete set of genes encoding major intrinsic proteins in arabidopsis provides a framework for a new nomenclature for major intrinsic proteins in plants. Plant Physiol 2001;126(4): 1358-69.

[67] Gustavsson S, Lebrun AS, Norden K, Chaumont F, Johanson U. A novel plant major intrinsic protein in Physcomitrella patens most similar to bacterial glycerol channels. Plant Physiol 2005; 139(1): 287-95.

[68] Danielson JAH, Johanson U. Unexpected complexity of the Aquaporin gene family in the moss *Physcomitrella* patens. BMC Plant Biol 2008; 8: 45.
[69] Sade N, Vinocur BJ, Diber A, *et al.* Improving plant stress tolerance and yield production: is the tonoplast aquaporin SlTIP2;2 a key to isohydric to anisohydric conversion? New Phytol 2009; 181(3): 651-61.
[70] Jaillon O, Aury JM, Noel B, *et al.* The grapevine genome sequence suggests ancestral hexaploidization in major angiosperm phyla. Nature 2007; 449(7161): 463-65.
[71] Shelden MC, Howitt SM, Kaiser BN, Tyerman D. Identification and functional characterisation of aquaporins in the grapevine, Vitis vinifera. Funct Plant Biol 2009; 36(12): 1065-1078.
[72] Fouquet R, Leon C, Ollat N, Barrieu F. Identification of grapevine aquaporins and expression analysis in developing berries. Plant Cell Rep 2008; 27(9): 1541-50.
[73] Bramley H, Turner NC, Turner DW, Tyerman SD. Roles of Morphology, Anatomy, and Aquaporins in Determining Contrasting Hydraulic Behavior of Roots. Plant Physiol 2009; 150(1): 348-64.
[74] Cochard H, Venisse JS, Barigah TS, *et al.* Putative role of aquaporins in variable hydraulic conductance of leaves in response to light. Plant Physiol 2007; 143(1): 122-33.
[75] Lovisolo C, Secchi F, Nardini A, Salleo S, Buffa R, Schubert A. Expression of PIP1 and PIP2 aquaporins is enhanced in olive dwarf genotypes and is related to root and leaf hydraulic conductance. Physiol Plantarum 2007; 130(4): 543-51.
[76] Parent B, Hachez C, Redondo E, Simonneau T, Chaumont F, Tardieu F. Drought and Abscisic Acid Effects on Aquaporin Content Translate into Changes in Hydraulic Conductivity and Leaf Growth Rate: A Trans-Scale Approach. Plant Physiol 2009; 149(4): 2000-12.
[77] Postaire O, Tournaire-Roux C, Grondin A, *et al.* A PIP1 aquaporin contributes to hydrostatic pressure-induced water transport in both the root and rosette of Arabidopsis. Plant Physiol 2010; 152(3): 1418-30.
[78] Sade N, Gebretsadik M, Seligmann R, Schwartz A, Wallach R, Moshelion M. The role of tobacco aquaporin1 in improving water use efficiency, hydraulic conductivity, and yield production under salt stress. Plant Physiol 2010; 152(1): 245-54.
[79] Vandeleur RK, Mayo G, Shelden MC, Gilliham M, Kaiser BN, Tyerman SD. The role of plasma membrane intrinsic protein aquaporins in water transport through roots: diurnal and drought stress responses reveal different strategies between isohydric and anisohydric cultivars of grapevine. Plant Physiol 2009; 149(1): 445-60.
[80] Bramley H, Turner NC, Turner DW, Tyerman SD. The contrasting influence of short-term hypoxia on the hydraulic properties of cells and roots of wheat and lupin. Funct Plant Biol 2010; 37(3): 183-93.
[81] Sadok W, Sinclair TR. Transpiration response of 'slow-wilting' and commercial soybean (Glycine max (L.) Merr.) genotypes to three aquaporin inhibitors. J Exp Bot 2010; 61(3): 821-9.
[82] Voicu MC, Zwiazek JJ. Inhibitor studies of leaf lamina hydraulic conductance in trembling aspen (Populus tremuloides Michx.) leaves. Tree Physiol 2010; 30(2): 193-204.
[83] Tyerman SD, Niemietz CM, Bramley H. Plant aquaporins: multifunctional water and solute channels with expanding roles. Plant Cell Environ 2002; 25(2): 173-194.
[84] Vandeleur R, Niemietz C, Tilbrook J, Tyerman SD. Roles of aquaporins in root responses to irrigation. Plant and Soil 2005; 274(1-2): 141-61.
[85] Maurel C, Santoni V, Luu DT, Wudick MM, Verdoucq L. The cellular dynamics of plant aquaporin expression and functions. Curr Opin Plant Biol 2009; 12(6): 690-98.
[86] Boursiac Y, Boudet J, Postaire O, Luu DT, Tournaire-Roux C, Maurel C. Stimulus-induced downregulation of root water transport involves reactive oxygen species-activated cell signalling and plasma membrane intrinsic protein internalization. Plant J 2008; 56(2): 207-18.
[87] Ye Q, Muhr J, Steudle E. A cohesion/tension model for the gating of aquaporins allows estimation of water channel pore volumes in Chara. Plant Cell Environ 2005; 28(4): 525-35.
[88] Picaud S, Becq F, Dedaldechamp F, Ageorges A, Delrot S. Cloning and expression of two plasma membrane aquaporins expressed during the ripening of grape berry. Funct Plant Biol 2003; 30(6): 621-30.
[89] Maurel C, Verdoucq L, Luu DT, Santoni V. Plant aquaporins: membrane channels with multiple integrated functions. Annu Rev Plant Biol 2008; 59: 595-624.
[90] Chen GP, Wilson ID, Kim SH, Grierson D. Inhibiting expression of a tomato ripening-associated membrane protein increases organic acids and reduces sugar levels of fruit. Planta 2001; 212(5-6): 799-807.
[91] Kim SH, Grierson D. Subcellular localisation and silencing of ripening-associated membrane protein (TRAMP) in tomato *Lycopersicon esculentum* Mill. Plant Science 2005; 169(6): 1022-9.
[92] Chervin C, Tira-umphon A, Terrier N, Zouine M, Severac D, Roustan JP. Stimulation of the grape berry expansion by ethylene and effects on related gene transcripts, over the ripening phase. Physiol Plantarum 2008; 134(3): 534-46.

[93] Tyerman SD, Bohnert HJ, Maurel C, Steudle E, Smith JAC. Plant aquaporins: their molecular biology, biophysics and significance for plant water relations. J Exp Bot 1999; 50: 1055-71.

[94] Barrieu F, Chaumont F, Chrispeels MJ. High expression of the tonoplast aquaporin ZmTIP1 in epidermal and conducting tissues of maize. Plant Physiol 1998; 117(4): 1153-63.

[95] Fetter K, Van Wilder V, Moshelion M, Chaumont F. Interactions between plasma membrane aquaporins modulate their water channel activity. Plant Cell 2004; 16(1): 215-28.

[96] Zelazny E, Borst JW, Muylaert M, Batoko H, Hemminga MA, Chaumont F. FRET imaging in living maize cells reveals that plasma membrane aquaporins interact to regulate their subcellular localization. *P Natl Acad Sci USA* 2007; *104*(30): 12359-64.

[97] Zhou Y, Setz N, Niemietz C, *et al.* Aquaporins and unloading of phloem-imported water in coats of developing bean seeds. Plant Cell Environ 2007; 30(12): 1566-77.

[98] Gomes D, Agasse A, Thiebaud P, Delrot S, Geros H, Chaumont F. Aquaporins are multifunctional water and solute transporters highly divergent in living organisms. Biochim Biophys Acta 2009; 1788(6) 1213-28.

[99] Vera-Estrella R, Barkla BJ, Bohnert HJ, Pantoja O. Novel regulation of aquaporins during osmotic stress. Plant Physiol 2004; 135(4): 2318-29.

[100] Schlosser J, Olsson N, Weis M, *et al.* Cellular expansion and gene expression in the developing grape (*Vitis vinifera* L.) Protoplasma 2008; 232(3-4): 255-65.

[101] Deluc LG, Grimplet J, Wheatley MD, *et al.* Transcriptomic and metabolite analyses of Cabernet Sauvignon grape berry development. BMC Genomics 2007; 8: 429.

[102] Bravdo B, Hepner Y, Loinger C, Tabacman H. Effect of irrigation and crop level on growth, yield and wine quality of Cabernet Sauvignon. Am J Enol Viticult 1985; 36: 132 - 39.

[103] Chaves MM, Zarrouk O, Francisco R, *et al.* Grapevine under deficit irrigation: hints from physiological and molecular data. Ann Bot 2010; 105(5): 661-76.

[104] Roby G, Harbertson JF, Adams DA, Matthews MA. Berry size and vine water deficits as factors in winegrape composition: anthocyanins and tannins. Aust J Grape Wine R 2004; 10: 100-7.

[105] Kennedy JA, Matthews MA, Waterhouse AL. Effect of Maturity and Vine Water Status on Grape Skin and Wine Flavonoids. Am J Enol Vitic 2002; 53(4): 268-74.

[106] Santos T, Lopes CM, Rodrigues ML, *et al.* Partial rootzone drying irrigation affects cluster microclimate improving fruit composition of 'Moscatel' field-grown grapevines. Scientia Horticulturae 2007; 112: 321–30.

[107] Spayd SE, Tarara JM, Mee DL, Ferguson JC. Separation of sunlight and temperature effects on the composition of *Vitis vinifera* cv. Merlot berries. Am J Enol Vitic 2002; 53(3): 171-82.

[108] Haselgrove L, Botting D, van Heeswijck R, *et al.* Canopy microclimate and berry composition: The effect of bunch exposure on the phenolic composition of *Vitis vinifera* L cv. Shiraz grape berries. Aust J Grape Wine R 2000; 6: 141-49.

[109] Smart RE, Smith SM, Winchester RV. Light Quality and Quantity Effects on Fruit Ripening for Cabernet Sauvignon. Am J Enol Vitic 1988; 39(3): 250-8.

[110] Keller M. Managing grapevines to optimise fruit development in a challenging environment: a climate change primer for viticulturists. Aust J Grape Wine R 2010; 16(supplement s1): 56-9.

[111] Lovisolo C, Perrone I, Carra A, *et al.* Drought-induced changes in development and function of grapevine (*Vitis* spp.) organs and in their hydraulic and non hydraulic interactions at the whole plant level: a physiological and molecular update. Funct Plant Biol 2010; 37: 98-116.

[112] Owen SJ, Lafond MD, Bowen P, Bogdanoff C, Usher K, Abrams SR. Profiles of abscisic acid and its catabolites in developing merlot grape (*Vitis vinifera*) berries. Am J Enol Vitic 2009; 60(3): 277-84.

[113] Wheeler S, Loveys B, Ford C, Davies C. The relationship between the expression of abscisic acid biosynthesis genes, accumulation of abscisic acid and the promotion of Vitis vinifera L. berry ripening by abscisic acid. Aust J Grape Wine R 2009; 15(3): 195-204.

[114] Coombe BG. The grape berry as a sink. Acta Horticulturae 1989; 239: 149-58.

[115] Pan QH, Li MJ, Peng CC, *et al.* Abscisic acid activates acid invertases in developing grape berry. Physiol Plantarum 2005; 125(2): 157-70.

[116] Gambetta GA, Matthews MA, Shaghasi TH, McElrone AJ, Castellarin SD. Sugar and abscisic acid signaling orthologs are activated at the onset of ripening in grape. Planta 2010; 232(1): 219-34.

[117] Conde C, Agasse A, Glissant D, Tavares R, Geros H, Delrot S. Pathways of glucose regulation of monosaccharide transport in grape cells. Plant Physiol 2006; 141(4): 1563-77.

[118] Shinozaki K, Yamaguchi-Shinozaki K, Mizoguchi T, *et al.* Molecular responses to water stress in Arabidopsis thaliana. J Plant Res 1998; 111: 345–51.
[119] Jang JY, Kim DG, Kim YO, Kim JS, Kang H. An expression analysis of a gene family encoding plasma membrane aquaporins in response to abiotic stresses in Arabidopsis thaliana. Plant Mol Biol 2004; 54(5): 713-25.
[120] Zhu C, Schraut D, Hartung W, Schaffner AR. Differential responses of maize MIP genes to salt stress and ABA. J Exp Bot 2005; 56(421): 2971-81.
[121] Aroca R, Ferrante A, Vernieri P, Chrispeels MJ. Drought, abscisic acid and transpiration rate effects on the regulation of PIP aquaporin gene expression and abundance in Phaseolus vulgaris plants. Ann Bot 2006; 98(6): 1301-10.
[122] Cabanero FJ, Martinez-Ballesta MC, Teruel JA, Carvajal M. New evidence about the relationship between water channel activity and calcium in salinity-stressed pepper plants. Plant Cell Physiol 2006; 47(2): 224-33.
[123] Kaldenhoff R, Ribas-Carbo M, Sans JF, Lovisolo C, Heckwolf M, Uehlein N. Aquaporins and plant water balance. Plant Cell Environ 2008; 31(5): 658-66.
[124] Yua Q, Hua Y, Lib J, Wua Q, Lin Z. Sense and antisense expression of plasma membrane aquaporin BnPIP1 from Brassica napus in tobacco and its effects on plant drought resistance. Plant Sci 2005; 169: 647-56.
[125] Katsuhara M, Koshio K, Shibasaka M, Hayashi Y, Hayakawa T, Kasamo K. Over-expression of a barley aquaporin increased the shoot/root ratio and raised salt sensitivity in transgenic rice plants. Plant Cell Physiol 2003; 44(12): 1378-83.
[126] Tyerman SD, Vandeleur RK, Shelden MC, *et al.* Water transport and aquaporins in grapevine. In: Roubelakis-Angelakis KA, Ed. Grapevine Molecular Physiology & Biotechnology. Springer, The Netherlands 2010; pp. 73-104.
[127] Ruger S, Netzer Y, Westhoff M, *et al.* Remote monitoring of leaf turgor pressure of grapevines subjected to different irrigation treatments using the leaf patch clamp pressure probe. Aust J Grape Wine R 2010; 16(3): 405-12.

CHAPTER 2

Mineral Compounds in the Grape Berry

V. Martins[1,2], A. Cunha[1,2], H. Gerós[1,2,*], M. Hanana[3] and E. Blumwald[4]

[1]Departamento de Biologia, Universidade do Minho, Campus de Gualtar, 4710-057 Braga, Portugal; [2]Centro de Investigação e de Tecnologias Agro-Ambientais e Biológicas (CITAB), Portugal; [3]Center of Biotechnology of Borj Cédria, BP 901, Hammam-Lif 2050, Tunisia and [4]Dept of Plant Sciences, University of California, One Shields Ave, Davis, CA 95616, USA

Abstract: Both organic and inorganic compounds are vital for grapevine health and maintenance of vigour suitable for fruit production. In particular, the mineral soil composition has an essential influence on grape quality and on the organoleptic properties of wine. Grape berries are very rich in K; however, other mineral elements such as N, Ca, P, Mg, S and several micronutrients affect berry development and maturation in ways that have not always been acknowledged. Some of these compounds directly affect berry set and yield, whereas others act indirectly by modulating vine physiology. In addition, the complex interaction between mineral elements and organic molecules also renders different outcomes in vine and grape berry. The present chapter provides an overview on the dynamics of the major mineral compounds in the grapevine, especially in the berry, and on the contribution of each element to berry quality and yield. Special emphasis is given to K, since it is the main cation in must and wine and plays a decisive role in berry development and wine quality.

Keywords: Calcium, Copper, Iron, Macronutrients, Magnesium, Micronutrients, Molybdenum, Nitrogen, Osmolyte, Phosphorus, Potassium, Potassium Transporters, Sodium, Sulfur, Zinc.

INTRODUCTION

Mineral elements are vital for plant development but a balanced mineral supply is of paramount importance in the vineyard to avoid excessive vegetative vigour or mineral deficiency, both needed to sustain plant equilibrium [1]. The assimilation of minerals in berries is highly dependent on the vine content for each element and may be affected by several factors such as vine water status [1] and soil tillage systems [2]. The roles of the major macronutrients in grapevine and their importance for fruit set as well as assimilation patterns are presented in Table **1**. K^+ is the major cation in ripe grape berries and has a pivotal role in the quality of the fruit, playing important roles in enzyme activation, membrane transport and osmotic regulation. For these reasons, the amount of information available in the literature regarding this ion is much higher than for other mineral elements that also perform important functions in berry physiology and development.

More than 97% of the plant body biomass is represented by four elements: C, H, O and N. In particular, N is essential for the formation of proteins, enzymes and coenzymes, nucleic acids, chlorophyll, vitamins and hormones (like auxins and cytokinins), and is also necessary for the production of oil, resins [3, 4] and some secondary metabolites, such as the alkaloids. However, contrarily to the first three elements that enter the plants in the form of water (H) or molecular gases (C and O), usually available in nature in generous amounts, N concentration in the soils is generally low, which is why it consists of one of the major limiting nutrients for plant development. In nature, N is present in different forms, including molecular N_2, volatile NH_3 or N oxides (NO_x), mineral (NO_3^- and NH_4^+) and organic N (free amino acids, proteins and other nitrogenated organic compounds). Amino acids play essential roles in berry quality.

After N, P is the second most frequently limiting macronutrient for plant growth. It makes up about 0.2% of a plant's dry weight and is a component of key molecules such as nucleic acids, phospholipids and ATP. Maintenance of stable cytosolic inorganic P concentrations is essential for many enzyme reactions [5].

***Address correspondence to H. Gerós:** Centro de Investigação e de Tecnologias Agro-Ambientais e Biológicas (CITAB), Portugal; E-mail: geros@bio.uminho.pt

Hernâni Gerós, M. Manuela Chaves and Serge Delrot (Eds)

Table 1: Essential macronutrients in grape berry.

	Major forms in plants	**Main translocation route to berries**	**Cell functions**	**Accumulation in grape berry**	**Importance in grape**
Potassium	K^+	Phloem	Enzyme activation; membrane transport processes; translocation of assimilates; maintenance of membrane potential (anion neutralization); osmotic potential regulation	Gradual increase at all stages, more prominent from veraison onwards	Major cation in ripe berries; determinant of berry pH; regulation of phloem transport rates and assimilation partitioning patterns
Nitrogen	mineral (NO_3^- and NH_4^+) and organic	Phloem and xylem	Constituent of proteins, amino acids and other nitrogenated organic compounds	Two phases of intense incorporation: before 'pea size' stage and at veraison	Decisive in balance between vegetative and reproductive growth and thus in grape quality and yield
Calcium	Ca^{2+}	Xylem	Cell signaling; counter-cation for both organic and inorganic anions in the vacuole; structural roles in the cell wall and membranes	Increase until veraison, followed by decrease in some varieties	Maintenance of fruit integrity; regulation of ripening timing; mechanical resistance against external aggressions
Phosphorus	Pi, P-esters, P-lipids and nucleic acids	Phloem	Component of key molecules such as nucleic acids, phospholipids and ATP	Continuous throughout the season, more prominent after veraison	Preponderant in initiation, maintenance and differentiation of clusters
Magnesium	Mg^{2+}	Phloem	Most abundant free divalent cation in the plant cytosol; central atom of chlorophyll tetrapyrrolic moiety ; cofactor of a wide range of enzymes involved in photosynthesis, respiration and formation of DNA and RNA; modulation of cell ionic fluxes	Before veraison; may increase or remain constant after veraison	Essential in photosynthesis and several metabolic processes
Sulfur	SO_4^{2-} and organic	Phloem	Involved in several catalytic or electrochemical functions of the molecules of which it is a component		Essential component of enzymes involved in defense mechanisms against pathogens

Like P, Ca concentrations in the cell must be finely regulated since it is an essential element in cell signalling. Ca^{2+} signals are a core regulator of plant cell physiology and cellular responses to the environment [6]. For this pivotal role in the signalling web, cytosolic Ca^{2+} concentrations must be low and roughly constant in steady-state conditions, but they also have to change rapidly in response to internal or external stimulus and to stresses, whether biotic or abiotic. Therefore, whenever necessary, Ca^{2+} is readily transported to the cytosol and subsequently taken back to storage organelles, mainly the vacuole [7], in order to start or stop the signal transduction processes [8].

As for other metal ions, the vacuole is the major organelle responsible for the maintenance of Mg^{2+} balance in the cytosol. As the most abundant free divalent cation in the plant cell, Mg is the central atom of chlorophyll tetrapyrrolic moiety and is the cofactor of a wide range of enzymes involved in photosynthesis, respiration and synthesis of DNA and RNA, in addition to its role in the modulation of cell ionic fluxes [9]. Although this macronutrient does not seem to have a direct effect on berry growth and quality, it greatly influences canopy density and yield, which is directly linked to the supply of nutrients to the fruits.

S is an essential macronutrient for all plant species and the earth atmosphere ensures that S exists predominantly in the form of SO_4^{2-} [10]. Also, it is the least abundant macronutrient found in plants, representing 0.1% of dry matter, compared to 1.5% and to 45% for N and C, respectively. Unlike these elements, S is rarely a structural component of biomolecules. Rather, it is nearly always directly involved in catalytic or electrochemical functions of the molecules of which it is a component [11]. Plants can readily assimilate inorganic SO_4^{2-} and reduce it to sulphide, which is assimilated into cysteine, mainly used for protein synthesis. Additionally, SO_4^{2-} may be directly incorporated into organic molecules by the action of adenosine phosphosulfate kinase [10, 12]. S is involved in several cellular processes such as those related to oxidative stress by being a compound of the tripeptide glutathione (GSH). It is also involved in heavy metal stress responses and in plant defence, as a component of cysteine-rich proteins with antimicrobial properties such as defensins and thionins [10]. Only scarce information concerning S dynamics in berries is available. However, it plays important physiological roles in vine and is widely used in vineyard management to combat grapevine diseases.

Although present in minor amounts in plants, micronutrients also play preponderant roles in grapevine physiology, consequently affecting fruit development and yield. Early studies supporting the vital functions of these elements in plants grown hydroponically have been conducted by Sommer and Lipman [13], contradicting the assumptions of previous researchers. Whereas trace amounts are required to obtain healthy and fully developed plants, high amounts of micronutrients impair normal growth. Like for other mineral elements, plants take up micronutrients from the soil and their bioavailability depends on several factors, such as soil pH, organic matter content, redox conditions and the amount of clays and hydrous oxides responsible for the adsorption or precipitation of metal ions [14].

POTASSIUM

K is an essential element for all plants, especially for grapevine. Grape berries are a strong sink for this cation where it is involved in several physiological processes, such as enzyme activation, cellular membrane transport processes and translocation of assimilates, anion neutralisation, which is essential in the maintenance of membrane potential, and osmotic potential regulation, which controls plant water relations, turgor maintenance and growth. Other cations such as Na^+, Ca^{2+}, Mg^{2+}, Mn^{2+} and copper may replace K^+ in some functions or have a synergistic effect as osmotic components, however, K^+ is a cell-friendly cation. Thus, it is usually highly abundant in plant tissues. Moreover, it permeates plant membranes easily [15] and has high mobility in the plant body, which makes it an easy recruitable ion, for example in situations of soil deficiencies in this mineral.

At the plant level, together with sugars and amino N compounds, K^+ constitutes an important osmolyte of the phloem sap, playing a determinant role on its transport rates and assimilation partitioning patterns [16]. Thus, by establishing an osmotic gradient between the leaves (source) and the berries (sink), it may be involved in sugar loading. During grape berry development and maturation, K^+, together with sugars and organic acids, decreases cell osmotic potential, contributing to the generation of a water potential gradient that causes water uptake into the berries and their subsequent expansion.

As the major cation in ripe berries (about 70% of the mineral cations) [17] as well as in must and wine (~900 mg/L) [18, 19], K^+ determines grape and wine quality since it greatly influences the grape berry pH. Its relationship with organic acids, especially with tartrate and malate, which are the most important organic acids in the grape berry, has been extensively described. To achieve good quality wines, a high tartrate/malate ratio is desired, since tartrate is significantly stronger and gives a crisp and fresh acid taste to the wine. Excessive accumulation of K^+ in the berry decreases free tartaric acid, due to exchange of tartaric acid protons with K^+, causing precipitation as potassium bitartrate, and thus increasing wine pH [19]. Indeed, direct correlation has been found between K^+ concentration and wine pH [20], whereas tartaric acid concentration and tartaric/malic acid ratio showed inverse correlation with this parameter. Grape juice with a high pH often results in unstable musts and wines that are more susceptible to oxidative and biological spoilage, and often produces a wine with low acidity and flat taste. High pH also reduces anthocyanin ionisation and decreases the colour quality of red wines [15]. The fermentation process is also affected since higher levels of malate to tartrate enhance malolactic fermentation [19, 21]. The deleterious effects of high K^+ levels are more evident in red wines since during fermentation the berry skin is left in the must for some period after crushing for anthocyanin extraction and more K^+ may also be extracted.

Patterns of Potassium Accumulation and Partitioning in the Grape Berry

In general, K^+ levels increase gradually during grape berry development, but a sharp increase occurs at the onset of ripening [15, 22, 23]. Rogiers *et al.* [23] have found that K^+ accumulation was slow during the pre-veraison phase, but increased 3.5 times during post-veraison enlargement, showing a steady upward trend from veraison to harvest, when the bunches contain the greatest proportion of K^+ of the vine [24]. K^+ levels may reach concentrations of 3 mg per g of berry fresh weight [15].

The accumulation of K^+ in the grape berry tissues throughout development has been described in detail in grapes from mature Muscat Gordo Blanco vines [22]. Among berry tissues, the skin presents the highest concentrations of K^+ at all developmental stages, accumulating up to 15 mg/g fw at the early ripe stage (17º Brix). K^+ concentrations in the skin are 2-7 and 2-4 times higher than in the pulp and in the seeds, respectively. Interestingly, the lowest concentrations are detected in the flesh tissues adjacent to the skin, whereas in the central tissues including ventral vascular bundles, K^+ levels are higher but still lower than in the skin [22].

Although K^+ content increases throughout development, K^+ concentration per unit fresh weight may increase or remain relatively constant, since berry growth resulting from water entry into grape berry flesh cells may occur at a similar rate than that of K^+ accumulation. Indeed, the flesh is the tissue where K^+ predominates, although at lower concentrations than in skin and seeds [20]. Vines of Grenache Noir grafted onto Richter 110 present a berry K^+ distribution of 37% in the skin, 60% in the flesh and 3% in the seeds [1], the predominant accumulation in skin and flesh reflecting the role of this cation in cellular expansion.

In general, longitudinal gradients are small, from the pedicel to the style remnant of the berries, but the brush develops concentrations in ripe and overripe berries midway between those of skin and flesh tissues. Other non-accumulating solutes such as sucrose, inorganic anions, phenols and tartrate also show high levels in the brush. Flesh tissue adjacent to ventral vascular bundles shows higher K^+ levels than flesh adjacent to the seeds, but K^+ values still exceed those found in more peripheral flesh. The longitudinal distribution pattern of K^+ changes throughout grape berry development. In green berries, K^+ concentrations are higher near the stylar end, whereas after veraison K^+ accumulation is more evident in the zone near the pedicel [15, 25]. The differences registered when comparing different berry tissues may be due to the specific characteristics of the cell types of each tissue and also to the role of K^+ in each cell type [15]. Differences are also noted at intracellular level, where K^+ concentration is 5 to 10 times higher in the cytoplasm than in the vacuole.

Interestingly, the rate of berry K^+ accumulation during ripening is related to plant water status [1]. Indeed, water deficit occurring from veraison onwards modifies the accumulation of this element in the different berry compartments. However, the percentage of K^+ of each tissue at maturity does not seem to be affected by water availability.

Translocation and Accumulation of Potassium in Grape Berries

In the initial phases of grape berry development, nutrients are supplied *via* both the phloem and the xylem, however, in some varieties the xylem flow is discontinued or reduced at the onset of ripening, perhaps not due to loss of function of the tracheids but rather to loss of an appropriate driving force (hydrostatic gradient) in the berry apoplast [26]. This may be caused by the low transpiration rate of the berry, resulting from changes such as decreased stomatal frequency and the deposit of epicuticular waxes. Thus, K^+ movement to the berries occurs mainly through the phloem.

The ability of berries to absorb K^+ is affected by several factors such as its availability in the soil, translocation from root to shoot and re-translocation from shoot to root, the amount of K^+ reserve, source-sink ratio, hormone levels and the developmental stage itself [15]. Zhenming *et al.* [27] have demonstrated that fruits are the main organs accumulating the foliar-applied K^+ analogue ^{86}Rb in three-year-old grapevines (*Vitis vinifera* L. x *V. labrusca* L. cv. Jingyou). Transport from leaves to fruits varied according to the stage of fruit development, being more intense in the latter stages, 20 days before harvest. This indicates that K^+ should be sufficiently and timely applied both at full bloom and fruit colouring stages to the table grape cultivars, to ensure fruit growth as well as high use efficiency of the K^+ fertiliser. In addition, changes in leaf/fruit ratios significantly affected K^+ influx into fruits from leaves, suggesting that a suitable leaf/fruit ratio should be maintained in grape production to ensure high quality fruits, with balanced source–sink capacity.

Correlation between K^+ levels in berries and in petioles was found in some grapevines [28]. In the Shiraz variety the sharp increase in berry K^+ levels until veraison coincided either with an abrupt fall or with a plateauing of K^+ content in petioles, depending on whether the vine was grown in its own roots or in rootstocks. This is consistent with the fact that between veraison and harvest, the rate of K^+ uptake by the vine decreases sharply but a steady increase is observed in the K^+ levels of bunches [24], indicating translocation of this element from reserve tissues to fruits.

Interestingly, high canopy densities causing shaded vines may promote a reduction in photosynthetic activity of mature leaves and thus speed up the process of leaf senescence even before fruits are fully developed. Hence, a greater translocation of solutes to fruits occurs before leaf fall, causing berries to have excessive amounts of K^+. Although leaf removal may seem like an option to reduce this effect, one must take into account the resulting reduced photosynthesis by the vine and the subsequent decrease in sugar content [29].

K^+ content of grape berries is greatly affected by the grapevine variety, the rootstock and the rootstock/scion combination. Early studies conducted by Downton [28] have demonstrated that, in general, the Shiraz, Cabernet Sauvignon, Sultana and Muscat Gordo Blanco varieties, present higher levels of K^+ in the berries when they grow on Salt Creek, Harmony and Schwarzmann rootstocks, than in non-grafted plants. The levels attained by harvest time appeared to be related to the K^+ status of the plant, since petioles of vines on rootstocks also contained more K^+ than those of own-rooted vines. Interestingly, Walker *et al.* [20] also found that, generally, ripe berries of ten-year-old vines of Muscat Gordo Blanco, Shiraz, Riesling and Cabernet Sauvignon (but not Chardonnay) grown on Ramsey (*Vitis champini*) rootstock contain higher levels of K^+ than those of non-grafted plants. Among all the scion/rootstock combinations, Shiraz in Ramsey contained the highest concentrations of K^+ ions in the pulp and in the whole berry. Chardonnay and Riesling had the highest proportion of K^+ in the berry skin, followed by Cabernet Sauvignon and Shiraz, and ultimately Muscat Gordo Blanco. Hence, berries of the latter variety had over 60% and Riesling berries over 40% of their K^+ in the pulp. The differences in K^+ concentration among berries of own-rooted and grafted grapevine varieties were noted in different tissues. K^+ concentrations were higher in the skin of own-rooted Muscat and Cabernet, in the pulp of grafted Muscat, Cabernet and Riesling and in the seeds of own-rooted Riesling. These differences will later influence K^+ content in must and wine throughout the fermentation process. Indeed, K^+ concentrations were proven to be higher in ferments of grafted varieties than in own-rooted, the biggest difference occurring in Shiraz. Among the studied varieties, Shiraz also results in wines with the highest K^+ concentrations, reflecting its higher concentration in the berry, especially in the pulp, as mentioned previously. Wines resulting from Cabernet, Chardonnay and Muscat varieties show intermediate levels of K^+ and the lowest concentration is found in Riesling wine. As a result, grapevine varieties, rootstock/scion combinations, as well as biological and environmental factors of berry growth and must fermentation, should be carefully managed in order to control K^+ levels and establish an optimal balance between vine yield and wine quality.

Transmembrane Transport of Potassium in Grape Berry Cells

The knowledge of the mechanisms involved in K^+ transport to grape berries and their regulation is of utmost importance for developing strategies to optimise the levels of this cation in berries and improve fruit quality, while assuring the production of wine with adequate pH. K^+ transport into plant cells occurs by multiple mechanisms, including K^+ channels that mediate passive low affinity transport across membranes and K^+ carriers that mediate energised high and low affinity uptake [19, 30]. In the grape berry, only a restricted number of K^+ transporters were identified to date. These are summarised in Table **2** and discussed below in more detail.

Table 2: Potassium transporters in grapevine.

	Type of transporter	Organs/tissues/cell types with detectable expression	Expression during berry development	Role in berry physiology/ripening	Reference
SIRK	Inward-rectifying channel	Berry pericarp, leaves and young stalks; guard cells	Low levels during the first stages of berry growth; drastic decrease after veraison	Regulation of grape berry water loss and/or K^+ loading before veraison	[31]
VvKUP1 VvKUP2	KUP/KT/HAK	Reproductive organs such as flowers and young berries; seeds and canes (VvKUP2); present especially in the skin	High levels prior to veraison; lower but significant post-veraison levels	K^+ uptake into berries or compartmentation into the skin cells during the first stages of berry development; K^+ homeostasis throughout fruit development	[30]
VvNHX1	NHX	Berry flesh, low levels in red skin; vacuolar localisation	High levels during veraison and after veraison	Vacuole and berry expansion	[32]

Pratelli *et al.* [31] have cloned a grapevine inward rectifying K^+ channel belonging to the Shaker family, *SIRK*, which is present in guard-cells and is implicated in ionic homeostasis of the berry. *SIRK* is expressed not only in grape berry pericarp, but also in leaves and young stalks. In *Arabidopsis*, the construct was also expressed in xylem parenchyma. A detailed analysis indicated that the protein exhibits the same structure as Shaker channels of the KAT type, containing a hydrophobic core with six transmembrane domains and the conserved pore region, P-domain, which contributes to K^+ conductivity [33]. In addition, semiquantitative reverse transcriptase-polymerase chain reaction analysis indicated that *SIRK* transcripts are present in low levels during the first stages of berry growth, suffer a drastic decrease by the time of veraison and are no longer detected after this phase. Thus, the encoded channel polypeptide appears to play an important role in the regulation of grape berry water loss and/or K^+ loading before veraison [31].

Two additional genes encoding KUP/KT/HAK-type berry K^+ transporters have been identified [30]. The cDNAs encoding *VvKUP1* and *VvKUP2* were able to complement the defect in K^+ uptake of a mutant *E. coli* for K^+ transporters. In berries of *V. vinifera* L. cv. Shiraz, both transporters are highly expressed in reproductive organs such as flowers and young berries, especially in the skin. Similar patterns were found in *V. vinifera* L. cv. Cabernet Sauvignon berries, the expression of *VvKUP1* being predominantly elevated prior to veraison, at the beginning of the lag phase. *VvKUP2* was also expressed in the seeds and, to a much lesser extent, in the canes of Shiraz berries. Since the higher transcript abundance was found in pre-veraison berries, it was suggested that these VvKUP transporters play an important role either in the uptake into the berries or in the compartmentation of K^+ into the skin cells during this period. Interestingly, lower but significant levels of expression of these transporters were detected in post-veraison berries, suggesting that VvKUP1 and VvKUP2 may contribute to K^+ homeostasis throughout fruit development.

Tonoplast transporters are crucial to ensure a fine control of the K^+ and Na^+ levels in the cytosol, while maintaining acid and sugar homeostasis in the berry. The berry vacuoles maintain an acidic pH, ranging from pH 2.5 in the green stage to pH 3.5 during ripening. Two H^+ pumps vacuolar ATPase (V-ATPase) and pyrophosphatase (V-PPase) generate and maintain a H^+ gradient across the vacuole membrane [34], which in turn is essential to energise solute compartmentation processes. The hydrolytic activity of these vacuolar pumps increases throughout development with an acceleration during ripening, the ratio of V-PPase activity to V-ATPase activity being always in favour of the former and reaching its maximum value at veraison [35]. Although several expressed sequence tags (ESTs) encoding putative vacuolar transporters have been identified in databases, there is no functional assessment of their putative role. Hanana *et al.* [32] have cloned and characterised an NHX-type vacuolar cation/H^+ antiporter from *V. vinifera* L. cv. Cabernet Sauvignon. This family of transporters couples the passive movement of H^+ out of the vacuole to the active movement of monovalent cations, mainly K^+ and Na^+, into the vacuole. After heterologous expression of *VvNHX1* in *Saccharomyces cerevisiae ena 1-4 nhx1* strains, there was a significant improve in the ability of the mutant to grow in the presence of NaCl and LiCl, indicating that this transporter is able to replace at least partially the yeast nhx1 function. The vacuolar localisation of VvNHX1 was assessed through transformation of yeast and berry cells with constructs bearing green fluorescent protein (GFP)-tagged VvNHX1. In contrast to untransformed cells or cells expressing GFP alone, in yeast cells expressing VvNHX1::GFP, the fluorescence was seen in the vacuole and in small vesicular and pre-vacuolar organelles, whereas in transformed berry suspension cells, the fluorescence clearly fitted with the vacuolar compartment. Immunoblots from yeast cells expressing a FLAG-tagged VvNHX1 also corroborated the vacuolar/pre-vacuolar localisation of the antiporter.

Bioinformatic analysis has revealed that VvNHX1 consists of two functional domains: (1) the transmembrane N-terminal part is hydrophobic, consists of 10 transmembrane regions and contains the pore for ion exchange, whereas (2) the C-terminal part is hydrophilic and seems to be essentially implicated in the regulation of the antiporter activity [36]. Indeed, VvNHX1 contains various interaction domains as well as several putative post-translational modification sites that are essential for regulating protein activity, influence structural stability or interact with other proteins or signalling molecules. The secondary structure of VvNHX1 is arranged as 56% α-helix, 37% linear form and only 7% β-sheet. The amino acid composition of this protein is particularly important since it determines the establishment of the α-helix structure and the charge distribution, which generates critical electrostatic interactions for the ionic flux. The presence of multiple α-helix structures composed of hydrophobic amino acids is favourable to the insertion of this protein in membranes.

Functional studies indicated that VvNHX1 displays low affinity for K^+/H^+ and Na^+/H^+ exchange activities (12.8 and 40.2 mM, respectively), and similar apparent V_{max} values suggest a higher affinity for K^+ than for Na^+ [32]. The expression of *VvNHX1* also varies throughout berry development. Transcripts were not detected in the early stages of berry development or in the green skin; high levels were detected in the berry flesh after veraison and low transcript amounts were detected in the red skin tissue. In addition, no transcripts were detected in leaves and roots, suggesting that *VvNHX1* expression is specific to the berry flesh. These results indicated that VvNHX1 plays a key role in vacuole expansion by mediating K^+ compartmentation. Together with the rapid accumulation of reducing sugars, this drives water entry in berry cells, leading to the well-known phenomenon of grape berry expansion, typical of veraison and post-veraison stages.

NITROGEN

Several factors affect N nutrition of grapevines, such as vine cultivar and rootstock, climate and season, N levels in the soil, cultural practices, canopy shading and microclimate [37]. Grapevines are able to absorb both NO_3^- and NH_4^+ ions from the soil. The former is actively transported across the plasma membrane by both low and high affinity H^+ symporters, and once in the root cells, it may have four fates: (1) be reduced to NH_4^+ and then incorporated into amino acids, (2) undergo efflux out of the cell, (3) be taken up and stored in the vacuole serving as a reservoir to sustain growth and participating in the general osmoticum or (4) move from cell to cell within the symplast of the root to be loaded in the xylem and transported to the aerial parts [38]. The reduction of NO_3^- is started by nitrate reductase, forming NO_2^-, which is then translocated to the chloroplast and reduced to NH_4^+ by nitrite reductase [39].

Nitrate reductase activity, indicating amino acid formation, is detected in root tips, green plant parts (mostly in the leaves) and even in berries. Thus, NO_3^- seems to be present in any part of the vine at any time [3] and has an important role in regulating C and N metabolic pathways, exerting complementary effects to sugars on gene expression [40]. Indeed, inorganic N sources seem to have a preponderant role in grapevine nutrition, to the detriment of the organic sources.

N reserves are located in the woody perennial parts of the grapevine, namely the roots, trunk and canes, consisting of an insoluble and a soluble fraction, the former being predominant [41]. Arginine and aspartic acid seem to be the main storage forms of soluble N.

Both NO_3^- and organic compounds are found in the xylem of the grapevine, where aspartic acid, glutamic acid and its amide and arginine are the major amino acids. Arginine has the lowest C/N ratio of all common amino acids, which is probably the reason why it was selected as the main translocable form of N other than NO_3^-. Curiously, the phloem was shown to transport the same N forms that are translocated in the xylem. The amount of NO_3^- in a tissue is influenced mainly by its uptake and reduction in the roots, the velocity of transport and the NO_3^- reduction activity of the leaves. Most of the N reaching the berries consists of soil-supplied NH_4^+ and of amino acids provided by leaves and shoots, with NO_3^- being present in small amounts during all stages of development of reproductive organs. However, it is interesting to see that the variation in NO_3^- and total N contents in the reproductive organs are similar, suggesting a relatively constant ratio between the anion and total N [3]. In the grape berry, the mineral N in the form of NH_4^+ can represent up to 80% of the total N before veraison, decreasing to 5-10% after maturation and even further after fermentation of the must. On the other hand, the content in NO_3^- and NO_2^- can be considered negligible, about 0.5-2 mg/L and 5-40 μg/L, respectively [18, 19].

A positive correlation has been found between the levels of supplied NH_4NO_3 to grapes of the Tempranillo cultivar and the N levels found in the leaf blades at bloom [42]. Although leaf composition was changed as a result of N treatments, vine yield, vigour and berry size were not affected, in contrast to the generally accepted concept that N fertilisation improves vineyard yield, especially when soils are deficient in this nutrient. However, high levels of N caused a substantial delay in the accumulation of sugars during ripening, which did not seem to be due to reduced photosynthesis, resulting from changes in canopy structure and/or the light microclimate since leaf area indices were not altered upon N fertilisation. However, studies conducted by Grechi *et al.* [43] in grapevines of Merlot cultivar have shown that leaf dry weight was increased with high N supply. In fact, N availability and light seem to interact to control vine biomass accumulation and root-to-shoot biomass partitioning is controlled to some extent by the internal C:N balance. Decreasing N availability leads to a higher root-to-shoot ratio, whereas decreased light regime causes the contrary effect, with more implications in biomass accumulation.

Hilbert *et al.* [44] have found that a high N supply leads to increased N in the grape berry from veraison to harvest and to higher pruning weight, cane number, leaf area and number of clusters per vine, while berry weight and number of berries per cluster decrease. Also, berries from vines with low N supply had higher sugar content and lower acidity, with no changes in fertility and morphology. Hence, N seems to have a decisive role in grape yield and quality since it apparently stimulates plant growth to the detriment of sugar accumulation in the grapes during ripening, with implications in fruit maturation.

Since berries are a strong sink for N, the presence or absence of fruits in vines regulates C and N partitioning, corroborating the findings presented above. Indeed, the amount of N found in the vegetative tissue of shoots of non-bearing vines of Cabernet Sauvignon cultivar was found to be greater than that in vegetative tissue of fruit-bearing vines [45].

It was also observed that fruit growth before veraison was favoured to the detriment of vegetative growth, in low N supply conditions. This effect, however, was not noted after veraison. Since low N constrains vegetative growth and enhances berry maturation, it is crucial that N be available in sufficient amounts to obtain well-developed berries with N levels that will sustain vinification.

Nitrogen Dynamics during Fruit Growth and Development

The translocation of N to grape berries seems to increase gradually throughout development. During berry ripening there is translocation of soil-supplied NH_4^+ and of amino acids provided by leaves, shoots and roots to fruits, most likely driven by source-sink mechanisms [3]. The concentration of total amino acids in grape and juice increases, whereas the NH_4^+ concentration declines, from veraison to harvest, as was demonstrated in berries from the cultivar Cabernet Sauvignon [37]. Thus, with the exception of NH_4^+, all major forms of N reach a maximum level of concentration at some point of the berry ripening process, depending on parameters such as cultural practices and environmental conditions. At harvest, N content may reach concentrations of 1800 μg per berry [2].

Generally, at budbreak there is a strong flow of N to clusters, mainly in the organic form, which is joined by the inorganic N forms at bloom. A significant N allocation to growing grape berries begins after this phase, causing the woody tissues to reach minimal levels of N. Although a small proportion of this macroelement is still translocated among the green parts of the plant and reserve tissues, N flow to clusters is predominant from bloom to harvest, when half of the N present in the annual structures of the grapevine is located in the reproductive parts. Thus, grape growth and N reserve formation seem to be antagonistic.

There are two phases of intense N incorporation corresponding to the peaks of grape dry mass formation. The first takes place during the two weeks before the 'pea size' stage of the berry, and the second starts one month later, at the onset of ripening, and lasts for another two weeks. Berries are able to accumulate about 50% of the final N mass during this period. Towards the end of fruit ripening large amounts of free amino acids are transported from roots to berries, increasing once more the concentration of soluble and total N [3].

The distribution of amino acids in grape berry tissues appears to vary according to cultural practices and climate conditions. However, it seems that the bulk of amino acids are found in the pulp and skin. Therefore, when berry skin is included in the must, more amino acids will be extracted [37].

Amino Acids in Grape Berry

As referred to previously, the organic N fraction of the translocation solutes is composed of only a limited number of amino acids, mainly arginine that represents up to 80% of the N reserves in the perennial structures. Additionally, it participates in the biosynthesis of other amino acids (such as proline, which seems to be mostly synthesised in the fruit), guanidines and polyamines. Amino acids can account for up to 90% of the N in grape juice [46, 47], arginine and proline generally making up the greatest proportion of the total amino acid concentration present in the berry. Correlations have been observed between the ability of plants to withstand osmotic stress and the degree of proline accumulation in response to that stress [47, 48].

Due to their buffering capacity, amino acids interfere with the sensation of acidity, thus contributing to the overall taste of wine. Seven amino acids are found in must in quantities above 100 mg/L: proline, arginine, glutamine, alanine, glutamate, serine and threonine [19]. Thus, the amino acid profile of wine can be used to differentiate wines according to vine variety, geographical origin and year of production [19, 49, 50]. Also, N composition of grape juice has implications in yeast growth and in the rate and duration of the fermentation process. NH_4^+ and the amino N of must, especially glutamine and aspargine, are largely assimilated by yeast during fermentation [46]. Hence, they constitute a fraction of the total yeast-assimilable N (YAN), together with other assimilable amino acids such as arginine, phenylalanine, histidine, valine, alanine, aspartate and glutamate. Lysine, proline and glycine are much less easily assimilated and, together with NO_3^-, are included in the yeast-non-assimilable N (YNAN). Throughout berry ripening, there are changes in the total proportion of YAN and YNAN as a percentage of total N and this is influenced by N application in the vineyard and subsequent vine N status [37].

Effects of Nitrogen Supply on the Secondary Metabolism of Grape Berries

In addition to its central role in the primary metabolism, N also plays a decisive role in the secondary metabolism, namely in the production of compounds essential for the quality of red grapes and for the bitterness and astringency of wine [19]. Adequate levels of N stimulate L-phenylalanine ammonia lyase (PAL), which is responsible for the initial step of the metabolism of L-phenylalanine. This amino acid is a major intermediate of the biosynthesis of phenolic compounds, such as anthocyanins. Thus, PAL activity in berry skin is closely related to anthocyanin accumulation during ripening [51]. Although N addition does have an impact on berry phenolic compound composition, trends of its effects are yet to be evident [37]. Some studies, however, have allowed some insight on this matter.

N fertilisation affects colour density of the must, total tannins and anthocyanins of the berry skin, according to the developmental stage. Although high levels of N are often associated with poor colour development of the fruit, Delgado *et al.* [42] have found that the supply of high levels of NH_4NO_3 to grapevines of the Tempranillo cultivar, at veraison, significantly improved the colour density of grapes. At harvest, however, results were inconclusive. Also, a stronger accumulation of tannins in grape berries with no N fertilisation was observed, whereas intermediate levels of N increased anthocyanin accumulation in the skin and significantly increased the colour density of the must, without causing changes in the pH. Thus, N could improve must colour density by intervening directly in the synthesis of anthocyanins and/or other phenols that influence colour by copigmentation.

Hilbert *et al.* [44] have found that inorganic N supply not only affects anthocyanin production but may also determine anthocyanin degradation profile. The highest anthocyanin content in the berry skin of Merlot cultivar was observed in vines supplied with low amounts of N shortly after veraison, whereas high levels of this nutrient inhibit their accumulation. Indeed, the rapid decrease in anthocyanin content during the late maturation phase suggested the establishment of a catabolic pathway activated by high N supply, reflected by the quantitative and qualitative delay in the biosynthesis of the anthocyanins and enhanced degradation. Hence, remarkably, berries had higher arginine and lower anthocyanin content, with relatively more abundant acylated anthocyanins, compared to berries supplied with low N levels.

Since some phenolic compounds produced in berries are involved in plant protection as biologically active growth inhibitors of other living systems such as fungi [19], it makes sense that N supply affecting phenolic accumulation will result in changes on grape berry susceptibility to pathogens. Indeed, Mundy and Beresford [52] have observed that yeast assimilable N (YAN) appears to influence the susceptibility of grape berries to *Botrytis cinerea*. However, this also has a strong connection with sugar concentration, which tends to be lower in berries with greater YAN, as reflected by the delay in berry maturation, as previously discussed. Thus, high levels of N seem to have an indirect effect in increased berry susceptibility, possibly through an increase in leaf canopy density and prolonged ripening period, both of which increase the risk of a favourable microclimate for botrytis bunch rot development. Also, N nutrition affects the cuticle development, decreasing berry mechanical defence against infection [53].

Botrytis infection leads to a reduction of total amino acid concentration of berries, in a cultivar dependent manner, and thus decreases the juice YAN concentration [37].

In conclusion, N fertilisation must be carefully controlled; on one hand it may contribute to an increase in berry quality and to sustain good fermentation conditions, whereas on the other hand it may alter the production of phenolic compounds, change berry susceptibility to pathogen infections and cause deleterious effects in both grape and wine quality [54, 55].

CALCIUM

As referred to in the Introduction, Ca is a central element in cell signalling and, in addition to its role as a counter-cation for both organic and inorganic anions in the vacuole, Ca^{2+} is required for structural roles in the cell wall and membranes [8], having a crucial role in fruit integrity [56]. Also, similarly to K^+, Ca^{2+} ions may also combine with tartaric acid, forming calcium tartrate crystals that precipitate during fermentation, affecting organoleptic properties and aging potential of wines [19].

In contrast to other elements, such as K and N, Ca is phloem immobile. Thus, it is uptaken by the roots and redistributed to the shoots through the xylem, both in the symplast and the apoplast, but it cannot be mobilised from older tissues and redistributed *via* the phloem. Fruits in which the phloem is the major route for nutrient supply must acquire Ca^{2+} directly from the xylem, but deficiency symptoms may occur, since the uptake is highly dependent on transpiration [8, 57], which is low in young leaves, in enclosed tissues and in fruits, and may be involved in the ceasing of the xylem flow after the veraison stage of grape berries in some cultivars [26]. This seems consistent with the fact that low concentrations of Ca^{2+} promote softening of fruit flesh, which is characteristic of berry ripening. However, as pointed out by Ferguson [56], there seems to be an apparent paradox. Maintenance of high tissue concentrations of Ca^{2+} either delays or retards senescence or ripening, ostensibly by maintaining normal cell function under conditions of organised progressive dysfunction. At the same time, normal cell function requires maintenance of low cytosolic Ca^{2+} concentrations, with the cell working to keep this cation out in the extracellular or extracytosolic environment. Perhaps the inhibitory effects of Ca^{2+} on senescence are primarily extracellular, acting on the cell walls and the external surface of the plasma membrane. Therefore, high levels of Ca^{2+} may be required to trigger the activity of enzymes involved in retarding senescence.

Since Ca^{2+} has a major role in the cross-linking of cell wall polysaccharides, it is naturally implicated in grape berry structural defence against pathogens. Interestingly, the loosening of cell walls and loss of cell cohesion associated with fruit softening seem to be partially due to a rise in polygalacturonase (PG) activity [58, 59]. Whereas the plant produces this enzyme naturally, fungi such as *Botrytis cinerea* also produce a set of PGs to degrade pectin molecules during plant tissue infection. Accordingly, the activity of these enzymes is inhibited by Ca^{2+} [60, 61], which occupies the sites of polygalacturonic acid, the substrate of PGs, or interacts with it, making it unavailable for the enzyme. Hence, plants seem to accumulate Ca^{2+} in large amounts in zones exposed to infection, increasing their resistance to pathogens.

Patterns of Calcium Accumulation in the Grape Berry

Similarly to other mineral compounds, the pattern of Ca^{2+} accumulation in grape berries varies according to grape cultivar and rootstock, environmental conditions and Ca^{2+} availability in the soil. Conradie [24] has described an active Ca^{2+} absorption by the vine about three weeks after budburst, which continued up to veraison, and a second less prominent increase during the six weeks before leaf fall. Amiri [62] has found that the rate of Ca^{2+} accumulation during berry development follows a typical sigmoid pattern.

However, Cabanne and Donèche [63] have found that in grape berries from the Sauvignon blanc, Semillon, Merlot and Cabernet Sauvignon cultivars, Ca^{2+} content increased in the pericarp until veraison and decreased afterwards, which is consistent with a reduction of xylem flow after this stage. A similar trend was observed in the flesh, Ca^{2+} levels reaching ~0.1 mg per berry before veraison, and decreasing gradually afterwards. In contrast, the skin and seeds have showed increasing Ca^{2+} concentrations throughout ripening, the former reaching a plateau around maturity and accumulating in concentrations slightly above those detected in the flesh. The whole berry could accumulate Ca^{2+} in concentrations of up to 0.5 mg and this value was also found for berries of the white Riesling cultivar [2]. In Asgari cultivar, a similar trend of Ca^{2+} accumulation in flesh and skin has been observed, with the former increasing until veraison and decreasing during ripening, and the latter increasing throughout all stages of development [62]. In Grenache Noir berries this trend was also observed [1]. Thus, during the first growth phase, Ca^{2+} is accumulated in the skin and flesh, and the continued accumulation in the skin appears to occur in parallel with the decrease in the flesh, supporting the assumption that during ripening Ca^{2+} is translocated from the flesh to the skin [63]. In addition, Ca^{2+} translocation seems to occur from the pericarp to the seeds, which is corroborated by the fact that the peripheral vascular system that leads to the berry pericarp disappears at veraison, thus shifting the Ca^{2+} supply towards the seeds, through the axial system which appears to remain intact. Therefore, the increase in berry Ca^{2+} during ripening was shown to be exclusively due to Ca^{2+} accumulation in the seeds. However, this trend does not seem to be universal for all cultivars since in Grenache Noir Ca^{2+} accumulation in the seeds halted after veraison [1].

Although these studies indicated that Ca^{2+} accumulation in the berry pericarp ceases after veraison, in berries of the Shiraz cultivar there is a steady accumulation of Ca^{2+} in this tissue throughout berry ripening, reaching concentrations of 0.66 mg per berry, at 115 days after flowering [23]. This suggested that the xylem conduits remain functional during post-veraison enlargement of berries or that this cation enters the berry through non-vascular connections. Furthermore, in conditions of non-restrictive water supply, Ca^{2+} content per berry could have a steady upward trend during post-veraison enlargement [1].

In accordance to the findings above, high levels of Ca^{2+} may be required during the initial phases of berry growth to inhibit premature fruit softening and colour changes, whereas after veraison the reduction in Ca^{2+} content allows fruit maturation. This theory, however, does not seem to apply to cultivars where Ca^{2+} flow to berries does not cease after veraison. Also, Conradie [24] has found that, in contrast to other elements such as K, N and P, bunches require no Ca between veraison and harvest, and Ca^{2+} is stored mainly in the leaves. Indeed, whereas bunches presented only 7.7% of total vine Ca^{2+} at harvest, the leaves contained the highest levels of all organs (46.4%).

The movement of Ca^{2+} in grape berry cells is an interesting target for research as numerous transporters are required for Ca^{2+} movement in plants. The tonoplast contains at least one vacuolar Ca^{2+} pump (Ca^{2+}-ATPase), Ca^{2+}/H^+ antiporters that are responsible for Ca^{2+} uptake and channels catalysing the efflux of Ca^{2+} from the vacuole [6]. The Ca^{2+} vacuolar pump mediates high-affinity Ca^{2+} transport, whereas the antiporters exhibit lower affinity but higher capacities for Ca^{2+} transport. These are driven by the transmembrane pH gradient generated by the V-ATPase and V-PPase [64]. Interestingly, pharmacological studies have demonstrated that the Ca^{2+}/H^+ antiporter is inhibited by several agents known to affect other Ca^{2+} transporters, including ruthenium red, verapamil, La^{3+}, Mn^{2+} and Cd^{2+}.

PHOSPHORUS

Plants cannot grow without a reliable supply of P. Although the total amount of P in the soil may be high, it is often present in unavailable forms or in forms that are only available outside the rhizosphere. Moreover, soil microbes release immobile forms of P to the soil solution and are also responsible for the immobilisation of this nutrient. Therefore, despite being a mobile element in plants, P immobility in soils may give rise to deficiency symptoms at any time of the growing season. Thus, it is essential that grapevines accumulate sufficient levels of P during vegetative growth in order to sustain reproductive growth.

The vegetative growth of vines is highly affected by P nutrition [65] and the effects vary according to the scion and rootstock [66] since the latter differ in their ability to take up P from the soil and to translocate it to the scion.

The initiation and maintenance of clusters and flowers within developing buds is inhibited under P depletion [65]. Depending on the timing of P supply throughout the season, there is an increase in berries per cluster with increasing duration of the supply of this nutrient. Accordingly, cluster weight and berry weight also depend upon the levels of P supply throughout the season. Therefore, a primary factor in P regulation of vine productivity is *via* maintenance and differentiation of initiated clusters, the former increasing with length of the P supply period.

Transport of Phosphorus to Grape Berries

P is taken up from soil by plants in the inorganic orthophosphate form, and grapevines have a large ability for P redistribution from bark, wood and roots to support growth during periods of high demand or low supply of this element. Similarly to K, the phloem is the main pathway for P transport from leaves to fruits, only low amounts being transported *via* the xylem when the main influx pathways of phloem are obstructed [67]. In addition, both the xylem and the phloem are responsible for P direct transport from roots to fruits, and horizontal transportation may also contribute to P dynamics. Interestingly, P absorption rates from leaves change during berry development, the highest rate being observed at the first stage of fruit growth.

Although the studies on the evolution of P levels throughout berry development are scarce, Conradie [24] has found that in Chenin blanc grafted with 99R, P levels in the bunches increase throughout development, reaching levels of 641 mg per vine at harvest. The roots contained more than 80% of vine P during the 27 days preceding budburst. Although between veraison and harvest the P content of the vine remained constant, a sharp increase was noted in bunches, which represented the largest sink for P accumulation. The correspondent loss of this element from the leaves corroborated the retranslocation of P from leaves to fruits during this period. In this study, two distinct periods of P absorption in the grapevine were observed:

the first one occurring between budding and veraison (ceasing at this phase), and the second one being less prominent, starting about five weeks after harvest and lasting until leaf fall. Thus, the post-harvest period seems to be of great importance for the accumulation of P reserves in the permanent parts of the vine, since no translocation from leaves to the permanent structure seems to occur during leaf fall.

Schaller [2] has described a pattern of P accumulation in the berry, with very high uptake rates at the beginning of berry development, which decrease continuously until veraison and increase again until harvest, when P content reaches almost 300 µg per berry.

The dynamics of P at the plant cell level is rather complex. The physiological functioning of cells requires that Pi concentrations in the cytosol be maintained within a narrow range [68] since its homeostasis is essential for many enzyme reactions. This homeostasis is achieved by a combination of membrane transport and exchange between various intracellular pools of P. These pools can be classified in a number of different ways, according to their location in physical compartments (such as the cytoplasm, vacuole, apoplast, and nucleus), to the chemical form of P (such as Pi, P-esters, P-lipids and nucleic acids) and to the physiological function, as metabolic, stored, and cycling forms. Therefore, the pH of each compartment determines the form of Pi and the proportion of the total P in each chemical form (except P in DNA) changes with tissue type and age and in response to P nutrition [5].

MAGNESIUM

According to Conradie [24], unlike Ca, K or P, Mg shows a single continuous absorption period by the grapevine. An increase in Mg^{2+} absorption by the vine was observed during the 45 days until the end of bloom, increasing further from this stage until veraison. However, most of Mg^{2+} accumulation was directed to the roots, shoots and leaves. Until harvest, a slight but significant increase in Mg^{2+} uptake occurred, but again, bunches were not the major target for this accumulation, containing only 15.4% of the total Mg^{2+}, whereas the leaves contained the major portion of this macronutrient. This pattern is mostly in accordance with the structural role of Mg^{2+} in chlorophyll molecules. However, other studies indicate that the accumulation of this element in the berry occurs in two successive phases, *i.e.* before and after veraison, and that the high rate of accumulation of this element after veraison is thus consistent with flow through the phloem into the grape berry after veraison. Indeed, Schaller [2] has described elevated uptake rates in the early growth phase with a significant reduction at veraison, followed by a second peak in the absorption before harvest; Mg^{2+} content reached almost 150 µg per berry.

As reported for Ca and K, vine water status also affects Mg accumulation in grape berries [1]. Indeed, in Grenache Noir berries, Mg^{2+} enters the berry mostly before veraison and its accumulation remains constant after this phase in non-irrigated vines. However, irrigation causes a shift in the accumulation of this element. By contrast, a decrease in Mg^{2+} concentration per unit of dry matter occurs after veraison. This change could be attributed directly to the gain in berry size and solute accumulation at post-veraison. Among berry tissues, Mg^{2+} content seems to be identical in flesh and seeds at veraison, the former tissue presenting much higher values than skin and seeds during ripening. However, a halt in Mg^{2+} accumulation is seen in seeds at veraison, explaining the drastically reduced values during ripening. Thus, there seems to be an isolation of the seeds after veraison in regard to Mg^{2+} accumulation; the flesh and skin remain unaffected as indicated by the continued accumulation of this element in these tissues.

SULFUR

Although studies on berry S content and translocation pathways are scarce, studies conducted by Rogiers *et al.* [69] show that this element is included in the group of phloem mobile elements, accumulating throughout berry growth and ripening. In addition, among berry tissues, the seeds were the strongest sink for S accumulation.

As already mentioned, S is involved in several cellular processes such as oxidative stress, heavy metal stress and plant defence. It has been used with high efficiency to prevent grapevine diseases caused by pathogens such as *Uncinula necator* (powdery mildew) and *Botrytis cinerea* [10]. Fumigation with SO_2 has been used in the storage of table grapes to prevent spoilage. However, repeated applications of SO_2 can damage berries by bleaching, which is especially critical in non-white cultivars. Also, this compound may cause premature browning of the stems, which decreases the marketability of the berries and increases the rate of water loss. The concentration of S dioxide required for control of the disease is very close to the levels that can be damaging to fruit. The application of 200 ppm S dioxide three times a week has been found suitable for controlling *Botrytis cinerea* on grape berries during storage, with reduced bleaching [70].

The protective effects of S are also found on field applications. At the ecological vine cultivation, the elemental S has been applied as a dust or as a wettable powder. However, these practices present risks, among which are phytotoxicity, inhibition of the microbial activity during fermentation of the grape must and the remaining of residues in the wine. Promising alternative approaches such as the incorporation into the soil of fine-granule S and *Thiobacillus* microorganisms revealed effective restriction of powdery mildew in infected berries and leaves by more than 80%, in comparison with 90% after treatment with a conventional fungicide used as control. Also, this approach affected positively the growth of the canes and shortened the time of ripening, increasing the provisional production [10, 71]. Fertilisation application strategies may stimulate S metabolic processes, inducing the natural resistance of plants against fungal pathogens [10].

The H^+/SO_4^{2-} co-transport in the plasma membrane of root cells is the first step for the uptake of SO_4^{2-} from the environment by plants. Further intracellular, cell-to-cell and long-distance transport must fulfil the requirements for SO_4^{2-} assimilation and source/sink demands within the plant [72]. Two kinds of transporters mediate the initial uptake and the distribution of SO_4^{2-} throughout the plant: one with low K_m (10 µM) assuring a high-affinity sulphate transport (HAST), and another with much higher K_m responsible for the low-affinity sulphate transport (LAST) [10].

Putative SO_4^{2-} transporters from *V. vinifera* (*VvST*) and *V. rupestris* (*VrST*) were first cloned by Tavares *et al.* [73] and identified as belonging to the high-affinity SO_4^{2-} transporters. In cell suspensions, there is a fast derepression of the transporters under conditions of S depletion. The enhanced influx rates and the upregulation of *VvST* and *VrST* were rapidly reversed by the addition of SO_4^{2-} to the medium. While *V. vinifera* showed a faster response at the protein level measured by influx rates, in *V. rupestris* the effect was observed primarily at mRNA level.

MACRONUTRIENTS INTERACTION IN THE GRAPEVINE

Although each of the grapevine macronutrients plays its essential role in vine growth and development, one must consider that these elements can interact with each other, producing different outcomes, which in turn will also depend on environmental conditions, scion-rootstock combinations and viticultural practices. For instance, Skinner and Matthews [74] have found that the translocation of Mg from roots to leaves is highly dependent upon P supply to the roots, the translocation of Mg being more sensitive than its uptake.

In a study conducted in table grapes, petiolar nutrient composition changed greatly in response to macronutrient supply to grapevines through drip irrigation, and changes in reproductive growth were also remarkable [75]. Indeed, N application alone increased shoot number per vine, shoot length and pruning weight, thus decreasing crop load efficiency due to excessive vegetative growth. However, when combined with K alone or K plus Mg, an increase in cluster number and weight was observed, resulting in greater yield per vine. Similarly, vines receiving both K and Mg fertilisation also displayed an increase in vine yield. Also, N alone or combined with K or Mg resulted in larger berries than those supplied solely with K or Mg or even with a combination of both of these nutrients. Vines supplied with the three elements at once also produced smaller berries. K seems to have more pronounced effects on various yield and quality components in table grapes, and its combination with N and/or Mg could create a balanced C/N ratio, resulting in a balance between vegetative and fruit production.

Another example has been shown in vines of the Tempranillo cultivar. Whereas a stronger accumulation of polyphenols occurred in the fruit of vines with no N fertilisation, the effects of different N treatments were reduced as the K fertilisation dose was increased. With average levels of K fertilisation, increased N supply caused a significant decrease in the content of total polyphenols, whereas in the presence of high levels of K, high N levels resulted in higher concentrations of these compounds [42]. Also, the degree of polymerisation of condensed tannins decreased when the N/K ratio was balanced and the amounts of both nutrients applied were high. Hence, a strong interaction between N and K is evident and optimal nutritional N/K ratios may enhance the phenolic features of grape berries, with no apparent changes in the vineyard vigour, yield and/or berry size.

ESSENTIAL ROLES OF MICRONUTRIENTS IN GRAPEVINE AND BERRIES

In grapevines, differences in size and depth of the roots influence the accumulation of metal ions [76] and grapevine rootstocks also determine the accumulation of micronutrients, such as Na^+ and Cl^-, in vine organs [28]. Interestingly, plants are able to influence the solubility and speciation of metals in the rhizosphere by exuding chelators and manipulating rhizosphere pH.

As other mineral elements, metal ions are taken up from the soil and distributed among plant organs and their movement in the xylem is essentially driven by the mass upward flow of water and thus dependent on transpiration rates. Copper, Mn^{2+} and Zn^{2+} are predominantly transported throughout the plant within the xylem rather than the phloem. However, as part of the regulation of free metal ion levels, the composition, pH and redox potential of the xylem sap seems to determine the movements of metal ion species. These elements usually travel associated with other molecules such as amino acids in the case of Cu, or organic acids such as citric and malic acid that render Zn^{2+} complex. Mn^{2+} seems to be the element with the greatest proportion of free ion form existing in the xylem sap, when compared to Cu and Zn^{2+} [14].

As referred to in the previous sections, micronutrient deficiency impairs plant growth and affects C assimilation processes, having implications in grapevine fruit formation and productivity since berry growth and ripening greatly depend on the import of photoassimilates.

Although the roots are the major vine organ for uptake of metal ions from the soil, metal translocation to aerial parts is tightly controlled, and thus the leaves play an important role in metal distribution in the plant. For instance, metabolically active mature leaves are responsible for guaranteeing effective distribution of Zn^{2+} within the grapevine [77]. Zn^{2+} supply increases pruning weight, cluster number, and thus the overall vine yield [75], whereas berry weight is apparently not affected. Remarkably, Fe seems to produce similar effects and when both nutrients are supplied at once, there is a resulting increase in cluster weight. Thus, these two micronutrients may have synergistic effects in vine yield. Bacha *et al.* [78] have also found that on grapevines of Thompson seedless and Roumy red grape cultivars (grown in calcareous soils), spraying Zn^{2+}, Mn^{2+} and Fe before bloom, after fruit set and/or during berry development caused an increase in vine yield. In particular, an increase in weight, size, length and diameter of the berries was observed, accompanied by significant changes in the concentration of some petiole nutrients. However, the extent of the differences varied among rootstock cultivars and was dependent on the number of nutrient applications per season.

Although Fe needs of fruit trees are relatively low, its deficiency represents the main constraint for successful cultivation of fruit tree crops in calcareous and alkaline soils [79], and grape is relatively susceptible to iron chlorosis. Therefore, Fe deficiency reduces yields and fruit quality and forces growers to adopt measures for controlling and preventing the development of iron chlorosis, such as the application of Fe fertilisers as synthetic chelates, which however do not represent a sustainable management approach due to high cost and potential pollution of the soil and water environments. The genetic approach to prevent chlorosis is based on the choice of tolerant rootstocks, which are known to activate mechanisms for improving Fe uptake under conditions of low availability. Unfortunately, for several fruit crops Fe tolerant rootstocks have adverse agronomic characteristics which make their adoption unlikely in modern fruit industry [79].

Supporting the previous statements, Mo deficiency in grapevines also causes disorders in berry growth and development. Upon Mo supply, the number of berries per cluster and berry weight may suffer significant alterations, depending on the developmental stage. Williams *et al.* [80] have found that the application of

Mo increased the number of bunches per vine at harvest. Additionally, there was an increase in size, number and weight of coloured grape berries at harvest, with a consequent improvement in vine yield. Interestingly, the increase in the percentage of coloured berries with one or more functional seeds was accompanied by a decrease in the proportion of green berries, indicating that Mo affects pollination and/or fertilisation and therefore berry development. Curiously, no visual symptoms of Mo stress on leaves or shoots or premature leaf senescence were observed. Hence, Mo seems to affect only a specific stage of development and not the overall vegetative vigour of the vines. However, Mo influences the petiolar content of other nutrients and the effects vary depending on the season, developmental stage and sampling area. Moreover, it affects the assimilation of inorganic N by being necessary for nitrate reductase activity [81]. Interestingly, depending on supply, changes in leaf Mo levels were observed between flowering and veraison, suggesting that the phloem mobility of this element is variable, which impairs the study of Mo requirements by the vine.

In contrast to other micronutrients, boron seems to have restricted mobility in most plant species, including the grapevine [82]. However, evidence for its retranslocation in the phloem has been found [83]. Like other micronutrients, boron deficiency has severe implications in vine health. Moreover, early studies pointed out the possible involvement of boron in the translocation of sucrose, in the form of sugar-borate complexes [84], indicating that it is linked to photoassimilate partitioning in the vine.

Studies on Cu effects on berry yield are scarce. Early studies have demonstrated the stimulating effect of this element in plant growth, when present in trace amounts [85]. Also, like S, this metal ion has been extensively used as the active principle of fungicides since the late 1800s, when the "Bordeaux mixture" was developed and its spectacular efficiency proved against fungal pathogens such as downy mildew [86]. Although initially it seemed to improve plant growth in unproductive lands, repeated use of fungicides based on Cu salts has led to the accumulation of large concentrations of this metal ion in vineyard soils and has raised concerns regarding negative effects on the environment, namely through toxicity to aquatic and soil organisms. This has led to several studies on Cu accumulation and distribution in vineyard soils and on its potential toxicity to grapevines [87-90]. Thus, Cu levels, as that of other micronutrients, must be carefully controlled in order to sustain plant growth while avoiding depletion or excess symptoms.

Within the plant cells, each of the micronutrients has particular and crucial functions, and most metal ions work as enzyme cofactors. Remarkably, due to different affinities of metalloproteins to metal ions, replacements among ions may occur, having serious implications in protein function and consequently destabilising cell environment [91]. Therefore, the concentrations of free metal ions must be finely adjusted. A small number of metal ion transporters of the plant vacuole have been described [64]. Notably, in some cases, metal ions consisting of either macro or microelements can enter the vacuole through common transporters, such as is the case of Mg^{2+} and Zn^{2+} which may share the same antiporter with H^+, with Zn^{2+} also entering through another H^+ antiporter shared with Mn^{2+}. More studies on micronutrient transport in berries are required to fully understand their dynamics during the season and optimise fertiliser application strategies, aiming at an increase in vine yield.

FUTURE PROSPECTS

Although several studies have provided solid knowledge on the accumulation, distribution and transport of mineral elements in grapevines and their partitioning in berries, there are still gaps to fill in, in order to achieve an optimal balance between vegetative and reproductive yield of the vines, aiming at the production of superior quality fruits. A great number of factors influence macro and micronutrient assimilation and transport in grapevines, among which the cultivar, the rootstocks/scion combination and soil and climate conditions. In addition, the interactions between these elements are complex, difficult to study, and introduce great variability in nutrient dynamics. Although some mineral elements are well characterised in the grape berry, essential information about other elements is still lacking. One great step towards the elucidation of mineral element dynamics in the berry is the unlocking of the mechanisms that rule macro and micronutrient transport between cell compartments, as well as the pathways of translocation throughout development and ripening, patterns of distribution and influence of external factors. Great advances have been made in the last decades and ongoing research has been providing profound knowledge on this subject. Its consolidation and rational application to the everyday life of vinicultural practice is thus of paramount importance to successfully improve berry and organoleptic quality.

ABBREVIATIONS

EST = Expressed sequence tag

GFP = Green fluorescent protein

GSH = Glutathione

HAK = High affinity K^+ transporter

HAST = High-affinity sulfate transport

KAT = *Arabidopsis* K^+ transporter

KUP = K^+ uptake (transporter)

KT = K^+ transporter

LAST = Low-affinity sulfate transport

PAL = Phenylalanine ammonia lyase

PG = Polygalacturonase

Pi = Inorganic phosphate

SIRK = Inward rectifying shaker-like K^+ channel

V-ATPase = Vacuolar- H^+-ATPase

V-PPase = Vacuolar- H^+-PPase (pyrophosphatase)

VrST = *Vitis rupestris* sulfate transporter

VvNHX1 = *Vitis vinifera* Na^+/H^+ *exchanger*

VvST = *Vitis vinifera* sulfate transporter

YAN = Yeast assimilable nitrogen

YNAN = Yeast-non-assimilable nitrogen

ACKNOWLEDGEMENTS

This work was supported in part by the Portuguese Foundation for Science and Technology (research project no. PTDC/AGR-ALI/100636/2008; grant no. SFRH/BD/64587/2009 to V.M).

REFERENCES

[1] Etchebarne F, Ojeda H, Deloire A. Grape berry mineral composition in relation to vine water status & leaf area/fruit ratio. In: Roubelakis-Angelakis KA, Ed. Grapevine Molecular Physiology & Biotechnology, 2nd ed. Springer 2009; pp. 53-72.

[2] Schaller K. Influence of different soil tillage systems on uptake of N, P, K, Mg, Ca, and organic N-compounds by grape berries during growth and development of the variety "white Riesling ". Bulletin OIV 1999; 72(823-824): 603-29.

[3] Wermelinger B. Nitrogen dynamics in grapevine: Physiology and modeling. In: Proceedings of the International Symposium on Nitrogen in Grapes and Wine. Seatle 1991; pp. 23-31.

[4] Williams LE, Miller AJ. Transporters responsible for the uptake and partitioning of nitrogenous solutes. Annu Rev Plant Physiol Plant Mol Biol 2001; 52: 659-88.

[5] Schachtman DP, Reid RJ, Ayling SM. Phosphorus uptake by plants: from soil to cell. Plant Physiol 1998; 116: 447-53.

[6] Dodd AN, Kudla J, Sanders D. The language of calcium signaling. Annu Rev Plant Biol 2010; 61: 593-620.

[7] Fontes N, Silva R, Vignault C, Lecourieux F, Gerós H, Delrot S. Purification and functional characterization of protoplasts and intact vacuoles from grape cells. BMC Research Notes 2010; 3:19.

[8] White PJ, Broadley MR. Calcium in Plants. Ann Bot 2003; 92: 487-511.

[9] Shaul O. Magnesium transport and function in plants: the tip of the iceberg. BioMetals 2002; 15: 309-23.

[10] Tavares S. Uptake and assimilation of sulfate in *Vitis vinifera*: A molecular and physiological approach. PhD Thesis, Instituto Superior de Agronomia, Universidade Técnica de Lisboa 2009.

[11] Leustek T, Martin MN, Bick JA, Davies JP. Pathways and regulation of sulfur metabolism revealed through molecular and genetic studies. Annu Rev Plant Physiol Plant Mol Biol 2000; 51: 141-65.

[12] Lee S, Leustek T. APS kinase from *Arabidopsis thaliana*: genomic organization, expression, and kinetic analysis of the recombinant enzyme *Biochem. Biophys.* Res Commun 1998; 247: 171-5.

[13] Sommer AL, Lipman CB. Evidence on the indispensable nature of zinc and boron for higher green plants. Plant Physiol 1926; 1: 231-49.

[14] Reichman SM. The responses of plants to metal toxicity: A review focusing on copper, manganese and zinc. Australian Minerals & Energy Environmental Foundation 2002.

[15] Mpelasoka BS, Schachtman DP, Treeby MT, Thomas MR. A review of potassium nutrition in grapevines with special emphasis on berry accumulation. Aust J Grape Wine Rev 2003; 9: 154-68.

[16] Lalonde S, Tegeder M, Throne-Holst M, Frommer WB, Patrick JW. Phloem loading and unloading of sugars and amino acids. Plant Cell Environ 2003; 26: 37-56.

[17] Hale CR. Relation between potassium and the malate and tartrate contents in grape berries. Vitis 1977*; 16*: 9-19.

[18] Blouin J, Cruège J. Analyse et composition des vins: Comprendre le vin. In: Editions La Vigne, Dunod. Paris, France 2003; p. 304.

[19] Conde C, Silva P, Fontes N, *et al.* Biochemical changes throughout grape berry development and fruit and wine quality. Food 2007; 1(1): 1-22.

[20] Walker RR, Clingeleffer PR, Kerridge GH, Rühl H, Nicholas P, Blackmore DH. Effects of the rootstock Ramsey (*Vitis champini*) on ion and organic acid composition of grapes and wine, and on wine spectral characteristics. Aust J Grape Wine Res 1998; 4: 100-10.

[21] Rühl EH. Effect of rootstock and K^+ supply on pH and acidity of grape juice. Acta Horticulturae 2000; 512: 31-7.

[22] Coombe BG. Distribution of solutes within the developing grape berry in relation to its morphology. Am J Enol Vitic 1987; 38(2): 120-7.

[23] Rogiers SY, Keller M, Holzapfel BP, Virgona JM. Accumulation of potassium and calcium by ripening berries on field vines of *Vitis vinifera* (L) cv. Shiraz *Aust.* J Grape Wine Res 2000; 6: 240-3.

[24] Conradie WJ. Seasonal uptake of nutrients by Chenin blanc in sand culture: II. Phosphorus, potassium, calcium and magnesium. S Afr J Enol Vitic 1981; 2(1): 7-13.

[25] Possner DRE, Kliewer WM. The localization of acids, sugars, potassium and calcium in developing grape berries. Vitis *1985; 24*: 229-40.

[26] Bondada BR, Matthews MA, Shackel KA. Functional xylem in the post-veraison grape berry. J Exp Bot 2005; 56(421): 2949-57.

[27] Zhenming N, Xuefeng X, Yi W, Tianzhong L, Jin K, Zhenhai H. Effects of leaf-applied potassium, gibberellin and source–sink ratio on potassium absorption and distribution in grape fruits. Sci Hortic 2008; 115: 164-7.

[28] Downton WJS. Influence of rootstocks on the accumulation of chloride, sodium and potassium in grapevines. Aust J Agric Res, 1977; 28: 879-89.

[29] Wood R, Parish M. The mechanisms and viticultural factors governing potassium accumulation in the grape berry - Part 1. In: The Australian & New Zealand Grapegrower & Winemaker. Annual Technical Issue 2003.

[30] Davies C, Shin R, Liu W, Thomas MR, Schachtman DP. Transporters expressed during grape berry (*Vitis vinifera* L.) development are associated with an increase in berry size and berry potassium accumulation. J Exp Bot 2006; 57(12): 3209-16.

[31] Pratelli R, Lacombe B, Torregrosa L, *et al.* A grapevine gene encoding a guard cell K^+ channel displays developmental regulation in the grapevine berry. Plant Physiol 2002; 128: 564-77.

[32] Hanana M, Cagnac O, Yamaguchi T, Hamdi S, Ghorbel A, Blumwald E. A grape berry (*Vitis vinifera* L.) cation/proton antiporter is associated with berry ripening. Plant Cell Physiol 2007; 48(6): 804-11.

[33] Czempinski K, Gaedeke N, Zimmermann S, Müller-Röber B. Molecular mechanisms and regulation of plant ion channels. J Exp Bot *1*999; 50: 955-66.

[34] Maeshima M. Tonoplast transporters: Organization and function. Annu Rev Plant Physiol Plant Mol Biol 2001; 52: 469-97.

[35] Terrier N, Sauvage F-X, Ageorges A, Romieu C. Changes in acidity and in proton transport at the tonoplast of grape berries during development. Planta 2001; 213: 20-28.

[36] Hanana M, Cagnac O, Mliki A, Blumwald E. Modèle topologique de la structure d'un antiport vacuolaire de type NHX chez la vigne cultivée (*Vitis vinifera*). Botanique 2009; 87: 339-47.

[37] Bell S-J, Henschke PA. Implications of nitrogen nutrition for grapes, fermentation and wine. Aust J Grape Wine Res 2005; 11: 242-95.

[38] Crawford NM, Glass ADM. Molecular and physiological aspects of nitrate uptake in plants. Trends Plant Sci 1998; 3(10): 389-95.

[39] Orsel M, Filleur S, Fraisier V, Daniel-Vedele F. Nitrate transport in plants: which gene and which control? J. Exp Bot 2002; 53(370): 825-33.

[40] Stitt M. Nitrate regulation of metabolism and growth. Curr Opin Plant Biol 1999; 2: 178-86.

[41] Conradie WJ. Distribution and translocation of nitrogen absorbed during late spring by two-year-old grapevines grown in sand culture. Am J Enol Vitic 1990; 41(3): 241-50.

[42] Delgado R, Martín P, del Álamo M, González M-R. Changes in the phenolic composition of grape berries during ripening in relation to vineyard nitrogen and potassium fertilisation rates. J Sci Food Agric, 2004; 84: 623-30.

[43] Grechi I, Vivin Ph, Hilbert G, Milin S, Robert T, Gaudillère J-P. Effect of light and nitrogen supply on internal C:N balance and control of root-to-shoot biomass allocation in grapevine. Environ Exp Bot. 2007; 59: 139-49.

[44] Hilbert G, Soyer JP, Molot C, Giraudon J, Milin S, Gaudillère JP. Effects of nitrogen supply on must quality and anthocyanin accumulation in berries of cv. Merlot. Vitis 2003; 42(2): 69-76.

[45] Rodriguez-Lovelle B, Gaudillère J-P. Carbon and nitrogen partitioning in either fruiting or non-fruiting grapevines: effects of nitrogen limitation before and after veraison. Aust J Grape Wine Res 2002, 8: 86-94.

[46] Roubelakis-Angelakis KA, Kliewer WM. Nitrogen metabolism in grapevine. In: Janick J, Ed. Horticultural Reviews. John Wiley & Sons, Inc 1992; pp. 407-12.

[47] van Heeswijck R, Stines AP, Grubb J, Moller I.S, Hoj P.B. Molecular biology and biochemistry of proline accumulation in developing grape berries. In: Roubelakis-Angelakis KA, Ed. Molecular Biology & Biotechnology of the Grapevine. Kluwer Academic Publishers 2001; pp. 87-107.

[48] Bates LS, Waldren RP, Teare ID. Rapid determination of free proline for water-stress studies. Plant Soil 1973; 39: 205-7.

[49] Garde-Cerdán T, Lorenzo C, Lara JF, Pardo F, Ancín-Azpilicueta C, Salinas MR. Study of the evolution of nitrogen compounds during grape ripening. Application to differentiate grape varieties and cultivated systems. J Agric Food Chem 2009; 57: 2410-19.

[50] Huang Z, Ough CS. Amino acid profiles of commercial grape juices and wines. Am J Enol Vitic 1991; 42(3): 261-7.

[51] Kataoka I, Kubo Y, Sugiura A, Tomana T. Changes in L-phenylalanine ammonia-lyase activity and anthocyanin synthesis during berry ripening of three grape cultivars. J Japan Soc Hort Sci 1983; 52(3): 273-9.

[52] Mundy DC, Beresford RM. Susceptibility of grapes to *Botrytis cinerea* in relation to berry nitrogen and sugar concentration. N Z Plant Protect 2007; 60: 123-7.

[53] Comménil P, Brunet L, Audran J-C. The development of the grape berry cuticle in relation to susceptibility to bunch rot disease. J Exp Bot 1997; 48(313): 1599-607.

[54] Waters EJ, Hayasaka Y, Tattersall DB, Adams KS, Williams PJ. Sequence analysis of grape (*Vitis vinifera*) berry chitinases that cause haze formation in wines. J Agric Food Chem 1998; 46: 4950-7.

[55] Waters EJ, Shirley NJ, Williams PJ. Nuisance proteins of wine are grape pathogenesis-related proteins. J Agric Food Chem 1996; 44: 3-5.

[56] Ferguson IB. Calcium in plant senescence and fruit ripening. Plant Cell Environ 1984; 7: 477-89.

[57] White PJ. The pathways of calcium movement to the xylem. J Exp Bot 2001; 52(358): 891-9.

[58] Li X, Xu C, Korban SS, Chen K. Regulatory mechanisms of textural changes in ripening fruits. Crit Rev Plant Sci 2010; 29: 222-43.

[59] Nunan KJ, Sims IM, Bacic A, Robinson SP, Fincher GB. Changes in cell wall composition during ripening of grape berries. Plant Physiol 1998, 118: 783-92.

[60] Cabanne C, Donèche B. Purification and characterization of two isozymes of polygalacturonase from *Botrytis cinerea*. Effect of calcium ions on polygalacturonase activity. Microbiol Res 2002; 157: 183-9.

[61] Chardonnet CO, Sams CE, Trigiano RN, Conway WS. Variability of three isolates of *Botrytis cinerea* affects the inhibitory effects of calcium on this fungus. Phytopathology 2000; 90(7): 769-74.

[62] Amiri ME. Accumulation of calcium by ripening berries on grape (*Vitis vinifera* L. cv. Asgari). In: The Proceedings of the International Plant Nutrition Colloquium XVI. UC Davis 2009.

[63] Cabanne C, Donèche B. Calcium accumulation and redistribution during the development of grape berry. Vitis 2003; 42(1): 19-21.

[64] Martinoia E, Maeshima M, Neuhaus HE. Vacuolar transporters and their essential role in plant metabolism. J Exp Bot 2007; 58(1): 83-102.

[65] Skinner PW, Matthews MA. Reproductive development in grape (*Vitis vinifera* L.) under phosphorus-limited conditions. Sci Hortic 1989; 38: 49-60.

[66] Grant RS, Matthews MA. The influence of phosphorus availability, scion, and rootstock on grapevine shoot growth, leaf area, and petiole phosphorus concentration. Am J Enol Vitic 1996; 47(2): 217-24.

[67] Xuefeng X, Zhenming N, Tianzhong L, Yi W, Jin K, Zhenhai H. Influx pathways and key stages of phosphorus and potassium transport into grape fruits. J Plant Nutr 2008; 31: 35-42.

[68] Smith FW, Mudge SR, Rae AL, Glassop D. Phosphate transport in plants. Plant Soil 2003; 248: 71-83.

[69] Rogiers SY, Greer DH, Hatfield JM, Orchard BA, Iieller M. Mineral sinks within ripening grape berries (*Vitis vinifera* L.). Vitis 2006; 45(3): 115-23.

[70] Marois JJ, Bledsoe AM, Gubler WD. Control of *Botrytis cinerea* on grape berries during postharvest storage with reduced levels of sulfur dioxide. The American Phytopathological Society 1986; 70(11): 1050-2.

[71] Bourbos VA, Skoudridakis MT, Barbopoulou E, Venetis K. Ecological control of grape powdery mildew (Uncinula necator). 2010 Nov 8. Available from: http://www.landwirtschaft-mlr.baden-wuerttemberg.de/servlet/PB//menu/1043197_l2/index1215611686464.html

[72] Buchner P, Takahashi H, Hawkesford MJ. Plant sulphate transporters: co-ordination of uptake, intracellular and long-distance transport. J Exp Bot 2004; 55(404): 1765-73.

[73] Tavares S, Sousa C, Carvalho LC, Amâncio S. Derepressed sulfate transporters are strongly and rapidly repressed after sulfate addition to sulfur-depleted *Vitis* cells. Int J Plant Sci 2008; 169(8): 987-97.

[74] Skinner PW, Matthews MA. A novel interaction of magnesium translocation with the supply of phosphorus to roots of grapevine (*Vitis vinifera* L.). Plant Cell Environ 1990; 13: 821-26.

[75] Amiri ME, Fallahi E. Influence of mineral nutrients on growth, yield, berry quality, and petiole mineral nutrient concentrations of table grape. J Plant Nutr 2007; 30: 463-70.

[76] Chopin EI, Marin B, Mkoungafoko R, *et al.* Factors affecting distribution and mobility of trace elements (Cu, Pb, Zn) in a perennial grapevine (*Vitis vinifera* L.) in the Champagne region of France. Environ Pollut 2008; 156: 1092-8.

[77] Volschenk CG, Hunter JJ, le Roux DJ, Watts JE. Effect of graft combination and position of application on assimilation and translocation of zinc in grapevines. J Plant Nutr 1999; 22(1): 115-9.

[78] Bacha MA, Sabbah SM, Hamady MA. Effect of foliar applications of iron, zinc and manganese on yield, berry quality and leaf mineral composition of Thompson Seedless and Roumy Red grape cultivars. J King Saud Univ Agric Sci 1417; 9(1): 127-40.

[79] Rombolà A.D, Tagliavini M. Iron nutrition of fruit tree crops. In: Barton LL, Abadía J, Eds. Iron Nutrition in plants and Rizhospheric Microorganisms. Springer 2005; pp. 61-83.

[80] Williams CMJ, Maier NA, Bartlett L. Effect of molybdenum foliar sprays on yield, berry size, seed formation, and petiolar nutrient composition of "Merlot" grapevines. J Plant Nutr 2004; 27(11): 1891-916.

[81] Mendel RR, Schwarz G. Molybdoenzymes and molybdenum cofactor in plants. Crit Rev Plant Sci 1999; 18(1): 33-69.

[82] Brown PH, Shelp BJ. Boron mobility in plants. Plant Soil 1997; 193: 85-101.

[83] Scott LE, Schrader AL. Effect of alternating conditions of boron nutrition upon growth and boron content of grape vines in sand culture. Plant Physiol 1947; 22(4): 526-37.

[84] Gauch HG, Dugger WM. The role of boron in the translocation of sucrose. Plant Physiol 1953; 28(3): 457-66.

[85] Sommer AL. Copper as an essential for plant growth. Plant Physiol 1931; 6(2): 339-45.

[86] Deacon JW. Principles and practice of controlling fungal growth. In: Deacon JW, Ed. Fungal Biology, 4th ed. Wiley-Blackwell, UK 2006; pp. 338-55.

[87] Brun LA, Maillet J, Richarte J, Herrmann P, Remy JC. Relationships between extractable copper, soil properties and copper uptake by wild plants in vineyard soils. Environ Pollut 1998; 102: 151-61.

[88] Nóvoa-Muñoz JC, Queijeiro JMG, Blanco-Ward D, Álvarez-Olleros C, Martínez-Cortizas A, García-Rodeja E. Total copper content and its distribution in acid vineyards soils developed from granitic rocks. Sci Total Environ 2007; *378*: 23- 7.

[89] Romeu-Moreno A, Mas A. Effects of copper exposure in tissue cultured *Vitis vinifera*. J Agric Food Chem 1999; 47: 2519-22.

[90] Toselli M, Baldi E, Marcolini G, *et al.* Response of potted grapevines to increasing soil copper concentration. Aust J Grape Wine Res 2009; 15: 85-92.

[91] Krämer U, Clemens S. Functions and homeostasis of zinc, copper, and nickel in plants. In: Tamás MJ, Martinoia E, Eds. Molecular biology of metal homeostasis and detoxification. Springer 2005; pp. 216-71.

CHAPTER 3

Source/Sink Relationships and Molecular Biology of Sugar Accumulation in Grape Berries

C. Davies[1,*], P. K. Boss[1], H. Gerós[2,3], F. Lecourieux[4] and S. Delrot[4]

[1]Commonwealth Scientific and Industrial Research Organisation, Plant Industry, P.O. Box 350, Glen Osmond, South Australia 5064, Australia; [2]Centro de Investigação e de Tecnologias Agro-Ambientais e Biológicas (CITAB), Quinta de Prados, 5001-801 Vila Real, Portugal; [3]Departamento de Biologia, Universidade do Minho, Campus de Gualtar, 4710-057 Braga, Portugal and [4]UMR 1287 Ecophysiologie et Génomique Fonctionnelle de la Vigne (EGFV), Université de Bordeaux, INRA, Institut des Sciences de la Vigne et du Vin (ISVV), 210 Chemin de Leysotte, CS 50008, 33882 Villenave d'Ornon, France

Abstract: Phloem transport of assimilates provides the materials needed for the growth and development of reproductive structures, storage and developing organs, and has long been recognised as a major determinant in crop yield. Thus, the understanding of the mechanisms and regulations of sugar transport into sink tissues has an important basic and applied relevance. The Grapevine is a good example of a crop where sugar accumulation in the fruit has an important economic role. Massive sugar transport and compartmentation into the grape berry mesocarp cells (up to 1 M glucose and fructose) start at veraison and continues until the harvest. Sucrose transported in the phloem is cleaved into hexoses by invertases and stored in the vacuole. The Sugar content determines the sweetness of table grapes, wine alcohol content, and regulates gene expression, including, for example, several genes involved in the synthesis of secondary compounds which contribute to grape and wine quality. Many viticultural practices affect source/sink relationships, thus altering sugar concentration in the berry. For instance, the rootstock used, which is a potential sink, has a strong impact on source activity, by affecting the morphology and activity of the aerial part of the plant. Molecular approaches have also provided major advances in grapevine research. Monosaccharide and disaccharide transporter genes have been recently identified and their products studied in heterologous systems. The sequencing of the grapevine genome and the development of grape microarrays have made a valuable contribution to the study of the biochemistry of grape berry development and ripening, for example, low affinity glucose uniporters identified in the genome may also be involved in the sugar uptake. In the present chapter, the routes of sugar import and storage in the grape cells are updated and discussed and a model with the main transport steps and biochemical pathways is proposed.

Keywords: Assimilates, Disaccharide transporter, Fructose, Glucose, Invertase, Monosaccharide transporter, Phloem transport, Phloem unloading, Sucrose, Sugars, Veraison.

INTRODUCTION

Sugar content and concentration in ripe berries are important parameters for both table grapes and wine making. Berry sugar concentration is the major determinant of the final alcohol content of the wine. A relationship between sugar and anthocyanin contents in grape skins [1, 2] and grape cells [3] has been shown, suggesting that the sugar status is not only important for the alcohol content, but also for the synthesis of secondary metabolites. Indeed, due to the relatively low amount of photosynthate produced by the berry itself, in the order of 10% of the carbon required for fruit development [4], imported sugars provide the starting material for most of berry metabolism. Sugars also provide the osmotic driving force for cell expansion as well as modulating gene expression through signalling mechanisms.

Prices are usually adjusted to the soluble solids present in the harvested grapes [5]. If the sugar content is too low, it may be increased by addition of exogenous sugars. This process, called chaptalisation is subject

***Address correspondence to C. Davies:** Commonwealth Scientific and Industrial Research Organisation, Plant Industry, P.O. Box 350, Glen Osmond, South Australia 5064, Australia; E-mail: christopher.davies@csiro.au

Hernâni Gerós, M. Manuela Chaves and Serge Delrot (Eds)

to legal constraints in most countries. For many decades, particularly in cooler regions, clonal selection and viticultural practices (row orientation, defoliation, cluster thinning) have been oriented towards high sugar concentration, which is considered as beneficial for wine quality. More recently, an excess of sugars has been observed in many vineyards around the world, which is thought to result from climate change. As a result, chaptalisation is used less frequently, and in some areas steps must be taken to avoid excessive sugar accumulation, which may be detrimental to wine quality and cause problems with both alcoholic and malolactic fermentations [6]. This is particularly the case for grapes grown under irrigation in hotter regions. Grapes of high °Brix may produce high alcohol wines which mask other quality components. An upper limit of 24°Brix is often used to indicate the proper maturity for quality wines in warmer regions [5].

Most of the viticultural practices used to control berry sugar content alter sink-source relationships in the plant. The molecular basis of sugar accumulation in the berries is now reasonably well understood, but the availability of the grape genome sequence offers new opportunities to investigate this process. In this context, this review will summarise the main physiological and molecular data related to sugar transport in grapevine and sugar accumulation in grape berries.

SUGAR TRANSPORT AND SINK/SOURCE RELATIONSHIPS

The ripening of grape berries is accompanied by a massive accumulation of hexoses, and by the synthesis and accumulation of a wide range of phenolic compounds and aroma precursors. The vacuoles are the main reservoir of sugars, organic acids, aromas, flavours, ions and water in grape berry cells during ripening (see the chapter by Fontes *et al.*, this edition).

The amount of sucrose imported by the berry depends on carbon allocation between the photosynthesising leaves and different types of sinks. To ensure vegetative growth, reproduction and acclimation to environmental stresses, vines must effectively allocate available annual resources to both vegetative and reproductive tissues. Source/sink relationships are important for total yield per vine, for sugar concentration in the berry and, to some extent, for berry and wine quality. Many viticultural practices directly (trellising, shoot trimming, cluster thinning) or indirectly (plantation density, row orientation) affect these relationships.

In grapevines, carbohydrate produced by photosynthesis is exported from the leaf as sucrose and is transported *via* the phloem to the berry [7, 8]. Grapevines accumulate and store carbohydrates necessary for support of initial growth the following spring and, in some cases, for acclimation to freezing temperatures even during the fruit maturation period. Each growing season therefore is not an event alone in time, but blends with and is influenced by those that came before and contributes to those which follow [9].

The Root System

Growth and developmental patterns of the below ground and aerial parts of the grapevine plant depend on complex exchanges of carbon and nitrogen, and on water availability and fluxes. Starch is accumulated in the wood of canes, roots and trunk at the end of the growth season. During winter, it is hydrolysed to osmotically active sugars which are then converted back to starch when spring approaches [10]. The stored starch is used as a starter to facilitate vegetative development during the early season. The ratio of above ground: below ground biomass increases during the season, and is negatively affected by moderate water deficits [11].

In fruit trees, the root system is the organ which is the most affected by the presence of fruits [12], and it is the main sink only when the plants are fruitless [13]. Eighty to 85% of the vineyards worldwide are grafted [14]. Complex relationships are established between the root and shoot system of grafted grapevines. The rootstock depends entirely on the scion for the supply of reduced carbon. The rootstock is therefore a potential sink and is an important component in the source/sink relationships. The grapevine rootstock may affect the photosynthetic capacities of the scion [15, 16]. These effects depend on the rootstock/scion combinations [17]. In the vineyard, the different photosynthetic activities induced by the rootstock are linked to the water status of the plants [18, 19].

Scion development strongly depends on the genotype of the rootstock. Rootstocks affect intensity and duration of individual shoot growth, leaf area, trunk size, pruning weight, bud fertility, yield and phenology [11]. Rootstock effects explain a high proportion of variance of shoot length at the beginning of each growth cycle [20]. The rootstock genotype also has a significant impact on scion gas exchange, water status, canopy growth and yield [21]. In conclusion, the rootstock is a potential sink, and it has a strong impact on source activity by controlling the morphology and activity of the aerial part of the plant.

Water stress reduces net photosynthesis and stomatal conductance with large differences between grafted and ungrafted vines [22]. Net photosynthesis, stomatal conductance and carboxylation efficiency were affected by the rootstock genotype combination, so that under water stress only some rootstock genotypes transferred drought tolerance to the scion [22]. Abscisic acid in leaves increased significantly after water stress, but the relations with the CO_2 assimilation rate, stomatal conductance, and carboxylation efficiency appeared to be affected by the genotype of the vegetative apparatus [22]. Under conditions of water stress, when less carbon is available, the fruits reach a better quality [5, 23].

The Aerial System

The relationship between yield, berry sugar content and perceived wine quality is extremely complex and, as yet, incompletely understood. This is probably due to there being a large number of factors that can affect the relationships between yield and sugar content and sugar content and wine quality. There is a general belief that high yields are not compatible with high quality and based on this belief top French appellations, for example, limit the yield below specified levels [5]. While there are a number of studies that support this belief (for example, [24, 25]) in other studies differences in fruit composition were found to be more dependent on seasonal differences than either yield or crop load [26].

Since the ultimate source of sugars is photosynthesis in leaves, the leaf area/fruit ratio is a critical factor for berry sugar level. The ratio below which sugar levels starts to decline is reported to be between 8 and 10 cm^2/g fruit, whereas increasing the ratio above this threshold hardly affects soluble solid levels [27]. For Tokay vines, the responses of berry growth, sugar accumulation and fruit colour to different amounts of leaf area/g berry were all saturated at about 12 cm^2 area/g fruit [28]. The wine sensory evaluation score of Sauvignon blanc decreased with decreasing leaf area, when the fruit weight ratio was below 18 cm^2/g and as crop load increased [25].

Shoot trimming minimises vegetative dominance and prevents physiological imbalance between sources and sinks. It has also been used to improve fruit set and maximise carbohydrate partitioning to fruit during ripening [29].

Sunlight-exposed fruits generally have higher levels of total soluble solids, anthocyanins and phenolics, while berry weight is lower compared to unexposed or canopy shaded fruits. Bergqvist *et al.* [30] showed that soluble solids initially increased with greater sunlight exposure, then declined when mid-day PAR exceeded 31 to 50 and 51 to 100 micromol m^{-2} sec^{-1}, respectively, for clusters on the north and south sides of the canopy. The increased temperature of fruits exposed beyond these levels may have inhibited ripening, since cluster temperatures above 37°C inhibit sugar accumulation.

THE MOLECULAR BIOLOGY OF SUGAR TRANSPORT AND ACCUMULATION IN THE GRAPE

As grape berries develop they change in both size and composition. Grapes exhibit a double sigmoid pattern of growth [31], the first rapid growth phase that occurs after fruit set being due to an increase in both cell numbers and cell expansion. In most cultivars the first expansion phase is followed by a lag phase during which little or no growth occurs. The commencement of the second growth phase, which follows the lag phase, coincides with the onset of ripening (veraison). In this review we will define veraison as the last time point before a significant increase in sugar accumulation and will reinterpret data in line with this definition where possible. One of the most important ripening-related changes that occurs at veraison is the beginning of a massive accumulation of sugars.

Our understanding of the mechanisms of carbon partitioning and accumulation in grapevines is improving due to the use of molecular techniques. The considerable progress made in this area in other, more tractable, plants can be used to aid our understanding of sugar metabolism in grape berries. There is convincing evidence that membrane-located sugar transporter proteins are heavily involved in the active transport and redistribution of sugars between cells and tissues (reviewed by Lalonde *et al.* [32]). Sugars that are ultimately stored in the cell vacuole must cross tonoplast and/or plasma membranes at some stage in their transport cycles whether the modes of loading and unloading are symplastic or apoplastic. Both mono- and disaccharide transporters have been identified and they appear to be important in sugar accumulation in grape berries.

Until veraison most of the sugar imported into the berry is metabolised and there is little storage. Just before veraison, the berry contains no more than 150 mM hexose, with a glucose/fructose ratio of two [33]. After veraison there is a steady rise in the levels of stored sugars and the mechanism of sugar uptake into the berry may alter at this time (see below). Cultivars of *Vitis vinifera* generally store little sucrose and accumulate sugars in the form of the hexoses glucose and fructose in roughly equal amounts. However, the levels of sugars stored and the sucrose to hexose ratio can vary somewhat depending on the cultivar [34, 35]. Some cultivars, in particular those derived from *V. labrusca*, store considerable amounts of sucrose, which suggests that they accumulate sugars in a different manner to *vinifera* cultivars.

Sugar concentrations in skin are generally lower than those present in the flesh tissue. Furthermore, there are longitudinal differences in the levels of glucose and fructose in the outer and central flesh tissues [36]. Hexose concentrations in the flesh increase from low near the brush to high at the stylar end. Sucrose concentration in the flesh is low throughout ripening, but increases in the skin when the berries reach 17-26° Brix and in the central tissues (including the brush) in overripe berries.

Phloem Unloading in Berries

Although phloem loading is obviously crucial to assimilate transport it will not be discussed here as this review deals with only those processes that occur within the berry. Assimilates imported in the berry are transported through the carpellary vascular bundles which are divided into the peripheral and central bundles. A dorsal bundle network extends at the periphery of the fruit, and central vascular bundles connected to the seeds irrigate the central flesh.

Phloem unloading is the process of assimilate movement from the sieve element/companion cells to the site of storage or utilisation. In broad terms unloading can occur by either a symplastic or apoplastic route. The symplastic route involves the passive movement of assimilate through plasmodesmata by diffusion possibly coupled with some bulk flow. The alternative apoplastic route requires the movement of assimilate across membranes at some stage in the unloading process. This apoplastic step can occur at or beyond the sieve element boundary [37, 38]. Very recently, it was observed that endocytosis may be involved in the direct uptake of hexoses from the apoplast to the vacuole [39]. The determination of which route predominates in a particular tissue has been based on such things such as anatomical features, the type of assimilate transported, the reaction to inhibitors, and through studying the movement of dyes, labelled viral movement proteins and radiolabelled assimilates. The study of the accumulation of the transcripts of various genes and their corresponding products thought to be involved in sugar metabolism and transport is also useful in determining possible modes of transport. In some species more amenable to manipulation, transgenic plants with altered sugar metabolism have also cast light on the processes involved. In most sinks the symplastic route is preferred [38]. In sinks which accumulate high levels of osmotically active assimilates, the apoplastic route predominates. A number of variations of the two general schemes of phloem unloading described above have been explained [37, 38]. Apoplastic phloem unloading appears to occur throughout apple fruit development [40]. There is also evidence that the mechanism of unloading can change between symplastic and apoplastic during the course of development. This occurs in tomato fruit during the phase of rapid hexose accumulation [41]. In addition to xylem conduits, membrane integrity, and thus cellular compartmentation, also apparently remain intact in ripening berries [42]. Concordantly, flow cytometry, bright-field, epifluorescence and confocal microscopy analyses have confirmed that pulp cells from ripe berries are structurally intact, in line with earlier results [43]. These cells are physiologically active as they are able to incorporate fluorescent sugars [39].

Mono- and disaccharide transporters comprise some of the many members of the major facilitator superfamily that conduct the transport of a range of molecules across membranes [44]. These proteins have similar structures as they contain 12 membrane-spanning regions. However, the different proteins within this family use a range of transport mechanisms. Some transport proteins, symporters and antiporters, conduct secondary active transport where the flux of the transported molecule is dependent on the movement of another molecular species in the same or opposite direction, respectively. Some are defined as uniport carriers which conduct facilitated diffusion where the flux is down the potential gradient but some selectivity is supplied by the uniport protein. Sugar transporters driven by the H^+ gradient across membranes are crucial in sugar distribution and accumulation and have been widely studied in a range of plants [32, 37, 38, 45, 46].

The mechanisms by which grape berries unload and take up sugars from the phloem and store them in the vacuole, predominantly in the form of fructose and glucose, have been studied by a number of groups using a range of techniques. Uptake and efflux studies with isolated skin suggested active uptake of D-glucose but not L-glucose. Surprisingly, active uptake of monosaccharides was measured in skin samples from both green unripe and ripe berries, whereas it was evident only in flesh tissues sampled from ripening berries [47]. However, contrary to what is observed in mesocarp tissues, a high proportion of the sugar absorbed by skin pieces is diffusive, whatever the sampling time.

Two recent studies have provided evidence for apoplastic unloading during berry ripening. Wang *et al.* [48] developed a novel experimental system in which peeled berries still attached to the vine were bathed in a buffer solution which allowed them to measure sugar efflux and to apply inhibitors. For berries tested during their ripening phase both sodium fluoride and *p*-chloromercuribenzenesulfonic acid (PCMBS) inhibited sugar efflux into the bathing solution. This suggested that plasma membrane transport was occurring and hence an apoplastic step is involved in sugar unloading.

More recent evidence has indicated that grape berries change their mode of unloading from the symplastic route to the apoplastic route at, or slightly before, the commencement of ripening [49]. Electron microscopy was used to show that there are plasmodesmata between the sieve elements and parenchyma cells in grape berries which should allow a symplastic connection. Approximately 70% of these plasmodesmata remain apparently intact throughout berry ripening. During ripening 5% of the plasmodesmata appear to be branched and are therefore likely to be inefficient, and 10-20% seemed to be blocked by electron dense material. However, even the 70% of plasmodesmata that appeared normal were shown to be unable to transport small molecules efficiently during ripening. Both the marker dye molecule carboxyfluorescein and the cucumber mosaic virus movement protein 3a tagged with GFP moved out freely from the phloem strands into flesh cells when injected into the berry rachis up until the time of veraison. In contrast, both markers were restricted to the phloem strands during the ripening phase suggesting that the berry flesh cells were symplastically isolated from the phloem cells. However, the unloading of [^{14}C] labelled assimilates into the flesh cells demonstrated that transport was still occurring. In addition, cell wall invertase activity and total soluble sugar levels in the apoplast were elevated after veraison (discussed below). These data led Zhang *et al.* [49] to conclude that there is a shift in the mode of unloading from the symplastic to the apoplastic pathway coincident with the commencement of berry ripening. As mentioned above a similar shift from symplastic to apoplastic phloem unloading during ripening has been described for a number of other fruit including tomato [38], apples [40], walnuts [50] and Jujube [51]. It has been suggested that the apoplastic route is favoured in sink tissues, such as fruit, where high concentrations of solutes are stored as it may be the only way uptake can be achieved against high concentration gradients [38, 49]. No transporter protein responsible for the efflux of sucrose into the apoplast was identified in these studies but recently some progress has been made in identifying transport proteins that may perform efflux functions in other plants [52-54]. For the carrier-type proteins described by Zhou *et al.* [53] to release sucrose from the phloem into the apoplast a sucrose gradient maintained by sucrose import into parenchyma storage cells by sucrose symporters or by the cleavage of sucrose by a cell wall invertase would have to be exist. Another mechanism for efflux from the phloem has been described for the ZmSUT1 sucrose/H^+ symporter from maize that can operate in a reverse mode to efflux sucrose (and protons) [52].

Grape Sugar Transporters

Molecular biology has been used to investigate the role of sugar transporters in the uptake of sugars into grape. By aligning the amino acid sequences from the sucrose transporters of a number of other plant species conserved regions can be identified. These conserved sequences can be used to design redundant oligonucleotide primers for use in PCR. Using this approach cDNAs were cloned from the *vinifera* cultivar Shiraz that encoded proteins with considerable amino acid sequence similarity to sucrose/H^+ symporters [55]. Three full-length cDNA clones, VvSUC11, VvSUC12 and VvSUC27, were isolated from a Shiraz berry cDNA library. The sequences of these three clones are approximately 50% identical to each other at the amino acid level. Another sequence was isolated from Uni blanc by Ageorges *et al.* [56] who named it *VvSUT1*. The deduced amino acid sequence of this clone is 99.4% identical to the VvSUC11 sequence from Shiraz and is therefore likely to represent the same transporter 'species' in a different cultivar. A forth distinct grape sucrose transporter sequence VvSUT2 (AAL32020) is more similar to VvSUC27 (Fig. **1**) [32, 46], its function has been tested and it seems to be lowly expressed in most grape tissues, particularly berries [57].

Functional analysis of VvSUC11 (VvSUT1), VvSUC12 and VvSUC27 has been carried out in yeast and confirmed that they are sucrose/H^+ symporters [56, 58, 59]. The uptake of sugars by all three transporters was pH dependent with maximal activity at, or below, pH 5 [58, 59]. The K_m values for VvSUC11 and VvSUC12, at pH 5, were 0.88 mM and 1.36 mM respectively [58]. Proton uncouplers greatly reduced the transport activity of these transporters in yeast and a significant reduction was observed upon the addition of the P-type H^+-ATPase inhibitor vanadate. The K_m values of around 1 mM indicate that these two transporters are likely to be of the High Affinity Low Capacity (HALC) type. VvSUC27 has a K_m of 8 at pH 5 and has been classified as a Low Affinity High Capacity (LAHC) type transporter [59]. The uptake activity of VvSUC27 in yeast was activated by glucose, fructose and mannose and inhibited by maltose and diethyl pyrocarbonate.

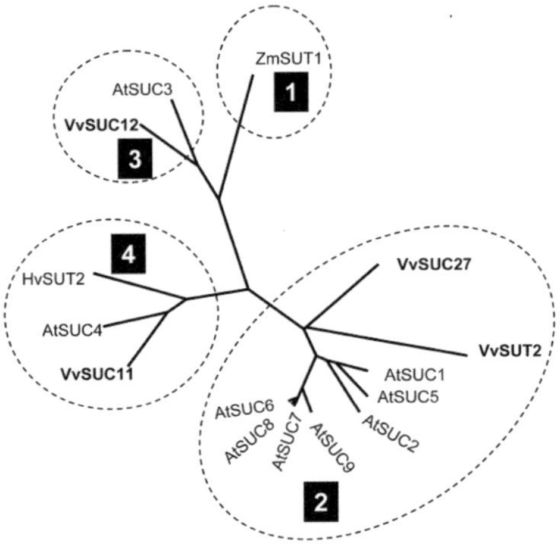

Figure 1: Unrooted dendrogram showing the relationship between various plant sucrose transport proteins. The clades are labelled as per Sauer [46]. The sequences used were Arabidopsis; AtSUC1 (Atg71880), AtSUC2 (At1g22710), AtSUC3 (At2g02860), AtSUC4 (At1g09960), AtSUC5 (At1g71890), AtSUC6 (At5g43610), AtSUC7 (At1g66570), AtSUC8 (At2g14670), AtSUC9 (At5g06170), Barley; HvSUT2 (CAB75881), Grape; VvSUC11 (AAF08329), VvSUC12 (AAF08330), VvSUC27 (AAF08331), VvSUT2 (AAL32020), Maize; ZmSUT2 (BAA83501). Protein sequences were aligned with CLUSTALW [117], distances calculated with PRODIST, tree developed with NEIGHBOUR and displayed with DRAWTREE [118].

The four distinct grape sequences mentioned above, along with the other sucrose transporters from both monocots and dicots, can be clustered into three or four groups of a phylogenetic tree (Fig. **1**) [32, 46]. The trees in these two reports are very similar to each other in structure and the more recent analysis by Sauer [46] will be referred to in the following discussion as it includes more sequences. Fig. **1** represents a modified version of this tree which contains all the Arabidopsis and grape sucrose transporter sequences and the additional sequences from maize and barley which are discussed below. Four distinct clades were described. Clade 1 contains only sequences from monocots and the member of interest in this clade, ZmSUT1 will be discussed later. VvSUC27 and VvSUT2 ('VvSUCy', AAL320020) cluster together in Clade 2. The AAL320020 sequence is derived from a genomic clone and contains a number of bases for which no call was made and so is incomplete. The sequences in this clade are all from dicots and many appear to be HALC type plasma membrane–located transporters. Some members of this clade have been shown to be localised to the phloem with proposed roles in sugar loading [32, 46]. Contrary to many of the members of this group that are of the HALC type, VvSUC27 has a K_m that suggests it is a LAHC transporter. From this it can be concluded that sequence similarity is of little use in predicting likely K_ms for these proteins and that it is possible that the members of a clade conduct a range of different functions.

VvSUC12 is located in Clade 3 that is distinguished by its members being larger, mainly due to having an extended cytoplasmic loop between two of the membrane-spanning regions [46]. An alignment of VvSUC11 (VvSUT1), VvSUC12 and VvSUC27 sequences showed that there is an insertion of 51 amino acids in the VvSUC12 sequence when compared to the VvSUC27 sequence. This insertion is in the hydrophilic loop between the sixth and seventh membrane spanning regions. Based on similarity to sensors from other organisms it has been proposed that this extended central loop may be involved in sugar sensing and that this group of proteins function as sugar sensors [60] but there is little evidence in support of this suggestion. It has also been suggested that this loop may be involved in the control of transporter activity [61]. However, the removal of the 55 'extra' amino acids from this loop in AtSUC3, which clusters in the same clade as VvSUC12, had little effect on its transport properties in yeast [62]. One function of the central loop in transporters may be in insertion into membranes as a loop of minimal length is required for correct insertion [63]. Some of the transporters in this clade have been localised to sieve elements but they are expressed in sink tissues as well and so may have a role in both phloem transport and unloading [46].

Clade 4, like Clade 2, appears to contain transporters of both LAHC and HALC types. For example, AtSUC4 (AtSUT4) has a K_m of 12 mM (LAHC) but VvSUC11, which also clusters in Clade 4, has a K_m of 0.88 mM and so appears to be a HALC sucrose transporter. A study by Endler *et al.* [64] has identified two members of this clade, HvSUT2 and AtSUT4 (AtSUC4) as tonoplast-localised sucrose transporters. This raises the interesting possibility that VvSUC11 might play a role in sucrose transport across the grape tonoplast. Such transporters have long been thought to exist on biochemical evidence but have proved difficult to identify. The mechanism by which transporters located in the tonoplast transport sugars against the proton gradient is unknown but a number of possible modes have been suggested [32]. However, the considerable difference in K_m between VvSUC11 and the tonoplast- located AtSUC4 may infer different roles and, therefore, localisation of the VvSUC11 transporter is required.

The overall message from these comparisons is that the membership of a protein in a cluster does not necessarily indicate its activity or likely function. The clades contain members with a range of K_ms indicating a wide range of affinities and potential function. Also, different members within a single clade have been reported to have different subcellular and tissue/organ distributions.

The expression pattern of the three Shiraz sucrose transporter genes was investigated by northern analysis [55]. Transcripts for all three genes were detected in a range of tissues including berries. The expression patterns of two of the genes, *VvSUC11* and *VvSUC12*, were very similar to each other. Neither was expressed to any great level in roots or tendrils but their transcripts were found in leaves, seeds, flowers and in berries throughout berry development. Their expression in berries was up-regulated at the time when the concentration of reducing sugars in the berry increased, the expression of *VvSUC12* was most significantly enhanced from veraison (Fig. **2**). Both of these transporters are HALC types. These data suggest these transporters may have a role in the import of sucrose into the ripening berry cells and in general they could be described as sink-related transporters. It appears that the expression of *VvSUC11* and especially *VvSUC12* is influenced by ethylene as 1-methylcyclopropene treatment reduced transcript levels in berries at around veraison [65]. Although all three of the grape sucrose transporters described above that have been functionally tested in yeast acted as sucrose symporters it is possible that one or more of them may be able to act in the reverse mode [52] and so may play a role in sucrose efflux from the phloem. The third gene, *VvSUC27*, was expressed in roots, tendrils, non-mature leaves, seeds, flowers and berries. *VvSUC27* was expressed early in berry development but the level of message decreased significantly after veraison so that no message was detected in the mid to late stages of ripening. A possibly different function for VvSUC27 is also supported by its having a much higher K_m than the other two grape sucrose transporters. It may be that this transporter is important in the delivery of sugars to non-storage sinks which would include pre-veraison berries. In addition to the three grape sucrose transporters discussed above, Afoufa-Bastien *et al.* [57] also investigated the expression of VvSUT2 in berries and showed that the levels of expression was much lower.

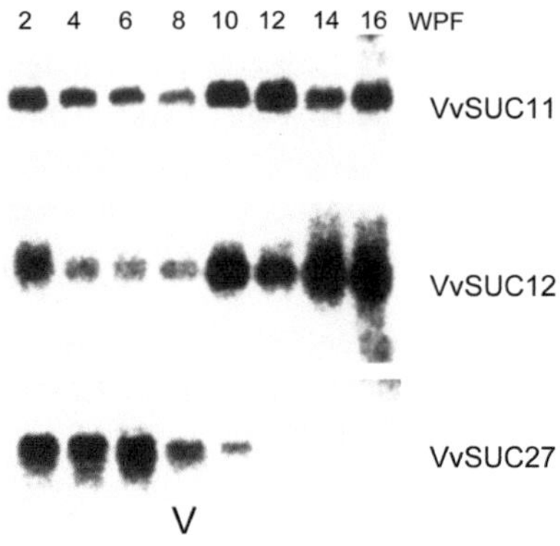

Figure 2: Northern blot analysis of Shiraz berry development series, probed with the three grape putative sucrose transporter clones. WPF, weeks post-flowering. The time of veraison is indicated by a 'V'. Adapted from Davies *et al.* [55].

Monosaccharide Transporters

Sucrose can be cleaved external to the plasma membrane and the resultant hexoses can be imported into the cell by monosaccharide specific transporters during grape berry development. This process, like the importation of sucrose by sucrose transporters described above, is consistent with the apoplastic mode of sugar accumulation after veraison [49]. Monosaccharide transporter (MST)-like genes have been cloned from a number of plant species and they comprise multigene families in plants. For example, 53 and 65 MST-like genes have been identified in Arabidopsis and rice, respectively [66]. The MST-like proteins from Arabidopsis can be clustered into seven groups having various functions [45].

Sequences of fifty nine putative MSTs have been identified in grapevine from the 8 x genomic sequence [57] (also see below). There appear to be at least ten distinct hexose transporters known to be transcribed in grapevine [57, 67- 69]. Whether any more of the putative MSTs are expressed or not remains to be seen. Two putative hexose transporters (VvHT1, VvHT2) were identified in grape berries by Fillion *et al.* [70]. RT-PCR was conducted using degenerate primers to amplify hexose transporters from cDNA made from Ugni blanc grape berry RNA. Two small but distinct clones were isolated which shared sequence similarity with hexose transporters from other species. One of these, *VvHT1*, was used as a probe to isolate a full-length clone from a cDNA library made from ripening berry RNA. The protein sequence deduced from this clone was closely related to other hexose transporters and contained the 12 membrane spanning regions characteristic of members of the major facilitator superfamily [44]. Northern blot analysis showed this gene was expressed most abundantly in young leaves and in berries [70]. The expression pattern for *VvHT1* in berries was biphasic. Expression was reported as elevated early in development (four weeks after flowering), after which it decreased to be at a much lower level at eight weeks after flowering and remained at this level until 12 weeks after flowering. In these berries sugar accumulation began sometime between four and eight weeks after flowering. The transcript level of VvHT1 then peaked again at 14 weeks after flowering but expression was virtually non-detectable at 15 weeks after flowering. This pattern of expression was confirmed using quantitative RT-PCR. The changes in transcript level of *VvHT1* do not coincide closely with the pattern of the increase in berry sugar concentration. Somewhat different expression patterns have been reported for *VvHT1* in other cultivars and it seems that the peak of expression observed in Ugni blanc at 14 weeks after flowering may be exceptional. In Cabernet Sauvignon berries the levels of *VvHT1* transcript declined steadily from early in berry development to be at low levels shortly after veraison; no later peak was observed [69, 71, 72]. The expression pattern of *VvHT2* was more closely related to the pattern of berry sugar accumulation than that of *VvHT1* [70]. *VvHT2* expression was low early in development but increased at about the same time as the increase in berry sugar levels occurred. A similar, but less marked, pattern of *VvHT2* expression was also observed in Cabernet Sauvignon [69]. VvHT1 functioned as a high affinity transporter (K_m 70 μM at pH 4.5) of glucose in yeast and transported fructose and sorbitol less well [67]. Galactose and mannose competed with glucose for transport but mannose was not transported (galactose transport was not tested) [67, 69]. An attempt was made to further explore the possible function of VvHT1 using tobacco plants transformed with sense and antisense CaMV 35S:*VvHT1* constructs. The strongest phenotype was in sense plants which were stunted with higher shoot/root ratios than control plants when grown *in vitro* [73]. The uptake of [^{3}H]glucose was significantly reduced in leaf discs from these plants. The level of 'host' MST transcripts was reduced but *VvHT1* transcripts were often present. These results show that hexose transporters are involved in plant development through affecting assimilate distribution, but the reason for the lack of complementation of the host function by *VvHT1* is unknown. Of the remaining three expressed MSTs, *VvHT11* was expressed moderately throughout berry development while *VvHT12* and *13* were expressed at low levels or not detected [57].

Studies in tomato give direct support to the notion that hexose transporters are involved in hexose accumulation in fruit. Tomato fruit with considerably reduced levels of *LeHT* expression, produced through RNAi-mediated knockdown, accumulated 55% less hexoses than the control line [74].

VvHT1 transcripts were localised mainly to the phloem in the peripheral vascular bundles in berries using *in situ* hybridisation [67]. VvHT1 protein was immunolocalised to the plasma membrane (and putative exocytic vesicles) of flesh and phloem cells and the plasma membrane of the sieve element/companion cell interface [67]. Due to the accumulation of *VvHT1* transcripts mainly before veraison [69, 71 ,72] and the presence of the VvHT1 protein mainly in green berries [68], it was thought that VvHT1 was not likely to be directly involved in post-veraison sugar accumulation but was more likely to be involved in the retrieval of glucose during cell growth and differentiation in the early stages of berry development.

Despite VvHT1 not being likely to be involved in the massive sugar accumulation that occurs after veraison, the manner in which its expression is controlled is of interest. In addition to the cDNA hexose transporter clones, Fillion *et al.* [70] isolated a genomic fragment containing approximately 2.5 kb of the promoter region of the *VvHT1* gene. Using computer analysis a considerable number of putative *cis* acting elements were identified. Some of these elements appeared likely to be involved in the suppression of transcriptional activity by sugars, but some were likely to be involved in the activation of expression by sugars. Sequence similarity was also observed between part of the *VvHT1* promoter region and the promoter region of grape alcohol dehydrogenase. Atanassova *et al.* [75] fused different lengths of the *VvHT1* promoter region to the GUS reporter gene and transformed these constructs into tobacco. They found GUS expression in a range of sink tissues. These constructs were also transformed into BY2 tobacco cells for further study. Both sucrose and glucose induced *VvHT1* expression but fructose did not. Palatinose, melibiose and turanose all induced gene GUS expression but lactulose did not, indicating that a glucosyl moiety is required. The increase in gene expression was shown not to be due to the osmotic effects of sugar addition. The regulation of *VvHT1* expression by sugars was also demonstrated in grape suspension cells (CSB, Cabernet Sauvignon Berry). In CSB cells *VvHT1* transcription is induced by sucrose and palatinose [75] but strongly repressed by high concentrations of glucose and 2-deoxy-D-glucose [68] (see below). It was therefore concluded that *VvHT1* gene expression can be both positively and negatively regulated by glucose and glucosyl-containing sugars depending on their concentration.

A deeper understanding of the mechanism involved in the control of *VvHT1* expression comes from yeast one–hybrid experiments. The proximal, 160-bp, part of the *VvHT1* promoter which contains two putative positive sugar-responsive motifs was used as the bait in the yeast one-hybrid system to clone *VvMSA,* an *ASR*-like (ABA-, stress-, and ripening-induced) gene [71]. In grape berries both *VvMSA* and *VvHT1* transcript levels were highest before veraison but were expressed at lower levels from veraison onwards. In berry cell suspensions *VvMSA* expression was induced by sucrose, and not by abscisic acid (ABA) alone, but ABA acted synergistically in the presence of sucrose. The time course of the induction of *VvMSA* and

VvHT1 transcript accumulation by sucrose in cell suspensions differed as *VvMSA* was induced transiently and before *VvHT1*. Gel-shift assays showed that the MSA protein bound to the sugar-responsive elements found in the *VvHT1* promoter. The interaction was also shown through co-expression studies in tobacco leaves using promoter/reporter gene fusions. The presence of a nuclear location signal was confirmed by the localisation of the MSA protein to the nucleus. Together these results suggest that ABA, with sucrose, can influence the cells hexose transporter activity through changing the level of VvMSA protein, which in turn interacts with the *VvHT1* promoter to control its activity. This provides an important link between sugars, hormones and sugar transport during the pre-veraison stage of berry development.

Three additional putative hexose transporters were cloned from Cabernet Sauvignon [69]. Two of these, VvHT4 (K_m 137 μM) and VvHT5 (K_m 89 μM) were shown to import [C^{14}]glucose when transformed into transport compromised yeast cells. These transporters had a lower transport rate and lower affinity for glucose than VvHT1. VvHT3 did not demonstrate transporter activity in the yeast system but a lack of functionality in yeast is not uncommon and there are a number of possible reasons for this. As seen for the sucrose transporters, clustering by sequence similarity does not allow the K_m of transporters to be accurately predicted as VvHT4 with a K_m of 137 μM, groups with AtSTP3 which has a K_m of 2 mM. However, in this case both transporters would be considered to fall into the HALC group. Uptake by VvHT1, 4 and 5 was shown to be protonophore and pH sensitive, which is consistent with their being H^+-dependent transporters. VvHT5 had a similar substrate specificity to VvHT1 except that fructose competed with glucose for transport slightly better in yeast transformed with VvHT5. Glucose transport by VvHT4 was largely unaffected by the competing sugars used (fructose, galactose, mannose, sucrose). VvHT3:GFP and VvHT4:GFP fusions driven by the CaMV 35S promoter were bombarded into onion cells to test transporter localisation. Both were found to be located in the plasma membrane, which is consistent with their proposed function. Real time PCR was used to study the transcript levels of *VvHT1-5* throughout Cabernet Sauvignon berry development. Little expression of *VvHT4* and *5* were observed [69]. *VvHT1* was expressed mainly during the pre-veraison period (Fig. **3**), in line with other reports [71, 72]. *VvHT2* was expressed throughout berry development while *VvHT3* expression was elevated early in berry development, declined during the lag phase and then increased again during the ripening phase (Fig. **3**). From this it could be proposed that *VvHT2* and *VvHT3* may be involved in the import of glucose after veraison but the functionality of neither of these two putative transporters has been demonstrated.

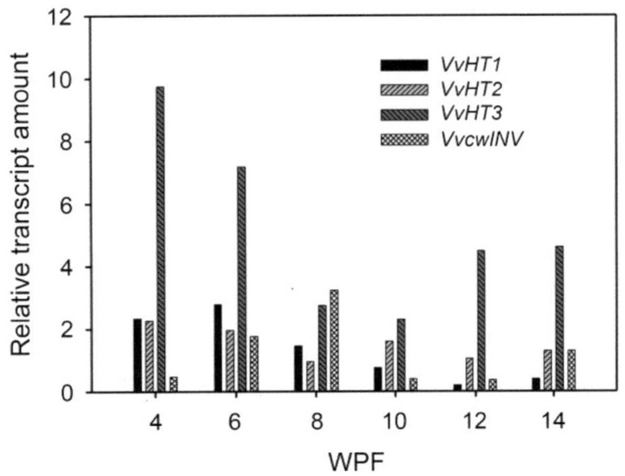

Figure 3: Quantitative RT-PCR analysis of hexose transporter (HT) and cell wall invertase (cwINV) expression during the development of Cabernet Sauvignon berries sampled at two weekly intervals from 4 to 14 wpf. Veraison was at 8 wpf. Adapted from Hayes *et al.* [69].

Conde *et al.* [68] studied glucose uptake and its regulation in some detail using grape suspension cells. A high affinity, broad specificity, H^+-dependent component and a diffusion- related component were detected. Fructose can inhibit glucose uptake but only at very high levels. Approximately 80% of glucose uptake was inhibited by the addition of a protonophore. In these cells, *VvHT1* was the most highly expressed MST, there was very little expression of *VvHT4* and *VvHT5* and low amounts of *VvHT2, 3, 6* and *7* (Fig. **4**). High glucose levels down-regulated *VvHT1* transcription and glucose uptake (Fig. **4A**, **B**), whereas low levels up-regulated both transcription and uptake. In this regard, higher plant cells resemble the green unicellular algae *Chlorella kessleri* [76]. The repression of transcription by high glucose was shown to be hexokinase-dependent (Fig. **4C**). High glucose also resulted in the activation of post-transcriptional control as protein levels and transport activity were reduced in some treatments where transcript levels were largely unaltered. These data were incorporated by the authors into a model for glucose uptake regulation under high and low glucose conditions. How these results relate to sugar uptake in ripening berries is unknown. As *VvHT1* expression in these cells is quite high, it would seem that they are in a different state of differentiation compared to the cells in ripening berries where the transcript level of *VvHT1* is low [69, 71, 72].

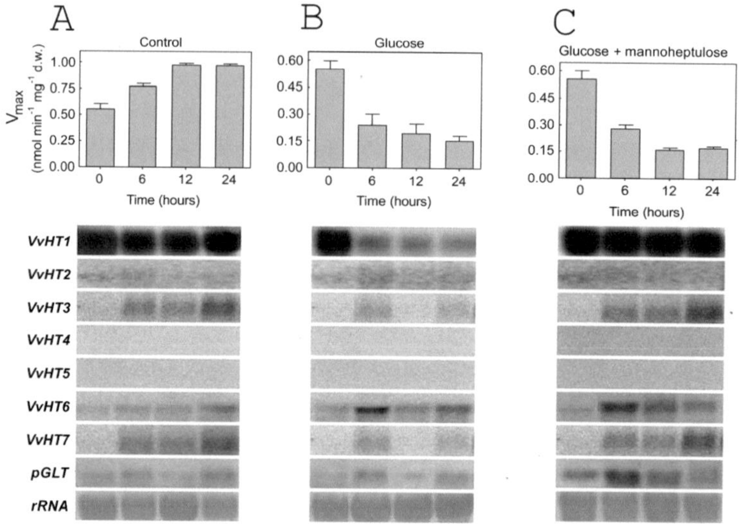

Figure 4: Transcript levels (measured by northern blotting) and glucose uptake of monosaccharide transporters in grape suspension cells treated with various sugars. D-[^{14}C]glucose uptake (V_{max}) and HT transcript levels were measured 0, 6, 12 and 24 hours after the addition of sugars. A) No addition ('control'), B) 150 mM glucose ('high glucose'), C) 150 mM glucose + 10mM mannoheptulose ('hexokinase inhibitor'). Reprinted from Plant Physiology, Vol. 141, Conde *et al.* 'Pathways of glucose regulation of monosaccharide transport in grape cells', p1563–1577, (2006) with permission from the American Society of Plant Biologists.

Recently a sugar-inducible protein kinase, VvSK1, that appears to regulate hexose and sugar accumulation in grapevine cells has been described [77], which gives us an insight into the overall control of sugar accumulation in grapevine and in particular in berries. In berries the expression of the corresponding gene was up-regulated at veraison and can be up-regulated by ABA and sucrose in grape cells. Overexpression of *VvSK1* in grape cells increased the expression of the hexose transporters *VvHT3, 4, 5, 6* and glucose uptake into these cells was increased, which suggests that those transporters may at least partly share common transcriptional regulations.

The transporters described above are in the main either proven, or predicted to be, plasma membrane located MSTs involved in hexose uptake from the apoplast. One vitally important aspect of sugar accumulation is the transport of sugars through the tonoplast, which has so far received little attention in grapes. Interestingly, there may be a possible candidate for a tonoplast hexose transporter among the grapevine sequences already published. A recent review of Arabidopsis MST-like genes by Büttner [45] contained a phylogenetic tree which divided the 53 members into seven groups based on the analysis of the full-length protein sequences. The sequence of VvHT6 closely matches that of three similar Tonoplast Monosaccharide Transporters (TMTs) in Arabidopsis (Fig. **5**). All four proteins are characterised by an extended loop between the transmembrane helixes six and seven [45, 69]. AtTMT1, 2 and a barley homolog (HvSTP1) were identified through proteomic analysis in vacuolar proteins from Arabidopsis and barley respectively [64, 78]. AtTMT1 and 3 have been localised to the tonoplast [79]. Both *AtTMT1* and *2* were expressed in a range of sink and source tissues and their expression was induced by stress. AtTMT1 was indirectly shown to transport glucose and fructose and so this group of proteins appears to comprise tonoplast monosaccharide transporters. Due to its similar sequence and putative protein structure VvHT6 may perform a similar function in grape. Microarray analysis has shown that *VvHT6* expression is low early in berry development but peaks at around the time of veraison before declining later in ripening [72, 80, 81]. This pattern of expression is consistent with VvHT6 having a role in post-veraison import of hexoses into the vacuole. The expression of two other putative tonoplast monosaccharide transporters has been reported during grape berry development. Two of these are significantly expressed during the ripening phase but there is no further functional data [57].

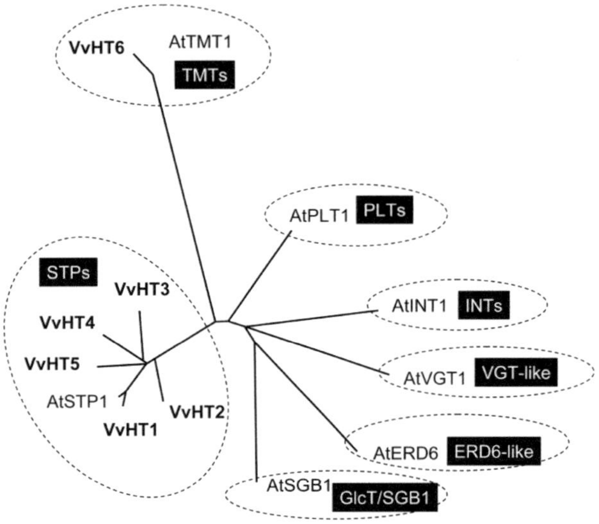

Figure 5: Unrooted dendrogram showing the relationship between various plant monosaccharide transport proteins. The clades are labelled as per Büttner [45]. The sequences used were Arabidopsis; AtERD6 (At1g08930), AtINT1 (At2g43330), AtPLT1 (At2g16120), AtSGB1 (At1g79820), AtSTP1 (At1g11260), AtTMT1 (At1g20840), AtVGT1 (At3g03090), grape; VvHT1 (CAA70777), VvHT2 (AY663846), VvHT3 (AAT09977), VvHT4 (AAT09978), VvHT5 (AAT09979), VvHT6 (AAX47312). Protein sequences were aligned with CLUSTALW [117], distances calculated with PRODIST, tree developed with NEIGHBOUR and displayed with DRAWTREE [118].

Using optical glucose sensors, Chen *et al.* [54] identified and characterised a new class of sugar transporters, named SWEETs, which are low-affinity glucose transporters functioning as uniporters. SWEETs are found in the genome of plants, animals and *Caenorhabditis elegans*. The SWEET family contains 17 members in *A. thaliana*, and the same number is found in the *V. vinifera* genome. It is tempting to speculate about the possible function of SWEET transporters in sugar accumulation in grape berries since a low-affinity diffusive component may be involved, as observed in sugar uptake experiments using grape suspension cells [68]. This linear component of glucose uptake is also observed in other biological systems and may suggest the involvement of an atypical glucose channel/transporter. Indeed, facilitated diffusion through a mercury-sensitive hydrophilic channel mediating glucose uptake was proposed for olive suspension-cultured cells [82]. The inhibitory effect of mercury indicated the involvement of a so far unidentified integral membrane protein because: (i) glucose transport linearly depended on sugar concentration up to 100 mM, a concentration for which it would be very unlikely to find a mediated transporter with physiological relevance, (ii) uptake rates of both D-glucose and L-glucose were equivalent, although a glucose analogue is not recognised by sugar permeases, (iii) glucose counter-transport, indicative of the activity of a membrane transporter, was absent in glucose-sufficient cells exhibiting linear sugar uptake, but present in sugar-starved cells displaying activity for a saturable-monosaccharide transport system, (iv) low activation energies were estimated from the initial glucose uptake at different temperatures by intact cells and plasma membrane vesicles, of 4 and 7 kcal mol^{-1}, respectively, similar to the values described for free diffusion of glucose in water and for the permeation of water through aquaporins and significantly lower than the activation energy for simple diffusion of glucose across phospholipids vesicles of 12 kcal mol^{-1}, and (v) propionic acid caused a sharp decrease in the diffusive uptake, suggesting that this putative protein can be regulated by cytosolic pH changes, much like the gating of plasma membrane intrinsic protein (PIP) in Arabidopsis roots during anoxia stress (reviewed by Conde *et al.* [83]). The involvement of channel-like protein was recently proposed for glucose uptake in Arabidopsis roots. Although the levels of soluble carbohydrates are low in the soil, plants readily take up glucose and sucrose when grown in axenic cultures. In Arabidopsis root tips, glucose and sucrose accumulation was insensitive to protonophores and was similar at pH 5.8, 6.8 and 7.8, suggesting that sugar influx may be mediated by H^+-independent transport systems, possibly channel-like [84]. According to the authors, high-resolution expression mapping in root tips showed that only a few H^+-dependent transporters of the STP (Sugar Transport Protein) and SUT/SUC (Sucrose Transporter) families are expressed in the external cell layers of root tips, meaning that the sugar uptake was due to other unknown mechanisms. These findings may suggest that the root tip expresses low-affinity transport mechanisms for glucose and sucrose, with a molecular nature yet to be determined [84] (reviewed by Conde *et al.* [83]). In another study, it was observed that *AtSWEET1* is highly expressed in Arabidopsis flowers, where the protein may supply nutrients to the gametophyte or nectaries, but not in roots [54]. Since the biochemical properties of AtSWEET1 are markedly similar to the unidentified transport activity characterised in roots using FRET sensors other AtSWEET paralogues for this function were proposed.

Apart from the role described above in the transport and storage of sugars, sugar transporters have a role in controlling gene expression and development. Several grape transporters whose activity is affected by sugars have been described above. The expression of a wide range of other genes can also be affected by sugars [85-89]. As sugar transporters are involved in the transport and partitioning of monosaccharides and sucrose they have the potential to modulate gene expression by altering either the concentration or compartmentation of sugars during development. A relatively small change in sugar transporter activity may exert a considerable influence on metabolism and development. It has also been suggested that some transporter-like proteins may exert a more direct influence by acting as sensors of sugar concentration, although none have yet been identified in grapes.

Based on the 8 x version of the genome, Agasse *et al.* [90] identified 59 putative hexose transporters and only four sucrose transporters and presented a diagram of evolutionary relationships of 111 taxa from *A. thaliana* and *V. vinifera*. More recently, also from the 8 x version of grapevine genome, Afoufa-Bastien *et al.* [57] identified 65 ORFs encoding putative sugar transporters and studied the expression of 20 genes in vegetative organs and during berry ripening. They also conducted *in silico* analysis of the promoter sequences of the sugar transporter genes. The 12 x version of the grapevine genome sequence released in 2010 showed significant differences from the 8 x. A joint effort of annotation is underway. A thorough study of the expression of all the transporters identified in various tissues and organs throughout the vegetative and reproductive cycles of grapevine will probably be helpful in reaching a better molecular understanding of assimilate partitioning in this plant.

Sucrose Cleavage

Invertases: Gene Expression and Activity

Invertases play an important part in sugar accumulation as photosynthate in grapes is transported from the leaf to the berry as sucrose and the hexoses, glucose and fructose, accumulate within the vacuole. The accumulation of roughly equal amounts of glucose and fructose suggest that the cleavage of sucrose may be catalysed by invertase (β-fructosidase; EC 3.2.1.26). The other enzyme commonly associated with sucrose conversion is sucrose synthase. Sucrose synthase is present in the cytoplasm and catalyses the reversible hydrolysis of sucrose into fructose and UDP-glucose. Hawker [8] measured the activity of a number of enzymes involved in sugar metabolism in grape berries during development and found that invertase activity was 200-300 times greater than that of sucrose synthase. Invertases hydrolyse sucrose and some other β-fructose containing oligosaccharides, and they can be divided into three main groups [91, 92]. Two forms have acidic pH optima. The soluble acid invertases have acidic pI's and are located in the vacuole. The insoluble (extracellular) invertases are ionically associated with the cell wall and have basic pI's. Members of the third group have neutral or alkaline pH optima and are thought to be located in the cytoplasm. The ability of invertases to cleave sucrose makes them powerful players in plant growth and development. Invertases are involved in processes such as the partitioning of carbon within the plant, controlling the composition of stored sugars (vacuolar invertases), providing hexoses for metabolism, osmoregulation and gene regulation/signalling [92]. The cleavage of sucrose by invertase is thought to produce the hexoses taken up by MSTs and to maintain sucrose gradients across membranes that may assist in transport (discussed above). Cleavage of sucrose in the apoplast by cell wall invertases provides hexoses for uptake by plasma membrane MSTs. It is also thought that the accumulation of hexoses resulting from the cleavage of sucrose by soluble invertases may be important in driving cell expansion during berry ripening. However, opinion seems divided on how important this may be in grape berries [93, 94].

Early evidence indicated that up to 25% of grape berry invertase activity was likely to be due to cell wall-associated forms of the enzyme as this activity was readily sedimented by the centrifugation of crude extracts [95]. However, the use of extraction media including ingredients such as borate buffer, Na_2CO_3, non-ionic detergents and PEG 4000 allowed almost complete solubilisation of invertase activity [96-98]. In contrast, recent studies have indicated that the levels of cell wall invertase activity and protein can be quite high at certain stages of development; indeed they exceed those of the soluble form after veraison in Kyoho grapes [49, 99].

The importance of soluble invertases in controlling the composition of sugars stored in fruit vacuoles is well demonstrated in tomato. Tomato species that store high levels of hexoses in preference to sucrose have high levels of soluble invertase activity [100, 101]. Conversely, other species that accumulate sucrose in preference to fructose and glucose have low levels of vacuolar invertase. This relationship is confirmed by the analysis of plants with reduced vacuolar invertase levels generated by classical and molecular breeding [102-104]. A similar relationship between the composition of stored sugars and acid invertase activity has been postulated in grapes. Steuben grapes store much higher levels of sucrose in their flesh cells than do Muscat Bailey A grapes. Acid invertase activity is much lower in the flesh of Steuben berries than in that of Muscat Bailey A [105].

Two cDNAs for genes encoding putative vacuolar grape berry invertases have been isolated and characterised [106]. The amino acid sequences deduced from these clones are similar to other plant invertases and are most closely related to vacuolar types. The theoretical isoelectric point of these two putative proteins is acidic, which is a further indication of their vacuolar nature. Both genes were expressed in a range of organs including flowers, seeds, leaves and berries [106]. The relative levels of expression of the two genes appear dissimilar, especially in berries where *VvGIN1* seems to be the more highly expressed of the two species. The transcript levels for both *VvGIN1* and *VvGIN2* were high early in berry development and declined rapidly at around veraison [49, 72, 80, 81, 106]. On a fresh weight basis soluble invertase activity was low at flowering and increased through the first phase of berry growth, reaching a peak at veraison. A similar pattern for soluble invertase activity was reported by Dreier *et al.* [93] and Zhang *et al.* [49]. After peaking at around veraison the level declined markedly. This is probably due to rapid fruit expansion during this period as the protein is stable in grapes and on a per berry basis the level of invertase activity remained relatively constant after veraison [106]. On a concentration basis the level of invertase protein declined in a manner similar to that of the transcript levels [49, 107, 108]. The parallels between invertase activity, protein level and expression of the invertase genes suggest that invertase activity is largely regulated at the transcriptional level in grapes.

Varying patterns of expression for the grape cell wall invertase gene *VvcwINV* (AY538262) have been reported. Hayes *et al.* [69] showed that in Cabernet Sauvignon berries transcript levels increased from low early in development to peak at veraison, after which they declined until another peak occurred late in development (Fig. **3**). Microarray data confirmed the increase in transcript levels later in ripening but not the increase during the pre-veraison stage [72]. According to Pan *et al.* [99] in Kyoho grapes cell wall invertase activity reached a peak some ten days prior to veraison, then declined until about 10 days, after veraison after which it increased again. A similar pattern was described by Zhang *et al.* [40] who showed that enzyme activity and protein levels were relatively low before veraison and increased after veraison so that they were considerably higher during the ripening phase [49]. It seems that there is likely to be considerable cell wall invertase activity in the berry during the sugar accumulation stage. The coordinated expression of cell wall invertase and hexose transporters indicates a possible apoplastic step during sugar accumulation in berries after veraison.

There is little information regarding the function of neutral invertases. They have neutral or alkaline pH-optima, neutral pI's and are localised to the cytoplasm [91]. The transcripts for one putative gene of this type have been detected in berries at the pea-size stage until ripeness [72]. Nine sequences with homology to neutral invertases have been identified from the 8 x version of the genomic assembly, and five of these are expressed in berries [109]. Their expression is relatively low and their precise role in sugar metabolism remains unknown.

Localisation of Vacuolar and Cell Wall Invertases in Developing Grape Berries

The distribution of cell wall and vacuolar invertases in Pinot noir berries up until the time of veraison was determined by immunohistochemistry [107]. In berries 10 days after flowering vacuolar invertase was found in pericarp cells free of crystals (there was a mosaic of crystal-containing and crystal-free parenchyma cells in the pericarp at this stage), in vascular bundles, surrounding the seed locule and in the developing seed coat. At veraison the vacuolar invertase was detected throughout the pericarp but was especially noticeable near the epidermis. The uneven distribution of soluble invertase protein indicates that the apparently homogeneous pericarp cells may consist of a mosaic of cells with differing specialisations. However, it might be that the variation is due to unsynchronised differentiation of the cells at an early stage of development. The presence of vacuolar invertase in most of the pericarp cells at veraison is consistent with it playing a role in controlling sugar composition in the vacuole during ripening. In contrast, cell wall invertase was observed in the crystal-containing parenchyma cells and especially in vascular tissue in berries at 10 days after flowering. Surprisingly, cell wall invertase was only detected in the cells directly under the epidermal layer in berries at veraison, with little signal detected in the pericarp cells [107]. The localisation of the cell wall invertase in berries at veraison seems to contrast with the increase in protein level and activity seen (see above) and further confuses the situation. Perhaps the physical and biochemical changes that occur at veraison may make detection of this protein difficult using this technique. More information will be needed to resolve this apparent conflict.

Hormonal Control of Invertases

There is some debate about the pattern of auxin accumulation during the early part of berry development (see the chapter by Davies and Böttcher, this edition), but studies using a synthetic auxin-like compound, benzothiazole-2-oxyacetic acid (BTOA), have indicated that expression of vacuolar invertase genes in grape may be related to auxin levels. By applying BTOA prior to ripening, the decrease in vacuolar invertase mRNA levels that normally occurs at about the time of veraison was considerably delayed [110]. Invertase transcript levels were maintained but the appearance of transcripts of other ripening-associated genes normally expressed at veraison was delayed. BTOA-treated fruit had higher invertase activity on a per gram basis than control fruit and there was a delay in the accumulation of reducing sugars. The expression of other genes was also affected by this treatment. BTOA treatment appeared to maintain the pre-veraison state of gene expression [110]. Therefore, it could be that the effect of BTOA on vacuolar invertase transcription is indirect as auxins may simply maintain the berry in the pre-veraison state, a state during which these invertases are normally expressed.

Pan *et al.* [99] demonstrated that the natural, cis-(+), form of ABA increased the activity of both soluble and cell wall acid invertases in Kyoho berry discs. The effect was greatest at pH 5.5, and both types of enzyme showed dose dependency effects although they differed in their sensitivity. A further experiment using intact isolated berries confirmed these effects, except that the pH and ABA analogue dependence was not tested. Although the transcript levels were not determined, their increase seems a likely cause of the elevated activities as the addition of a transcriptional inhibitor nullified the ABA-mediated induction. For all of the above experiments the changes in activity were reflected by corresponding changes in protein level as measured by western blotting. However, other experiments showed that invertase activity can also be controlled through post-translational modification. The protein kinase inhibitor quercetin suppressed the

ABA induction of both cell wall and soluble invertase activity in berry discs without substantially altering the amount of enzyme (which had been increased by the prior addition of ABA). The effect was less evident in enzyme extracts but it appears that phosphorylation was involved in the control of invertase activity. This conclusion is supported by the decrease in activity of both forms of invertase when acid phosphatase enzyme was added to the enzyme extract. There may also have been a slight stimulation of activity of both types of invertase by other protein kinase inhibitors but these results are less clear cut. In summary, it is likely that a complex series of reversible phosphorylation events are involved in the control of invertase activity in addition to an ABA-induced increase in enzyme level which probably derives from increased transcription. These results are important as they provide evidence of a link between ABA, invertase activity and sugar metabolism. ABA-induced increases in invertase activity could be an important part of the initiation of grape berry ripening.

Invertases are central to metabolism and the control of plant development and, as such, are highly controlled enzymes. The above discussion touches on some of the possible ways in which invertase function is controlled. There are, however, a number of other ways in which invertase activity is regulated in plants. These include various post-translational pathways involving kinases, diverse modes of control including proteinaceous inhibitors of invertase activity and the compartmentalisation of vacuolar invertases in precursor protease vesicles [111] have been reported in other species. It is likely, therefore, that as our knowledge increases the control exerted over invertases during berry ripening will be shown to be increasingly complex.

Possible Model for Sugar Accumulation During Grape Berry Ripening

The dynamics of sugar accumulation in ripening berries may be simulated with mechanistic models which may predict the effects of source/sink ratio alterations [112, 113]. The current physiological evidence, detailed above, suggests that sugar accumulation during the ripening phase occurs by the apoplastic pathway. Some of this is direct evidence as inhibitors of energy driven transport proteins inhibit sucrose efflux after veraison [48]. The movement of carboxyfluorescein and a viral movement protein is restricted to the phloem during the ripening phase [49]. In line with this, cell wall invertase activity and the expression of some sugar transport genes are up-regulated from veraison [49, 55, 69, 70]. Although the role of the cell wall invertase seems obvious, the role of each of the sugar transporter proteins so far described remains uncertain.

Fig. **6** is a speculative scheme describing the possible routes for sugar import and accumulation during berry ripening and the proteins that could be involved. Efflux of sucrose from the phloem (1 and 2). The scheme shows efflux across the membrane occurring at the sieve element/storage cell boundary, but as suggested by Patrick [38] and Lalonde *et al.* [37] sucrose may be unloaded into the parenchyma cells surrounding the phloem and the transport across membranes into storage parenchyma cells may occur subsequently from these cells. Wherever the apoplastic step occurs it might be *via* 1) pH-independent facilitators as described by Zhou *et al.* [53] or 2) through H^+-coupled sucrose symporters acting in 'reverse' as described for ZmSUT1 [52]. Both modes require a sucrose gradient across the membrane. In the case of a symporter acting in reverse, the gradient has to be large and the proton motif force sufficient for efflux to occur. The cleavage of sucrose by cell wall invertase and the uptake of residual sucrose by symporters appear to maintain low sucrose levels and high hexose levels in the apoplast [114]. The hexoses resulting from sucrose cleavage by invertase may be transported across the plasma membrane of storage cells by MSTs (3). VvHT2 and VvHT3 are candidates for this role in grape berries, however the function of neither has been defined and there may be other MSTs as yet unidentified that may perform this role. Importation of sucrose across the plasma membrane may also occur *via* sucrose symporters (4). The possibility that apoplastic sugars might be incorporated by endocytosis (5) has been recently suggested [39]. It is also tempting to speculate that a low-affinity diffusive component (6) may be involved in glucose uptake by fruit cells, as observed in suspension-cultured cells [68, 82] and Arabidopsis [54, 84]. Our knowledge of the action of monosaccharide antiporters in the tonoplast (7) which import hexoses into the vacuole is limited. The protein sequence of VvHT6 is quite dissimilar to the other reported grape MSTs [69] but it is very closely related to TMT proteins (Fig. **5**). TMTs are reported to be involved in vacuolar monosaccharide uptake in Arabidopsis [79]. It is also possible that sucrose is taken up directly into the vacuole (8), as suggested by transport data with radiolabeled sugars in tonoplast vesicles and intact vacuoles from grape cells. A K_m of 60 mM sucrose was estimated for this facilitated diffusion (N. Fontes, H. Gerós and S. Delrot, unpublished). Not shown on the schematic is the possibility that sugars stored in the vacuole might be exported back out to be utilised for metabolism.

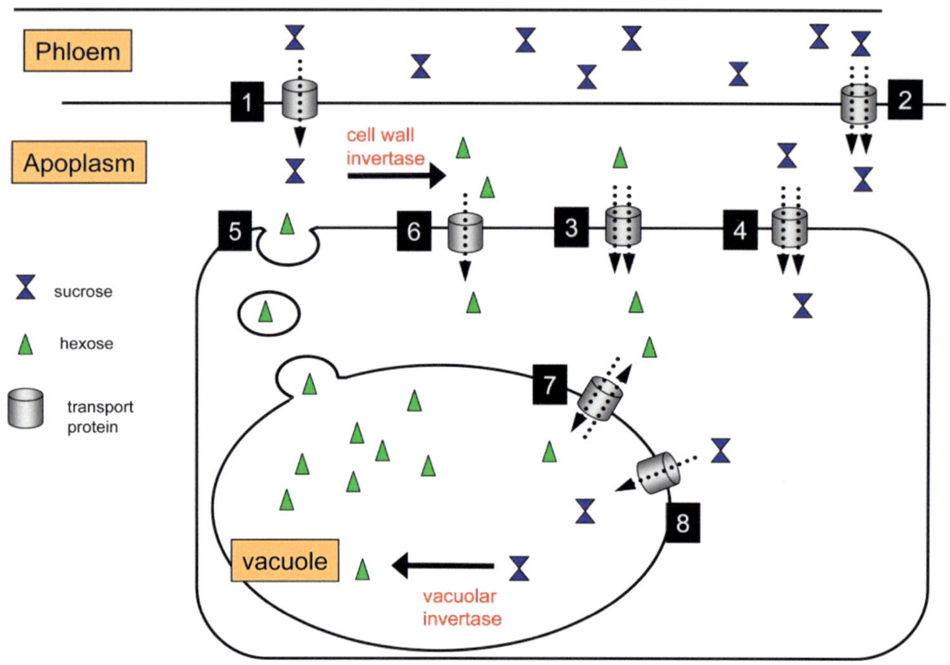

Figure 6: Speculative model for sugar uptake into berries after veraison. Single, green triangles represent hexose molecules, double, blue triangles represent hexoses. Transporter molecules are represented by cylinders crossing the lines, which represent membranes. The numbers are referred to in the text.

Three sucrose transporters have been described for grapevine and their sequences and properties have been discussed above. Given the lack of localisation it is difficult to assign roles to these proteins as they could be involved in retrieval of sucrose back into the sieve elements or conceivably be involved in efflux from the phloem (2) or import of sugars across the plasma or vacuolar membranes (4), (6). VvSUC11 clusters with AtSUC4 (Fig. **1**) which is tonoplast located [64] but in contrast to AtSUC4 it has a low K_m [58]. VvSUC12 has a long central loop but a function for this sequence has not yet been determined. The striking pattern of VvSUC27 expression suggests that it is involved in a process which shuts down at veraison. It is curious that vacuolar invertase activity remains high throughout ripening as if sucrose is cleaved in the apoplast by cell wall invertase. What role can vacuole located invertases play in such a scheme unless there is sucrose resynthesis in the cytoplasm? Of course it does not have to have a role and may just be left over from the preveraison stage of development.

In summary, although much progress has been made there are still large gaps in our knowledge of grape berry sugar metabolism during ripening. There is an obvious need to test the function and localisation of a number of the transporters which have only been partly characterised. Although sequence comparisons can be helpful there seems to be a great deal of variation in the K_m, location and function of various transporters within many of the clades formed through protein sequence clustering. Therefore, too much reliance cannot be placed on this data as the mechanism of transport can be changed by small changes in amino acid sequence. The way in which the expression of genes and the activities of the corresponding proteins involved in sugar transport and storage are controlled as well as the study of transgenic plants modified in the expression of these genes will be required to further refine their role in grape berry development and sugar accumulation.

CONCLUDING REMARKS

We now have a better understanding of sugar accumulation, a process that is vital to the grape industries. Our knowledge concerning both the function of genes involved in this process and the manner in which they are controlled has improved considerably as a result of the use of molecular techniques. In particular, the cloning of key genes and their expression in both homologous and heterologous systems have proven to be powerful tools and the sequencing of the grapevine genome [115] and the development of high throughput techniques such as microarrays and deep sequencing for gene expression analysis will enable rapid progress on these subjects in the future. These aspects of berry development have increased importance given the changes in environmental conditions predicted as a result of climate change and their likely effects on vine and berry development [6, 116].

ABBREVIATIONS

ABA = Abscisic acid

BTOA = Benzothiazole-2-oxyacetic acid

CaMV = *Cauliflower Mosaic Virus*

CSB = Cabernet Sauvignon Berry (cells)

cwINV = *Cell wall invertase*

FRET = *Fluorescence Resonance Energy Transfer*

GFP = Green Fluorescent Protein

HALC = High Affinity Low Capacity type transporter

HT = *Hexose Transporter*

LAHC = Low Affinity High Capacity type transporter

MST = Monosaccharide Transporter

PCMBS = *p-chloromercuribenzenesulfonic acid*

PCR = Polymerase Chain Reaction

RT-PCR = Reverse Transcriptase PCR

STP = Sugar Transport Protein

SUT/SUC = Sucrose Transporter/Sucrose transporter

TMTs = Tonoplast Monosaccharide Transporters

Wpf = *Weeks post-flowering*

ACKNOWLEDGEMENTS

We would like to acknowledge Prof. John Patrick for stimulating discussions regarding sugar transport mechanisms in fruit and Dr Christine Böttcher and Ms Katie Spackman for helpful comments on the manuscript. The Portuguese Foundation for Science and Technology (research project PTDC/AGR-ALI/100636/2008) has also given an important contribution.

REFERENCES

[1] Pirie A, Mullins MG. Interrelationships of sugars, anthocyanins, total phenols, and dry weight in the skin of grape berries during ripening. Am J Enol Vitic 1977; 28: 204-9.

[2] Hunter JJ, De Villiers OT, Watts JE. The effect of partial defoliation on quality characteristics of Vitis vinifera L. cv. Cabernet Sauvignon grapes. II. Skin color, skin sugar, and wine quality. Am J Enol Vitic 1991; 42: 13- 8.

[3] Larronde F, Krisa S, Decendit A, Chèze C, Deffieux G, Mérillon JM. Regulation of polyphenols production in *Vitis vinifera* cell suspension cultures by sugars. Plant Cell Reports 1998; 17: 946-50.

[4] Ollat N, Gaudillere J-P. Carbon balance in developing grapevine berries. Acta Hort 2000; 526: 345-50.

[5] Jackson DI, Lombard PB. Environmental and management practices affecting grape composition and wine quality- A review. Am J Enol Vitic 1993; 44: 409-30.

[6] Mira de Orduña R. Climate change associated effects on grape and wine quality and production. Food Res Int 2010; 43: 1844-55.

[7] Swanson CA, Elshishiny EDH. Translocation of sugars in the Concord grape. Plant Physiol 1958; 33: 33-7.

[8] Hawker JS. Changes in activities of enzymes concerned with sugar metabolism during development of grape berries. Phytochemistry 1969; 8: 9-17.

[9] Howell SS, Candolfi-Vasconcelos MC, Koblet W. Response of Pinot noir grapevine growth, yield, and fruit composition to defoliation the previous growing season. Am J Enol Vitic 1994; 45: 188–91.

[10] Winkler AJ, Williams WO. Starch and sugars of *Vitis vinifera*. Plant Physiol 1945; 20: 412-32.

[11] Tandonnet JP, Cookson S, Vivin P, Ollat N. Scion genotype controls biomass allocation and root development in grafted grapevine. Aust J Grape Wine Res 2010; 16: 290-300.

[12] Williams LE. Grape. In: Zamski E, Schaffer AA, Marcel Dekker, Eds. Photoassimilate distribution in plants and crops. In, New York 1996; pp. 851-81.

[13] Giorgessi F, Bortolin C, Sansone L, Giulivo C. Stock and scion growth relationships in *Vitis vinifera*. Acta Hort 1996; 427: 295-8.

[14] Smith JP. Investigations into the mechanism underlying grapevine rootstock effects on scion vigor and yield. Ph.D. Thesis, Charles Stuart University, Wagga Wagga, NSW, Australia 2004; pp. 196.

[15] Keller M, Kummer M, Carmo-Vasconcelos M. Soil nitrogen utilisation for growth and gas exchange by grapevines in response to nitrogen supply and rootstock. Aust J Grape Wine Res 2001; 7: 2-11.

[16] Padgett-Johnson M, Williams LE, Walker MA. The influence of *Vitis riparia* rootstock on water relations and gas exchange of *Vitis vinifera* cv. Carignane scion under non-irrigated conditions. Am J Enol Vitic 2000; 51: 137-43.

[17] During H. Photosynthesis of grafted and ungrafted grapevines : effects of rootstock genotype and plant age. Am J Enol Vitic 1994; 3: 297-9.

[18] Climaco P, Chaves MM, Castro R. Influence du porte-greffe sur le rendement et sur la réponse écophysiologique du cépage Périquita. CR GESCO 1998; 10: 150-6.

[19] Tandonnet JP, Ollat N. Etude de l'influence du porte-greffe et de l'intergreffe sur la physiologie de la vigne. CR GESCO 2001; 12: 377-83.

[20] Lefort PL, Léglise N. Quantitative stock-scion relationships in vine. Preliminary investigations by the analysis of reciprocal graftings. Vitis 1977; 16: 149-61.

[21] Soar CJ, Dry PR, Loveys BR. Scion photosynthesis and leaf gas exchange in *Vitis vinifera* L. cv. Shiraz: mediation of rootstock effects *via* xylem sap ABA. Aust J Grape Wine Res 2006; 12: 82-96.

[22] Iacono F, Buccella A, Peterlunger E. Water stress and rootstock influence on leaf gas exchange of grafted and ungrafted grapevines. Scientia Hort 1998; 75: 27-39.

[23] Keller M, Smithyman RP, Mills LJ. Interactive effects of deficit irrigation and crop load on Cabernet Sauvignon in an arid climate. Am J Enol Vitic 2008; 59: 221-34.

[24] Balasubrahmanyam VR, Eifert J, Diofasi L. Vine behaviour and wine composition in Italian Riesling grapes as influenced by differential cropping levels. Vitis 1979; 18: 122-6.

[25] Naor A, Gal Y, Bravdo B. Shoot and cluster thinning influence vegetative growth, fruit yield and wine quality of "Sauvignon Blanc" grapevines. J Amer Soc Hort Sci 2002; 127: 628-34.

[26] Keller M, Mills LJ, Wample RL, Spayd SE. Cluster thinning effects on three deficit-irrigated Vitis vinifera cultivars. Am J Enol Vitic 2005; 56: 91-103.

[27] Kaps ML, Cahoon GA. Growth snd fruiting of container-grown Seyval blanc grapevines modified by changes in crop level, leaf number and position, and light exposure. Am J Enol Vitic 1992; 43: 191-9.

[28] Kliewer WM, Weaver RJ. Effect of crop level and leaf area on growth, composition, and coloration of Tokay grapes. Am J Enol Vitic 1971; 22: 172-7.

[29] Vasconcelos MC, Castagnoli S. Leaf canopy structure and vine performance. Am J Enol Vitic 2000; 51: 390-6.

[30] Bergqvist J, Dokoozlian N, Ebisuda N. Sunlight exposure and temperature effects on berry growth and composition of Cabernet Sauvignon and Grenache in the central San Joaquin valley of California. Am J Enol Vitic 2001; 52: 1-7.

[31] Coombe BG. Research on development and ripening of the grape. Am J Enol Vitic 1992; 43: 101-10.

[32] Lalonde S, Wipf D, Frommer WB. Transport mechanisms for organic forms of carbon and nitrogen between source and sink. Ann Rev Plant Biol 2004; 55: 341-72.

[33] Findlay N, Oliver KJ, Nii N, Coombe B. Solute accumulation by grape pericarp cells. IV. Perfusion of pericarp apoplast *via* the pedicel and evidence for xylem malfunction in ripening berries. J Exp Bot 1987; 38: 668-79.

[34] Hawker JS, Ruffner HP, Walker RR. Sucrose content of some Australian grapes. Am J Enol Vitic 1976; 27: 125-9.

[35] Lott RV, Barrett HC. The dextrose, levulose, sucrose, and acid content of the juice from 39 grape clones. Vitis 1967; 6: 257-68.

[36] Coombe BG. Distribution of solutes within the developing grape berry in relation to its morphology. Am J Enol Vitic 1987; 38: 120-7.

[37] Lalonde S, Tegeder M, Throne-Holst M, Frommer WB, Patrick JW. Phloem loading and unloading of sugars and amino acids. Plant Cell Environ 2003; 26: 37-56.

[38] Patrick JW. Phloem unloading: Sieve element unloading and post-sieve element transport. Ann Rev Plant Physiol Plant Mol Biol 1997; 48: 191-222.

[39] Fontes N, Côrte-Real M, Gerós H. New observations on the integrity, structure and physiology of flesh cells from fully ripened grape berry. Am J Enol Vitic 2011. In press (doi: 10.5344/ajev.2011.10126)

[40] Zhang L-Y, Peng Y-B, Pelleschi-Travier S, *et al.* Evidence for apoplasmic phloem unloading in developing apple fruit. Plant Physiol 2004; 135: 574-86.

[41] Ruan YL, Patrick JW. The cellular pathway of postphloem sugar-transport in developing tomato fruit. Planta 1995; 196: 434-44.

[42] Keller M, Smith JP, Bondada BR. Ripening grape berries remain hydraulically connected to the shoot. J Exp Bot 2006; 57: 2577-87.

[43] Krasnow M, Matthews M, Shackel K. Evidence for substantial maintenance of membrane integrity and cell viability in normally developing grape (*Vitis vinifera* L.) berries throughout development. J Exp Bot 2008; 59: 849-59.

[44] Marger MD, Saier MH. A major superfamily of transmembrane facilitators that catalyze uniport, symport and antiport. Trends Biochem Sci 1983; 18: 13-20.

[45] Büttner M. The monosaccharide transporter(-like) gene family in Arabidopsis. FEBS Lett 2007; 581: 2318-24.

[46] Sauer N. Molecular physiology of higher plant sucrose transporters. FEBS Lett 2007; 581: 2309-17.

[47] Coombe BG, Matile P. Sugar accumulation by grape berry pericarp cells. I. Sugar uptake by skin segments. Biochem Physiol Pflanzen 1980; 175: 369-81.

[48] Wang Z-P, Deloire A, Carbonneau A, Federspiel B, Lopez F. An *in vivo* experimental system to study sugar phloem unloading in ripening grape berries during water deficiency stress. Ann Bot 2003; 92: 523-8.

[49] Zhang XY, Wang XL, Wang XF, *et al.* A shift of phloem unloading from symplasmic to apoplasmic pathway is involved in developmental onset of ripening in grape berry. Plant Physiol 2006; 14: 220-32.

[50] Wu GL, Zhang XY, Zhang LY, Pan QH, Shen YY, Zhang DP. Phloem unloading in developing walnut is symplastic in the seed pericarp and apoplasmic in the fleshy pericarp. Plant Cell Physiol 2004; 45: 1461-70.

[51] Nie P, Wang X, Hu L, *et al.* The predominance of the apoplastic phloem-unloading pathway is interrupted by a symplasmic pathway during Chinese jujube fruit development. Plant Cell Physiol 2010; 51: 1007-18.

[52] Carpaneto A, Geiger D, Bamberg E, Sauer N, Fromm J, Hedrich R. Phloem-localized, proton-coupled sucrose carrier ZmSUT1 mediates sucrose efflux under the control of the sucrose gradient and the proton motive force. J Biol Chem 2005; 280: 21437-43.

[53] Zhou Y, Qu H, Dibley KE, Offler CE, Patrick JW. A suite of sucrose transporters expressed in coats of developing legume seeds includes novel pH-independent facilitators. Plant J 2007; 49: 750-64.

[54] Chen LQ, Hou BH, Lalonde S, *et al.* Sugar transporters for intercellular exchange and nutrition of pathogens. Nature 2010; 468(7323): 527-32.

[55] Davies C, Wolf T, Robinson SP. Three putative sucrose transporters are differentially expressed in grapevine tissues. Plant Sci 1999; 147: 93-100.

[56] Ageorges A, Issaly R, Picaud S, Delrot S, Romieu C. Identification and functional expression in yeast of a grape berry sucrose carrier. Plant Physiol Biochem 2000; 38: 177-85.

[57] Afoufa-Bastien D, Medici A, Jeauffre J, *et al.* The Vitis vinifera sugar transporter family: phylogenetic review and macroarray expression profiling. BMC Plant Biol 2010; 10: 245.

[58] Manning K, Davies C, Bowen HC, White PJ. Functional characterization of two ripening-related sucrose transporters from grape berries. Ann Bot 2001; 87: 125-9.

[59] Zhang YL, Meng QY, Zhu HL, *et al.* Functional characterization of a LAHC sucrose transporter isolated from grape berries in yeast. Plant Growth Regul 2008; 54: 71-9.

[60] Barker L, Kuhn C, Weise A, *et al.* SUT2, a putative sucrose sensor in sieve elements. Plant Cell 2000; 12: 1153-64.

[61] Wilson TH, Yunker PL, Hansen CL. Lactose transport mutants of *Escherichia-coli* resistant to inhibition by the phospherase system. Biochim Biophys Acta 1990; 1029: 113-6.

[62] Meyer S, Melzer M, Truernit E, *et al.* AtSUC3, a gene encoding a new Arabidopsis sucrose transporter, is expressed in cells adjacent to the vascular tissue and in a carpel cell layer. Plant J 2000; 24: 869-82.

[63] Weinglass AB, Kaback HR. The central cytoplasmic loop of the major facilitator superfamily of transport proteins governs efficient membrane insertion. Proc Natl Acad Sci USA 2000; 97: 8938-43.

[64] Endler A, Meyer S, Schelbert S, *et al.* Identification of a vacuolar sucrose transporter in Barley and Arabidopsis mesophyll cells by a tonoplast proteomic approach. Plant Physiol 2006; 141: 196-207.

[65] Chervin C, Terrier N, Ageorges A, Ribes F, Kuapunyakoon T. Influence of ethylene on sucrose accumulation in grape berry. Am J Enol Vitic 2006; 57: 511-3.

[66] Johnson DA, Thomas MA. The monosaccharide transporter gene family in Arabidopsis and Rice: A history of duplications, adaptive evolution, and functional divergence. Mol Biol Evol 2007; 24: 2412-23.

[67] Vignault C, Vachaud M, Cakir B, *et al.* VvHT1 encodes a monosaccharide transporter expressed in the conducting complex of the grape berry phloem. J Exp Bot 2005; 56: 1409-18.

[68] Conde C, Agasse A, Glissant D, Tavares R, Geros H, Delrot S. Pathways of glucose regulation of monosaccharide transport in grape cells. Plant Physiol 2006; 141: 1563-77.

[69] Hayes MA, Davies C, Dry IB. Isolation, functional characterization, and expression analysis of grapevine (*Vitis vinifera* L.) hexose transporters: differential roles in sink and source tissues. J Exp Bot 2007; 58: 1985-97.

[70] Fillion L, Ageorges A, Picaud S, *et al.* Cloning and expression of a hexose transporter gene expressed during the ripening of grape berry. Plant Physiol 1999; 120: 1083-93.

[71] Cakir B, Agasse A, Gaillard C, Saumonneau A, Delrot S, Atanassova R. A grape ASR protein involved in sugar and abscisic acid signaling. Plant Cell 2003; 15: 2165-80.

[72] Deluc LG, Grimplet J, Wheatley MD, *et al.* Transcriptomic and metabolite analyses of Cabernet Sauvignon grape berry development. BMC Genomics 2007; 8:doi:10.1186/1471-2164-8-429.

[73] Leterrier M, Atanassova R, Laquitaine L, Gaillard C, Coutos-Thevenot P, Delrot S. Expression of a putative grapevine hexose transporter in tobacco alters morphogenesis and assimilate partitioning. J Exp Bot 2003; 54: 1193-204.

[74] McCurdy DW, Dibley S, Cahyanegara R, Martin A, Patrick JW. Functional characterization and RNAi-mediated suppression reveals roles for hexose transporters in sugar accumulation by tomato fruit. Mol Plant 2010; 3: 1049-63

[75] Atanassova R, Leterrier M, Gaillard C, *et al.* Sugar-regulated expression of a putative hexose transport gene in grape. Plant Physiol 2003; 131: 326-34.

[76] Komor E, Haass D, Tanner W. Unusual features of the active hexose uptake system of *Chlorella vulgaris*. Biochim Biophys Acta 1972; 266: 649-60.

[77] Lecourieux F, Lecourieux D, Vignault C, Delrot S. A sugar-inducible protein kinase, VvSK1, regulates hexose transport and sugar accumulation in grapevine cells. Plant Physiol 2009; 152: 1096-106.

[78] Carter C, Pan S, Zouhar J, Avila EL, Girke T, Raikhel NV. The vegetative vacuole proteome of *Arabidopsis thaliana* reveals predicted and unexpected proteins. Plant Cell 2004; 16: 3285-303.

[79] Wormit A, Trentmann O, Feifer I, *et al.* Molecular identification and physiological characterization of a novel monosaccharide transporter from Arabidopsis involved in vacuolar sugar transport. Plant Cell 2006; 18: 3476-90.

[80] Terrier N, Glissant D, Grimplet J, *et al.* Isogene specific oligo arrays reveal multifaceted changes in gene expression during grape berry (*Vitis vinifera* L.) development. Planta 2005; 222: 832-47.

[81] Pilati S, Perazzolli M, Malossini A, Cestaro A, *et al.* Genome-wide transcriptional analysis of grapevine berry ripening reveals a set of genes similarly modulated during three seasons and the occurrence of an oxidative burst at veraison. BMC Genomics 2007; 8:doi:10.1186/1471-2164-8-428.

[82] Conde C, Silva P, Agasse A, Tavares R, Delrot S, Gerós H. An Hg-sensitive channel mediates the diffusional component of glucose transport in olive cells. Biochim Biophys Acta Biomembr 2007; 1768: 2801-11.

[83] Conde A, Diallinas G, Chaumont F, Chaves M, Gerós H. Transporters, channels or simple diffusion? Dogmas, atypical roles and complexity in transport systems. Int J Biochem Cell Biol 2010; 42: 857-68.

[85] Halford NG, Paul MJ. Carbon metabolite sensing and signalling. Plant Biotechnol J 2003; 1: 381-98.

[84] Chaudhuri B, Hörmann F, Lalonde S, *et al.* Protonophore- and pH-insensitive glucose and sucrose accumulation detected by FRET nanosensors in Arabidopsis root tips. Plant J 2008; 56: 948–62.

[86] Koch KE. Carbohydrate-modulated gene expression in plants. Ann Rev Plant Physiol Plant Mol Biol 1996; 47: 509-40.

[87] Koch K. Sucrose metabolism: regulatory mechanisms and pivotal roles in sugar sensing and plant development. Curr Opin Plant Biol 2004; 7: 235-46.

[88] Rolland F, Baena-Gonzalez E, Sheen J. Sugar sensing and signalling in plants: Conserved and novel mechanisms. Ann Rev Plant Biol 2006; 57; 675-709.

[89] Rook F, Hadingham SA, Li Y, Bevan MW. Sugar and ABA response pathways and the control of gene expression. Plant Cell Environ 2006; 29: 426-34.

[90] Agasse A, Vignault C, Kappel C, Conde C, Gerós H, Delrot S. Sugar transport and sensing in grape. In: Roubelakis-Angelakis K, Ed. Grapevine Molecular Physiology and Biotechnology. Springer, Dordrecht, Heidelberg, London, New York 2009; pp. 105-40.

[91] Roitsch T, González MC. Function and regulation of plant invertases: sweet sensations. Trends Plant Sci 2004; 9: 606-13.

[92] Sturm A. Invertases. Primary structures, functions, and roles in plant development and sucrose partitioning. Plant Physiol 1999; 121: 1-7.

[93] Dreier LP, Hunter JJ, Ruffner HP. Invertase activity, grape berry development and cell compartmentation. Plant Physiol Biochem 1998; 36: 865-72.

[94] Pérez FJ, Gómez M. Possible role of soluble invertase in the gibberellic acid berry-sizing effect in Sultana grape. Plant Growth Regul 2000; 30: 111-6.

[95] Arnold WN. Beta-Fructofuranosidase from grape berries. Biochim Biophys Acta 1965; 110: 134-47.

[96] Hawker JS. Insoluble invertase from grapes - an artifact of extraction. Phytochemistry 1969; 8: 337-44.

[97] Ishikawa N, Nakagawa H, Ogura N. Isoforms of invertase in grape berries. Agric Biol Chem 1989; 53: 837-8.

[98] Ruffner HP, Hurlimann M, Skrivan R. Soluble invertase from grape berries-purification, deglycosylation and antibody specificity. Plant Physiol Biochem 1995; 33: 25-31.

[99] Pan QH, Li MJ, Peng CC, *et al.* Abscisic acid activates acid invertases in developing grape berry. Physiol Plant 2005; 125: 157-70.

[100] Stommel JR. Enzymatic components of sucrose accumulation in the wild tomato species *Lycopersicon-peruvianum.* Plant Physiol 1992; 99: 324-8.

[101] Yelle S, Chetelat RT, Dorais M, Deverna JW, Bennett AB. Sink metabolism in tomato fruit. 4. Genetic and biochemical-analysis of sucrose accumulation. Plant Physiol 1991; 95: 1026-35.

[102] Chetelat RT, Deverna JW, Bennett AB. Effects of the *Lycopersicon-chmielewskii* sucrose accumulator gene (SUCR) on fruit yield and quality parameters following introgression into tomato. Theor Appl Genet 1995; 91: 334-9.

[103] Klann EM, Hall B, Bennett AB. Antisense acid invertase (TIV1) gene alters soluble sugar composition and size in transgenic tomato fruit. Plant Physiol 1996; 112: 1321-30.

[104] Ohyama A, Ito H, Sato T, Nishimura S, Imai T, Hirai M. Suppression of acid invertase activity by antisense RNA modifies the sugar composition of tomato fruit. Plant Cell Physiol 1995; 36: 369-76.

[105] Takayanagi T, Yokotsuka K. Relationship between sucrose accumulation and sucrose-metabolizing enzymes in developing grapes. Am J Enol Vitic 1997; 48: 403-7.

[106] Davies C, Robinson SP. Sugar accumulation in grape berries - Cloning of two putative vacuolar invertase cDNAs and their expression in grapevine tissues. Plant Physiol 1996; 111: 275-83.

[107] Famiani F, Walker RP, Tecsi L, Chen ZH, Proietti P, Leegood RC. An immunohistochemical study of the compartmentation of metabolism during the development of grape (*Vitis vinifera* L.) berries. J Exp Bot 2000; 51: 675-83.

[108] Giribaldi M, Perugini L, Sauvage FX, Schubert A. Analysis of protein changes during grape berry ripening by 2-DE and MALDI-TOF. Proteomics 2007; 7: 3154-70.

[109] Nonis A, Ruperti B, Pierasco A, *et al.* Neutral invertases in grapevine and comparative analysis with *Arabidopsis*, poplar and rice. Planta 2008; 229: 129-42.

[110] Davies C, Boss PK, Robinson SP. Treatment of grape berries, a nonclimacteric fruit with a synthetic auxin, retards ripening and alters the expression of developmentally regulated genes. Plant Physiol 1997; 115: 1155-1161.

[111] Huang LF, Bocock PN, Davis JM, Koch KE. Regulation of invertase: a 'suite' of transcriptional and post-transcriptional mechanisms. Funct Plant Biol 2007; 34: 499-507.

[112] Dai ZW, Vivin P, Barrieu F, Ollat N, Delrot S. Physiological and modelling approaches to understand water and carbon fluxes during grape berry growth and quality development: a review. Aust J Grape Wine Res 2010; 16: 70-85.

[113] Dai ZW, Vivin P, Robert T, Milin S, Li SH, Génard M. Model-based analysis of sugar accumulation in response to source-sink ratio and water supply in grape (*Vitis vinifera*) berries. Functional Plant Biol 2009; 36: 527-40.

[114] Wada H, Shackel KA, Matthews MA. Fruit ripening in Vitis vinifera: apoplastic solute accumulation accounts for pre-veraison turgor loss in berries. Planta 2008; 227: 1351-61.

[115] Jaillon O, Aury JM, Noel B, *et al.* The grapevine genome sequence suggests ancestral hexaploidization in major angiosperm phyla. Nature 2007; 449: 463-5.

[116] Salazar Parra C, Aguirreolea J, Sanchez-Diaz M, Irigoyen JJ, Morales F. Effects of climate change senarios on Tempranillo grapevine (*Vitis vinifera* L.) ripening: response to a combination of elevated CO^2 and temperature, and moderate drought. Plant Soil 2010; 337: 179-91.

[117] Thompson JD, Higgins DG, Gibson TJ. CLUSTAL-W - improving the sensitivity of progressive multiple sequence alignment through sequence weighting, position-specific gap penalties and weight matrix choice. Nucleic Acids Res 1994; 22: 4673-80.

[118] Felsenstein J. PHYLIP - Phylogeny Inference Package (Version 3.2). Cladistics 1989; 5: 164-6.

CHAPTER 4

The Biochemistry of Organic Acids in the Grape

C. M. Ford*

School of Agriculture, Food and Wine, The University of Adelaide, 5005 Australia

Abstract: Grape berries are commonly perceived to be composed principally of high concentrations of fermentable sugars, accompanied by a complex suite of polyphenolic compounds responsible for colour and 'mouthfeel' properties. The organic acid composition of the berry, which is principally a reflection of the metabolism of tartaric and malic acids during development and ripening, has several important consequences for the use of grapes in winemaking. Early research showed two unusual features of acid metabolism in grapes – the occurrence of significant concentrations of tartaric acid, and a marked decrease in the concentration of malic acid as berries enter the ripening stage. Despite a few 'false starts', a synthetic pathway that led to the formation of tartaric acid from ascorbic acid was identified, and in the past few years it has been proved possible to confirm this by biochemical and molecular biological approaches. Evidence for the synthetic route to malate formation in the berry proved simpler to identify, but the greater metabolic versatility of malate compared with tartaric acid, and the post-veraison breakdown of malate by a variety of pathways have ensured a continuing interest in this component of the berry's acid complement. Modern 'post-genomic' approaches have been increasingly used to enable the measurement and analysis of berry metabolism. These approaches, and the advent of a readily transformable 'model' grapevine system, will undoubtedly continue to play a major role in the development of the understanding needed for the rational modification of berry acid composition in response to changing environmental and cultural practices.

Keywords: Ascorbic acid, Malate dehydrogenase, Malate synthesis, Malic acid, Organic acid, Oxaloacetate, PEP carboxylase, Tartaric acid, Tartrate synthesis.

INTRODUCTION

Berries of the cultivated grapevine *Vitis vinifera* are unusual among plants in the accumulation of significant concentrations[1] of two organic acids, tartaric and malic, during berry development and ripening. The importance of these metabolic events for the use of grapes by man cannot be overstated – immature berries are preserved from early predation by their unpalatability to birds and other aggressors, microbial spoilage is reduced and oxidation lessened by the low-pH conditions of grape juice, must and wine. Finally, acid composition gives wine its desirable sensory 'zing' to balance sugar and alcohol derived sweetness in wine styles from Champagne to full-bodied new-world reds, from the Rieslings of Watervale in South Australia to the Barolo of northern Italy.

Perhaps surprisingly therefore, the study of how these and various other organic acids such as citric, ascorbic and oxalic, which occur in lower concentrations in the berry, are formed, distributed and further consumed during the growth and development of the berry has received only modest attention from researchers working in the field of grapevine science. Much of our present knowledge can be traced back to research published in the period spanning the late 1950's to the early 1980's, during which time the pathways by which tartaric acid synthesis is believed to occur, and by which malic acid is both made and subsequently consumed, were elaborated. Recent advances have largely focused on the identification of

***Address correspondence to C. M. Ford:** School of Agriculture, Food and Wine, The University of Adelaide, 5005 Australia; E-mail: christopher.ford@adelaide.edu.au

[1] The comparison of concentrations of solutes such as acids in berry tissue, expressed usually as mg acid per g fresh (or dry) weight of berry tissue, tells us something about possible changes in metabolic activities; if data is also provided to inform of differences in berry weight it is possible also to quantify differences in total amounts of acid. Alternatively, data may be presented as amount of acid per berry, expressed usually as mg acid per berry, in which case metabolic changes can only be identified if berry weight data is provided.

genes and the confirmation of their putative activity by *in vitro* assay, and attempts are now underway to gain an understanding of the processes by which these metabolic events are regulated.

This chapter will provide a brief overview of the status of our knowledge of grapevine organic acid biochemistry at the start of 2011. Much of the work referred to has been the subject of earlier reviews, and the reader is directed in particular to the seminal works of Ruffner [1, 2], Frank Loewus' review in Phytochemistry [3], an excellent book chapter by Terrier and Romieu [4], and recent reviews that we have published from my laboratory (DeBolt, Melino and Ford [5]; Sweetman *et al.* [6]).

ORGANIC ACID COMPOSITION OF GRAPE BERRIES

Amongst the earliest complete analyses of grapevine tissue organic acid composition, the work of Amerine [7] (1956) and Stafford [8] (1959) stands out. Amerine reported that in berries, tartaric, malic, citric, ascorbic, phosphoric and tannic acid were present; Stafford confirmed the occurrence of all but ascorbic and tannic acids in grapevine leaves, and included oxalic acid, in the form of idioblast crystals of calcium oxalate. The later reports of Webb [9] and co-workers, and DeBolt *et al.* [10] were to confirm the occurrence of calcium oxalate raphide crystals in grape berries. In perhaps the broadest survey, Kliewer [11] identified no fewer than 23 acids in berries, although most of these were found only in trace amounts.

By far the predominant acids are tartaric and malic acid, which together may account for over 90% of the total acidity[2] in the berry, and which also contribute the greatest to the pH of the juice, must and wine during vinification and subsequent ageing of wines.

Tartaric and malic acids are diprotic, with two dissociable protons per molecule, and in each case it is the first dissociations, with pKa's of around 2.98 and 3.46 respectively that represent the 'important' acid properties in a winemaking context. At a typical wine pH of 3.4, tartaric acid will contribute around three times the number of protons, or in other words will be three times as acidic as malic acid.

From a winemaking perspective, however, the acid anions, specifically the bitartrate and bimalate monoanions, have important organoleptic roles to play in wine taste. Malic acid has an unpleasantly harsh, metallic taste, while the taste attributed to tartaric acid is more frequently referred to as being of a 'mineral' or citrus-like quality. Malic acid 'lost' in the late stages of berry ripening may be compensated with the addition of tartaric acid at crush or thereafter to provide the winemaker with control of pH and titratable (*i.e.* 'taste') acidity. Importantly in this respect, tartaric acid, unlike malic acid, is not a metabolic substrate for microbial growth of species of lactic acid bacteria. The malolactic fermentation that results from growth of LAB in wine converts malic acid to lactic acid, which is softer on the palate but is a weaker acid, thereby causing wine pH to rise and an increased risk of oxidative and microbial spoilage.

TARTARIC ACID SYNTHETIC PATHWAYS

L-Tartaric acid (2*R*,3*R*)-2,3-dihydroxybutanedioic acid is the naturally occurring stereoisomer of tartaric acid; its enantiomeric partner D-Tartaric acid, (2*S*,3*S*)-2,3-dihydroxybutanedioic acid is generally thought to be formed only by some microbial species (Fig. **1**). It was the distinct stereochemistry of tartaric acid and its salts that led Louis Pasteur to his explanation of chirality in chemical compounds, including famously sorting crystals of racemic tartaric acid (in which the L and D forms are present in equal amounts) into two populations based on their shapes when viewed under the microscope. Solutions prepared from recrystallised samples of each of these two crystal forms rotated polarised light in opposite directions, thus

---

[2] There have been a confusing number of ways in which the concentrations of acids are reported in the literature, including mg/L, mmol/L (mM), μmol/L (μM) and mEq/L, and that is before considering the thorny issue of 'titratable' and 'total' acidity, measures that are of great value to the winemaker but liable to mislead otherwise. This does nothing to simplify the business of reading these papers. In the limited instances reported here, acid concentrations are described in mg/L.

confirming the link between the earlier observation (by Jean Baptiste Biot, in 1832) of chirality in solutions of tartaric acid, and the asymmetric shapes of the crystallised forms of the compound.

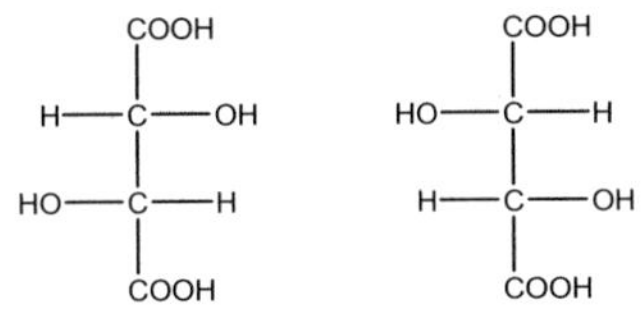

Figure 1: The Fischer projections of L (left) and D (right) tartaric acid.

From a botanical perspective, tartaric acid[3] is uncommon, found only in a few species, mostly in low concentrations and with no known function. In berries of the cultivated grapevine, however, tartaric acid accumulates to significant concentrations, reaching concentrations around 7.5 g/L. Accumulation occurs in the vacuole of the ripening berry, following synthesis within the berry itself, which begins at the earliest stages of berry formation and continues until approximately 40-50 days after anthesis (flowering). Once formed, there is generally no net loss of tartaric acid as berries ripen further, although some early reports suggested that it may be catabolised when environmental temperatures exceed 30ºC [12][4].

Despite the very obvious close structural similarity of tartaric acid to L-malic acid (hydroxybutanedioic acid; Fig. **2**), it was clearly demonstrated in 1958 that different metabolic pathways led to the formation of these two acids in grapevines. Frank Loewus and Helen Stafford, working during the former's summer vacation, exposed excised vine leaves to $^{14}CO_2$ for varying times under light or dark conditions. After 3 hours of exposure to the radioactive CO_2, this was flushed out and the plants left for a further 21 hours prior to extraction of metabolites and analysis. Two critical outcomes were observed: first, at least 75% of the applied radiolabel was recovered as malic acid, while tartaric acid, although present in the leaves at approximately four-fold greater concentration, contained less than 0.1% of the applied radiolabel. Dark-incubated leaves accumulated over 90% of the applied radioactivity as malic acid. Second, analysis of the pattern of radiolabelling (*i.e.*, the specific carbon atoms of tartaric and malic acids that were labelled by incorporation of the ^{14}C) showed substantial differences, suggesting that the tartaric acid and malic acid synthetic pathways were not intimately linked with one another [13].

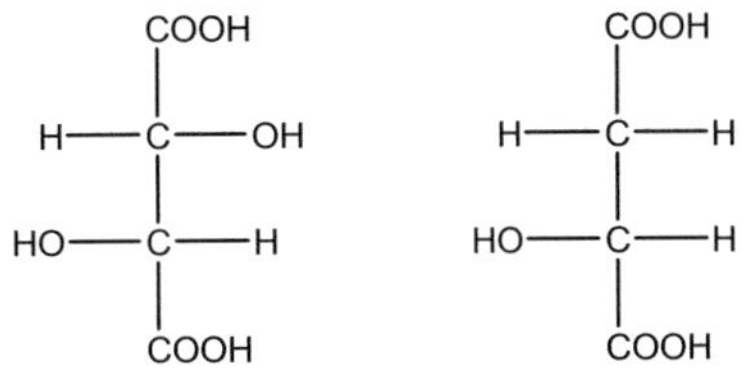

Figure 2: The Fischer projections of L tartaric acid (left) and L malic acid (right).

Earlier suggestions concerning the precursor of tartaric acid, based on the common configuration of internal carbon atoms and their hydroxyl groups, proposed that cleavage of 5-keto-D-gluconate between carbon atoms 4 and 5 would result in the formation of tartaric acid [14] (Fig. **3**). As we shall see, this proved to be a prescient observation, albeit one for which much evidence remains to be accumulated. Having established that tartaric and malic acids have different origins in the grape, the search for the likely precursor continued, which was reported in a series of publications from research groups in Japan and the USA. Common to all of these experimental advances were the use of radiolabelled compounds applied to tissues of the vine for varying times, followed by analysis of metabolites by fractionation and chromatographic processes, and rigorous quantitation of applied and recovered radioactivities.

---

[3] 'tartaric acid' is used to denote the naturally occurring enantiomer L-tartaric acid; 'malic acid' denotes L-malic acid

[4] This intriguing suggestion is corroborated by our observations of apparent loss of tartaric acid from berries exposed to extreme heat events in South Australia, when daytime temperatures can exceed 40ºC and night temperatures 30ºC for periods of several days.

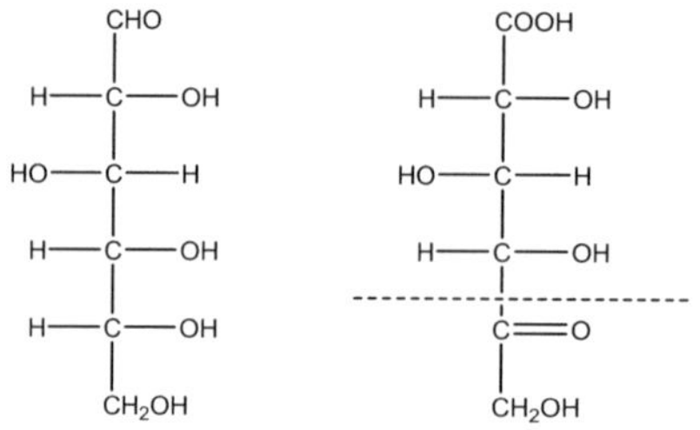

Figure 3: The Fischer projections of D-glucose, 5-keto D-gluconic acid and L-tartaric acid, showing the common configuration of carbon atoms 1 to 4 and the consequent potential for tartaric acid formation from the cleavage of 5-keto D-gluconic acid between carbon atoms 4 and 5 (shown by dashed line).

Ironically, the first of these studies, by Loewus and Stafford in 1958, was shown in retrospect to be a case of so near, yet so far. Building on earlier observations that suggested the co-occurrence of tartaric acid with hexulosonic acids and L-ascorbic acid (vitamin C) [15], the authors proposed ascorbate[5] as a precursor of tartaric acid, with a cleavage of the 6-carbon backbone of ascorbate acid between carbon atoms 2 and 3 (Fig. **4**). Ascorbate labelled on carbon 6 (L- [6-^{14}C] ascorbic acid) was provided to an excised vine leaf *via* the petiole, and after an 8-hour period of uptake and a further 17 hours incubation in the dark, extractions and analysis were undertaken. No radioactivity was found associated with tartaric acid at a level felt to be 'of any significance as concerns the possible metabolic conversion of ascorbate to tartaric acid in the grape leaf', and it was concluded that there was no evidence for a 'direct pathway' forming tartaric from ascorbate.

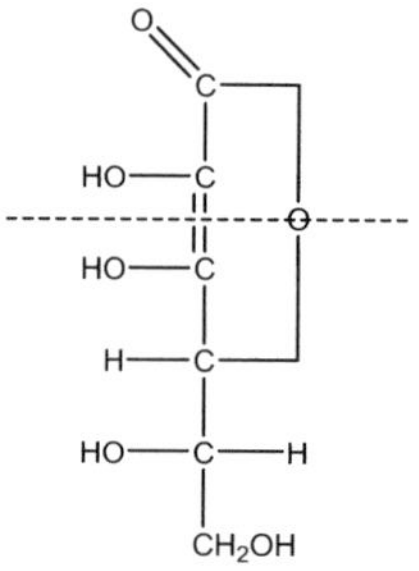

Figrue 4: L-ascorbic acid, with a dashed line showing the proposed cleavage site between carbons 2 and 3.

Subsequently, data was presented confirming distinct synthesis pathways of malic acid and tartaric acid, and showing that glucose might function as the biological precursor of tartaric acid [16]. Importantly, this paper also presented data that showed greater incorporation of radiolabel into tartaric acid following feeding with glucose labelled on carbon atom 1 compared to that arising in tartaric acid from berries fed with glucose radiolabelled on carbon atom 6. The synthesis of tartaric acid by immature berries exposed to $^{14}CO_2$ indicated that comparable metabolism could be expected in both tissues [17]. An alternative pathway for the formation of tartaric acid in grapes was also suggested around this time, based on the proposed reaction of two molecules of glycolate to form an intermediate, oxaloglycolate, which is subsequently reduced to form tartaric acid [18]. In addition to L-tartaric acid, both the D-enantiomer and meso forms of tartaric acid were also isolated in this study. Subsequently, detailed metabolic investigation using radiolabelled compounds revealed that only L-tartaric acid was formed when immature berries of *V. labrusca* were fed ascorbate or glucose [19], thus suggesting that the glycolate-based pathway proposed earlier did not function in these tissues.

The landmark paper of Saito and Kasai, published in 1969 [20], provided the by now long awaited breakthrough in the identification of the biological precursor of tartaric acid, and remains the foundation

---

[5] L-ascorbic acid occurs *in vivo* as the monoanion ascorbate; this term will be used in the text.

upon which our current understanding is based. After working with excised clusters (bunches) of immature berries of *Vitis labrusca* cv Delaware[6] exposed to $^{14}CO_2$ for 10 minutes in either light or dark conditions, followed by removal of the radiolabel and sampling of berries over the following 2 hours, a clear pattern of tartaric acid accumulation emerged. In berries exposed to light during and after $^{14}CO_2$ treatment, tartaric acid appeared as a radiolabelled compound within 20 minutes, whereas berries that were dark-treated did not show any radioactive tartaric acid accumulation for 8 hours. Malic acid accumulated as a radioactive product within 20 minutes of exposure to $^{14}CO_2$ in the light and dark treatments. This result strongly suggested that 'dark metabolism' of berries was consistent with patterns seen in other plants (*i.e.* fixation of CO_2 by non-photosynthetic activities such as PEP carboxylase), and more importantly, that the formation of tartaric acid arose only following the synthesis of radiolabelled sugars, observed almost immediately in the light-exposed treatment but only after 2 hours in the dark-exposed treatment.

The next experiments described were key to our understanding of the process, in suggesting not only the likely biological precursor to tartaric acid, but also proposing the basics of the pathway by which the transformation may be achieved. Berries were 'fed' radiolabelled sugars or other six-carbon compounds, and the proportion of the applied radioactivity recovered as tartaric acid was computed. 'Simple' sugars (glucose, fructose, sucrose) showed no increased efficiency as tartaric acid 'substrates' than was observed for $^{14}CO_2$ and in all cases they were preferentially incorporated into malic acid. *Myo*-Inositol, a 6-carbon carbohydrate compound, was shown to be more readily incorporated into tartaric than into malic acid, and as a consequence, uptake studies using the related compounds, D-glucurono-γ-lactone and D-glucuronic acid, each radiolabelled with ^{14}C at position 6 were undertaken. Radiolabel was recovered in tartaric acid to the extent of 25% of that applied as D-glucurono-γ-lactone, but only as 5% of that applied as D-glucuronic acid. It was established previously that D-glucurono-γ-lactone is an intermediate in the synthesis of ascorbic acid, and in a key experiment, L-[1-^{14}C] ascorbic acid was fed to immature berries in the same way used for the other compounds. In this case however, over 70% of the applied radioactivity was recovered as tartaric acid, providing strong circumstantial evidence that ascorbate *is* the preferred precursor. The key lay in the choice of radiolabelled ascorbate used by Saito and Kasai – critically theirs contained the ^{14}C at position 1 of the carbon backbone, in contrast to position 6 of the ascorbate used in the work by Loewus and Stafford 10 years earlier. When considered in the light of the observations of Ribereau-Gayon (1968) and Loewus and Stafford (1958b), the role of ascorbate, specifically the four-carbon moiety derived from carbon atoms 1 to 4 as the biological precursor of tartaric acid in grapevines, emerged.

Intriguingly, the elucidation of the identity of the precursor of tartaric acid in grapevines was quickly followed by the realisation that in other plants, ascorbate can fulfil the precursor role in a different way, namely by cleavage between carbons 2 and 3 to yield the 2-carbon compound oxalic acid, and thereafter tartaric acid formation from carbon atoms 3 to 6. Studies in *Pelargonium crispum* using L-[6-^{14}C] and L-[1-^{14}C] ascorbic acid fed to excised leaf tips demonstrated a high conversion efficiency of the applied radioactivity into oxalic and tartaric acid. Critically, when fed with L-[1-^{14}C] ascorbic acid, around 12% of the radioactivity recovered in the acid-soluble fraction was in the form of oxalic acid, and only 0.4% recovered as radiolabelled tartaric acid. However, when the leaves were fed however with L-[6-^{14}C] ascorbic acid, the outcome was reversed, and over 32% of radioactivity was recovered as tartaric acid and only 0.03% as oxalic acid [21]. Clearly, in this species, tartaric acid synthesis was directed from the 'alternative' terminal 4-carbon moiety of ascorbate, namely that derived from carbon atoms 3 to 6.

Wagner and Loewus [22] (1974) extended the findings of Saito and Kasai to cover additional members of the Vitaceae, showing unambiguously that only L-[1-^{14}C] ascorbic acid feeding resulted in recovery of radiolabel in tartaric acid. They also studied the metabolic fate of the two-carbon fragment representing

---

[6] The initial paper recorded this species as *V. labruscana* B cv Delaware; subsequent papers modified the name to *V. labrusca* L cv Delaware. The specific name *labruscana* has been suggested as a name for hybrid species formed from crosses between *V. vinifera* and *V. labrusca,* although the authors note that 'These cannot however be considered as different species in the true sense of the word.' (Thomas and vanHeeswijck, 2004 in P.R. Dry and B.G. Coombe Viticulture Volume 1. Winetitles. South Australia: p119-131.)

carbons 5 and 6 of ascorbate, by following the radiolabelled compounds formed when leaves and berries were fed with L-[6-^{14}C] ascorbic acid. In these experiments, between 15 and 45% of the applied radiolabel was recovered in the neutral fraction, representing its incorporation into sugars and suggesting that the two-carbon fragment derived from ascorbate is recycled into the general hexose metabolism of the vine. Subsequent efforts to determine the possible involvement of the glycolate pathway, by which the two-carbon fragment of ribulose-1,5-bisphosphate is recycled as a component of the photorespiratory cycle, in metabolism of carbon atoms 5 and 6 of ascorbate suggested that this did not occur [23]. Further experiments confirmed the role of the four-carbon fragment derived from ascorbate carbons 1-4 in the synthesis of tartaric acid in grapevines, and moreover provided confirmation of the production of oxalic acid by cleavage of ascorbate between carbon atoms 2 and 3 in *Pelargonium sp.* [24, 25].

The synthetic pathway by which ascorbate forms tartaric acid was elucidated by a series of studies performed using radiolabel incorporation and analysis of the metabolites formed following a period of incubation. In the earliest of these reports, Saito and Kasai [26] followed the metabolism of L-[1-^{14}C] ascorbic acid 'fed' to immature berry slices over a 5-hour period. In addition to radiolabelled tartaric acid, three radiolabelled compounds were identified in extracts prepared following incubation, L-idonic acid, L-idono-γ-lactone and 2-keto-L-idonate (Fig. **5**). When the grape slices were treated with iodoacetic acid, an inhibitor of enzymes with catalytically important cysteine residues, there was no accumulation of radiolabelled tartaric acid, and the radioactivity previously recovered in tartaric acid was recovered in the three intermediate compounds. This suggested a specific inhibition of the enzyme converting L-idonate to the next intermediate.

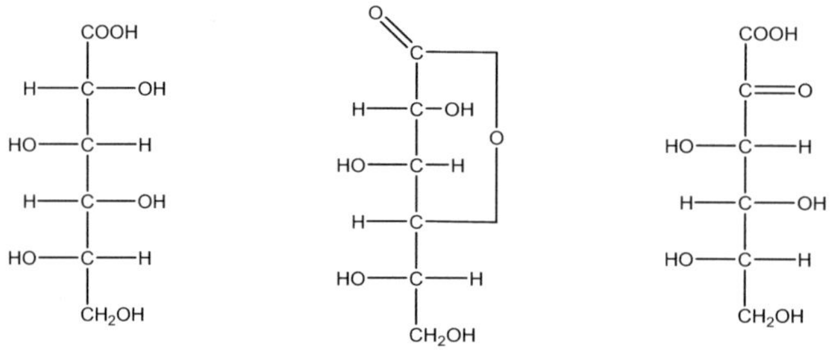

Figure 5: L-idonic acid, L-idono-γ-lactone and 2-keto-L-idonate, intermediates in the formation of tartaric acid from ascorbate.

Further investigation revealed the presence of an additional compound, identified as 5-keto-D-gluconic acid, in berry slices fed with L-[1-^{14}C] ascorbic acid. Recall that almost 30 years before, it had been suggested that tartaric acid could arise by cleavage of this compound between carbon atoms 4 and 5 – in fact, Saito and Kasai cited the first report of this possibility in plants from a publication in 1949! When berry slices were incubated with radiolabelled preparations of dehydroascorbate (the reversibly oxidised form of ascorbate), 2-keto-L-idonate, L-idonic acid and 5-keto-D-gluconic acid (all labelled with ^{14}C at position 1), very high levels of radiolabelled tartaric acid were formed – in all cases greater than 80% of the acid-soluble radioactivity was recovered as tartaric acid. Examination of the radiolabelled products formed upon addition of the individual intermediate compounds resulted in the proposal of a metabolic pathway for the formation of tartaric acid from ascorbate in grape berries as follows: Ascorbic acid/dehydroascorbate → 2-keto-L-idonate → L-idonic acid → 5-keto-D-gluconic acid → L-tartaric acid (Fig. **6**).

The identification that 5-keto-[1-^{14}C]-D-gluconic acid, when fed to leaves of *Pelargonium sp.,* resulted in the recovery of radiolabel in tartaric acid, led Saito and Kasai [27] to suggest that plants in which tartaric acid synthesis occurs *via* the cleavage of ascorbate between carbons 2 and 3 may contain a third pathway of tartaric acid synthesis, which has glucose as its precursor. In this proposal, glucose would form 5-keto-D-gluconic acid without ascorbate as an intermediate, a pathway that has been characterized in species of *Gluconobacter* (Fig. **7**). Subsequently, it would be shown that this pathway might contribute with a minor portion of tartaric acid synthesis in grapevines also, although it is unlikely that it represents a major synthetic route in this species [28, 29].

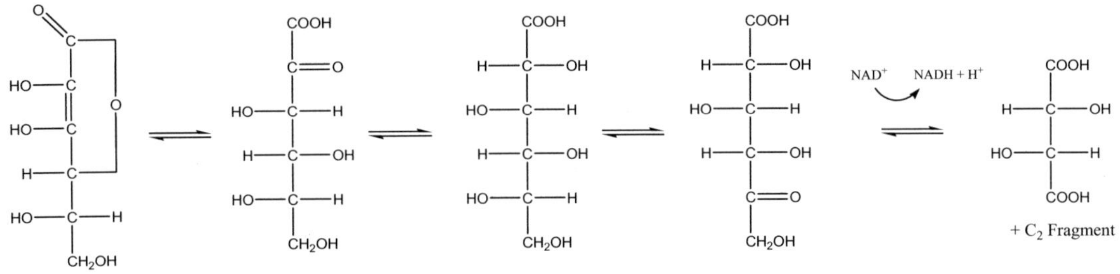

Figure 6: The metabolic pathway for the formation of tartaric acid from L-ascorbate, showing L-ascorbate, 2-keto L-idonate, L-idonate, 5-keto D-gluconate and L-tartaric acid. Note (i) that the identity of the redox acceptor between 2-keto L-idonate and L-idonate remains undetermined, and (ii) that an intermediate, tartaric acid semialdehyde, has been proposed to occur arising from cleavage of 5-keto D-gluconate, prior to the formation of tartaric acid.

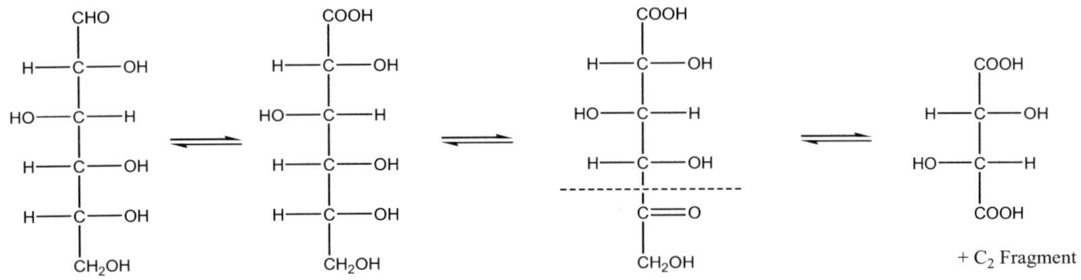

Figrue 7: Intermediates identified in the 'glucose pathway' of tartaric acid synthesis: D-glucose, D-gluconic acid and 5-keto D-gluconic acid. Figure also shows the result of cleavage of 5-keto D-gluconic acid between carbon atoms 4 and 5.

The role of ascorbate as the biological precursor to tartaric acid in plants had not been at first universally accepted. In a paper published in 1973, Ruffner and Rast [30] suggested that the tartaric acid incorporation data reported in Saito and Kasai (1969) was likely to have been derived from non-enzymatic breakdown of ascorbic acid to form tartaric acid. The authors proposed instead that glucose is the precursor of tartaric acid, and presented distinct pathways for tartaric acid formation in leaf and berry tissue. Much careful work by Loewus and co-workers in the succeeding years, however, gave growing support to the findings reported by Saito and Kasai in 1969. In a review published in 1982, Ruffner reiterated the doubts surrounding the role of ascorbate as the precursor of tartaric acid. These arguments were based on the very high conversion of applied radiolabelled ascorbic acid to tartaric acid reported by Saito and Kasai, an apparent loss of radiolabelled tartrate as the period of feeding increased, and the failure to detect other than a minimal amount of un-metabolised ascorbate when using mature grapes, in which there was no synthesis of tartaric acid occurring [1].

In 1987, strong evidence supporting both the role of ascorbate as the biological precursor of tartaric acid in grapevines, and the sequence of intermediates in the transition from ascorbate to tartrate, was published [31]. Time-course experiments were conducted using either ascorbic acid (L-[1-^{14}C] ascorbic acid), $^{14}CO_2$ or [U-^{14}C]-sucrose, fed to immature and mature grapevine leaves. The sequential accumulation of radiolabel was observed following L-[1-^{14}C] ascorbic acid feeding of young leaves, into 2-keto-L-idonate, L-idonate, 5-keto-D-gluconate and tartaric acid. There was a greater recovery of radiolabel from L-idonate than other intermediates, further supporting earlier suggestions that the rate-limiting step in the pathway lay between L-idonate and 5-keto-D-gluconate. In older leaves fed in the same way, accumulation of radiolabel was restricted to 2-keto-L-idonate and L-idonate. Moreover, when young leaves were fed with $^{14}CO_2$ or [U-^{14}C]-sucrose, the accumulation of radiolabel could be followed over time into glucose, ascorbate, 2-keto-L-idonate, L-idonate, 5-keto-D-gluconate and finally into tartaric acid, in relative amounts that indicated a strong precursor-product relationship as predicted by the proposed synthesis pathway. One final piece of evidence for the role of ascorbate as the precursor to tartaric acid was presented, in which the specific activities of ascorbate and tartaric acid were shown to be closely correlated. In the words of the authors, 'it is concluded that ascorbic acid must be considered a true physiological intermediate in grapevine tartrate biogenesis.'

An arguably 'overlooked' contributor to the organic acid composition of grapevine tissues, oxalic acid provided the basis for work published in 2004 [10]. Over the preceding 90 years, a number of reports had been elaborated concerning the presence and composition of crystals ('biomineralisation') within grapevine tissues. The most common form of biomineralisation in plants involves calcium oxalate crystal formation, and indeed crystals composed of calcium oxalate were identified in leaves of *Vitis sp.* as early as 1914 [32]. Further reports of calcium oxalate, in the form of both needle-like 'raphide' or star-shaped 'druse' crystals, in grapevine berries and roots [33, 34], were accompanied however by others, in which the composition of the crystal forms was suggested to be calcium tartrate or potassium hydrogen tartrate.

During the attempted preparation of protoplasts from berry tissues, an abundance of needle-shaped crystals was observed when the cells were examined by light microscopy. Upon purification and analysis by X-ray powder diffraction, these crystals were shown to be exclusively composed of calcium oxalate. No evidence was found either microscopically or by powder diffraction, for the presence of any tartrate in any of the crystal fractions analysed [10].

This intriguing finding suggested that grape berries may contain at least two distinct pathways of ascorbate catabolism – leading *via* cleavage between carbons 2 and 3 to the formation of oxalate and a four-carbon fragment[7], and the 'established' pathway of tartaric acid synthesis in which ascorbate is the precursor for a carbon 4-5 cleavage leading to the formation of tartaric acid. Experiments were conducted to test this, in which immature berries of *V. vinifera* cv Cabernet Sauvignon were fed with L-[1-^{14}C] ascorbic acid in four 16-hour pulses over a 96-hour period (with a 16-8 light dark regime) while still attached to the vine. Analysis of the organic acid composition of these berries by HPLC and radioactivity measurement (1.5M phosphoric acid was used to solubilise all acid salts present in the homogenised berry preparations, including calcium oxalate crystals) was conclusive – ^{14}C was detected in oxalate and tartaric acid peaks eluting from the HPLC. In accordance with earlier data, around 52% of the applied radiolabel was recovered in tartaric acid, and approximately 20% as radiolabelled oxalic acid.

Calcium oxalate raphide crystal formation occurs in grapevine tissues within specialised idioblast cells [35] and has been shown in other plant systems to be directed by as yet uncharacterised proteins [36]. It is therefore probable that ascorbate metabolism within idioblast cells in grapevine tissues proceeds by a distinct pathway to that resulting in tartaric acid formation.

Efforts to further characterise the biochemical basis of tartaric acid synthesis in grapevines culminated in the identification of L-idonate dehydrogenase, using a combination of molecular and biological approaches. Grapevine expressed sequence tags (ESTs) were used to identify cDNAs, and thence genes were expressed in the tissues and at the developmental stages, appropriate for tartaric acid synthesis. Initially, 87 cDNA candidates were identified, all expressed differentially in tartaric acid accumulating tissues. A closer examination of protein motifs and functional domains suggested four of these candidates contained motifs associated with oxidoreductase enzymes, and that one of these was absent from a tartaric acid-lacking grapevine species, *Ampelopsis aconitifolia*. Full-length cloning of the cDNA sequence, which was named Vv-L-IdnDH and characterisation followed. This sequence encoded a protein with similarity to an *Escherichia coli* enzyme catalysing the conversion of L-idonate to 5-keto-D-gluconate – previously shown to be the likely rate-limiting step in tartaric acid synthesis in grapevines. Functional studies of the activity of the recombinant enzyme *in vitro* confirmed its activity and co-factor specificity. Detailed analysis of the transcriptional activity of Vv-L-IdnDH confirmed that transcripts accumulate in the early stages of berry development, before declining – a result consistent with the observed pattern of tartaric acid accumulation in the developing berry [37]. These data have been confirmed subsequently, using detailed microarray analyses determined at multiple stages throughout berry development [38]. Interestingly, when tissue sections from developing berries were similarly examined, transcripts encoding Vv-L-IdnDH were more

---

[7] Earlier work, beginning with the radiolabelling studies of Loewus and Stafford in 1958, had suggested that tartaric acid found in grapevine leaves did not arise from the 4-carbon fragment derived from carbons 3 to 6 of ascorbate.

represented in seed-derived preparations than in mesocarp (pulp) or pericarp (skin) preparations [39]. Earlier work had shown tartaric acid levels in developing berries increase towards the centre of the berry, [40] thus perhaps providing a rationale for the observed distribution of L-IdnDH transcript.

To date, the identification and analysis of additional tartaric acid synthesis genes and their encoded enzymes awaits further discoveries.

MALIC ACID – SYNTHESIS AND CATABOLISM

The Biochemistry of Malate Synthesis

L-malic acid, or hydroxybutanedioic acid, is the second major organic acid that accumulates during grape berry development. Unlike tartaric acid, malic acid is ubiquitously distributed in the plant kingdom, and moreover fulfils a number of distinct and important metabolic and physiological roles. Furthermore, also in contrast to the fate of tartaric acid, malic acid formed in the berry pre-veraison is broken down during a brief period around veraison. This process, which is in some part regulated by environmental temperatures as well as developmental cues, results in marked changes in the acid composition of grapes at harvest.

Pathways of malic acid synthesis and breakdown and the details of its participation in a wide range of plant metabolic processes have been determined for many plant species [41, 42]. Early research indicated that grapevines use analogous pathways and enzymes to achieve the synthesis and catabolism of malate to those identified and characterised in other plants, and these will be briefly reviewed below. An interesting facet of this lies in the differences in malate metabolism that exist between grapevine leaves and berries. This is reflective in part of the multiple roles played by malate in plant metabolism, and also of the properties of leaves as 'source' and berries as 'sink' tissues. However, malate is principally of interest from the perspective of its arguably central role in plant metabolism – much has been written that details the numerous biochemical and physiological processes in which malate participates. The accumulation of malate in pre-veraison berries, and its loss during subsequent ripening has been viewed principally from a phenomenological perspective in much of the writing on this topic. In part this is unsurprising: the outcomes of acid metabolism are primarily of applied interest in terms of understanding the acid composition of grapes at harvest. In New World winemaking, the management decisions consequent upon these data in terms of acid additions to ensure microbial stability form a key part of the ongoing winemaking process. However, an examination of the physiological basis of malate accumulation and subsequent loss in grape berries, particularly in the 21st century context of warming climates, is overdue. The 'physiological biochemistry' of malate metabolism, particularly in regard to the effects of temperature, will be discussed in a subsequent section.

In many early papers, attempts were made to ascertain the metabolic origins of tartaric and malic acids in grapevines. As we have seen, the search for tartaric acid synthesis processes led to the identification of several 'novel' and seemingly dedicated metabolic pathways. However, it was apparent from these same experiments and others that malate is made by common processes of hexose metabolism, and the recognised pathways of carbon assimilation [43, 44]. Analysis of the patterns of radiolabelling showed that the radioactive carbon atom was incorporated predominantly at positions 1 or 4 (*i.e.* at the ends) of the molecule when short labelling periods were used or if the labelling period was followed by incubation of the leaves in the dark. When labelling was followed by continued light exposure, distribution of the ^{14}C was seen also on the internal carbons of the malate molecule. These data suggested that malate may be formed by at least two routes: as a consequence of the metabolism of hexoses, or by a direct carboxylation reaction. The contributions of these alternative pathways were shown to differ between leaf and berry tissues and also to be under developmental control [16].

Malic acid formation in grape berries is primarily conducted by the cytoplasmic enzyme PEP carboxylase, which catalyses the β-carboxylation of phospho-enol-pyruvate (PEP) arising from glycolysis, forming oxaloacetic acid (OAA; Fig. **8**) [45, 46]. Sometimes referred to as 'dark fixation' of carbon dioxide (a confusing term since the reaction proceeds equally well in the light), this reaction uses PEP arising from the catabolism of sucrose, which is imported to the immature berry, and carbon dioxide in the form of bicarbonate HCO_3^-.

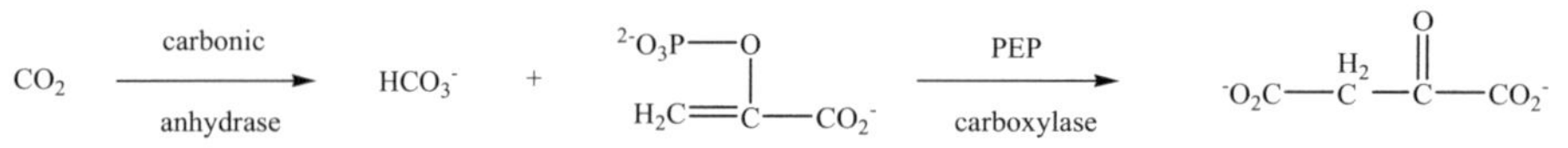

Figure 8: The reaction of PEP carboxylase.

PEP carboxylase (PEPC) is one of the most widely studied plant enzymes, and in consequence much is known about its activity, and moreover its regulation in the plant. Among the key features of the enzyme are its oligomeric quaternary structure, and complex reaction kinetics including its inhibition by L-malate, the principle end product of the carboxylation of pyruvate (see below) [47, 48].

PEPC is intimately involved in C4 and CAM photosynthetic carbon fixation as the main catalyst of CO_2 fixation following its uptake and reaction with carbonic anhydrase to produce bicarbonate. It is further implicated in a number of non-photosynthetic carbon fixation reactions in all plants [42]. In C3 plants such as grapevine, it has been shown that leaves undergo carbon fixation by the Calvin cycle, with ribulose 1,5-bisphosphate mediated formation of triose phosphates (see the review by Ruffner, 1982, for further details) [2]. Berries, which possess photosynthetic capacity only up to veraison appear to use PEPC-mediated fixation of carbon into malate, leading some authors to suggest 'C4-like metabolism' for these organs. Further discussion of this matter will be found subsequently. In C4 and CAM plants, the activity of PEPC is regulated by post-translational modification in the form of phosphorylation by a specific PEPC kinase, which acts to reduce the inhibitory effect of malate and to increase the stimulatory effect of glucose-1-phosphate [49].

The second reaction in PEPC-mediated formation of malate is catalysed by malate dehydrogenase (MDH). Several isoforms of MDH have been identified in plants[8], located in the cytosol and various organelles, with numerous metabolic roles, some photosynthetic and others associated with reactions in the TCA cycle, photorespiration and cellular transport processes.

Thermodynamic measurements indicate that the reactions of PEPC forming OAA from PEP ($\Delta G^{0'}$ -30 kJ mol^{-1}) and of malate dehydrogenase to form malate from OAA ($\Delta G^{0'}$ -28 kJ mol^{-1}) are both strongly in the direction of malate formation. Indeed, it has been shown that malate formation from PEP is more thermodynamically favoured (standard free energy change under biological conditions, $\Delta G^{0'}$ -58 kJ mol^{-1}) than is the pyruvate kinase mediated ATP-generating formation of pyruvate from PEP ($\Delta G^{0'}$ -31.4 kJ mol^{-1}) [42], the 'default' glycolytic reaction.

The pathways of malate formation in grape berries that are derived from hexose metabolism may include its abstraction from the TCA cycle following synthesis by fumarase, the carboxylation of pyruvate by NADP-dependent (cytoplasmic) malic enzyme or as a product of the glyoxylate cycle. For a detailed treatment of the proposed role of these pathways in malate synthesis, the reader is directed to the recent review by Sweetman *et al.* (2009) [6].

The Biochemistry of Malate Breakdown

In contrast to the fate of tartaric acid, malic acid is subject to catabolic reactions (breakdown) during the later stages of berry development. Multiple pathways have been identified by which malate is catabolised in

---

[8] A further complication arises with respect to the naming of malate dehydrogenases. The EC (Enzyme Commission) number of malate dehydrogenase is 1.1.1.37, also listed as malic dehydrogenase. Many isoforms of this enzyme exist, widely distributed in cellular compartments. EC 1.1.1.38 is also called malate dehydrogenase, specifically oxaloacetate decarboxylating, NAD+, or NAD-dependent malic enzyme, which is associated with mitochondria. EC 1.1.1.39 is analogous to 1.1.1.38, but does not decarboxylate exogenosly added OAA. EC 1.1.1.40, is malate dehydrogenase (oxaloacetate decarboxylating), NADP+, or NADP-dependent malic enzyme, which is associated with the cytosol.

plants, reflecting the widely recognised roles of malate in many facets of plant metabolism. Grape berries have been shown to contain several enzymes involved in malate breakdown, including those associated with gluconeogenesis, respiration and fermentation [6]. Generally the net breakdown of malate is reached as berries attain veraison, and this is reflected in the specific activities[9] of malate metabolising enzymes measured at various stages of berry development.

Early experiments showed a clear distinction in the timing between malate synthesis (measured as PEP carboxylase activity) and malate breakdown, measured as the activities of malic enzyme and PEP-carboxykinase in crude protein extracts during berry ripening [16, 45, 50]. Enzyme activity associated with the onset of malate breakdown reflected a rapid decrease in the activity of PEP carboxylase, and an increase in the measured activities of malic enzyme and PEP carboxykinase. Subsequent work, following the pattern of labelling by ^{14}C fed to berries as 2,3 [^{14}C] fumarate, confirmed the participation of malate in gluconeogenic formation of sucrose post-veraison [51]. The activity of PEP carboxykinase, responsible for the direct ATP-dependent decarboxylation of OAA to PEP, was seen in berries from around 4 weeks post-flowering [52]. This coincides with the period of malate accumulation (indicated by the presence of PEP carboxylase) and it was suggested by the authors that PEP carboxykinase may have a role also in the synthesis of malate from PEP. Subsequent molecular studies, in which the transcription of two PEPCK transcripts was individually observed to increase in ripening (post-veraison) fruit, suggested that the story of PEP carboxykinase may yet prove to be more complex than at first thought [6, 53].

Ripening berries can respire exogenously added malate through the TCA cycle, and although the grape is conventionally classified as a non-climacteric fruit (*i.e.*, grapes do not evidence a strong respiratory burst associated with the onset of ripening) it is clear that the respiratory pathway of malate dissimilation is active in these tissues. The physiological context for this will be discussed in a later section. The main enzyme activities associated with malate oxidation are malate dehydrogenase (MDH) and cytosolic NADP-dependent malic enzyme. Early experiments suggested that the activities of these enzymes increased during grape berry ripening [45].

As noted earlier, many isoforms of MDH exist within plant cells, and the assignment of activities has relied largely upon experiments in which cells are homogenised and, in some cases, then fractionated using differential centrifugation techniques. It was shown that the ripening of grapes coincided with the appearance of multiple isoforms of MDH [54]. Subsequent experiments in which fractionation techniques were used suggested that up to 80% of the activity measured was derived from cytosolic MDH, and both cytosolic and mitochondrial fractions contained multiple isoforms of the enzyme [55]. Further analysis of the MDH isoforms isolated from these fractions led the authors to suggest that mitochondrial MDH (mMDH), which showed a greater thermostability than the cytosolic isoforms, may function principally to oxidise malate at later stages of berry ripening, when the fruit is likely to be exposed to higher ambient temperatures [56]. Molecular analyses of MDH expression have further revealed the complexity of the role of MDH and its isoforms in malate oxidation. Work in which transcription of mMDH isoforms was measured at five stages during berry development by Northern hybridization confirmed the expression of mMDH at later stages, but was unable to entirely rule out cross-reactivity with cytosolic MDH material [57]. A more detailed analysis of DNA microarray data, in which mitochondrial and cytosolic MDH isoforms were distinguished, provided clear evidence for increased transcription of at least one isoform of mMDH at later stages of berry ripening [6].

---

[9] The concept of the specific activity of an enzyme is more commonly encountered in discussion of protein purification. It is broadly defined as the activity of the enzyme divided by the concentration of protein in the fraction assayed. In the context of changes in the extent to which an enzyme is expressed in a particular tissue or at a specific developmental stage, an indication of its abundance may be inferred by comparing specific activity data. To account for growth of the plant, an indication of the amount of tissue from which the extract was made is critical, albeit often concealed within the detail of the methods sections. In this respect, the use of 'activities' in sentences such as 'the activity of malic enzyme was increased post-veraison' infers that there was more malic enzyme at this time, rather than that pre-existing malic enzyme became more active post-veraison.

The direct formation of pyruvate from malate by the activity of malic enzyme is widely believed to play a role in the breakdown of malate during berry ripening. Cytosolic NADP-dependent malic enzyme activity was shown to increase with berry ripening, and yet to be present during the period of malate accumulation as well. The reversibility of the reaction catalysed by the enzyme led to speculation that it may play a role also in malate synthesis [45, 58], but careful analysis of the fate of radiolabelled alanine (which is rapidly metabolised to pyruvate) clearly showed that the malic enzyme-catalysed carboxylation of pyruvate does not play a significant role in malate formation at any stage of berry development [59]. Studies with a highly purified preparation of the enzyme, indicating co-operativity in the kinetics of the malate decarboxylating activity, suggested that malic enzyme may play a regulatory role in the breakdown of malate [60]. Moreover, the NADP-dependent malic enzyme catalysed breakdown of malate was shown to be sensitive to the cytosolic NADPH to NADP+ ratio, with higher rates of enzyme activity associated with conditions of low NADPH concentration. The requirement of many biosynthetic reactions for reducing equivalents in the form of NADPH led the authors to propose a model in which malate fulfils a role as a store of 'reducing power' that can be drawn upon in the later stages of berry ripening, at which time the 'normal' glycolytic metabolism of hexose sugars has ceased. Subsequent analysis of the catalytic activity of NADP-dependent malic enzyme purified from berries confirmed the sensitivity to NADPH levels [61].

A mitochondrial isoform of malic enzyme, dependent on NAD+ as its cofactor, has been reported in many plants but has only been detected immunohistochemically in grape berries. It has been suggested that the production of NADH arising from malate oxidation by NAD-dependent malic enzyme under conditions of malate abundance (*i.e.* post-veraison) may be sufficient to activate the non-phosphorylating pathway of respiration [6]. A consequence of this would be the reduction in levels of reactive oxygen species *via* activity of the alternative oxidase. Although the grape is not considered to be a climacteric fruit, hydrogen peroxide release from berries of Pinot Noir undergoing veraison was observed, concomitant with an increase in transcription of the grapevine alternative oxidase gene [62].

MINOR ACIDS IN THE GRAPE BERRY

As noted earlier, more than 20 organic acids have been identified in grape berry analyses. Of the minor acids, only oxalic and ascorbic acids have been the subject of research to investigate aspects of metabolism. Oxalic acid biochemistry in the grape was discussed earlier, in the context of tartaric acid synthesis and the 'dual fate' of ascorbate in berry tissues. Ascorbic acid, in addition to being the precursor of tartaric acid, has cellular antioxidant roles in plant tissues, including the grape berry. Results of investigations into its metabolism will be described below.

Ascorbic Acid – Synthesis and Recycling in the Grape

Ascorbic acid functions as a major soluble antioxidant in a wide number of physiological and developmental situations in plants, providing regenerable protection against damage caused by Reactive Oxygen Species (ROS; for a review of this topic the reader is directed to recent reviews by Smirnoff [63, 64] and Hancock & Viola [65]). Its function as an antioxidant in many biochemical reactions is enhanced by the presence of an enzyme-based system of regeneration, by which the monodehydroascorbate anion is converted back to ascorbate by the activity of monodehydroascorbate reductase, thereby providing fine control over the cellular redox state.

There is continuing strong interest in the metabolism of ascorbate, no doubt largely due to its anthropogenic incarnation as Vitamin C, with much of the research in the area focussed upon strategies whereby increased concentrations of vitamin C may be achieved in a range of foodstuffs, most notably fruits [66]. The grape berry is however not among those fruits with a high vitamin C content (ca 11 mg per 100 g fresh weight; compare this to ca 200 mg per 100 g fresh weight for blackcurrants and ca 50 mg per 100 g fresh weight for lemons) [67] and is unusual among plants in using ascorbate also as a precursor for the synthesis of tartaric and oxalic acids [5].

Evidence has been presented for three distinct metabolic pathways by which ascorbate synthesis can occur in plants [65]. Predominant among these is the 'Smirnoff-Wheeler pathway', shown to be present in many different tissues, and for which a number of the enzymatic reactions have been characterised in recent years [63]. Two further pathways have been proposed, the first in which myo-inositol is oxidised *via* D-glucuronic acid, entering the Smirnoff-Wheeler pathway at the level of L-gulono-1,4-lactone (the immediate precursor of ascorbate), and a synthetic route *via* D-galacturonic acid, shown to be up-regulated in ripening strawberries and associated with the breakdown of cell walls that accompanies softening processes [68].

Early research showed that immature berries fed with radiolabelled ascorbate were capable of its further metabolism (see earlier section on tartaric acid synthesis), but provided no evidence for synthesis of ascorbate within the berry. Recent data from molecular analysis of developing and ripening berries, combined with measurements of the levels of ascorbate in berries during development and ripening, indicated the operation of two distinct ascorbate synthesis pathways, which may be under developmental regulation. In immature, developing berries transcripts for several genes associated with the Smirnoff-Wheeler pathway were identified that were absent or down-regulated from berries sampled at post-veraison time points [69, 70]. Analysis of the concentration of ascorbate and its oxidised form, dehydroascorbate, provided strong evidence that in developing berries the increase in ascorbate concentration correlates with increased transcription of the Smirnoff-Wheeler pathway genes encoding GDP-D-mannose-3,5-epimerase, GDP-L-galactose phosphorylase and L-galactose dehydrogenase. Berries sampled at these times also showed a low ascorbate:dehydroascorbate ratio, concomitant with a low level of transcription of the genes encoding the ascorbate recycling enzymes, monodehydroascorbate reductase and dehydroascorbate reductase [70]. Intriguingly, in post veraison berries, transcription of Smirnoff-Wheeler pathway genes was down-regulated, and instead the berries showed increased transcription of *GalUR*, encoding D-galacturonic acid reductase, a key component of the ascorbate synthesis pathway in ripe strawberries. Further investigations added a physiological dimension to these studies, with the revelation that under high light conditions, berry ascorbate levels increased. At the same time, the transcription of Smirnoff-Wheeler pathway genes was unaffected but transcription of *GalUR* increased, suggesting a possible basis for increased ascorbate in light-exposed berries (see below for discussion of tartaric acid synthesis responses to light exposure and light exclusion) [69]. Our recent data suggests that total light exclusion during berry development has only a transient effect on the transcription of ascorbate synthesis genes from the Smirnoff-Wheeler pathway, which is reflected in reduced ascorbate and tartaric acid concentrations in berries that persisted throughout development and ripening [90].

DEVELOPMENTAL AND PHYSIOLOGICAL BIOCHEMISTRY OF ORGANIC ACID METABOLISM IN GRAPE BERRIES

INTRODUCTION

As noted earlier, research on the metabolism of organic acids during berry development has often been focussed on increasing our understanding of fruit composition at harvest. The changes in acid metabolism during berry development, and the impact of environmental conditions during the growing season upon the composition and concentration of acids have been widely studied. The viticultural perspective, reflecting the requirement to achieve the optimal balance of acids with other desirable parameters including yield, sugar levels and flavour/tannin composition, has focussed largely on the effects of canopy temperature and light, both of which may to a degree be varied by appropriate cultural practices. The following sections will briefly review this literature, and attempt to describe the likely biochemical events associated with the metabolic outcomes observed. As will be shown, our present understanding of the mechanisms that underlie the regulation of acid metabolism at genetic, developmental and physiological levels remains poorly developed.

Genetic Variation in Organic Acid Metabolism

It has been long known that species within the genus *Vitis* and individual varieties of the cultivated grapevine *V. vinifera* show wide variation in the natural acidity of berries. Analyses of the acid composition in developing and ripe berries of 26 species of *Vitis* and 78 varieties of vinifera (28 of table grapes and 50 of wine grapes) showed differences in levels of tartaric and malic acid [71, 72]. Tartaric acid levels in wine grapes varied from a maximum of 1.08 g/100 ml (Emerald Riesling) to below 0.6 g/100 ml in cultivars including Mataro, Pinot Gris and Semillon. The values changed little for fruit harvested 'early' or 'late' in the growing season. Malic acid levels in the same samples however were, as expected, very different in 'early' and 'late' samples. Within the wine grapes sampled, the maximum 'early' sampled malic acid level was 0.85 g/100 ml, recorded for Pinot St. George, and the lowest for White Riesling, 0.26 g/100 ml. Several varieties showed 'late' malate levels below 0.15 g/100 ml. More importantly from the perspective of acid metabolism, some varieties showed greater than 50% loss of malate between early and late sampling, while others (White Riesling and Grey Riesling for instance) retained malate at comparable levels across the sampling period [72].

Consideration and attempted extrapolation of these data to a broader context indicates the great difficulties encountered when analysing acid metabolism during berry development. In the work described, berry sampling was not achieved under conditions of 'equivalent development' for each variety studied. For instance, in the research described by Kliewer *et al.*, at both time points the total soluble solids (TSS) of the berries differed between varieties, reflecting the chronological range over which ripening occurs, the differences in berry size and volume, and the extent to which sugar accumulation rates vary between varieties. It is not known to what extent if any malate metabolism is regulated in relation to other parameters of berry ripening – in the case of the two Riesling varieties for example, was the comparatively small loss of malate between early and late sampling dates reflecting a low rate of malate loss overall, or that malate loss had already occurred before the first sampling date, or was it that in these varieties there is in fact a continued synthesis of malate after veraison? These questions can be answered only by 'developmental sampling' of berries across multiple time points in the growing season, and the subsequent comparison of acid levels in the light of changes in other measurable parameters of growth and development. It is in this respect that 'omics' approaches are already proving to be of great value [73, 74].

Developmental and Physiological Biochemistry of Tartaric Acid

Tartaric acid levels in ripe berries reflect the extent to which its synthesis occurred during the early stages of berry development. Tartaric acid can be detected in grapevine flowers, and levels in the berry increase for at least the first 4 weeks post anthesis (flowering) [1, 75]. Thereafter, its synthesis stops, and further changes in concentration are due to increasing berry volume in the later stages of ripening, leading to dilution of tartaric acid. Although early reports suggested that tartaric acid occurs in the leaf, and that its accumulation in the berry is a consequence of translocation, the work of Hale and Hardy in the early 1960s provided conclusive evidence that the berry itself is the site of synthesis. Although there has not yet been any study of the distribution of tartaric acid synthesis enzymes within the berry, as noted in an earlier section, molecular evidence suggests that the transcript encoding L-IdnDH, the only characterised enzyme in the tartaric acid synthesis pathway, occurs at higher frequency towards the centre of the berry.

A detailed investigation was reported in which berries sampled at 10-day intervals during development and ripening were dissected into four concentric zones of tissue, representing skin, outer pulp, intermediate pulp and inner pulp fractions [40]. Tartaric and malic acid concentrations were recorded for each fraction, along with data for sugars and some cations. The results showed clearly that tartaric acid distribution through the berry was uneven, and that as development and ripening progressed, tartaric acid concentration in the skin and outer pulp fractions decreased, while remaining relatively unchanged within the two inner regions of the berry. The decline in tartaric acid concentration in skin and outer pulp fractions, although apparent from ca 20 days after anthesis, was most marked in the immediate period post-veraison, following which there seemed to be a levelling off of the changes in the relative concentrations. It is unclear however to what extent any of these changes reflected increases in cell size with berry growth and expansion – it may be supposed that the pulp fractions would have increased in weight during development more than was the case for the skin.

Intriguing data concerning developmental changes in the distribution of tartaric acid suggested that initially high tartaric acid levels in the inner and outer pulp and the skin are followed, post-veraison, by a significant decrease in its concentration in the outer mesocarp layers immediately beneath the skin [76]. Measurement of acid composition was made using acid extraction to solubilise any salts and it was therefore not possible to examine the extent to which tartaric acid was present in the berries as the neutral (un-dissociated) acid or in the anion forms, known to predominate as the berry imports potassium ions that are exchanged for protons in the vacuole [77]. The authors suggested that metabolism of tartaric acid may be occurring post-veraison in tissues close to or associated with the vascular tissue of the berry.

A number of investigations have looked at the effects upon the acid composition of varying the temperature or light that reaches the berry. The most pronounced effects are seen with respect to malic acid (see next section) – generally speaking, the tartaric acid concentration in ripe berries is unaffected by temperature. Nonetheless, evidence from a sampling regime conducted in South Australia during the spring and summer of 2005-6, during which time a period of significant high seasonal temperature followed a very dry winter and early spring, suggested that the situation may be more complex. It was observed in this unreplicated trial that tartaric acid, expressed as amount per berry, decreased beyond 80 days after flowering (Melino and Ford, unpublished observation). Earlier reports alluded to a 'respiration' of tartaric acid at temperatures above 30°C (cited in [12]) although it may be questioned, in view of the continued success of grape growing in many warm-climate regions, whether this is an entirely accurate comment.

The effects of light on the synthesis of tartaric acid have been studied from the perspective of potential modifications to the bunch microclimate through the application of canopy management techniques. Studies of vines in which 'shading treatments' have been imposed through the use of various screening strategies reported little or no significant change in the total acidity of berries in shaded or control replicates [78, 79]. In later investigations that compared the composition of berries from 'sun-exposed' and 'shaded' parts of the canopy, for which the individual acids were analysed by HPLC, it was seen that an increase in the concentration of tartaric acid associated with growth in the sun-exposed berries was due to a reduction in berry size – the amount of tartaric acid per berry was unchanged [80]. An alternative approach, in which bunches were totally excluded from light at specified periods during berry development by enclosing them in foil bags, provided no evidence for any light-mediated changes in tartaric acid composition [81].

A confounding effect in experiments of this nature arises from the difficulties of separating the effects of light from those of heating associated with increased exposure to the sun. The use of black plastic boxes placed around the developing bunches at anthesis allowed for the exclusion of light while maintaining ambient temperatures within. Berries grown in this way were shown to have reduced levels of tartaric acid on a per berry basis and, at the same time, to be somewhat smaller than berries grown outside of the boxes [82]. Berries from bunches on treatment vines within the trial in which increased sun exposure was achieved by the use of cane positioning did not show elevated tartaric acid levels when expressed on a per berry basis compared to control vines (in which the bunches were moderately exposed by the natural leaf cover). In the first report of a specific molecular effect on tartaric acid synthesis, levels of the transcript encoding L-IdnDH were down-regulated in berries grown within these light-excluding boxes [37].

Developmental and Physiological Biochemistry of Malic Acid

Malate levels in the berry reflect not just its synthesis, but also its subsequent participation in a wide range of metabolic activities at later stages of ripening. Malate in immature grapes is first detectable by HPLC about 7 days after flowering, when berries are around 2-4 mm in diameter (modified E-L stages 27-29) [83]. Its synthesis continues over the first 45-60 days of berry development, but at the onset of the period of berry softening (and colour formation in some varieties), a rapid loss of malate is seen. Experiments in which bunches were vascularly isolated from the surrounding plant following a period of exposure of the adjacent leaf to $^{14}CO_2$ by 'girdling' confirmed that malate was synthesised in the berry and not imported from another part of the vine [43].

The research described earlier in which the distribution of tartaric acid within the berry was discussed also provided some clues concerning malate. Malate accumulation was recorded within the outer, intermediate and inner pulp fractions throughout the period of berry development. The rate of its accumulation was highest in the inner pulp and decreased towards the periphery of the developing berry [40]. The lowest pre-veraison concentration of malate was recorded in the skin, with a sudden burst of accumulation seen in the week preceding veraison at a rate greater than seen earlier in other berry fractions. Intriguingly, the rate of malate degradation immediately after veraison was greatest in the central tissues of the berry (following a short, ca 10-day lag in the inner pulp), gradually decreasing towards the peripheral tissues to the extent that in the fully ripe berries, the highest malate concentrations were recorded in the skin.

A biochemical perspective to these findings was provided by measurements of NADP-dependent malic enzyme activity in different sections of the developing and ripening berry. Earliest experiments, in which malic enzyme activity was assayed in tissue fractions similar to those described by Possner and Kliewer from green (immature) and ripening berries, showed a broad negative correlation between the 'activity' of malic enzyme (*i.e.* activity per unit protein concentration) and the concentration of malate in the ripening berry tissue [84]. Although there was no comparable correlation seen in pre-veraison berries, as would be expected if malic enzyme is principally associated with malate catabolism, its activity was nonetheless clearly detectable. The authors suggested that the catalytic properties of the purified enzyme described earlier (see the section on malic acid synthesis above) may reflect the physiological mechanism of enzyme regulation. As the vacuole 'releases' malate into the cytosol, the pH will decrease and the decarboxylating activity of malic enzyme will be enhanced through a combination of higher affinity for the substrate and higher catalytic rate at lower pH.

These observations were extended in the later work of Gutiérrez-Granda and Morrison [76], who studied malate and malic enzyme levels across 9 developmental time points spanning an eight-week sampling period from two weeks before to six weeks after veraison. There was no clear correlation, however, between the concentration of malate and the activity of malic enzyme – in skin samples post-veraison malic enzyme and malate levels were higher than in the other tissue sections. It was suggested that this 'paradox' may be solved by considering the mobilisation of malate within the cells of ripening berries: the peripheral region of the berry is more vascularised, and malate from the inner parts of the berry may be translocated towards the outer part, where it is then readily broken down by malic enzyme (present at higher activities than in the internal 'pulp' cells) prior to transport *via* the berry's vascular structures. As proposed by Coombe, the consequences of such a rapid catabolism of malate would be reflected in the establishment of a gradient of malate, which would drive the continual diffusion of further malate from the internal zones of the berry outwards [85].

Similarly to tartaric acid, following its synthesis malate is transported to the vacuole of the cell during the period of acid accumulation. The processes responsible for the transport of organic acids within developing berries remain to be fully elucidated – vacuolar proton pumps have been identified and characterised (for a review see [4]) but the identification and functional characterisation of the anion channels awaits.

The physiological basis of malate metabolism in berries has been studied for many years. Both temperature and light have been suggested to play roles in these processes, and although early reports of increased rates of acid loss in berries grown at high temperatures did not identify the acids affected (see [86] and references therein) it is safe to assume that malate synthesis was affected. Interestingly this study, in which berries were heated to ca 32°C at night and allowed to reach ambient temperatures by day (which reached on occasion 45°C), showed no effect on final berry concentration of acid, but a significant decrease in the order of 30% in berry weight in the heated berries compared to the control treatment. These data provide some of the earliest indications of the likely effects of increased environmental temperatures arising from anthropogenic climate change, predicted to affect many areas where grapes are presently grown.

A detailed study was conducted by Kliewer, following the fate of $^{14}CO_2$ uptake by excised shoots and its subsequent metabolic fate in berries grown under a range of temperatures [87]. These data showed that immature and mature berries differed greatly in their response to temperature with respect to malate metabolism. Pre-veraison berries incubated at temperatures around 10°C showed reduced malate accumulation compared with the controls (incubated at ca 25 °C), while temperatures between 10 and 37°C appeared to have little effect on malate synthesis in the 24 hours following exposure to $^{14}CO_2$. In post-veraison berries, however, a significant decrease in malate accumulation was seen in berries sampled from shoots incubated at temperatures greater than 10°C. At the lower temperature, 24 hours after exposure to $^{14}CO_2$ over 80% of the radioactivity fixed into organic acids was malate. This was halved in berries incubated for the same period at 37°C, suggesting that higher temperatures lead to reduced malate synthetic capacity in mature berries. Later results from trials in which pot-grown cuttings of Cabernet Sauvignon were treated with a range of temperature regimes at specified developmental stages confirmed that high temperatures during the pre-veraison period of malate synthesis led to lower levels of malate accumulation, while high temperatures post-veraison resulted in greater levels of malate degradation [88].

Investigations into the effects of light exposure and exclusion on malate metabolism have been in many cases confounded by the inter-relatedness of light and temperature effects. Although widely recognised, these effects are difficult to disentangle entirely, most particularly when the effects of greater than normal canopy light levels are considered (as for instance may occur with shoot-positioning treatments designed to maximise exposure to the sun). Experiments conducted in a phytotron, in which temperatures were maintained within narrow limits, examined differences in berry malate concentrations in two cultivars, Pinot Noir and Cabernet Sauvignon [81]. Shading of the berries at differing stages resulted in reduced malate levels compared to the full-light control when measured during development and ripening, but the increased levels of malate degradation in light-exposed fruit compared to the shaded treatments resulted in comparable levels at harvest. In studies using light-excluding boxes designed to minimise temperature variations with the external environment to study the effects of light, it was seen that malate levels appear to be little affected by the light regime, in contrast to the effects seen on tartaric acid and ascorbate [37, 82].

An important consideration in the full interpretation of data from experiments examining effects such as light and temperature upon berry acid composition is the complex interplay of different pathways of carbon metabolism. For instance, in immature berries, acid synthesis occurs and sugars are largely used as a source of energy imported to the berry from the leaves. Beyond veraison, the berry turns into a 'sink' for sugars, which accumulate and are no longer solely used as a source of energy in the berry. Thus carbon metabolism, in the context of the fate of organic acids, proceeds in a series of different 'metabolic environments', which will determine the extent and directionality of acid synthesis and breakdown reactions at different stages of the berry's ripening and development.

CONCLUSIONS AND FUTURE PERSPECTIVES

The organic acid composition of the ripe grape berry reflects the outcomes of several inter-related metabolic activities, involving synthesis by 'conventional' (malate) and 'grape-specific' (tartaric) pathways, intracellular transport and storage in the vacuole (both acids), and a range of catabolic processes in the post-veraison period (malate)[10]. Evidence points to genetic and environmental components in the determination of acid synthesis and breakdown, but further research is needed to more clearly define the specific details.

It can be suggested that the present 'unknowns' relating to the biochemistry of organic acids in the grape berry be placed into two broad groups. First, completion of the gaps in our understanding of the enzymatic steps in the various synthetic and catabolic pathways, including the elucidation of the processes of intracellular transport both into and out of the vacuole. The identification of these activities and their characterisation by, *inter alia*, enzyme purification, functional assays and inhibitor studies will complete the work that was started over 50 years ago.

Despite a 'slow start', the tools of the 21st century biologist are increasingly being applied to the grapevine. The completion of at least three genome sequencing projects and the continual annotation of the information contained within, and the accumulation of large repositories of array data, as well as significant

---

[10] Recall also that early reports and 'anecdotal' data from several laboratories including the author's, point to the potential breakdown of tartaric acid following periods of high environmental temperature by as-yet uncharacterized pathways.

metabolite and proteomics-based analyses attest to the growing importance of post-genomics-based approaches in the study of the grapevine. The recent development of the grape microvine will allow for the first time the transformation, regeneration and fruiting within a 12-18-month time timespan [89]. A further 'tool' is provided by large-scale automated growth and non-destructive sampling facilities, in which plants can be grown under specific temperature and light regimes with regular analysis by imaging technologies, such as the recently completed Plant Accelerator in Australia (http://www.plantaccelerator.org.au/).

These technologies and others will prove essential in solving the second group of presently unknown facets of organic acid metabolism in grape berries – what may for simplicity be called regulatory mechanisms. It is likely for example that transcriptional control plays a role in tartaric acid synthesis, but as yet there is no data concerning the specific transcription factors involved. Similarly, regulation of the well-characterised switch between synthesis and breakdown of malate at the time of veraison remains largely uncharacterised. These and other localised control mechanisms will undoubtedly be revealed in the coming years as the data held in genome sequence datasets becomes more fully interpreted.

The nature of organic acid metabolism in the developing and ripening grape arguably provides us with a very amenable experimental model to study some fundamental processes of plant development. Tartaric acid synthesis is an unusual fate for L-ascorbate, which in many plant systems fulfils a primarily antioxidant function. The rapid and extensive switch in ascorbate metabolism towards tartaric acid synthesis at the earliest stages of berry development, and its reversal as veraison approaches, represents a readily followed modification to the carbon flux in the fruit. An understanding of the mechanisms underlying this process may yield important insights to the larger story of fruit development. Similarly, the apparent responsiveness of malate to environmental temperatures (indeed, it may be suggested that the most immediately negative impact of climate change for grape composition will be in the loss of malate levels at harvest) provides a sensitive model of metabolic response to environmental parameters, which we are now in a position test in both laboratory and field-based investigations. It may therefore be that the organic acids, long overlooked in compositional importance compared with the sugars and polyphenolics, will yet play a role in our deeper understanding of the mysteries of berry metabolism.

ABBREVIATIONS

ATP = Adenosine triphosphate

CAM = Crassulacean acid metabolism

cDNA = Complementary DNA

EST = Expressed sequence tag

GDP = Guanosine diphosphate

HPLC = High performance liquid chromatography

MDH = Malate dehydrogenase

NAD = Nicotine adenine dinucleotide

NADP = Nicotine adenine dinucleotide phosphate

OAA = Oxaloacetate

PEP = Phospho enol pyruvate

PEPC = PEP carboxylase

ROS = Reactive oxygen species

TSS = Total soluble sugars

Vv-L-IdnDH = *Vitis vinifera* L-idonate dehydrogenase

ACKNOWLEDGEMENTS

Work in the author's laboratory is supported by Australia's grapegrowers and winemakers through their investment body, the Grape and Wine Research and Development Corporation, with matching funds from the Australian Government. The University of Adelaide is part of the Wine Innovation Cluster, based on the Waite Campus. The author wishes to thank his past and present PhD students, collaborators and colleagues for their support, and enthusiastic willingness to sample grapes in unreasonably hot conditions. Drs. Nicole Cain and Peter Valente (University of Adelaide) are gratefully acknowledged for assistance with the preparation of figures.

REFERENCES

[1] Ruffner HP. Metabolism of Tartaric and Malic Acids in *Vitis*: A Review-Part A. Vitis 1982; 21: 247-59.

[2] Ruffner HP. Metabolism of Tartaric and Malic Acids in *Vitis*: A Review-Part B. Vitis 1982; 21: 346-58.

[3] Loewus F. Biosynthesis and Metabolism of Ascorbic Acid in Plants and of Analogs of Ascorbic Acid in Fungi. Phytochemistry 1999; 52(2): 193-210.

[4] Terrier N, Romieu C, Roubelakis-Angelakis KA. Grape Berry Acidity. In: Molecular Biology and Biotechnology of the Grapevine. 2001; pp. 35-57.

[5] DeBolt S, Melino V, Ford C. Ascorbate as a Biosynthetic Precursor in Plants. Annals of Botany 2007; 99(1): 3-8.

[6] Sweetman C, Deluc LG, Cramer GR, Ford CM, Soole KL. Regulation of malate metabolism in grape berry and other developing fruits. Phytochemistry 2009; 70(11-12): 1329-44.

[7] Amerine MA. The Maturation of Wine Grapes. Wines and Vines 1956; 37(11): 53-5.

[8] Stafford H. Distribution of Tartaric Acid in the Leaves of Certain Angiosperms. American Journal of Botany 1959. 46(5): 347-52.

[9] Webb A, Cavaletto M, Carpita C, Lopez E, Arnott J. The Intravacuolar Organic Matrix Associated with Calcium Oxalate Crystals in Leaves of *Vitis*. Plant Journal 1995; 7(4): 633-48.

[10] Debolt S, Hardie J, Tyerman S, Ford CM. Composition and synthesis of raphide crystals and druse crystals in berries of *Vitis vinifera* L. Cv. Cabernet Sauvignon: ascorbic acid as precursor for both oxalic and tartaric acids as revealed by radiolabelling studies. Australian Journal of Grape and Wine Research 2004; 10(2): 134-42.

[11] Kliewer W. Sugars and Organic Acids of *Vitis vinifera*. Plant Physiology 1966; 41: 923-31.

[12] Peynaud E, Maurié A. Synthesis of Tartaric and Malic Acids by Grape Vines. American Journal of Enology and Viticulture 1958. 9(1): 32-6.

[13] Stafford H, Loewus F. The Fixation of C14O2 into Tartaric and Malic Acids of Excised Grape Leaves. Plant Physiology 1958; 33(3): 194-9.

[14] Vickery H, Palmer J. The Metabolism of the Organic Acids of Tobacco Leaves VII: Effect of Culture of Excised Leaves in Solutions of (+)-Tartrate. The Journal of Biological Chemistry 1954; 207(1): 275-285.

[15] Hough L, Jones JK. The Biosynthesis of the Monosaccharides. In: Advances in Carbohydrate Chemistry. 1956; 185-262.

[16] Ribéreau-Gayon G. Etude des mecanismes de synthese et de transformation de l'acide malique, de l'acide tartrique et de l'acide citrique chez *Vitis vinifera* L. Phytochemistry 1968; 7(9): 1471-82.

[17] Saito K, Kasai Z. Accumulation of Tartaric Acid in the Ripening Process of Grapes. Plant and Cell Physiology 1968; 9(3): 529-37.

[18] Maroc J. La conversion du glycolate en glucose et ses relations avec la biogenes de l'acide tartrique. Physiologie Vegetale 1967; 5(1): 37-46.

[19] Wagner G, Yang J, Loewus F. Stereoisomeric characterization of tartaric acid produced during L-ascorbic acid metabolism in plants. Plant Physiology 1975; 55(6): 1071.

[20] Saito K, Kasai K. Tartaric Acid Synthesis from L-Ascorbic Acid-1-14C in Grape Berries. Phytochemistry 1969; 8(11): 2177-82.

[21] Wagner G, Loewus F. The Biosynthesis of (+)-Tartaric Acid in *Pelargonium crispum*. Plant Physiology 1973; 52(6): 651-4.

[22] Wagner G, Loewus F. L-ascorbic acid metabolism in vitaceae: conversion to (+)-tartaric acid and hexoses. Plant Physiology 1974; 54(5): 784.

[23] Saito K, Loewus F. The Metabolism of L-[6-^{14}C] Ascorbic Acid in Detached Grape Leaves. Plant and Cell Physiology 1979; 20(8): 1481-8.

[24] Williams M, Loewus F. Biosynthesis of (+)-Tartaric Acid From L-[4-^{14}C] Ascorbic Acid in Grape and Geranium. Plant Physiology 1978; 61(4): 672-4.

[25] Williams M, Saito K, Loewus F. Ascorbic Acid Metabolism in Geranium and Grape. Phytochemistry 1979;. 18(6): 953-6.

[26] Saito K, Kasai Z. Conversion of L-Ascorbic Acid to L-Idonic Acid, L-Idono-γ-Lactone and 2-Keto-L-Idonic Acid in Slices of Immature Grapes. Plant and Cell Physiology 1982; 23(3): 499-507.

[27] Saito K, Kasai Z. Synthesis of L-(+)-tartaric acid from L-ascorbic acid *via* 5-Keto-D-gluconic acid in grapes. Plant Physiology 1984; 76(1): 170-174.

[28] Saito K. Specific transfer of 3H from D-[3-3H] gluconic acid into L-tartaric acid in vitaceous plants. Phytochemistry 1994; 37(4): 1017-22.

[29] Saito K, Loewus F. Formation of Tartaric Acid in Vitaceous Plants: Relative Contributions of L-Ascorbic Acid-Inclusive and Acid-Noninclusive Pathways. Plant and Cell Physiology 1989; 30(6): 905-10.

[30] Ruffner H, Rast D. The Biogenesis of Tartaric Acid in the Grapevine. Zeitschrift fur Pflanzenphysiologie 1974; 73(1): 45-55.

[31] Malipiero U, Ruffner HP, Rast DM. Ascorbic To Tartaric Acid Conversion In Grapevines. Journal of Plant Physiology 1987; 129(1-2): 33-40.

[32] Haberlandt G. Physiological Plant Anatomy. Macmillan and Co. London 1914.

[33] Arnott HJ, Webb MA. Twinned raphides of calcium oxalate in grape (*Vitis*): Implications for crystal stability and function. International Journal of Plant Sciences 2000; 161(1): 133-42.

[34] Storey R, Jones RGW, Schachtman DP, Treeby MT. Calcium-accumulating cells in the meristematic region of grapevine root apices. Functional Plant Biology 2003; 30: 719-27.

[35] Hardie W, O'Brien T, Jaudzems V. Morphology, anatomy and development of the pericarp after anthesis in grape, *Vitis vinifera* L. Australian Journal of Grape and Wine Research 1996; 2(2): 97-142.

[36] Nakata P. Advances in our understanding of calcium oxalate crystal formation and function in plants. Plant Science 2003; 164(6): 901-9.

[37] DeBolt S, Cook D, Ford C. L-Tartaric acid synthesis from vitamin C in higher plants. Proceedings of the National Academy of Sciences of the United States of America 2006; 103(14): 5608-13.

[38] Deluc LG, Grimplet J, Wheatley MD, *et al.* Transcriptomic and metabolite analyses of Cabernet Sauvignon grape berry development. BMC Genomics 2007; 8(1): 429.

[39] Grimplet J, Deluc LG, Tillett RL, *et al.* Tissue-specific mRNA expression profiling in grape berry tissues. BMC Genomics 2007; 8(1): 187.

[40] Possner DRE, Kliewer W. The localisation of acids, sugars, potassium and calcium in developing grape berries. Vitis 1985; 24: 229-40.

[41] Fernie AR, Martinoia E. Malate. Jack of all trades or master of a few? Phytochemistry 2009; 70(7): 828-32.

[42] Lance C, Rustin P. The Central Role of Malate in Plant Metabolism. Physiologie Vegetale 1984; 22(5): 625-41.

[43] Hale CR. Synthesis of Organic Acids in the Fruit of the Grape. Nature 1962; 195(4844): 917-8.

[44] Loewus F, Stafford H. Observations on the Incorporation of C14 into Tartaric Acid and the Labeling Pattern of D-Glucose from an Excised Grape Leaf Administered L-Ascorbic Acid-6-C14. Plant Physiology 1958; 33(2): 155-6.

[45] Hawker J. Changes in the activities of malic enzyme, malate dehydrogenase, phosphopyruvate carboxylase and pyruvate decarboxylase during the development of a non-climacteric fruit (the grape). Phytochemistry 1969; 8(1): 19-23.

[46] Lakso A, Kliewer W. The influence of temperature on malic acid metabolism in grape berries: I. Enzyme responses. Plant Physiology 1975; 56(3): 370-2.

[47] Chollet R, Vidal J, Oleary M. Phosphoenolpyruvate carboxylase: A ubiquitous, highly regulated enzyme in plants. Annual Review of Plant Physiology and Plant Molecular Biology 1996; 47: 273-98.

[48] O'Leary M. Phosphoenolpyruvate carboxylase: an enzymologist's view. Annual Review of Plant Physiology 1982; 33: 297-315.

[49] Vidal J, Chollet R. Regulatory phosphorylation of C4 PEP carboxylase. Trends in Plant Science 1997; 2(6): 230-7.

[50] Ruffner H, Hawker J, Hale C. Temperature and Enzymic Control of Malate Metabolism in Berries of *Vitis vinifera*. Phytochemistry 1976; 15(12): 1877-80.

[51] Ruffner H, Koblet W, Rast D. Gluconeogenesis in ripening grapes (*Vitis vinifera* L.). Gluconeogenese in reifenden Beeren von *Vitis vinifera*. Vitis 1975; 13(4): 319-28.

[52] Ruffner HP, Kliewer W. Phosphoenolpyruvate Carboxykinase Activity in Grape Berries. Plant Physiology 1975; 56(1): 67-71.

[53] Terrier N, Glissant D, Grimplet J, *et al.* Isogene specific oligo arrays reveal multifaceted changes in gene expression during grape berry (*Vitis vinifera* L.) development. Planta 2005; 222(5): 832-47.

[54] Dal Peruffo A, Pallavicini C. Enzymatic changes associated with ripening of grape berries. Journal of the Science of Food and Agriculture 1975; 26(5): 559-566.

[55] Taureilles-Saurel C, Romieu CG, Robin J-P, Flanzy C. Grape (*Vitis vinifera* L.) malate dehydrogenase. I. Intracellular compartmentation of the isoforms. American Journal of Enology and Viticulture 1995; 46(1): 22-8.

[56] Taureilles-Saurel C, Romieu CG, Robin J-P, Flanzy C. Grape (*Vitis vinifera* L.) malate dehydrogenase. II. Characterization of the major mitochondrial and cytosolic isoforms and their role in ripening. American Journal of Enology and Viticulture 1995;. 46(1): 29-36.

[57] Or E, Baybik J, Sadka A, Saks Y. Isolation of mitochondrial malate dehydrogenase and phosphoenolpyruvate carboxylase cDNA clones from grape berries and analysis of their expression pattern throughout berry development. Journal of Plant Physiology 2000; 157(5): 527-34.

[58] Meynhardt JT. Biosynthesis of dicarboxylic acids through carbon dioxide fixation by an enzyme extract from Barlinka grape berries. South African Journal of Agricultural Science 1965; 8: 381-7.

[59] Ruffner HP, Possner D, Brem S, Rast DM. The Physiological Role of Malic Enzyme in Grape Ripening. Planta 1984; 160(5): 444-8.

[60] Possner D, Ruffner HP, Rast DM. Isolation and biochemical characterisation of grape malic enzyme. Planta 1981; 151(6): 549-54.

[61] Franke KE, Adams DO. Inhibition of malic enzyme from grape berries by sulfhydryl reagents and oxalic acid. American Journal of Enology and Viticulture 1992; 43(2): 153-8.

[62] Pilati S, Perazzolli M, Malossini A, Cestaro A, *et al.* Genome-wide transcriptional analysis of grapevine berry ripening reveals a set of genes similarly modulated during three seasons and the occurrence of an oxidative burst at veraison. BMC Genomics 2007; 8(1): 428.

[63] Smirnoff N, Conklin PL, Loewus FA. Biosynthesis of Ascorbic Acid in Plants: A Renaissance. Annual Review of Plant Physiology and Plant Molecular Biology 2001; 52: 437-67.

[64] Smirnoff N. Ascorbic Acid: Metabolism and Functions of a Multi-Facetted Molecule. Current Opinion in Plant Biology 2000; 3(3): 229-35.

[65] Hancock R, Viola R. Biosynthesis and catabolism of L-ascorbic acid in plants. Critical Reviews in Plant Sciences 2005; 24: 167-88.

[66] Poiroux-Gonord F, Bidel LP, Fanciullino AL, Gautier H, Lauri-Lopez F, Urban L. Health Benefits of Vitamins and Secondary Metabolites of Fruits and Vegetables and Prospects To Increase Their Concentrations by Agronomic Approaches. Journal of Agricultural and Food Chemistry 2010; 58: 12065-82.

[67] Davey Mw, Van Montagu M, Inze D, Sanmartin M, Kanellis A, Smirnoff N, Benzie Ijj, Strain Jj, Favell D, Fletcher J. Plant L-ascorbic acid: chemistry, function, metabolism, bioavailability and effects of processing. Journal of the Science of Food and Agriculture 2000; 80: 825-60.

[68] Agius F, González-Lamothe R, Caballero JL, Muñoz-Blanco J, Botella MA, Valpuesta V. Engineering increased vitamin C levels in plants by overexpression of a D-galacturonic acid reductase. Nature Biotechnology 2003; 21(2): 177-81.

[69] Cruz-Rus E, Botella MA, Valpuesta V, Gomez-Jimenez MC. Analysis of genes involved in l-ascorbic acid biosynthesis during growth and ripening of grape berries. Journal of Plant Physiology 2010; 167(9): 739-48.

[70] Melino VJ, Soole KL, Ford CM. Ascorbate metabolism and the developmental demand for tartaric and oxalic acids in ripening grape berries. BMC Plant Biology 2009; 9(1): 145.

[71] Kliewer W. Concentration of Tartrates, Malates, Glucose and Fructose in the Fruits of the Genus *Vitis*. American Journal of Enology and Viticulture 1967; 18(2): 87-96.

[72] Kliewer W, Howarth L, Omori M. Concentrations of Tartaric Acid and Malic Acids and Their Salts in *Vitis vinifera* Grapes. American Journal of Enology and Viticulture 1967; 18(1): 42-54.

[73] Cramer GR, Ergül A, Grimplet J, Tillett RL, *et al.* Water and salinity stress in grapevines: early and late changes in transcript and metabolite profiles. Functional and Integrative Genomics 2006. DOI 10.1007/s10142-006-0039-y.

[74] Grimplet J, Wheatley MD, Jouira HB, Deluc LG, Cramer GR, Cushman JC. Proteomic and Selected Metabolite Analysis of Grape Berry Tissues Under Well-Watered and Water-Deficit Stress Conditions. Proteomics 2009; 9(9): 2503-28.

[75] Kliewer W. Changes in the Concentration of Malates, Tartrates, and Total Free Acids in Flowers and Berries of *Vitis vinifera.* American Journal of Enology and Viticulture 1965; 16(2): 92-100.

[76] Gutierrez-Granda M, Morrison J. Solute Distribution and Malic Enzyme-Activity in Developing Grape Berries. American Journal of Enology and Viticulture 1992; 43(4): 323-8.

[77] Boulton R. The General Relationship Between Potassium, Sodium and pH in Grape Juice and Wine. American Journal of Enology and Viticulture 1980; 31(2): 182-6.

[78] Kliewer W, Schultz HB. Influence of Environment on Metabolism of Organic Acids and Carbohydrates in *Vitis vinifera.* II. Light. American Journal of Enology and Viticulture; 1964. 15(3): 119-29.

[79] Schultz H, Lider LA. Modification of the light factor and heat load in vineyards. American Journal of Enology and Viticulture 1964; 15: 87-92.

[80] Crippen D, Morrison J. The Effects of Sun Exposure on the Compositional Development of Cabernet Sauvignon Berries. American Journal of Enology and Viticulture 1986; 37(4): 235-42.

[81] Dokoozlian N, Kliewer W. Influence of Light on Grape Berry Growth and Composition Varies During Fruit Development. Journal of the American Society for Horticultural Science 1996; 121(5): 869-74.

[82] DeBolt S, Ristic R, Iland PG, Ford CM. Altered light interception reduces grape berry weight and modulates organic acid biosynthesisduring development. HortScience 2008; 43(3): 957-61.

[83] Coombe BG. Adoption of a System for Identifying Grapevine Growth Stages. Australian Journal of Grape and Wine Research 1995; 1(3): 104-10.

[84] Possner D, Ruffner H, Rast D. Regulation of malic acid metabolism in berries of *Vitis vinifera.* Acta Horticulturae Fruit Set and Development, XXI IHC 139 1983; 139: 117-22.

[85] Coombe B. Distribution of Solutes within the Developing Grape Berry in Relation to Its Morphology. American Journal of Enology and Viticulture 1987; 38(2): 120-7.

[86] Radler F. The effect of temperature on the ripening of Sultana grapes. American Journal of Enology and Viticulture 1965; 16(1): 38-41.

[87] Kliewer W. Influence of environment on metabolism of organic acids and carbohydrates in *Vitis vinifera.* I. Temperature. Plant Physiology 1964; 39(6): 869-80.

[88] Buttrose M, Hale C, and Kliewer WM. Effect of temperature on the composition of Cabernet Sauvignon'berries. American Journal of Enology and Viticulture 1971; 22(2): 71-5.

[89] Chaïb J, Torregrosa L, Mackenzie D, Corena P, Bouquet A, Thomas MR. The grape microvine - a model system for rapid forward and reverse genetics of grapevines. The Plant Journal 2010; 62(6): 1083-92.

[90] Melino VJ, Hayes MA, Soole KL, Ford CM. The role of light in the regulation of ascorbate metabolism during berry development in the cultivated grapevine *Vitis vinífera* L. Journal of the Science of Food and Agriculture 2011; 91(9): 1712-21.

CHAPTER 5

Phenolics in Grape Berry and Key Antioxidants

S. D. Castellarin[1,*], L. Bavaresco[2,3], L. Falginella[1], M. I. V. Z. Gonçalves[3], G. Di Gaspero[1]

[1]Dipartimento di Scienze Agrarie e Ambientali, University of Udine, Italy; [2]Centro di Ricerca per la Viticoltura, Consiglio per la Ricerca e la Sperimentazione in Agricoltura, Conegliano, Italy and [3]Istituto di Frutti-Viticoltura, Università Cattolica del Sacro Cuore, Piacenza, Italy

Abstract: Phenolic compounds are important components of the grape berry in the determination of wine style and quality. In the past decade, significant advances towards a better understanding of the genetics, biochemistry, and physiology governing the synthesis of this class of secondary metabolite have been made. This deeper knowledge of phenylpropanoid and flavonoid metabolism strengthens the foundation for practical applications in the vineyard; investigations involving cultural practices and manipulation of environmental effects can help viticulturists deliver grapes to winemakers that are better suited to particular enological objectives, as well as possibly enriching the product in health-promoting compounds.

Keywords: Anthocyanins, Antioxidants, Flavan-3-ols , Flavonoid pathway, Flavonoids, Glutathione, Hydroxycinnamic acids, Phenolic compounds, Phenylalanine ammonia lyase, Proanthocyanidin, Resveratrol, Salicylic acid, Stilbenes.

INTRODUCTION

The term 'phenolic' is used to define any substance that possesses one or more hydroxyl (OH) substituents bonded to a C6 aromatic ring. This class of compounds is named after the substance with the simplest molecular structure, phenol, which consists of a phenyl ($-C_6H_5$) group bonded to an OH group. Polyphenols are phenolics with more than one phenol unit per molecule.

Phenolics are a diverse group of secondary products with a broad range of biological functions. These functions include roles in plant reproduction through attraction of pollinators and seed dispersers, in protection against biotic and abiotic stresses, and in responses to diverse environmental cues.

Grape berries accumulate a vast array of phenolic compounds, mostly polyphenols. Phenolic composition is highly diversified among different varieties and the environments where they are grown. Berry phenolics contribute to wine quality and have beneficial effects in many aspects of human health [1, 2]. Three main classes of phenolic compounds are synthesised in grape berry: phenolic acids, stilbenes, and flavonoids. In this chapter, we will review the chemical composition of phenolics in grape berry and the pattern of their accumulation, the extent of diversity in phenolic composition of the grapevine germplasm, and the current understanding of the molecular basis of their synthesis, including the effect of environmental factors.

PHENOLICS RELEVANT TO BERRY QUALITY

Phenolic Acids

Simple phenols are molecules with a single C6 ring, which include benzoic acids (C6-C1) with a carbon bonded to the phenyl group, and hydroxycinnamic acids (C6-C3) with a C3 skeleton attached to the aromatic ring.

Benzoic acids in grape berry include gentisic, salicylic, and gallic acid. Gentisic acid is accumulated at very low levels. Salicylic acid (SA) and its volatile ester methyl salicylate are signalling molecules mediating plant response to pathogen infections. SA has anti-inflammatory properties, but it is present in traces of 40

***Address correspondence S. D. Castellarin:** Dipartimento di Scienze Agrarie e Ambientali, University of Udine, Italy; E-mail: simone.castellarin@uniud.it

Hernâni Gerós, M. Manuela Chaves and Serge Delrot (Eds)

µg per kg of berry fresh weight (FW), which makes SA intake from grape products almost negligible. Gallic acid is present in a range of ~2 to ~13 mg per kg of berry FW, predominantly in seeds where it frequently forms esters with catechins.

Hydroxycinnamic acids are the major class of phenolic acids in grape berry. In terms of concentration, *p*-coumaric, caffeic, and ferulic acids are predominant. These three hydroxycinnamic acids differ by the type and number of substituents of the aromatic ring (Fig. **1**). They are present primarily as *trans* isomers, but traces of *cis* isomers have been detected. Hydroxycinnamic acids are esterified with tartaric acid, and thus named coutaric acid (*trans-p*-coumaroyl-tartaric acid), caftaric acid (*trans*-caffeoyl-tartaric acid), and fertaric acid (*trans*-feruloyl-tartaric acid). Esterification of *p*-coumaric acid with anthocyanin-3-glucosides can also occur in the skin of red varieties.

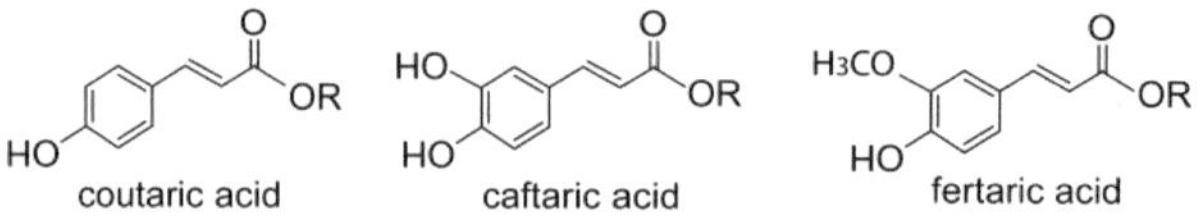

Figure 1: Structures of hydroxycinnamates in grape berries. R is tartaric acid.

Hydroxycinnamic acids accumulate in all berry tissues, predominantly in flesh. They are the major class of phenolics in grape juice and white wines [3, 4]. Caftaric acid is the main hydroxycinnamate ester in both white and red grapes [5]. Synthesis of hydroxycinnamates occurs mainly before veraison. During ripening, the content of hydroxycinnamates per berry remains almost constant, while their concentration decreases with increasing fruit size and dilution of solutes [6]. The average concentration of hydroxycinnamates in ripe berries is dependent on the variety, and is on average ~150 mg per kg of berry FW. Hydroxycinnamates do not play any direct role in the taste of wines, but eventually they undergo oxidation into quinones, an enzymatic process driven by polyphenol oxidase. uinones react with other compounds causing browning.

Stilbenes

Stilbenes occur naturally in a few edible plants. Several species of the genus *Vitis*, including *V. vinifera*, are capable of producing stilbenes [7-9] (Fig. **2**). Stilbenes are constitutively synthesised in healthy grape berries [10]. Stilbene content increases from veraison to ripening, and significant differences in final content exist among varieties. Their synthesis also increases upon pathogen infection and in response to abiotic stresses. Some stilbenes, particularly resveratrol, have long been known for their benefits to human health. High resveratrol producers include Pinot noir, Lemberger, and Marsanne, while low levels of resveratrol accumulate in Aglianico, Schiava Grossa, and Nebbiolo grapes.

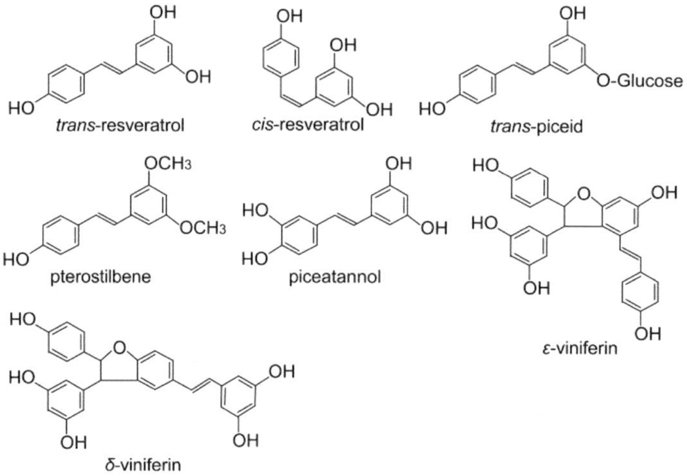

Figure 2: Structures of stilbenes in grape berries.

Trans-resveratrol (3, 5, 4'-trihydroxystilbene) is the stilbene with the simplest molecular structure. *Trans*-resveratrol itself has antimicrobial activity, and it is the parent stilbene that is converted into other compounds through various modifications of the stilbene unit. *Cis*-resveratrol is an isomer of *trans*-resveratrol. This compound is less stable than its precursor, due to steric hindrance between the aromatic rings.

Glycosylation of stilbenes is functional to storage, translocation, modulation of antifungal activity and protection from oxidative degradation. Glycosylated derivatives of resveratrol include piceid, *trans*- and *cis*-resveratrol-3-*O*-*β*-*D*-glucopyranoside, as well as astringin which is a 3'-OH-*trans*-piceid.

Methylation is a modification of the stilbene unit by the addition of methyl groups. Pterostilbene (3,5-dimethoxy-4'-hydroxystilbene) is a dimethylated derivative of resveratrol with enhanced fungitoxic activity with respect to its precursor.

Oligomerisation is another common process that converts stilbene units into more complex compounds. Many stilbenes are dimers, trimers, and tetramers that originate from oxidative coupling of resveratrol or resveratrol derivatives. A major group of resveratrol oligomers consists of viniferins. Viniferins are produced through oxidation of the basic stilbene by 4-hydroxystilbene peroxidases. The most important viniferins are *α*-viniferin, a cyclic dehydrotrimer of resveratrol, *β*-viniferin, a cyclic dehydrotetramer of resveratrol, *γ*-viniferin a more polymerised oligomer of resveratrol, *δ*-viniferin, an isomer of the resveratrol dehydrodimer, and ε-viniferin, a cyclic dehydrodimer of resveratrol. Among them, ε-viniferin is the major stilbene synthesised in berries infected with *Botrytis cinerea* [11]. Furthermore, synthesis of ε-viniferin and *δ*-viniferin is induced upon *Plasmopara viticola* infection or UV irradiation [12].

Several other stilbenes accumulate in grape berry. Piceatannol (*trans*-3, 4, 3', 5'-tetrahydroxystilbene) is a hydroxylated derivative of resveratrol, which has antioxidant properties and participates in reactive oxygen species (ROS) scavenging. Isomeric and glycosylated forms of the compounds mentioned above, such as hopeaphenol, resveratroloside, resveratrol-4'-*O*-*β*-*D*-glucopyranoside, and resveratrol di- and tri-glucoside derivatives have also been identified in trace amount.

Flavonoids

Flavonoids are C6-C3-C6 polyphenolic compounds, in which two hydroxylated benzene rings, A and B, are joined by a three carbon chain that is part of a heterocyclic C ring (Fig. **3**). Flavonoids are divided into structural classes that include flavonols, flavan-3-ols, and anthocyanins, according to the oxidation state of the C ring. The number of structural variants within each class is increased by modifications of stereochemistry, the position and type of substituents (including hydroxyl, methyl, galloyl, and glycosyl groups, and aliphatic and aromatic acids), and the degree of polymerisation of basic units. Total content and composition of flavonoids is highly variable across genotypes, and it is also modulated to some extent by biotic and abiotic factors.

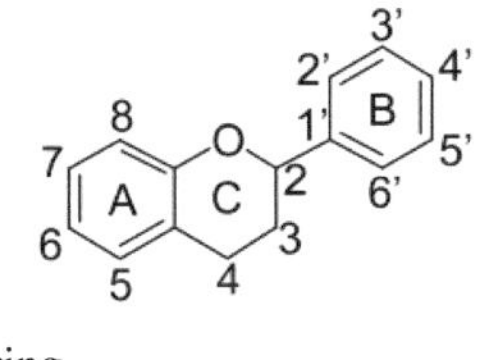

Figure 3: Flavonoid ring structure and numbering.

Flavonols

Flavonols are a class of flavonoids with a 3-hydroxyflavone backbone. The flavonols synthesised in the grape berry are kaempferol, quercetin, myricetin, and their methylated forms isorhamnetin, laricitrin, and syringetin [13, 14] (Fig. **4**). They differ by the number and type of substituents on the B ring, and occur normally as glucosides, galactosides, rhamnosides, and glucuronides. The sugar is bonded to the 3 position of the flavonoid skeleton.

In the grape berry, flavonols are synthesised only in the skin, where they serve as UV-protecting agents. Synthesis of flavonols begins in flower buttons, it levels off during early fruit development, and it resumes during fruit ripening. Flavonols reach the peak amount per berry a few weeks after veraison [15]. As for their concentration per kg of FW, the highest values occur a few days after blooming and decrease during fruit development due to a dilution effect following the increase in berry size.

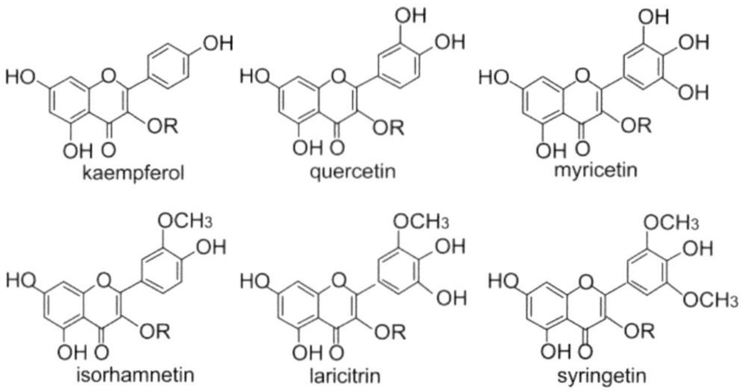

Figure 4: Structures of flavonols in grape berries. R is glucose, galactose, rhamnose, or glucuronic acid.

Flavonol concentration varies extensively among varieties, ranging from 1 to 80 mg per kg of berry FW. Flavonol content is on average higher in red varieties than in whites. However, flavonol content can be strongly affected by environmental factors, particularly sunlight exposure [15]. As for flavonol composition, white varieties synthesise only the mono-hydroxylated kaempferol, the di-hydroxylated quercetin, and its methoxylated form isorhamnetin, while red grapes also accumulate the tri-hydroxylated myricetin and its methoxylated forms laricitrin and syringetin [13]. Quercetin is the main flavonol in white grapes and in some red varieties, such as Pinot noir and Nebbiolo. Other reds such as Cabernet Sauvignon and Tempranillo predominantly accumulate myricetin. Methoxylated flavonols are present in low amounts.

Flavonols are yellow pigments that contribute to the colour of white wines. In reds, flavonols can conjugate to anthocyanins, a process called copigmentation that reinforces the stability of wine colour [16].

Flavan-3-ols and Proanthocyanidins

The most abundant class of phenolics in grape berries is polymeric flavan-3-ols, also known as proanthocyanidins or condensed tannins, which originate from monomeric flavan-3-ols. Flavan-3-ols, also known as catechins, are characterised by the presence of a hydroxyl group at the 3 position of the C ring (Fig. **5**). The five flavan-3-ols in grapes are (+)-catechin and its isomer (-)-epicatechin, (+)-gallocatechin, (-)-epigallocatechin, and (-)-epicatechin-3-*O*-gallate [17]. Flavan-3-ols are detectable in every part of the berry, with the highest concentration in seeds, followed by skin, and to a much lesser extent in flesh [18, 19].

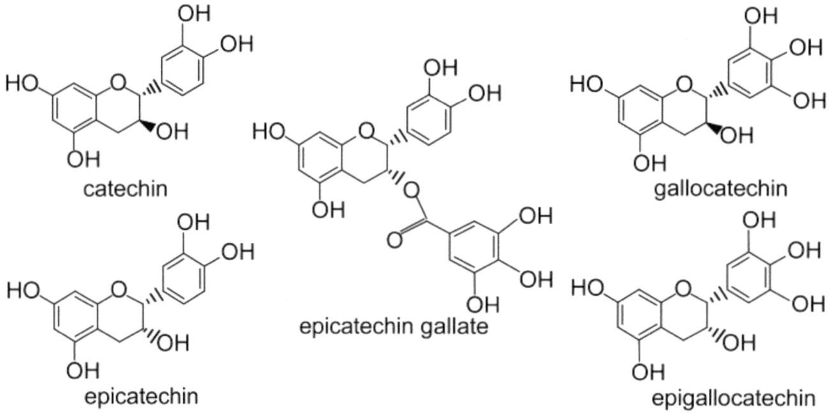

Figure 5: Structures of flavan-3-ols in grape berries.

Proanthocyanidins are polymers composed of flavan-3-ol subunits that are linked *via* 4-6 and 4-8 interflavan bonds (Fig. **6**). Proanthocyanidins are a very diverse group of compounds. They vary in size depending on the number of subunits, ranging from dimers to polymers with more than 40 subunits [20, 21]. Seed and berry skin have striking differences in the composition of flavan-3-ols and proanthocyanidins. Di-hydroxylated flavan-3-ols, catechin, epicatechin, and the galloylated form, epicatechin-3-*O*-gallate, predominate in seeds. Berry skins also contain these forms, as well as tri-hydroxylated flavan-3-ols, gallocatechin, and epigallocatechin.

Composition of flavan-3-ols and proanthocyanidins in seeds and berry skin is diversified among varieties but relatively stable within each variety, suggesting a prominent genotype effect [17]. Concentration of proanthocyanidins is much higher than flavan-3-ols. In seeds and skins of Syrah grapes, the concentration of flavan-3-ols is ~3 and ~0.1 g per kg of berry FW, respectively, while proanthocyanidin concentration in seed and skin is ~25 and ~5 g per kg of berry FW, respectively.

Synthesis of flavan-3-ols and proanthocyanidins commences in flower buttons and proceeds in developing seeds and skins during fruit growth. Their synthesis is completed by veraison in skins and a few weeks after veraison in seeds. As a result, the content of flavan-3-ols and proanthocyanidins remains stable or decreases slightly during fruit ripening [20-24]. Flavan-3-ols and proanthocyanidins are predominantly found in the hypodermal cell layers of berry skin and in the soft parenchyma of the seed coat, within the vacuole or bound to cell wall polysaccharides [25-27].

Flavan-3-ols, proanthocyanidins, and their complexes with polysaccharides have a significant impact on wine quality. All of them contribute to mouth-feel and smooth sensory perceptions [28]. In red wines, proanthocyanidins can also conjugate to anthocyanins in the process of copigmentation [16].

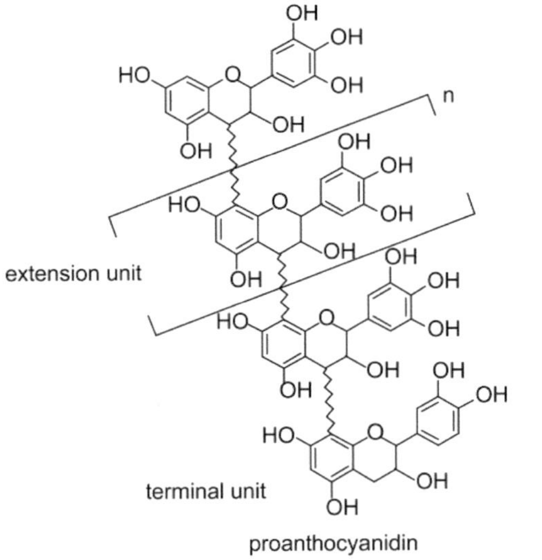

Figure 6: Structure of a hypothetical proanthocyanidin.

Proanthocyanidin average size is also known as mean degree of polymerisation (mDP). Proanthocyanidin mDP fluctuates during fruit development and ripening [20, 23, 29]. In Syrah, the mDP of skin proanthocyanidins may increase from 7.3 before veraison to 27 in ripe berries. In another study, Syrah grapes reached the highest mDP value of 40 six weeks before veraison and then decreased with the progression of ripening to a value of 20. The mDP of seed proanthocyanidins followed a similar trend, but with values much lower than in skins. In ripe Pinot noir grapes, the mDP ranged from 5.9 to 9.4 in seeds, and from 27 to 42 in skins across three seasons. Correlation between proanthocyanidin content and mDP in harvested berries and the same parameters measured in wine is poor. The mDP in wines is normally lower than in berries, since only a variable fraction of flavan-3-ols and proanthocyanidins present in grapes is extracted during vinification [17, 23].

Anthocyanins

Anthocyanins are responsible for red, purple, and blue pigmentation of grape berries. Anthocyanins are derivatives of the flavylium cation. They are synthesised from anthocyanidins by glycosylation at the 3 and 5 positions of the C ring. *V. vinifera* has only 3-*O*-monoglucosides due to two disruptive mutations in the anthocyanin 5-*O*-glucosyltransferase gene, producing a nonfunctional form of the enzyme that normally performs 5-glycosylation in other grapevine species [30].

The basic structure of the aglycone is a C6-C3-C6 skeleton. There are 17 naturally occurring aglycones, but only six are reported in grapevine: cyanidin, peonidin, malvidin, delphinidin, and petunidin are invariably found in all varieties, and traces of pelargonidin have been detected in Pinot noir and Cabernet Sauvignon [31] (Fig. **7**). The differences in chemical structure of the aglycones occur at the 3' and 5' positions of the B ring, due to hydroxyl or methoxyl substitutions. Additional structural variation is caused by the nature of aliphatic or aromatic acids (acetic acid, coumaric acid, caffeic acid) linked to the 6' position of the glucose bonded to the 3 position of the C ring. The grape berries of, for example, Cabernet Sauvignon accumulate 18 anthocyanins, including the six 3-*O*-monoglucosides, five acetylated anthocyanins, six *p*-coumaroylated anthocyanins, and one caffeoylated anthocyanin. All of them are derived from either the di-hydroxylated cyanidin or the tri-hydroxylated delphinidin through subsequent steps of methylation and acylation.

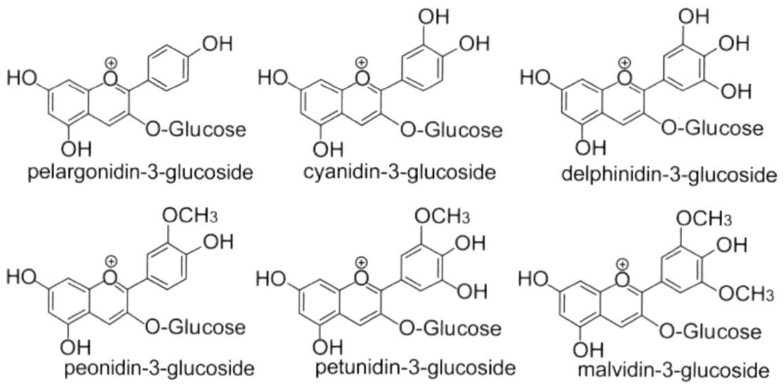

Figure 7: Structures of anthocyanin-monoglucosides in grape berries.

Anthocyanin composition is particular to each variety and relatively stable under different growing conditions [13, 32]. Most red varieties have a predominance of tri-hydroxylated anthocyanins, and within this subclass the di-methoxylated malvidin is the prevalent form. A few varieties, including Muscat Rouge de Madere, Moscato Rosa, and Nebbiolo, accumulate predominantly di-substituted anthocyanins. Mono-substituted anthocyanins are present only in traces in the form of the monoglucoside pelargonidin. The extent of the occurrence and the type of acylation of the glucose moiety also vary among varieties [13]. Monoglucoside anthocyanins with no acylation usually predominate over acetylglucosides (formed by acylation with an acetyl group) and *p*-coumarilglucosides (acylation with a *p*-coumaril group) (Fig. **8**). In extreme cases, such as the varieties Ciliegiolo and Sangiovese, the acylated anthocyanin content is extremely low, and they are barely detectable in Pinot noir, Pinot gris, Muscat Rouge de Madere, and Gewürztraminer.

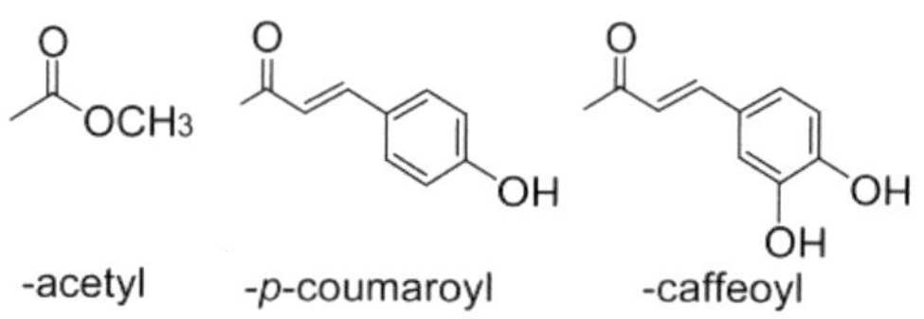

Figure 8: Acyl groups linked to anthocyanin-monoglucosides in grape berries.

Anthocyanins are accumulated in berry skins of red varieties from veraison until full maturity, when the rate of synthesis levels off. Anthocyanins are synthesised in the cytosol of berry epidermal cells and then stored in the vacuole. In a few teinturier varieties, accumulation in berry skin is paralleled by accumulation in flesh. The anthocyanin content in ripe grapes varies substantially with the genotype: in a set of 64 red varieties, anthocyanins ranged from ~26 mg per kg of berry FW in Muscat Rouge de Madere to ~6300 mg per kg of berry FW in Casetta, a locally grown variety in Italy [13]. However, anthocyanin content is more sensitive than anthocyanin composition to environmental effects.

Every anthocyanin has a particular absorbance spectrum and thus has a particular hue ranging from red to blue. Delphinidin derivatives are associated with blueness, cyanidin derivatives are reddish. Methylation may affect colour stability, as anthocyanin-*O*-methylation reduces the chemical reactivity of phenolic hydroxyl groups [33]. The combination of quantity of anthocyanins and anthocyanin profile contributes to the intensity and hue of the colour of fruit and wines [34].

BIOSYNTHESIS OF PHENOLICS

Phenolics are all synthesised from the amino acid phenylalanine through the phenylpropanoid and flavonoid pathways [35]. Phenylalanine is a product of the shikimate pathway, which links carbohydrate metabolism with the biosynthesis of aromatic amino acids and secondary metabolites.

Biosynthesis of Phenylpropanoids

The general phenylpropanoid pathway converts phenylalanine into 4-coumaroyl-CoA through three successive enzymatic reactions. The first reaction involves deamination of phenylalanine by the enzyme phenylalanine ammonia lyase (PAL), which produces cinnamic acid. In the second step, cinnamic acid is converted into *p*-coumaric acid by a hydroxylation at the 4 position by cinnamate-4-hydroxylase (C4H). Then, the enzyme 4-coumaroyl:CoA-ligase (4CL) catalyses the esterification of *p*-coumaric acid with CoA, which produces 4-coumaroyl-CoA. Intermediates of the general phenylpropanoid pathway give rise to phenolic acids. The end-product of the phenylpropanoid pathway, 4-coumaroyl-CoA, is the substrate of two enzymes, chalcone synthase (CHS) and stilbene synthase (STS), which control the enzymatic reactions at the entrance of the flavonoid pathway and stilbene pathway, respectively (Fig. **9**).

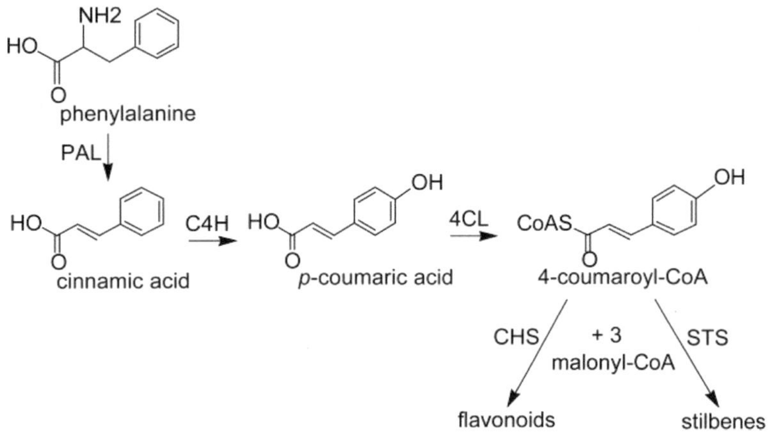

Figure 9: General phenylpropanoid pathway.

Modifications to intermediates of the phenylpropanoid pathway generate hydroxycinnamic acids. Caffeic acid originates from 3-hydroxylation of *p*-coumaric acid. Caffeic acid can be converted by 3-methylation into ferulic acid. In grape berries, hydroxycinnamic acids form an ester with tartaric acid. Esterification occurs *via* the activated CoA phenols of the corresponding hydroxycinnamic acid: caftaric acid originates from caffeoyl-CoA, coutaric acid from *p*-coumaroyl-CoA, and fertaric acid from feruloyl-CoA [3]. Benzoic acids may be derived from phenylpropanoid intermediates. Formation of gallic acid, the most relevant benzoate, occurs through shortening of the side chain of cinnamic acid through β-oxidation and, as an alternative to the phenylpropanoid route, by dehydrogenation of shikimic acid.

Biosynthesis of Flavonoids

The flavonoid pathway leads to the synthesis of different classes of metabolites: flavonols, flavan-3-ols, proanthocyanidins, and anthocyanins (Fig. **10**). This biosynthetic pathway is highly conserved in higher plants. All flavonoids stem from tetrahydroxychalcone, which is synthesised by the enzyme chalcone synthase (CHS) using one molecule of 4-coumaroyl-CoA, the end-product of the phenylpropanoid pathway, and three molecules of malonyl-CoA. CHS catalyses the condensation and the partial cyclisation of three acetate units onto 4-coumaroyl-CoA, forming the A and B rings. Chalcone isomerase (CHI) accelerates the formation of the C ring to produce naringenin, the earliest intermediate with the typical flavonoid core.

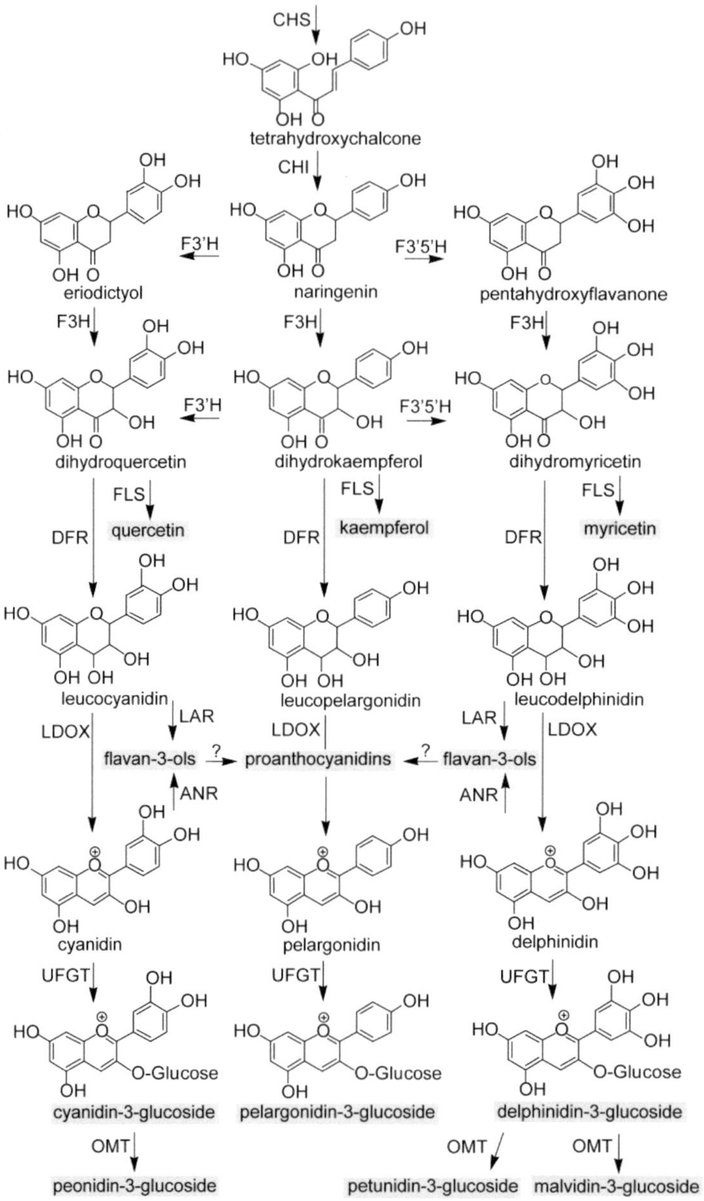

Figure 10: Flavonoid pathway.

Naringenin is a substrate for several enzymes. It may undergo hydroxylation at the 3 position by the activity of flavanone-3-hydroxylases (F3H), forming dihydrokaempferol. It may also undergo hydroxylation at the 3' and 3',5' positions by the activity of flavonoid-3'-hydroxylase (F3'H) and flavonoid-3',5'-hydroxylase (F3'5'H), which catalyse the conversion into eriodictyol and pentahydroxyflavanone, respectively. Eriodictyol and pentahydroxyflavanone are then converted into dihydromyricetin or dihydroquercetin by the activity of F3H. Dihydrokaempferol is also a substrate for F3'H and F3'5'H, which catalyse the hydroxylation into dihydromyricetin or dihydroquercetin.

At this point in the pathway, the three dihydroflavonols, *i.e.* dihydrokaempferol, dihydroquercetin, and dihydromyricetin, are substrates for a number of enzymes which partition them into different downstream pathways. Dihydroflavonols may become substrates for the enzyme flavonol synthase (FLS), which catalyses the formation of the flavonols kaempferol, quercetin, and myricetin. Flavonols are normally glycosylated at the 3 position of the C ring. Substrate specific 3-glycosyl transferases have not been isolated so far, but a UDP-glucose:flavonoid 3-*O*-glucosyl transferase (UFGT) with a high affinity for anthocyanidins also accepts flavonol substrates.

Dihydroflavonols may also enter the pathways leading to the formation of anthocyanins and flavan-3-ols. The first committed step is the reduction of dihydroflavonols to flavan-3,4-ols, which is catalysed by a dihydroflavonol reductase (DFR) [36]. The activity of DFR on 3'-OH dihydrokaempferol sends the product into the branch of the pathway that leads to pelargonidin, the activity of DFR on 3',4'-OH dihydroquercetin leads to cyanidin, and the activity of DFR on 3',4',5'-OH dihydromyricetin leads to delphinidin. Flavan-3,4-ols, also known as leucoanthocyanidins, are converted into anthocyanidins by the enzyme leucoanthocyanidin dioxygenase (LDOX), also known as anthocyanidin synthase [37]. Leucoanthocyanidins and anthocyanidins may enter the flavan-3-ol pathway. Two enzymes, leucoanthocyanidin reductase (LAR) and anthocyanidin reductase (ANR), convert leucoanthocyanidins and anthocyanidins into (+)-catechin and (-)-epicatechin, respectively. *ANR* is also named *BAN* after its *BANYULS* homolog in Arabidopsis. Flavan-3-ols can then polymerise into proanthocyanidins. This process has important implications for wine quality, but the underlying biochemistry of polymerisation remains largely unknown.

Anthocyanidins with a hydroxyl group at the 3 position of the C ring are unstable. Instead of undergoing reduction by ANR into epicatechin, anthocyanidins can be also be stabilised by the addition of a glucose at the 3 position of the C ring, resulting in anthocyanin production. The formation of glucosides is catalysed by UFGT. This enzyme accepts both anthocyanidins and flavonols as substrates, but its catalytic efficiency is much higher for anthocyanidins [38].

UFGT controls the entry point into the terminal part of the flavonoid pathway, leading to the formation of anthocyanins [39, 40]. The primary di-hydroxylated and tri-hydroxylated anthocyanins that are synthesised by UFGT are cyanidin-3-*O*-glucoside and delphinidin-3-*O*-glucoside. The 3'-OH of cyanidin and delphinidin and their 3-*O*-glucosides, and eventually the 5'-OH of delphinidin-3-*O*-glucoside, can be methylated by an *O*-methyltransferase (OMT) generating peonidin, petunidin, and malvidin, and their 3-*O*-glucosides [41, 42]. The observation that acylated anthocyanins are present in most red grapes suggests the presence of a still elusive anthocyanin acyltransferase.

Anthocyanins are synthesised in the cytosol and delivered into the vacuole, where they are stored as coloured coalescences called anthocyanic vacuolar inclusions (see the chapter by Fontes *et al.*, this edition). Vacuolar uptake may depend on multiple mechanisms, either mediated by tonoplast transporters or based on vesicular trafficking. Transporter-mediated uptake may rely on two mechanisms, for both of which some experimental evidence has been provided. MATE-type proteins are localised to the tonoplast and function as vacuolar H^+-dependent transporters of acylated anthocyanins [43]. ATP-binding cassette proteins are glutathione *S*-conjugate pumps involved in the uptake of glycosylated flavonoids, independent of the presence of an H^+ gradient. Glutathione S-transferases (GST) are thus expected to participate in vacuolar trafficking. Among them, a member of the *GST* gene family has overlapping patterns of transcription with anthocyanin accumulation [44, 45]. Another putative anthocyanin carrier with high similarity to mammalian bilitranslocases has been isolated from grape berries [46]. Although evidence exists for a number of transporters, it remains unknown whether anthocyanins enter the vacuole as single molecules and then aggregate, or if cytoplasmic vesicles containing coalesced anthocyanins interact with the tonoplast.

Biosynthesis of Stilbenes

The first committed step of the stilbene pathway is controlled by stilbene synthase (STS). STS accepts the same substrates as chalcone synthase: 4-coumaroyl-CoA, the end-product of the phenylpropanoid pathway, and 3 molecules of malonyl-CoA. In a similar way as CHS, STS carries out three reactions of condensation that produce resveratrol, a molecule with two aromatic rings. In the STS reaction, the terminal carboxyl group is removed prior to closure of the A ring, which causes different ring-folding in resveratrol compared to tetrahydroxychalcone, the product of CHS. CHS and STS thus compete for the same substrates and control the entry point into the flavonoid pathway and the stilbene pathway, respectively.

Flavonoids are ubiquitous in higher plants. In contrast, only a few phylogenetically unrelated species produce stilbenes. STS and CHS are closely related enzymes sharing up to 79% amino acid identity, but specifically control flavonoid or stilbene biosynthesis. This limited occurrence in nature has raised the question as to whether STS has independently evolved from the ubiquitous CHS in diverse plant lineages, or if it is more ancient and has been retained only by the stilbene-producing species. STS evolved from CHS several times in higher plants and a few amino acid exchanges are sufficient to switch from chalcone to stilbene synthesis [47]. After its formation in the grapevine lineage, the STS gene has massively proliferated in the grapevine genome mostly by local duplications, giving rise to a family of 43 members. It is currently unknown to what extent each gene copy is redundant or if any specialisation has occurred. In contrast, only three copies of CHS are present in the grapevine genome, with diversified patterns of expression in the grape berry [48].

Little is known about other enzymatic steps that convert resveratrol into downstream derivatives. A resveratrol *O*-methyltransferase (ROMT) catalyses the conversion of resveratrol into the highly fungitoxic pterostilbene [49]. A glucosyltransferase is capable of producing glucosides of *cis*- and *trans*-resveratrol, and also has residual glucosyl activity on hydroxycinnamic acids and some flavonoids [50].

Regulation of the Flavonoid Pathway

Regulation of the Flavonol Pathway

Flux of flavonoid intermediates towards flavonol synthesis is controlled by flavonol synthases (FLS) [15, 51]. Five *FLS* genes are present in the grapevine genome. Two isoforms sharing 80% nucleotide identity are transcribed in skins of ripening berries, with expression profiles that mirror the accumulation of flavonols.

Expression of *FLS* is specifically activated by the R2R3-MYB transcription factor VvMYBF1. The promoter of *VvMYBF1* contains light regulatory elements, which renders *VvMYBF1* expression extremely sensitive to light induction [52].

Glucosylation of flavonols is partially attributed to the aspecific activity of glucosyltransferases that primarily use similar flavonoid substrates. UFGT glucosylates quercetin with the maximum activity, and to a lesser degree myricetin and kaempferol. However, when compared to anthocyanidins, quercetin is glucosylated with a k_{cat} 48 times lower than cyanidin [38]. In a similar way, another glucosyltransferase that preferentially glucosylates resveratrol is capable of producing glucosides of quercetin and kaempferol, but it does not accept the anthocyanidin cyanidin as a substrate [50].

Differences in flavonol composition across genotypes and in response to light are also attributed to the differential activity of flavonoid 3'- versus 3'5'-hydroxylase enzymes, which dictates the relative proportion of precursors of either quercetin or myricetin. Methylated flavonols are generated by the activity of an *O*-methyltransferase that exhibits equivalent kinetic parameters with quercetin 3-*O*-glucoside and the aglycones quercetin and myricetin, as well as with anthocyanins [41]. Also, two *O*-methyltransferases characterised for their capability of converting hydroxypyrazine precursors into methoxypyrazines have an even greater activity on quercetin [53].

Regulation of Flavan-3-ol and Proanthocyanidin Pathways

Leucoanthocyanidin reductase (*LAR*) and anthocyanidin reductase (*ANR*) control the branching points of the flavonoid pathway that lead to the formation of flavan-3-ols. In the grapevine genome, there are two highly similar copies of *LAR*, while *ANR* is a single-copy gene.

LAR1, *LAR2*, and *ANR* are expressed in berry skin from blooming until veraison, consistent with the accumulation pattern of flavan-3-ols [19, 54]. Transcription of *LAR1*, *LAR2*, and *ANR* declines by the time the synthesis of flavan-3-ols is complete and before flavonoid metabolism shifts to the synthesis of anthocyanins. This temporal separation prevents competition between LAR/ANR and UFGT for common substrates in red grapes. Trends of expression are not dissimilar in seeds, where flavan-3-ol accumulation continues a bit longer (until 2 to 4 weeks after veraison) and transcripts of *LAR2* persist throughout veraison [54]. In flesh, which marginally contributes to berry flavan-3-ols, *LAR2* and *ANR* are weakly expressed before veraison [19].

Expression of *LAR1* and *ANR* is regulated by the Myb transcription factors VvMybPA1 and VvMybPA2 [55, 56]. VvMybPA1 and VvMybPA2 activate the promoters of *LAR1* and *ANR* triggering their expression, along with the expression of early genes of the flavonoid pathway, but they do not activate the promoter of *LAR2* (the other terminal step in flavan-3-ols synthesis) or the *UFGT* promoter. However, expression patterns of *VvMybPA1* overlap more precisely with the expression of *LAR1*, *LAR2*, and *ANR* in seeds, while expression patterns of *VvMybPA2* closely match the expression of *LAR1*, *LAR2*, and *ANR* in berry skin. Another MYC-like basic Helix-Loop-Helix transcription factor, *VvMYC1*, participates in the control of flavan-3-ol synthesis [57]. *VvMYC1* can independently activate the *ANR* promoter, but it also physically interacts with *VvMybPA1* and, in this combination, activation of the *ANR* promoter increases 8-fold compared with *VvMYC1* alone. In general, expression profiles of *VvMYC1* follow flavan-3-ol synthesis patterns in both seeds and skins, as well as the pattern of anthocyanin synthesis. In seeds, where only flavan-3-ols are synthesised, *VvMYC1* expression is 10 times greater than in skins.

Regulation of the Anthocyanin Pathway

Expression of *UFGT* is developmentally regulated at the onset of ripening, and it is required for anthocyanin pigmentation. A functional copy of *UFGT* is present in the genome of red and white grape varieties, although whites do not synthesise anthocyanins in the berry.

Transcription of *UFGT* is regulated by the additive activity of two MybA-type transcription factors, *VvMybA1* and *VvMybA2*, which reside within a MybA gene cluster of four copies. This locus accounts for 62 % of the variance in anthocyanin content in linkage mapping experiments [58]. In a highly diverse core collection of grapevine varieties, five DNA polymorphisms within *VvMybA1*, *VvMybA2*, and *VvMybA3* account for 84% of the variance in anthocyanin content in those varieties. The allelic variation caused by the insertion of the *Gret1* Gypsy-type retrotransposon into the promoter of *VvMybA1* alone explains 59 % of that variation. *Gret1* prevents *VvMybA1* transcription [59]. Reversion of this *VvMybA1* silencing can occur, albeit infrequently, in red somatic mutants of white seedlings due to the excision of *Gret1*. The *Gret1* insertion has occurred rather recently in the evolutionary history of the grapevine lineage, in a haplotype that had already undergone two disruptive mutations in the coding sequence of the adjacent *VvMybA2* [60, 61]. Unlike the wild-type functional allele, which has a G nucleotide at a critical position of the coding sequence in the R2R3 recognition domain, a G→T mutated allele of *VvMybA2* encodes a protein with a non-conservative amino acid change, rendering it unable to trigger *UFGT* transcription [58, 61]. Genotypes homozygous for the haplotype carrying *Gret1 VvMybA1* and the mutated T *VvMybA2* do not synthesise berry anthocyanins. Genotypes with at least one wild-type haplotype carrying *VvMybA1* with no *Gret1* insertion and the functional G *VvMybA2* allele are able to fully activate the promoter of *UFGT*. The haplotype with no *Gret1* insertion but containing the mutated T *VvMybA2* allele is able to partially bind the promoter of *UFGT*. In a few white table grape varieties of eastern Mediterranean origin, *e.g.* Sultanina, lack of anthocyanin synthesis is caused by an unknown additional mutation [62, 63].

The functional role of wild-type *VvMybA1* and *VvMybA2* in the regulation of anthocyanin synthesis has been confirmed by white-skinned bud sports of heterozygous red cultivars, such as Pinot noir and Cabernet

Sauvignon, whose phenotype is caused by large deletions removing *VvMybA1* and *VvMybA2* from the wild-type haplotype [64, 65].

VvMybA1 regulates expression of UFGT and other early genes of the flavonoid pathway and, much more strongly, a narrow set of genes relevant to anthocyanin metabolism: *F3'5'H*, *OMT*, *GST*, and a vacuolar anthocyanin MATE-transporter [66, 67].

F3'5'H and F3'H compete with each other for the delivery of flavonoid precursors into the parallel pathways leading to synthesis of tri-hydroxylated and di-hydroxylated anthocyanins, respectively [68]. F3'5'H is encoded by a transcriptionally specialised family of 16 genes, while a single F3'H is expressed in grape berry [69]. Dark red grapes, which are more rich in tri-hydroxylated anthocyanins, transcribe more copies of *F3'5'H* and do so at a higher rate than reddish grapes. *F3'5'H*s are not expressed in white cultivars, which also fail to accumulate tri-hydroxylated flavonols, since their synthesis departs from the core flavonoid pathway downstream of the F3'5'H/F3'H branching point.

OMT methylates cyanidin and delphinidin into peonidin, petunidin, and malvidin [41, 42]. Activation of *OMT* expression is particularly intense in grapes that accumulate mostly methylated anthocyanins, such as Pinot gris, especially when normalised to the rate of anthocyanin synthesis using the rate of *UFGT* transcription [45].

GST has long been suspected to control glutathione conjugation of anthocyanins, a process that should favour vacuolar uptake through ATP-binding cassette transporters. GST is encoded by a gene family of nearly a hundred members, 55 of which are expressed at veraison in red grapes, and several of which are strongly induced at that stage compared with early fruit development [70].

Knowledge from studies of model systems would predict the participation of basic Helix-Loop-Helix and WD-repeat proteins in the transcriptional regulatory complex with Myb factors that activates anthocyanin synthesis. This is known to occur between the VvMYC1 transcription factor and VvMybA1 through physical interaction. The promoter of *UFGT* is induced ~40-fold after the co-expression of *VvMYC1* and *VvMybA1* [57]. An ectopically expressed WDR protein was found to enhance anthocyanin accumulation in Arabidopsis, and its expression pattern in grape skins follows those of *VvMYC1*, the *VvMybA*s, and *UFGT* [71]. *VvMYC1* also interacts with *Myb5a*/*Myb5b,* enhancing their activation of early genes in the flavonoid pathway [72, 73].

Regulation of the Stilbene Pathway

STS controls the entry point into the stilbene pathway. STS is encoded by a gene family of 43 highly similar paralogues. All *STS* genes are transcribed in healthy berries. This suggests a basal activation of stilbene synthesis during early fruit growth, in addition to a developmental induction at veraison [10, 70, 74]. *STS* transcription increases as ripening progresses and higher levels of *STS* expression are detected in varieties that are high resveratrol producers [10]. *STS* is expressed predominantly in berry skins, but it is not absent from the flesh. At the subcellular level, STS proteins are located along the plasma membrane, in the cell wall, in chloroplasts, and on the outer face of the endoplasmic reticulum [75, 76].

Expression of at least some *STS* copies is under the control of an R2R3 MYB transcription factor, which was named *VvMYB14* after its most similar homologue, *AtMYB14* in Arabidopsis [77]. Supporting evidence for the role of *VvMYB14* is provided by the observation that *VvMYB14* has overlapping patterns of expression with *STS* upon elicitation, and *VvMYB14* silencing significantly reduces *STS* transcription.

Less is known about the regulation of resveratrol conversion into downstream derivatives. The methylation of resveratrol into pterostilbene is controlled by a ROMT [49]. A glucosyltransferase participates in the formation of resveratrol glucosides [50].

THE INFLUENCE OF ENVIRONMENTAL FACTORS AND CULTURAL PRACTICES ON BERRY PHENOLICS

Because it is a quantitative trait, phenolic accumulation in grape berry is determined to a variable extent by genetic factors and by effects of the environment, and to some degree by the interaction between them.

For some flavonoids, synthesis largely depends on the variants of genetic determinants that are present in a given genotype and is little affected by external factors. This is the case for anthocyanin composition in ripe berries, a biochemical marker used for differentiating cultivars before the advent of DNA fingerprinting. In contrast, flavonol synthesis is highly related to the level of sunlight exposure, an external factor that predominates over genetic factors in the determination of flavonol content.

Environmental factors include all external stimuli, the most influential of which for phenolic synthesis are light and temperature, as well as others such as water status, nutritional status, and pathogenesis. These factors modulate grapevine physiology and may influence vine vigour, crop load, as well as the balance between photosynthetic carbon fixation and partitioning of assimilates to ripening berries. Environmental factors also include modulation of vine physiology imposed by cultural practices such as trellis systems, shoot positioning, defoliation, and cluster thinning. The impact of some external factors (*e.g.* water deficits, pathogen infections) on berry physiology and phenolic synthesis is mediated by endogenous plant growth regulators (abscisic acid, ethylene, MeJA, SA), whose roles have been studied alone and in relation to external stimuli.

Experimental designs carried out in the field are complicated by difficulties in separating multiple effects from each other, in order to study specific environmental and viticultural factors. This entails the careful disjoining of a network of interactions at the vineyard site. On the other hand, trials in growth chambers and greenhouses facilitate a better understanding of single processes using a well-controlled, simplified system, but artificial growing conditions fail to provide a full understanding of the relevance of a given factor in the natural context of the vineyard.

Phenolic accumulation is also influenced in vineyards by genotype × environment interaction. Some grape varieties ripen with similar phenolic composition under variable sites of cultivation, while others only perform well in narrow viticultural zones that guarantee optimal environmental conditions. Generally speaking, phenolic synthesis is part of a coordinated suite of changes that accompanies berry ripening. Thus, many endogenous factors and external cues that promote or interfere with ripening have a side-effect on phenolic synthesis.

Environmental Factors and Stilbene Synthesis

Stilbene synthesis is enhanced in response to several abiotic and biotic cues. Abiotic factors include UV-radiation, wounding, ozone, anoxia, and metal ions. Several microorganisms, their elicitors, and some endogenous factors released upon pathogen infection can serve as biotic stimuli.

Abiotic factors are usually tested under experimental conditions that are less likely to occur in the field. Exposure to UV light induces the accumulation of stilbenes in grape berry through the induction of *STS* expression [76, 78]. In berries, this is dependent on the developmental stage, since unripe berries respond to UV irradiation to a greater extent. Ozone treatments promote accumulation of pterostilbene and resveratrol. Short-term anoxia increases resveratrol content [79].

Several stilbenes have antimicrobial activity *in vitro* at concentrations normally detected in planta, and are claimed to reinforce defence responses against *B. cinerea* and *P. viticola*. Pathogen infections strongly induce transcription of *STS* and synthesis of resveratrol [12, 80, 81]. Antifungal activity is modulated by the subsequent conversion into a number of derivatives with variable toxicity. The accumulation of resveratrol begins shortly after pathogen inoculation. Grapevines that are more resistant to infections are capable of converting resveratrol into the more toxic viniferins with a higher efficiency than susceptible plants, in which resveratrol is preferentially glycosylated into the non-toxic form, piceid. An inverse correlation does exist between sporangial density of *P. viticola* and concentration of resveratrol, δ-viniferin, and ε-viniferin across different genotypes [82]. The main stilbenes involved in the response to *B. cinerea* are resveratrol, ε-viniferin, and α-viniferin [11].

Phytohormones that participate in defence-signalling normally induce stilbene synthesis. Exogenous MeJA induces accumulation of resveratrol and ε-viniferin [83]. SA stimulates resveratrol synthesis to a variable extent depending on the variety. Pre-harvest treatment with benzothiadiazole may increase resveratrol content by 40% due to the general induction of the phenylpropanoid pathway [84]. A number of other molecules such as chitosan oligomers, beta-aminobutyric acid, and emodin, as well as some microorganisms (*e.g. Trichoderma viride*, *Rhizopus stolonifer*, and *Aspergillus carbonarius*) may elicit stilbene synthesis.

Several viticultural factors are potential modulators of stilbene metabolism. These include soil type, altitude of the cultivation site, heat stress, defoliation, mineral supply, and controlled development of noble rot, all of which have shown some influence. However, consistent and substantial effects were only obtained by severely impacting the plant water status. Water deficits, induced either by growing vines with a limited water supply or by imposing a prolonged grape withering, have been shown to promote stilbene accumulation [85-88].

Environmental Factors and Flavonol Synthesis

Synthesis of flavonols is a light-dependent process. Sealing grape bunches in light-excluding boxes from before flowering until harvest completely inhibits flavonol synthesis [89]. If shading is applied later in fruit development, flavonol content is reduced and no further accumulation is detected past the initiation of light deprivation. Light directly affects the expression of *FLS*. Light induction of *FLS* transcription is mediated by the transcription factor VvMYBF1, which contains light-responsive regulatory elements in its promoter [52, 71]. UV radiation, particularly UV-B, most efficiently activates flavonol synthesis, which is a consequence of the biological role of these compounds in UV screening. In the vineyard, any cultural practices that favour the exposure of grape bunches to sunlight boosts flavonol accumulation. This occurs equally in white and red grapes. Flavonol synthesis is scarcely affected by other external stimuli.

Environmental Factors and Synthesis of Flavan-3-ols and Proanthocyanidins

Among the flavonoids, flavan-3-ols and proanthocyanidins are those with the most stability under diverse growing conditions. This especially holds true for the accumulation of these compounds in seeds.

In skins, sunlight may have a positive impact on the quantity of flavan-3-ols and proanthocyanidins, as well as on their mean degree of polymerisation [90]. Though these changes are detectable, the differences may not be statistically significant. Sunlight exposure more substantially impacts the proportion of proanthocyanidin subunit type by increasing the relative abundance of the tri-hydroxylated gallocatechins at the expense of the di-hydroxylated catechins. This effect is a consequence of light stimuli on the balanced expression of *F3'5'H*/*F3'H*, which affects the flavonoid pathway upstream of the flavan-3-ol branching point [91].

Heat summation over fruit development is linearly related to proanthocyanidin accumulation, consistent with the general belief that hot climates are conducive to optimal tannin profiles, particularly in reds. In the vineyard, cultural practices that favour the radiative heating of grape bunches tend to increase accumulation of proanthocyanidins [92].

Environmental Factors and Anthocyanin Synthesis

Temperature and light, and the combination of these two factors, which are not easily separated under natural conditions, have a significant impact on anthocyanin biosynthesis.

In growth chambers, optimal conditions for anthocyanin accumulation occur when grapes are exposed to diurnal fluctuations that simulate cool nights (15°C in the dark) and warm, but not excessively hot, days (25°C under light). Higher temperatures (30-35°C) not only inhibit transcription of anthocyanin synthetic genes, but also promote the degradation of existing anthocyanins, which results in a significant net loss [93, 94]. Temperature regimes of 37°C day/32 °C night imposed on red grape bunches resulted in the reduction of berry anthocyanin content by ~35% in 5 days, and by ~59% in 20 days.

Reddening is normally promoted by solar radiation [95]. The anthocyanin-promoting effect of light exposure is more apparent in pale red varieties, such as Tokay, Emperor, and Sultanina Rose, which completely fail to redden if shaded during ripening [96]. Exposing grape bunches to sunlight induces anthocyanin synthetic genes, once they have been developmentally activated. This exposure is associated with an increase in anthocyanin accumulation, until the point when excessive heat causes berry temperature to become detrimental. Thus, the benefits of light exposure depend on the extent to which berry temperature is elevated by increased solar radiation. Berry temperature increases linearly with incident radiation, and is a result of the flux density of absorbed radiation and convective heat loss. Depending on air flow within the canopy and on berry skin colour, solar radiation may elevate berry skin temperature by up to 17°C above air temperature. Thus, canopy management (*i.e.* shoot positioning, vegetative vigour, and timing and intensity of basal defoliation) remains a critical factor in achieving the most appropriate level of light exposure for red bunches. This requires fine tuning that varies according to variety, site, and season. In general, increasing sunlight exposure for red grapes is detrimental in warm viticultural areas and beneficial in cool climates.

Spayd *et al.* [97] were able to separate the effects of temperature and solar radiation on anthocyanin metabolism in a field experiment by comparing sun-exposed bunches, bunches that were sun-exposed but chilled to the temperature of shaded bunches, shaded bunches, and shaded bunches that were heated to the temperature of sun-exposed bunches. Cool temperatures in sun-exposed clusters maximised anthocyanin concentration.

Anthocyanin composition is also affected by the degree of incident radiation. The proportion of tri-hydroxylated anthocyanins increases over di-hydroxylated derivatives as sunlight exposure increases from veraison until ripening, as a result of altered *F3'5'H/F3'H* gene expression [91].

Water status is another major element of plant physiology that affects anthocyanin metabolism [98-103]. Water deficits increase anthocyanin accumulation by promoting the expression of anthocyanin synthetic genes, regardless of when the deficit is applied. On a per berry basis, the quantity of anthocyanins increases consistently among red varieties. The positive effect of water shortage on anthocyanin synthesis is imposed upon the adverse effect on fruit growth, which together highly increase anthocyanin concentration. Severe and prolonged water deficits also cause substantial increases in the proportion of tri-hydroxylated anthocyanins over di-hydroxylated derivatives, and a predominance of methylated forms over the parent hydroxylated anthocyanins. These metabolite changes stem from the increased ratio of *F3'5'H/F3'H* gene expression and the induction of an *OMT* [100].

Viticultural practices that modify crop load and carbon source-sink relationships may affect anthocyanin accumulation under some circumstances. In Nebbiolo grapes, cluster thinning increased anthocyanin concentration, while it had no effect on Cabernet Sauvignon [104, 105]. The intensity of nitrogen fertilisation and resulting vine vigour are inversely correlated with anthocyanin accumulation.

Plant growth regulators involved in fruit ripening also modulate anthocyanin synthesis. Abscisic acid and ethylene-releasing compounds increase anthocyanin content through a transient induction of several anthocyanin synthetic genes, including *VvMybA* and *UFGT* [106-110]. MeJA, in combination with sucrose, stimulates the expression of several anthocyanin biosynthetic genes and increases the accumulation of anthocyanins. This is frequently observed in MeJA-mediated defence in response to pathogen infections [111]. In contrast, auxins repress expression of anthocyanin synthetic genes and reduce anthocyanin accumulation [112].

OTHER KEY ANTIOXIDANTS

Many classes of phenolics have been regarded as effective antioxidants *in planta* and in humans following dietary consumption of fresh and transformed grape products. The most striking examples are resveratrol, which has been linked to the reduced incidence of coronary heart disease among moderate wine drinkers, and proanthocyanidins, which possess an antioxidant power 20 times greater than vitamin E and 50 times greater than vitamin C. However, other non-phenolic compounds normally present in the grape berry are also associated with antioxidant activity.

Glutathione

Glutathione is a tripeptide thiol formed from glutamate, cysteine, and glycine by two sequential reactions controlled by γ-glutamylcysteine synthase and glutathione synthase.

Glutathione accumulates in grape berry from veraison until ripening, and participates in the prevention of oxidative stress by detoxification of ROS [113]. In crushed berries, it reacts with hydroxycinnamate quinones, reducing oxidative browning of white must. Glutathione also conjugates to other molecules in reactions catalysed by GST. Glutathione may also enter the synthetic pathway that leads to cysteinylated precursors of mercaptohexan-1-ol, a volatile thiol that contributes to the flavours of grapefruit and passion fruit, distinctive of Sauvignon blanc and other whites. Glutathione content in the grape berry decreases with UV light exposure and temperature stresses. This reduction is paralleled by an increase in glutathione conjugates. Expression of several copies of the *GST* gene is induced under those conditions. Free glutathione in intact berries of table grapes and the fraction of non-oxidised glutathione in juice and wine may serve as a source of antioxidants in the human diet.

Ascorbic Acid (vitamin C)

Grapes are not as high in ascorbic acid as other fruits such as citrus and kiwifruit. This is because compared with other fleshly fruit, grapes are unusually rich in tartaric acid, which is the end-product of irreversible ascorbic acid catabolism [114] (see the chapter by Ford C, this edition). In crushed berries, ascorbic acid is readily available for energetically favourable oxidation, thus its content falls very rapidly to undetectable levels after oxygen exposure. The ascorbic acid detected in finished wines is a food additive used by the wine industry.

Tocochromanols (vitamin E)

Tocochromanols are molecules with vitamin E activity which prevent lipid oxidation. Depending on the nature of their isoprenoid chain, they are distinguished into tocopherols (those containing a phytyl chain) or tocotrienols (those with a geranylgeranyl chain). Within each group, molecules are further differentiated into α-, β-, γ-, and δ-forms according to the number and position of methyl groups on the chromanol ring. Tocopherols accumulate in seeds, flesh, and skins of grape berries, in a concentration ranging from 5 to 100 μg per g dry weight, decreasing as ripening progresses [115]. The α-tocopherol is the main tocopherol found in grapes. Tocotrienols are only detected in seeds at a concentration that increases during fruit ripening, with a maximum of 54 μg per g of seed dry weight. The most abundant form of tocotrienols in the grape is γ-tocotrienol.

ABBREVIATIONS

4CL	=	4-coumaroyl:CoA-ligase
ANR	=	Anthocyanidin reductase
C4H	=	Cinnamate 4-hydroxylase
CHI	=	Chalcone isomerase
CHS	=	Chalcone synthase
DFR	=	Cihydroflavonol reductase
F3'5'H	=	Flavonoid-3',5'-hydroxylase
F3'H	=	Flavonoid-3'-hydroxylase

F3H = Flavanone-3-hydroxylases

FLS = Flavonol synthase

FW = Fresh weight

GST = Glutathione *S*-transferases

LAR = Leucoanthocyanidin reductase

LDOX = Leucoanthocyanidin dioxigenase

mDP = Mean degree of polymerisation

MeJA = Methyl-jasmonate

OMT = *O*-methyltransferase

PAL = Phenylalanine ammonia lyase

ROMT = Resveratrol *O*-methyltransferase

ROS = Reactive oxygen species

SA = Salicylic acid

STS = Stilbene synthase

UFGT = UDP-glucose:flavonoid 3-*O*-glucosyl transferase

ACKNOWLEDGEMENT

We thank Courtney Coleman for the assistance provided in the preparation of the manuscript.

REFERENCES

[1] Corder R, Mullen W, Khan NQ, Marks SC, Wood EG, Carrier MJ, Crozier A. Oenology: red wine procyanidins and vascular health. Nature 2006; 444(7119): 566.

[2] Jang M, Cai L, Udeani GO, *et al.* Cancer chemopreventive activity of resveratrol, a natural product derived from grapes. Science 1997; 275(5297): 218-20.

[3] Adams DO. Phenolics and ripening in grape berries. Am J Enol Viticult 2006; *57*(3): 249-56.

[4] Waterhouse AL. Wine phenolics. Ann Ny Acad Sci 2002; 957; 21-36.

[5] Singleton VL, Zaya J, Trousdale EK. Caftaric and coutaric acids in fruit of Vitis. Phytochemistry 1986; 25(9): 2127-33.

[6] Ong BY, Nagel CW. Hydroxycinnamic acid tartaric acid ester content in mature grapes and during maturation of white riesling grapes. Am J Enol Viticult 1978; 29(4): 277-81.

[7] Langcake P, Pryce RJ. The production of resveratrol by *Vitis vinifera* and other members of the *Vitaceae* as a response to infection or injury. Physiol Plant Pathol 1976; 9: 77-86.

[8] Jeandet P, Douillt-Breuil AC, Bessis R, Debord S, Sbaghi M, Adrian M. Phytoalexins from the vitaceae: biosynthesis, phytoalexin gene expression in transgenic plants, antifungal activity, and metabolism. J Agr Food Chem 2002; 50(10): 2731-41.

[9] Bavaresco L, Fregoni C, van Zeller de Macedo Basto Gonçalves MI, Vezzulli S. Physiology & Molecular Biology of Grapevine Stilbenes: An update. In: Kalliopi A, Roubelakis-Angelakis KA, Ed. Grapevine Molecular Physiology and Biotechnology, 2nd ed. Springer 2009; pp. 341-63.

[10] Gatto P, Vrhovsek U, Muth J, *et al.* Ripening and genotype control stilbene accumulation in healthy grapes. J Agr Food Chem 2008; 56(24): 11773-11785.

[11] Bavaresco L, Petegolli D, Cantu E, Fregoni M, Chiusa G, Trevisan M. Elicitation and accumulation of stilbene phytoalexins in grapevine berries infected by Botrytis cinerea. Vitis 1997; 36(2): 77-83.

[12] Pezet R, Gindro K, Viret O, Spring JL. Glycosylation and oxidative dimerization of resveratrol are respectively associated to sensitivity and resistance of grapevine cultivars to downy mildew. Physiol Mol Plant P 2004; 65(6): 297–303.

[13] Mattivi F, Guzzon R, Vrhovsek U, Stefanini M, Velasco R. Metabolite profiling of grape: Flavonols and anthocyanins. J Agr Food Chem 2006; 54(20): 7692-702.

[14] Downey M, Rochfort S. Simultaneous separation by reversed-phase high-performance liquid chromatography and mass spectral identification of anthocyanins and flavonols in Shiraz grape skin. J Chromatogr A 2008; 1201(1): 43-7.

[15] Downey MO, Harvey JS, Robinson SP. Synthesis of flavonols and expression of flavonol synthase genes in the developing grape berries of Shiraz and Chardonnay (*Vitis vinifera* L.). Aust J Grape Wine R 2003; 9(2): 110-21.

[16] Boulton R. The copigmentation of anthocyanins and its role in the color of red wine: A critical review. Am J Enol Vitic 2001; 52(2): 67-87.

[17] Mattivi F, Vrhovsek U, Masuero D, Trainotti D. Differences in the amount and structure of extractable skin and seed tannins amongst red grape varieties. Aust J Grape Wine R 2009; 15(1): 27-35.

[18] Kennedy JA, Matthews MA, Waterhouse AL. Changes in grape seed polyphenols during fruit ripening. Phytochemistry 2000; 55(1): 77-85.

[19] Verries C, Guiraud JL, Souquet JM, Vialet S, Terrier N, Olle D. Validation of an extraction method on whole pericarp of grape berry (Vitis *vinifera* L. cv. Shiraz) to study biochemical and molecular aspects of flavan-3-ol synthesis during berry development. J Agr Food Chem 2008; 56(14): 5896-904.

[20] Downey MO, Harvey JS, Robinson SP. Analysis of tannins in seeds and skins of Shiraz grapes throughout berry development. Aust J Grape Wine R 2003; 9(1): 15-27.

[21] Hanlin RL, Downey MO. Condensed tannin accumulation and composition in skin of Shiraz and Cabernet Sauvignon grapes during berry development. Am J Enol Viticult 2009; 60(1): 13-23.

[22] De Freitas VAP, Glories Y, Monique A. Developmental changes of procyanidins in grapes of red Vitis vinifera varieties and their composition in respective wines. Am J Enol Viticult 2000; 51(4): 397-403.

[23] del Rio JLP, Kennedy JA. Development of proanthocyanidins in *Vitis vinifera* L. cv. Pinot noir grapes and extraction into wine. Am J Enol Viticult 2006; 57(2): 125-132.

[24] Harbertson JF, Kennedy JA, Adams DO. Tannin in skins and seeds of Cabernet Sauvignon, Syrah, and Pinot noir berries during ripening. Am J Enol Viticult 2002; 53(1): 54-9.

[25] Amrani-Joutei K, Glories Y, Mercier M. Localization of tannins in grape berry skins. Vitis 1984; 33: 133-8.

[26] Geny L, Saucier C, Bracco S, Daviaud F, Glories Y. Composition and cellular localization of tannins in grape seeds during maturation. J Agr Food Chem 2003; 51(27): 8051-4.

[27] Hanlin RL, Hrmova M, Harbertson JF, Downey MO. Review: Condensed tannin and grape cell wall interactions and their impact on tannin extractability into wine. Aust J Grape Wine R 2010; 16(1): 173-88.

[28] Cheynier V, Prieur C, Guyot S, Rigaud J, Moutounet M. The structures of tannins in grapes and wines and their interactions with proteins. Acs Sym Ser 1997; 661: 81-93.

[29] Kennedy JA, Hayasaka Y, Vidal S, Waters EJ, Jones GP. Composition of grape skin proanthocyanidins at different stages of berry development. J Agr Food Chem 2001; 49(11): 5348-55.

[30] Janvary L, Hoffmann T, Pfeiffer J, Hausmann L, Topfer R, Fischer TC, Schwab W. A Double mutation in the anthocyanin 5-O-glucosyltransferase gene disrupts enzymatic activity in *Vitis vinifera* L. J Agr Food Chem 2009; 57(9): 3512-18.

[31] He F, He JJ, Pan QH, Duan CQ. Mass-spectrometry evidence confirming the presence of pelargonidin-3-O-glucoside in the berry skins of Cabernet Sauvignon and Pinot Noir (*Vitis vinifera* L.). Aust J Grape Wine R 2010; 16(3): 464-8.

[32] Pomar F, Novo M, Masa A. Varietal differences among the anthocyanin profiles of 50 red table grape cultivars studied by high performance liquid chromatography. J Chromatogr A 2005; 1094(1-2): 34-41.

[33] Sarni P, Fulcrand H, Souillol V, Souquet JM, Cheynier V. Mechanisms of anthocyanin degradation in grape must-like model solutions. J Sci Food Agr 1995; 69(3): 385-91.

[34] Mazza G. Anthocyanins in grapes and grape products. Crit Rev Food Sci Nutr 1995; 35(4): 341-71.

[35] Sparvoli F, Martin C, Scienza A, Gavazzi G, Tonelli C. Cloning and molecular analysis of structural genes involved in flavonoid and stilbene biosynthesis in grape (*Vitis vinifera* L.). Plant Mol Biol 1994; 24(5): 743-55.

[36] Gollop R, Even S, Colova-Tsolova V, Perl A. Expression of the grape dihydroflavonol reductase gene and analysis of its promoter region. J Exp Bot 2002; 53(373): 1397-409.

[37] Gollop R, Farhi S, Perl A. Regulation of the leucoanthocyanidin dioxygenase gene expression in Vitis vinifera. Plant Sci 2001; 161(3): 579-88.

[38] Ford CM, Boss PK, Hoj PB. Cloning and characterization of *Vitis vinifera* UDP-glucose:flavonoid 3-O-glucosyltransferase, a homologue of the enzyme encoded by the maize Bronze-1 locus that may primarily serve to glucosylate anthocyanidins *in vivo*. J Biol Chem 1998; 273(15): 9224-33.

[39] Boss PK, Davies C, Robinson SP. Analysis of the expression of anthocyanin pathway genes in developing *Vitis vinifera* L cv Shiraz grape berries and the implications for pathway regulation. Plant Physiol 1996; 111(4): 1059-66.

[40] Boss PK, Davies C, Robinson SP. Expression of anthocyanin biosynthesis pathway genes in red and white grapes. Plant Mol Biol 1996; 32(3): 565-9.

[41] Hugueney P, Provenzano S, Verries C, *et al.* A novel cation-dependent *O*-methyltransferase involved in anthocyanin methylation in grapevine. Plant Physiol 2009; 150(4): 2057-70.

[42] Lucker J, Martens S, Lund ST. Characterization of a *Vitis vinifera* cv. Cabernet Sauvignon 3 '5 '-*O*-methyltransferase showing strong preference for anthocyanins and glycosylated flavonols. Phytochemistry 2010; 71(13): 1474-84.

[43] Gomez C, Terrier N, Torregrosa L, *et al.* Grapevine MATE-type proteins act as vacuolar H+-dependent acylated anthocyanin transporters. Plant Physiol 2009; 150(1): 402-15.

[44] Ageorges A, Fernandez L, Vialet S, Merdinoglu D, Terrier N, Romieu C. Four specific isogenes of the anthocyanin metabolic pathway are systematically co-expressed with the red colour of grape berries. Plant Sci 2006; 170(2): 372-83.

[45] Castellarin SD, Di Gaspero G. Transcriptional control of anthocyanin biosynthetic genes in extreme phenotypes for berry pigmentation of naturally occurring grapevines. BMC Plant Biol 2007; 7: 46.

[46] Braidot E, Petrussa E, Bertolini A, *et al.* Evidence for a putative flavonoid translocator similar to mammalian bilitranslocase in grape berries (Vitis vinifera L.) during ripening. Planta 2008; 228(1): 203-13.

[47] Tropf S, Lanz T, Rensing SA, Schroder J, Schroder G. Evidence that stilbene synthases have developed from chalcone synthases several times in the course of evolution. J Mol Evol 1994; 38(6): 610-8.

[48] Goto-Yamamoto N, Wan GH, Masaki K, Kobayashi S. Structure and transcription of three chalcone synthase genes of grapevine (*Vitis vinifera*). Plant Sci 2002; 162(6): 867-72.

[49] Schmidlin L, Poutaraud A, Claudel P, *et al.* A stress-inducible resveratrol *O*-methyltransferase involved in the biosynthesis of pterostilbene in grapevine. Plant Physiol 2008; 148(3): 1630-9.

[50] Hall D, De Luca V. Mesocarp localization of a bi-functional resveratrol/hydroxycinnamic acid glucosyltransferase of Concord grape (*Vitis labrusca*). Plant J 2007; 49(4): 579-91.

[51] Fujita A, Goto-Yamamoto N, Aramaki I, Hashizume K. Organ-specific transcription of putative flavonol synthase genes of grapevine and effects of plant hormones and shading on flavonol biosynthesis in grape berry skins. Biosci Biotechnol Biochem 2006; 70(3): 632-8.

[52] Czemmel S, Stracke R, Weisshaar B, *et al.* The grapevine R2R3-MYB transcription factor VvMYBF1 regulates flavonol synthesis in developing grape berries. Plant Physiol 2009; 151(3): 1513-30.

[53] Dunlevy JD, Soole KL, Perkins MV, *et al.* Two O-methyltransferases involved in the biosynthesis of methoxypyrazines: grape-derived aroma compounds important to wine flavour. Plant Mol Biol 2010.

[54] Bogs J, Downey MO, Harvey JS, Ashton AR, Tanner GJ, Robinson SP. Proanthocyanidin synthesis and expression of genes encoding leucoanthocyanidin reductase and anthocyanidin reductase in developing grape berries and grapevine leaves. Plant Physiol 2005; 139(2): 652-63.

[55] Terrier N, Torregrosa L, Ageorges A, Vialet S, Verries C, Cheynier V, Romieu C. Ectopic expression of VvMybPA2 promotes proanthocyanidin biosynthesis in grapevine and suggests additional targets in the pathway. Plant Physiol 2009; 149(2): 1028-41.

[56] Bogs J, Jaffe FW, Takos AM, Walker AR, Robinson SP. The grapevine transcription factor VvMYBPA1 regulates proanthocyanidin synthesis during fruit development. Plant Physiol 2007; 143(3): 1347-61.

[57] Hichri I, Heppel SC, Pillet J, *et al.* The basic Helix-Loop-Helix transcription factor MYC1 is involved in the regulation of the flavonoid biosynthesis pathway in grapevine. Mol Plant 2010; 3(3): 509-523.

[58] Fournier-Level A, Le Cunff L, Gomez C, *et al.* Quantitative genetic bases of anthocyanin variation in grape (*Vitis vinifera* L. ssp *sativa*) berry: A Quantitative Trait Locus to Quantitative Trait Nucleotide integrated study. Genetics 2009; 183(3): 1127-39.

[59] Kobayashi S, Goto-Yamamoto N, Hirochika H. Retrotransposon-induced mutations in grape skin color. Science 2004; 304(5673): 982.

[60] Fournier-Level A, Lacombe T, Le Cunff L, Boursiquot JM, This P. Evolution of the VvMybA gene family, the major determinant of berry colour in cultivated grapevine (*Vitis vinifera* L.). Heredity 2010; 104(4): 351-62.

[61] Walker AR, Lee E, Bogs J, McDavid DA, Thomas MR, Robinson SP. White grapes arose through the mutation of two similar and adjacent regulatory genes. Plant J 2007; 49(5): 772-85.

[62] Azuma A, Kobayashi S, Mitani N, *et al.* Genomic and genetic analysis of Myb-related genes that regulate anthocyanin biosynthesis in grape berry skin. Theor Appl Genet 2008; 117(6): 1009-19.

[63] Lijavetzky D, Ruiz-Garcia L, Cabezas JA, *et al.* Molecular genetics of berry colour variation in table grape. Mol Genet Genomics 2006; 276(5): 427-35.

[64] Walker AR, Lee E, Robinson SP. Two new grape cultivars, bud sports of Cabernet Sauvignon bearing pale-coloured berries, are the result of deletion of two regulatory genes of the berry colour locus. Plant Mol Biol 2006; 62(4-5): 623-35.

[65] Yakushiji H, Kobayashi S, Goto-Yamamoto N, *et al.* A skin color mutation of grapevine, from black-skinned Pinot Noir to white-skinned Pinot Blanc, is caused by deletion of the functional VvmybA1 allele. Biosci Biotech Bioch 2006; 70(6): 1506-1508.

[66] Cutanda-Perez MC, Ageorges A, Gomez C, *et al.* Ectopic expression of VlmybA1 in grapevine activates a narrow set of genes involved in anthocyanin synthesis and transport. Plant Mol Biol 2009; 69(6): 633-48.

[67] Poudel PR, Azuma A, Kobayashi S, Goto-Yamamoto N. Coordinate induction of anthocyanin biosynthetic pathway genes by VvMybAs In: *Proceedings of the 10th International Conference on Grapevine Breeding and Genetics.* Geneva, 1-5 August 2010; 2010.

[68] Bogs J, Ebadi A, McDavid D, Robinson SP. Identification of the flavonoid hydroxylases from grapevine and their regulation during fruit development. Plant Physiol 2006; 140(1): 279-91.

[69] Falginella L, Castellarin SD, Testolin R, Gambetta GA, Morgante M, Di Gaspero G. Expansion and subfunctionalisation of flavonoid 3',5'-hydroxylases in the grapevine lineage. BMC Genomics 2010; 11: 562.

[70] Zenoni S, Ferrarini A, Giacomelli E, *et al.* Characterization of transcriptional complexity during berry development in *Vitis vinifera* using RNA-Seq. Plant Physiol 2010; 152(4): 1787-95.

[71] Matus JT, Aquea F, Arce-Johnson P. Analysis of the grape MYB R2R3 subfamily reveals expanded wine quality-related clades and conserved gene structure organization across *Vitis* and *Arabidopsis* genomes. BMC Plant Biol 2008; 8: 83.

[72] Deluc L, Barrieu F, Marchive C, *et al.* Characterization of a grapevine R2R3-MYB transcription factor that regulates the phenylpropanoid pathway. Plant Physiol 2006; 140(2): 499-511.

[73] Deluc L, Bogs J, Walker AR, Ferrier T, *et al.* The transcription factor VvMYB5b contributes to the regulation of anthocyanin and proanthocyanidin biosynthesis in developing grape berries. Plant Physiol 2008; 147(4): 2041-53.

[74] Pilati S, Perazzolli M, Malossini A, Cestaro A, *et al.* Genome-wide transcriptional analysis of grapevine berry ripening reveals a set of genes similarly modulated during three seasons and the occurrence of an oxidative burst at veraison. BMC Genomics 2007; 8: 428.

[75] Fornara V, Onelli E, Sparvoli F, *et al.* Localization of stilbene synthase in *Vitis vinifera* L. during berry development. Protoplasma 2008; 233(1-2): 83-93.

[76] Pan QH, Wang L, Li JM. Amounts and subcellular localization of stilbene synthase in response of grape berries to UV irradiation. Plant Sci 2009; 176(3): 360-6.

[77] Vannozzi A, Boss PK, Walker AR, Lucchin M, Dry IB. Identification of an R2R3 MYB transcription factor involved in the regulation of the stilbene synthase pathway in grapevine In: Proceedings of the 10th International Conference on Grapevine Breeding and Genetics. Geneva, 1-5 August 2010; 2010.

[78] Douillet-Breuil AC, Jeandet P, Adrian M, Bessis N. Changes in the phytoalexin content of various *Vitis* spp. in response to ultraviolet C elicitation. J Agr Food Chem 1999; 47(10): 4456-61.

[79 Jiménez J, Orea J, Ureña A, Escribano P, Osa P, Guadarrama A. Short anoxic treatments to enhance trans-resveratrol content in grapes and wine. Eur Food Res Technol 2007; 224(3): 373-8.

[80] Vezzulli S, Battilani P, Bavaresco L. Stilbene synthase gene expression after *Aspergillus carbonarius* infection in grapes. Am J Enol Viticult 2007; 58(1): 132-4.

[81] Polesani M, Bortesi L, Ferrarini A, *et al.* General and species-specific transcriptional responses to downy mildew infection in a susceptible (*Vitis vinifera*) and a resistant (*V. riparia*) grapevine species. BMC Genomics 2010; 11: 117.

[82] Langcake P. Disease resistance of *Vitis* spp. and the production of the stress metabolites resveratrol, ε-viniferin, α-viniferin and pterostilbene. Physiol Plant Pathol 1981; 18(2): 213-26.

[83] Vezzulli S, Civardi S, Ferrari F, Bavaresco L. Methyl jasmonate treatment as a trigger of resveratrol synthesis in cultivated grapevine. Am J Enol Viticult 2007; 58(4): 530-3.

[84] Iriti M, Rossoni M, Borgo M, Faoro F. Benzothiadiazole enhances resveratrol and anthocyanin biosynthesis in grapevine, meanwhile improving resistance to Botrytis cinerea. J Agr Food Chem 2004; 52(14): 4406-13.

[85] Deluc LG, Decendit A, Papastamoulis Y, Merillon JM, Cushman JC, Cramer GR. Water deficit increases stilbene metabolism in Cabernet Sauvignon berries. J Agr Food Chem 2011; 59(1): 289-97.

[86] Rizzini FM, Bonghi C, Tonutti P. Postharvest water loss induces marked changes in transcript profiling in skins of wine grape berries. Postharvest Biol Tec 2009; 52(3): 247-53.

[87] Versari A, Parpinello GP, Tornielli GB, Ferrarini R, Giulivo C. Stilbene compounds and stilbene synthase expression during ripening, wilting, and UV treatment in grape cv. Corvina. J Agr Food Chem 2001; 49(11): 5531-6.

[88] Zamboni A, Minoia L, Ferrarini A, Tornielli GB, Zago E, Delledonne M, Pezzotti M. Molecular analysis of post-harvest withering in grape by AFLP transcriptional profiling. J Exp Bot 2008; 59(15): 4145-59.

[89] Downey MO, Harvey JS, Robinson SP. The effect of bunch shading on berry development and flavonoid accumulation in Shiraz grapes. Aust J Grape Wine R 2004; 10(1): 55-73.

[90] Cortell JM, Kennedy JA. Effect of shading on accumulation of flavonoid compounds in (*Vitis vinifera* L.) pinot noir fruit and extraction in a model system. J Agr Food Chem 2006; 54(22): 8510-20.

[91 Koyama K, Goto-Yamamoto N. Bunch shading during different developmental stages affects the phenolic biosynthesis in berry skins of 'Cabernet Sauvignon' grapes. J Am Soc Hortic Sci. 2008; 133(6): 743-53.

[92] Cohen SD, Tarara JM, Kennedy JA. Assessing the impact of temperature on grape phenolic metabolism. Anal Chim Acta 2008; 621(1): 57-67.

[93] Kliewer WM. Influence of temperature, solar radiation and nitrogen on coloration and composition of Emperor grapes. Am J Enol Vitic 1977; 28(2): 96-103.

[94] Mori K, Goto-Yamamoto N, Kitayama M, Hashizume K. Effect of high temperature on anthocyanin composition and transcription of flavonoid hydroxylase genes in 'Pinot noir' grapes (*Vitis vinifera*). J Hortic Sci Biotech 2007; 82(2): 199-206.

[95] Dokoozlian NK, Kliewer WM. Influence of light on grape berry growth and composition varies during fruit development. J Am Soc Hort Sci 1996; 121(5): 869-74.

[96] Weaver RJ, Mccune SB. Influence of light on color development in *Vitis vinifera* grapes. Am J Enol Vitic 1960; 11(4): 179-84.

[97] Spayd SE, Tarara JM, Mee DL, Ferguson JC. Separation of sunlight and temperature effects on the composition of *Vitis vinifera* cv. Merlot berries. Am J Enol Viticult 2002; 53(3): 171-82.

[98] Bucchetti B, Matthews MA, Falginella L, Peterlunger E, Castellarin SD. Effect of water deficiton Merlot grape tannins and anthocyanins across four seasons. Sci. Hortic.-Amsterdam 2011; in press.

[99] Castellarin SD, Matthews MA, Di Gaspero G, Gambetta GA. Water deficits accelerate ripening and induce changes in gene expression regulating flavonoid biosynthesis in grape berries. Planta 2007; 227(1): 101-12.

[100] Castellarin SD, Pfeiffer A, Sivilotti P, Degan M, Peterlunger E, Di Gaspero G. Transcriptional regulation of anthocyanin biosynthesis in ripening fruits of grapevine under seasonal water deficit. Plant Cell Environ 2007; 30(11): 1381-99.

[101] Matthews MA, Anderson MM. Fruit ripening in *Vitis vinifera* L - Responses to seasonal water deficits. Am J Enol Viticult 1988; 39(4): 313-20.

[102] Ojeda H, Andary C, Kraeva E, Carbonneau A, Deloire A. Influence of pre- and postveraison water deficit on synthesis and concentration of skin phenolic compounds during berry growth of *Vitis vinifera* cv. Shiraz. Am J Enol Viticult 2002; 53(4): 261-7.

[103] Roby G, Harbertson JF, Adams DA, Matthews MA. Berry size and vine water deficits as factors in winegrape composition: Anthocyanins and tannins. Aust J Grape Wine R 2004; 10(2): 100-7.

[104] Guidoni S, Allara P, Schubert A. Effect of cluster thinning on berry skin anthocyanin composition of *Vitis vinifera* cv. Nebbiolo. Am J Enol Viticult 2002; 53(3): 224-6.

[105] Keller M, Mills LJ, Wample RL, Spayd SE. Cluster thinning effects on three deficit-irrigated *Vitis vinifera* cultivars. Am J Enol Viticult 2005; 56(2): 91-103.

[106] Gambetta GA, Matthews MA, Shaghasi TH, McElrone AJ, Castellarin SD. Sugar and abscisic acid signaling orthologs are activated at the onset of ripening in grape. Planta 2010; 232(1): 219-34.

[107] Jeong ST, Goto-Yamamoto N, Kobayashi S, Esaka A. Effects of plant hormones and shading on the accumulation of anthocyanins and the expression of anthocyanin biosynthetic genes in grape berry skins. Plant Sci 2004; 167(2): 247-52.

[108] Koyama K, Sadamatsu K, Goto-Yamamoto N. Abscisic acid stimulated ripening and gene expression in berry skins of the Cabernet Sauvignon grape. Funct Integr Genomic 2010; 10(3): 367-81.

[109] Peppi MC, Walker MA, Fidelibus MW. Application of abscisic acid rapidly upregulated UFGT gene expression and improved color of grape berries. Vitis 2008; 47(1): 11-4.

[110] El-Kereamy A, Chervin C, Roustan JP, *et al.* Exogenous ethylene stimulates the long-term expression of genes related to anthocyanin biosynthesis in grape berries. Physiol Plantarum 2003; 119(2): 175-82.

[111] Kortekamp A. Expression analysis of defence-related genes in grapevine leaves after inoculation with a host and a non-host pathogen. Plant Physiol Bioch 2006; 44(1): 58-67.

[112] Davies C, Boss PK, Robinson SP. Treatment of grape berries, a nonclimacteric fruit with a synthetic auxin, retards ripening and alters the expression of developmentally regulated genes. Plant Physiol 1997; 115(3): 1155-61.

[113] Kobayashi H, Takase H, Suzuki Y, Tanzawa F, Takata R, Fujita K, Kohno M, Mochizuki M, Suzuki S, Konno T. Environmental stress enhances biosynthesis of flavor precursors, S-3-(hexan-1-ol)-glutathione and S-3-(hexan-1-ol)-L-cysteine, in grapevine through glutathione S-transferase activation. J Exp Bot 2010; doi 10.1093/jxb/erq376.

[114] Melino VJ, Soole KL, Ford CM. Ascorbate metabolism and the developmental demand for tartaric and oxalic acids in ripening grape berries. BMC Plant Biol 2009; 9: 145.

[115] Horvath G, Wessjohann L, Bigirimana J, *et al.* Accumulation of tocopherols and tocotrienols during seed development of grape (*Vitis vinifera* L. cv. Albert Lavallee). Plant Physiol Bioch 2006; 44(11-12): 724-31.

CHAPTER 6

Aroma and Aroma Precursors in Grape Berry

P. Darriet*, C. Thibon and D. Dubourdieu

Université de Bordeaux, Institut des Sciences de la Vigne et du Vin Unité de recherche Oenologie, EA 4577, USC INRA, 210 chemin de Leysotte, CS 50008, 33882 Villenave d'Ornon cedex, France

Abstract: The grape berry is the site of biosynthesis and accumulation of compounds that are likely to contribute in wines some of their aromatic characteristics. These compounds have the aromatic potential, and exist in part as volatile forms but mainly as non-volatile aroma precursors that can be released through chemical and biochemical reactions during vinification and ageing. The chemistry of aromas has gradually identified a number of key volatile compounds and their precursor forms. Among these compounds, we have considered methoxypyrazines, monoterpenes and C13-norisoprenoids, which are derivatives of carotenoids and sulfur compounds possessing a thiol group. They represent a large diversity of flavour nuances (herbaceous, fruity, floral, empyreumatic, *etc.*), often at trace concentrations (in the nanogram per litre range). The reactivity of these compounds in enological conditions, and the state of the knowledge on their biosynthesis in grapes, both in volatile form, possibly odorous, and in precursor form (glycosylated or conjugation of cysteine and glutathione) are described.

Keywords: Aroma precursors, C13-norisoprenoids, Methoxypyrazines, Monoterpenes, Odoriferous compounds, S-conjugates, ß-damascenone, Terpenes, Thiols.

INTRODUCTION

The aromas of fine wines, with their finesse, complexity and uniqueness, are a source of great pleasure. Bouquet and flavour are obviously related to the expertise of the winemaker and the techniques s/he uses, but primarily reflect grape composition and especially varietal character, as well as its particular expression depending on climate and soil, and their impact on vine physiology. Grape and wine smells vary from the floweriness of Muscat to the boxwood, tomato leaf and tropical fruit nuances of Sauvignon Blanc; and from the green pepper, blackcurrant, and smoky aromas of Cabernet Sauvignon to the violet, meaty aromas of Syrah. All of these nuances form the wine's taste profile and are initially found in the aromatic potential of the grapes.

How can the chemistry of flavours explain what our senses perceive? For instance, from one to several hundred compounds are present in the headspace of a wineglass. The concentration of these compounds, as well as their olfactory detection threshold values, can vary from several hundred milligrams per litre to trace concentrations in the picogram per litre range. A relatively small number, often present in trace concentrations, are considered as key volatile compounds. These compounds contribute to the typical odours of wine aroma. However, these alone cannot explain all the nuances of wine aroma and the "olfactory images" are the result of perceptual interactions produced at the brain level amongst the volatile compounds [1], particularly implicating key varietal compounds. Knowledge of these essential pieces in the puzzle of wine olfactory images is very important, especially to understand how wine aromas develop and the chemical and biochemical mechanisms involved.

The study of grape and wine aromas is laborious because the most important grape aroma compounds are often present in very low concentrations. The development of instrumental analysis techniques is thus crucial to progress in this field. A large part of the volatile compounds from grape contributing to wine aroma exist in the must under non-volatile form of precursors, which are released through alcoholic fermentation and ageing. Natural factors (climate, soil), viticultural parameters and enological modalities

***Address correspondence to P. Darriet:** Université de Bordeaux, Institut des Sciences de la Vigne et du Vin Unité de recherche Oenologie, EA 4577, USC INRA, 210 chemin de Leysotte, CS 50008, 33882 Villenave d'Ornon cedex, France; E-mail: philippe.darriet@oenologie.u-bordeaux2.fr

Hernâni Gerós, M. Manuela Chaves and Serge Delrot (Eds)

have a great impact on these compounds. The first studies on grape flavours date back to the early 1950s when gas chromatography made it possible to separate volatile components in the vapour phase. Since then, many volatile compounds have been identified, especially through the coupling of gas chromatography with olfactometry and mass spectrometry.

In the first part of this chapter, the compounds known at present which, due to their power and odorant concentrations found in grapes (and possibly wine), are likely to participate in the aroma. In the second part, we build our state of knowledge on the odourless forms and non-volatile flavour precursors that lead to a perception of flavour.

METHOXYPYRAZINES

Methoxypyrazines are nitrogen heterocycle compounds belonging to the pyrazine group, largely represented in both animals and plants [2]. Among the various methoxypyrazines, some alkylated methoxypyrazines, such as 2-methoxy-3-isobutylpyrazine (IBMP), 2-methoxy-3-sec-butylpyrazine and 2-methoxy-3-isopropylpyrazine (IPMP) are extremely volatile, with very low odour thresholds in the nanogram per litre range in water (Table **1**). The odours of these methoxypyrazines are vegetable-like, reminiscent of pea pods, green peppers and, depending on the concentration and the compound, earthy nuances. These same substances have also been identified in bell peppers, pea pods, potatoes and carrots, as well as blackcurrants, raspberries, and blackberries [3-5].

Table 1: 2-methoxy-3-alkylpyrazines identified in grapes and wines.

Name	Structure	Descriptors	Detection threshold (ng/L)
	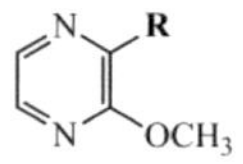		
2-methoxy-3-isobutylpyrazine	R: $CH_2CH(CH_3)_2$	greenpeper/ pea pod	1 [a]
2-methoxy-3-isopropylpyrazine	R: $CH(CH_3)_2$	greenpeper/ pea pod/earthy	2 [a]
2-methoxy-3-sec-butylpyrazine	R: $CH(CH_3)CH_2CH_3$	greenpeper/ pea pod	1 [a]
2-methoxy-3-ethylpyrazine	R: CH_2CH_3	greenpeper/ pea pod	40 [b]
2-methoxy-3-methylpyrazine	R: CH_3	greenpeper/ pea pod	4000 [c]

[a] Determined at Faculté d'Oenologie, ISVV, University Bordeaux Segalen.

[b] Concentrations from reference [206].

[c] Concentrations from reference [207].

These compounds, particularly IBMP, have been identified in various grape varieties such as Cabernet Sauvignon, Cabernet Franc, Sauvignon Blanc [6-9], Merlot, Carmenère and Verdejo [5, 10-12]. 2-methoxy-3-isobutylpyrazine has also been found in Pinot Noir, Chardonnay, Riesling, Chenin Blanc, Traminer, Syrah and Pinotage [13, 14], but in very low concentrations. Furthermore, other methoxypyrazines such as 2-methoxy-3-methylpyrazine [7] and 2-methoxy-3-ethylpyrazine [15, 16] have been identified in Sauvignon Blanc grapes, but these compounds are much less odiferous than IBMP (Table **1**).

In Cabernet Sauvignon, the pea pod, pepper-like aroma of IBMP contributes to a herbaceous expression of grape aromas, which can have a negative effect on wine flavour at concentrations as low as 15 nanograms per litre [17]. IBMP can also affect the flavour of Tempranillo and Grenache grown in Spain [18]. On the other hand, these same aromas can also contribute, to some extent, to desirable varietal aromas in Sauvignon Blanc wines [19, 20]. As noticed by Escudero *et al.* [21] and Pineau *et al.* [22], perceptual phenomena can also reflect interaction between volatile compounds to modulate the perceived herbaceous flavour of this compound by tasters.

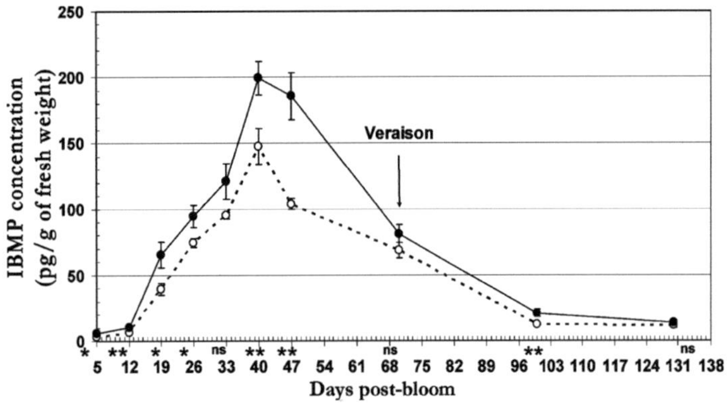

Figure 1: 2-methoxy-3-isobutylpyrazine concentrations of shaded (●) and exposed berries (○-) of *Vitis vinifera* L cv Cabernet Franc during the growing season (reprinted with permission from the Journal of Agricultural and Food Chemistry. Copyright 2011 American Chemical Society) [32].

Levels of IBMP in Grapes and Wines

After the Allen group, who first developed a sensitive and reproducible method for assaying methoxypyrazines [7-9], numerous works have been published concerning the content of methoxypyrazines in grapes and wines, and their development in relation to enological, viticultural and natural parameters.

IBMP is proportionally the most abundant compound in grapes, grape juice and wine. Its concentration in Sauvignon Blanc grape juice varies from 0.5 to 40 ng/L. Levels of 2-methoxy-3-isopropylpyrazine (between 1 and 6 ng/L) are 7-8 times lower than those of 2-methoxy-3-isobutylpyrazine in grapes. The latter concentration is generally less than ng/L [8]. Suffice to say that among these three methoxypyrazines, 2-methoxy-3-isobutylpyrazine, the most abundantly found, is also the most likely to contribute to the grassy aroma of Sauvignon Blanc grapes. As for Verdejo, concentrations range from 5 to 15 ng/L [12]. IBMP was also quantified in Cabernet Sauvignon grapes and wines at levels between 0.5 ng/l and 100 ng/l [9, 17, 23]. The highest content is usually found in Carmenère grapes and wines, which are frequently characterised by herbaceous flavours [5, 11].

Effect of Winemaking on Methoxypyrazine Concentration in Wine

In ripe grapes, IBMP is mainly present in the skins [24, 25] and seeds, not the pulp, with a notable difference in the concentration between the pre-veraison and harvest periods [24, 25]. During vinification, IBMP is easily extracted during pressing or pre-fermentation maceration [24, 25]. Neither alcoholic fermentation nor ageing modify the original concentration in the must by more than 30% [24, 25, 33]. This means that conventional winemaking techniques do not decrease the IBMP content of grapes in the finished wine. However, the settling of must during the fermentation of white or rosé wines can reduce IBMP levels by about 50% [24, 25]. As for red wines, thermovinification can lead to a decrease of IBMP by evaporation [24, 25]. During agéing, and because this compound is non-oxidisable, oxygenation has no impact on its concentration [26]. Moreover, IBMP content is very stable during ageing and can affect wine aroma for many years [5].

Methoxypyrazine Biosynthesis in Grapes and the Impact of Vine Physiology

The biosynthesis of methoxypyrazines in grapes, particularly IBMP, has not yet been characterised, and the proposed biosynthetic pathway refers to microbial synthesis [4, 27]. Biosynthesis begins with an amino acid (*i.e.* leucine for IBMP), which is transformed into an amide (leucinamide) and reacts with a 1-2-dicarbonyl compound (glyoxal, for example) to form a pyrazinone. The final step concerns the methylation of the pyrazinone *via* an ortho-methyltransferase, leading ultimately to methoxypyrazine (i.e methylating 2-hydroxy-3-isobutylpyrazine (IBHP) to form 2-methoxy-3-isobutylpyrazine) [28]. One O-methyltransferase

(OMT) catalysing such reactions was partially purified in grapes, but its specificity towards the substrate (IBHP) is not very good [29, 30]. Two cDNA encoding OMT having homology with the purified enzyme by Hashizume [29, 30] were recently characterised and expressed in *Escherichia coli* [31] but the question remains whether these genes constitute the key step in IBMP formation in grapes.

Over the past 20 years, several authors have investigated IBMP formation along with cluster development and the parameters involved. IBMP appears in clusters a few days after anthesis [32] and concentrations increase at least until closure of the cluster, and even until 2 to 3 weeks before veraison [5, 28, 32]. The concentration decreases during the ripening period and the content at harvest drops according to the degree of maturity. The amount of IBMP in ripe grapes is closely correlated with the maximum determined during the bunch closure period [33]. While some enzyme activities and postulated genes have been characterised, the parameters of grape and plant physiology governing IBMP development in grapes (accumulation, decrease) have not been clarified. It should be noted that IBMP was detected both in stems and leaves [23, 24, 28]. The authors found much higher concentrations of IBMP in the basal leaves, proximal to the cluster, and experiments using labelled IBMP showed that this molecule can move from the basal leaves to the cluster and then to the berries [5, 23]. Moreover, whereas IBMP concentrations decrease continuously in berries from veraison to maturity, IBMP biosynthesis continues to take place in basal leaves [5].

The biosynthesis of methoxypyrazines in grapes, particularly IBMP, has not yet been characterised, and the proposed biosynthetic pathway refers to microbial synthesis [4, 27]. Biosynthesis begins with an aminoacid (*i.e.* leucine for IBMP), which is transformed into an amide (leucinamide), and reacts with a 1-2-dicarbonyl compound (glyoxal, for example) to form a pyrazinone. The final step concerns the methylation of the pyrazinone realized by an ortho-methyltransferase, leading ultimately to methoxypyrazine (*i.e.* methylating 2-hydroxy-3-isobutylpyrazine (IBHP) to form 2-methoxy-3-isobutylpyrazine) [28]. One O-methyltransferase (OMT) catalyzing such reactions was partially purified in grapes, but its specificity towards the substrate (IBHP) is not very good [29, 30]. Two cDNA encoding OMT having homology with the purified enzyme by Hashizume [29, 30] were recently characterized and expressed in *Escherichia coli* [31] but the question remains whether these genes constitute the key step in IBMP formation in grapes.

Over the past 20 years, several authors have investigated IBMP formation along with cluster development, and the parameters involved. IBMP appears in clusters a few days after anthesis [32] and concentrations increase at least until closure of the cluster, and even until 2 to 3 weeks before veraison [5, 28, 32]. The concentration decreases during the ripening period and the content at harvest drops according to the degree of maturity (Fig. **1**). The amount of IBMP in ripe grapes is closely correlated to the maximum determined during the bunch closure period [33]. While some enzyme activities and postulated genes have been characterized, the parameters of grape and plant physiology governing IBMP development in grapes (accumulation, decrease) have not been clarified. It should be noted that IBMP was detected both in stems and leaves [23, 24, 28]. The authors found much higher concentrations of IBMP in the basal leaves, proximal to the cluster, and experiments using labeled IBMP showed that this molecule can move from the basal leaves to the cluster and then to the berries [5, 23]. Moreover, whereas IBMP concentrations decrease continuously in berries from veraison to maturity, IBMP biosynthesis continues to take place in basal leaves [5].

Also, numerous studies have been published relative to the impact of natural (climate, soil) and viticultural factors on IBMP concentration in grapes and wines. Allen *et al.* [13] first noticed lower IBMP concentrations in wines obtained from grapes ripened at a higher temperature, whereas Falcão *et al.* [34] discovered higher IBMP levels in wines obtained from grapes grown at different altitudes. The sensitivity of IBMP to UV light and its degradation leading to the forming of 2-methoxy-3-methylpyrazine, a much less odiferous compound, as a main product [5, 35, 36] should be noted. The variation of IBMP content in grapes and wines relative to climate is thus a key factor [17, 28, 32, 37].

Paradoxically, the impact of light is not related to the kinetics of post-veraison IBMP degradation proportional to basal leaf removal, as proved by Ryona *et al.* [32] and subsequently Scheiner *et al.* [38], but instead, according to these authors, to the limiting of the accumulation of IBMP from fruit set to veraison (Fig. **1**). Roujou *et al.* [5] had indirectly demonstrated this by showing that IBMP concentration in grapes was lower when the basal leaves were removed at an early stage (berry set), and not at veraison. However, in light of the abundance of IMBP in basal leaves, it is possible that the more marked decrease in the concentration of this compound in grapes was due to elimination of the source.

Moreover, other physiological parameters of grape vines, such as yields and the availability of water and nitrogen [5, 39] can have an impact on IBMP development. IBMP content in ripe grapes can also vary significantly depending on the clonal origin of the vines [11]. In addition, the detection of IPMP in wines can be related to contamination of grapes during harvesting by the Asian Lady Beetle, *H.axyridis*, an insect introduced for the biological control of aphids [40].

Moreover, other physiological parameters of grape vines, such as yields and the availability of water and nitrogen [5, 39] can have an impact on IBMP development. IBMP content in ripe grapes can also vary significantly depending on the clonal origin of the vines [11]. In addition, the detection of IPMP in wines can be related to contamination of grapes during harvesting by the Asian Lady Beetle, *H.axyridis*, an insect introduced for the biological control of aphids [40].

TERPENIC COMPOUNDS

Terpenes are a very large and widespread group of compounds. The name "terpene", invented by Kekule (1866), comes from the German "turpentine" in connection with the richness of the terpene composition of oil of turpentine.

Within this family, terpenes with 10 carbon atoms (monoterpenes) formed from two isoprene units of 5-carbon terpenes, as well as others with 15 carbon atoms (sesquiterpenes) formed from 3 isoprene units are present in grapes and wine, and contribute to interesting odours. Based on an early hypothesis by Cordonnier [42], many studies have been undertaken by various enological research teams to learn more about these compounds [42-47].

There are forty main monoterpene compounds in grapes. The most important from the point of view of odour are certain monoterpene alcohols and oxides such as linalool, geraniol, citronellol, (E)-hotrienol, cis or (Z)-rose oxide and nerol (3,7-dimethyl-2(Z),6-octadien-1-ol), which develop floral aromas. The detection thresholds of these compounds are quite low, ranging from tens to hundreds of micrograms per litre (Table **2**). As early as 1970, Ribéreau-Gayon and Boidron showed that a mixture of the main terpenols had a significantly lower odour detection threshold than each one taken separately, thereby highlighting a synergistic action between compounds [42]. In addition, the detection threshold of linalool and geraniol is lowered by the presence of ß-damascenone [48].

The monoterpene alcohols mentioned above play a major role in the aroma of grapes and wines from the Muscat family (Muscat de Frontignan, also called Muscat à Petits Grains or Muscat d'Alsace; Muscat of Alexandria, also called Muscat à Gros Grains; Muscat d'Ottonel; White Muscat from Piemonte, *etc.*) as well as crosses between Muscat of Alexandria and other varietals, including such varietals as Muscat de Hambourg (crossed with Frankenthal [49]), and the Argentinian variety Torrontes (a cross with Mission, also called Pais variety [50]). The concentrations of the main monoterpenes in these grapes and wines are much greater than the odour detection threshold of these compounds. Monoterpenes also have a more or less pronounced impact on the flavour of Gewürztraminer, Albariño, Scheurebe and Auxerrois grapes and wines, and to some extent on those of Riesling, Muscadelle and some clones of Chardonnay. These monoterpene alcohols are often found in many varieties (Sauvignon, Syrah, Cabernet Sauvignon, *etc.*) at levels generally below the olfactory perception threshold. In addition, seeing as several monoterpenes have an asymmetric carbon, varying degrees of enantiomeric forms of monoterpenes are found in grapes [51]. For example, linalool, hotrienol, cis and trans-rose oxide are predominantly present (88-97%) in a single enantiomeric form in the various varieties of Muscat that were analysed (*S* form for linalool and (E)-hotrienol and *(2S, 4R)* and *(2R, 4S)* forms for (Z)-rose oxide, respectively). The abundance of one or the other enantiomer may contribute to modulate the strength of these odorous compounds in grape juice and wine, as well as aromatic expression. *R*-(-)-linalool, with an odour threshold of 0.8 µg/L, is described as having floral notes and overtones of woody lavender. It is more odiferous than *S*-(+)-linalool, which presents a floral, sweet scent with a detection threshold of 17 µg/L [52]. However, the proportions of enantiomeric forms may change over the fermentation process, the most fragrant enantiomer, cis-*(2S, 4R)*-rose oxide or (Z)-(-)-rose oxide, being for example between 38 and 76% in wine [53, 54].

Table 2: Characteristics of some monoterpenes identified in grapes and wines.

Name	Structure	Descriptors	Detection threshold [a] (µg/L)	Concentrations in Muscat wines (Min-Max) [b] (µg/L)
linalol 3,7-dimethyl-1,6-octadien-3-ol	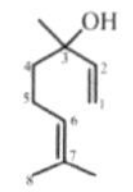	coriander seed- rose	20/3 [c]	170-815
geraniol 3,7-dimethyl-2(E),6-octadien-1-ol	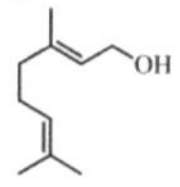	rose	90/9	223-1056
citronellol 3,7-dimethyl-6-octen-1-ol	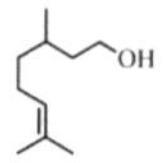	rose –lemongrass	20/15	nd-20
(E)-hotrienol 3,7-dimethyl-1,5(E)7-octatrien-3-ol		linden-rose	300/70	5-300
(2S, 4R)-(-)-cis-rose oxide	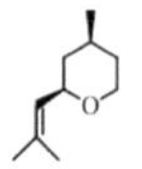	floral-green	8 [d]/0.2 [e]	0.2-10 [e]

[a] Determined at Faculté d'Oenologie, ISVV, University Bordeaux Segalen.

[b] Concentrations from references [43-46].

[c] Detection threshold in model solution and water respectively.

[d] Determined on racemic mixture.

[e] Values from references [52, 53].

Aldehydes (geranial, linalal), acids (trans-geranic), monoterpene diols, triols [43, 55] and some menthene diols [56] derived from oxidation of alpha-terpineol have been identified in grapes, while esters (geranyl and neryl acetate) have been found in wines [57]. However, except for aldehydes and geranyl acetate, these compounds are not very odiferous. Other than rose oxide, oxides present in grapes, such as the oxides of linalool and nerol, have little olfactory impact (*i.e.* high perception thresholds of 1 to 5 mg/L). Also, the presence of linalool oxide contributes to the increased perception of linalool [42].

Monoterpenes Biosynthesis in Grapes

The first step in the pathway of terpene biosynthesis is the formation of isopentenyl pyrophosphate (IPP). This compound is biosynthesised either in the cytosol (mevalonate pathway) or in plastids (1-deoxy-D-xylulose-5-phosphate/2*C*-methyl-D-erythritol-4-phosphate or DXP pathway) as first shown by Rohmer [58]. Geranyl diphosphate is then formed by the condensation of IPP and its isomer DMAPP (dimethylallyl pyrophosphate). Luan *et al.* [59] proved that the DXP pathway is the dominant route for geraniol or linalool biosynthesis in grape berry exocarp and mesocarp. Once GPP is synthesised, terpene synthases such as linalool synthase [60] or geraniol synthase [61] can compete for this substrate leading to the formation of either linalool or geraniol, and then other monoterpenes and high-oxydation-state intermediates (polyhydroxylated monoterpenes (diendiols, trien-diols), furanic and pyranic linalool oxides) [62]. For example, the enzymatic hydroxylation of linalool, due to cytochrome P 450 monooxygenase, forms various diols (diendiol-1, diendiol-2) from linalool or even 8-hydroxylinalool ((Z) and (E) forms) and 8-hydroxygeraniol from geraniol [62, 63]. According to the same mechanism, the hydroxylation of citronellol (3,7-dimethyl-6-octen-1-ol) leads to the formation of 3,7-dimethyl-5-octen-1,7-diol, a precursor of rose oxide [54]. Moreover, sequencing of the grapevine genome has revealed the complexity of the terpene synthase gene family, as numerous putative terpene synthase genes have been identified [64]. Among these genes, one code for an alpha-terpineol synthase has been characterised [65]. Recently, genetic analysis of monoterpene production in grapes has led to the identification of 2 major Quantitative Trait Loci influencing total monoterpenol content and linalool concentrations, respectively [66, 67]. The deoxy-D-xylulose-phosphate synthase gene is therefore a very strong candidate to account for the genetic variability of berry terpenol content [66, 67].

Monoterpene biosynthesis occurs during grape ripening, starting from berry set. These compounds are mainly located in the berry skin [68-70]. Depending on the authors, there is either a continuous accumulation of grape monoterpenes [71] or a decrease of free monoterpenes before the full accumulation of sugars in the grapes [43, 72]. Ripening conditions and especially temperature and light exposure during the maturation period account for various levels of monoterpenes in grapes, non-excessive exposure being favourable to increased volatile monoterpene levels at grape ripeness [73].

Botrytis cinerea development on grapes can also alter the composition of grape monoterpenes by degrading the main monoterpene alcohols and their oxides [45, 74, 75] into generally less odorous components. Laboratory experiments have established that linalool is converted into lilac aldehyde and lilac alcohol, and 8-hydroxylinalool may also be formed by the action of enzymes secreted by *B. cinerea* [45, 75].

Rearrangements of Monoterpene Compounds during Alcoholic Fermentation

Fermentation significantly alters the monoterpene composition of grapes through chemical and microbiological processes. The most major transformation concerns the degradation of nerol and geraniol by the yeast *Saccharomyces cerevisiae via* an enzymatic reduction to form citronellol, alpha-terpineol and linalool [57, 76]. The proportion of the above compounds depends on the yeast strain and grape juice composition [77]. During fermentation, the enzymatic reduction of 3,7-dimethyl-2,5-octadien-1,7-diol or geranyldiol leads to the formation of diendiol (3,7-dimethyl-5-octen-1 ,7-diol), a precursor of rose oxide due to a dehydration reaction under acidic conditions [54]. This diol is also derived from the enzymatic hydroxylation of citronellol in grapes [54]. These results illustrate the deep changes in the flavour of grapes that can result from the fermentative metabolism of yeasts.

Chemical Rearrangements in Grape Juice and Wine

The terpenols themselves may undergo rearrangements in acid to produce other monoterpene alcohols [78]. This is a classic chemical reaction involving the dehydration of alcohols in an acid medium. Thus, concentrations of geraniol and nerol, which constitute part of the aroma of young wines, can rapidly decrease after 2 to 3 years of bottle ageing (for instance, in Muscat), and no longer contribute in the same way to wine aroma. Linalool is more stable. Concentrations of this compound may even actually increase at the beginning of ageing since they are formed from geraniol and nerol [79]. More specifically, the cyclisation of nerol produces alpha terpineol; nerol is transformed into alpha-terpineol and linalool. Linalool also turns into terpine hydrate over time [79]. These results account, at least partially, for the fact that the very intense young character of Muscat wine disappears during ageing, acquiring a resinous odour. Some monoterpene polyols (diols and triols) present in grapes can be transformed into other monoterpenes. Heating accelerates this transformation. Thus, the alcohol dehydratation of 3,7-dimethylocta-1,5-dien-3,7-diol in the wine acidic medium yields (E)-hotrienol, while (E)-2,6-dimethyl-6-hydroxyocta-2,7-dienoic acid has recently been identified as the precursor of the *wine-lactone via* stereoselective cyclisation [80].

Sesquiterpenoids

Sesquiterpenoids ((+)-aromadendrene, α-humulene, α-bisabolol, dehydro-aromadendrene, *etc.*) are secondary metabolites of grapes [81] that do not generally contribute directly to grape and wine flavour as their concentrations are usually in the μg/L range, *i.e.* below the olfactory perception threshold of these compounds. Nevertheless, the characterisation of pepper nuances in Syrah wines succeeded in identifying (-)-rotundone, a powerful sesquiterpene with an olfactory perception threshold in the ng/L range (Table **3**).

Concentrations in Syrah range from 50-600 ng/L, and it is assumed that this compound can contribute to the black pepper flavour of this variety [82, 83].

Table 3: (-)-rotundone a powerful sesquiterpenoid in wine [82, 83].

Name	Structure	Detection threshold (µg/L)	Concentrations in wines (µg/L)
(-)-rotundone	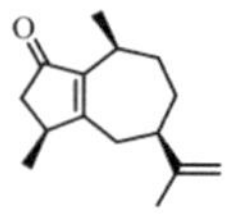	0.016/0.008	0.05-0.6

C13-NORISOPRENOIDS DERIVATIVES

The oxidative degradation of carotenoids, which belong to the family of terpenes with 40 carbon atoms (tetraterpenes), leads to many derivatives, including nor-isoprenoids with 13 carbon atoms (C13-norisoprenoids) that may contribute to the aroma of wines. These compounds were originally studied in tobacco, where they are abundant, but this family of compounds is also the subject of study in grapes since the 1970s [84, 85].

Table 4: Characteristics of some C13norisoprenoids identified in grapes and wines.

Name	Structure	Descriptors	Detection threshold [a] (ng/L)	Concentrations in wines (Min-Max) [b] (ng/L)
ß-damascenone	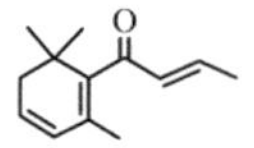	apple sauce - rose	60/2 [c]	100-2500
ß-ionone	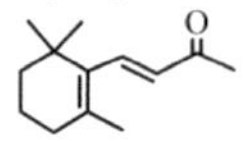	violet	800/120	nd -2415
TPB	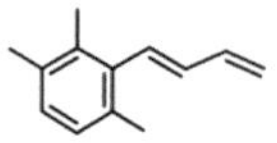	geranium leaf	24/20	nd -120
TDN	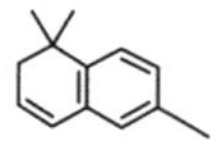	kerosene-like	20 000/3000	nd -30000

[a] Determined at Faculté d'Oenologie, ISVV, University Bordeaux Segalen.

[b] Concentrations from references [22, 88, 206].

[c] Detection threshold in model solution and water, respectively.

Three major groups, each containing various volatile odiferous compounds, are concerned. The oxygenated megastigmane group includes powerful compounds such as beta-damascenone, whose name, invented by Demole, refers to the identification of ketone in the Bulgarian rose (*Rosa Damascena*) [86]. The smell of ß-damascenone is reminiscent of apple sauce and tropical fruit. ß-damascenone has a very low odour threshold in water (2 ng/L) and in model solution (60 ng/L) (Table **4**). This compound, initially identified in grape juice from the Riesling and Scheurebe varieties by Schreier [87] and then in many other varieties [85], has maintained the myth of a major contribution to wine aromas given its very low olfactory threshold in water (2 ng/L). In fact, the perception threshold of ß-damascenone in wine is between 2 and 7 µg/L, although this compound, usually found in concentrations of between 700 ng/L and 2.5 µg/L [22, 88], is rarely the only one involved in the aromatic component of wines. However, ß-damascenone can contribute, *via* synergistic phenomena, to lower the level of ethyl hexanoate [22] and linalool [47]. ß-ionone, with a distinctive smell of violet, has a perception threshold of 120 ng/L in water, 800 ng/L in model solution and 4 µg/L in white wine, and its influence has been demonstrated in various grape and wine varieties [85, 87, 89] . Other C13-oxygenated norisoprenoids, such as 3-oxo-alpha-ionol (tobacco), 3-hydroxy-beta-damascone (tea, tobacco) and ß-damascone (tobacco, fruit), can provide only a very weak potential contribution to wine aroma. The second group consists of non-oxygenated megastigmane compounds, with 1-(2,3,6-trimethylphenyl)buta-1,3-diene (TPB) as a major representative. The detection threshold of this compound, presenting a typical geranium leaf odour, is 40 ng/L in wine and 20 ng/L in water. Concentration ranges in some old Semillon wines can afford 200 ng/L [90-92]. The third group, composed of non-megastigmanes, includes some odorous compounds as TDN (1,1,6-trimethyl-1,2-dihydronaphthalene), which smells like kerosene and has a detection threshold of 20 µg/L [93]; (E) and (Z)-vitispirane, which have camphor/woody nuances; -riesling acetal (fruity descriptor); and actinidol (woody). TDN is considered to account in large part for the "petroleum" aromas of aged Riesling wines [85], while (E) and (Z)-vitispirane, riesling acetal and actinidol are considered to have a limited contribution to wine aroma, particularly Riesling, as their concentrations are usually much lower than their detection threshold [94].

While megastigmane compounds from the first group can be detected in grape must and are present in the young wine, the representatives of the two other groups are only formed during wine ageing.

Genesis of C13-norisoprenoid in Grape and Wine

As previously indicated, C13-norisoprenoid, mainly located in the skin of the berries, originates from grape carotenoid degradation as carotene, lutein, neoxanthin and violaxanthin during grape maturation [85, 95]. In particular, grape exposure to light increases degradation [96]. These compounds undergo first cleavage by grape dioxygenase, followed by reduction and glycosylation, to constitute precursor forms [97, 98]. At the end of maturation, and then during vinification and wine ageing, C13-norisoprenoid compounds are formed in acidic media, through sometimes complex chemical mechanisms from several volatile and non-volatile precursors [84, 85]. For example, ß-damascenone originates from neoxanthin through the following steps: oxidative cleaveage of neoxanthin to produce grasshopper ketone; successive reduction and alcohol dehydratation steps during maturation and vinification then lead to megastigma-4,6,7-trien-3,9-diol and subsequently megastigma-3,4-dien-7-yn-9-ol, which are both transformed into ß-damascenone [99, 100] (Fig. **2**). TDN, as vitispirane, can originate from various enzymatic and chemical steps, typically reduction, cyclisation and alcohol dehydration reactions from various volatile non-odoriferous precursors [101, 102]. Note from an enological point of view that TDN concentrations in wines and spirits are also increased by the bruising and crushing of grapes prior to fermentation [103]. Also, the nitrogen fertilisation of the vine (*V.vinifera* L. var. Riesling) was shown to limit TDN concentrations in the wines during ageing [104].

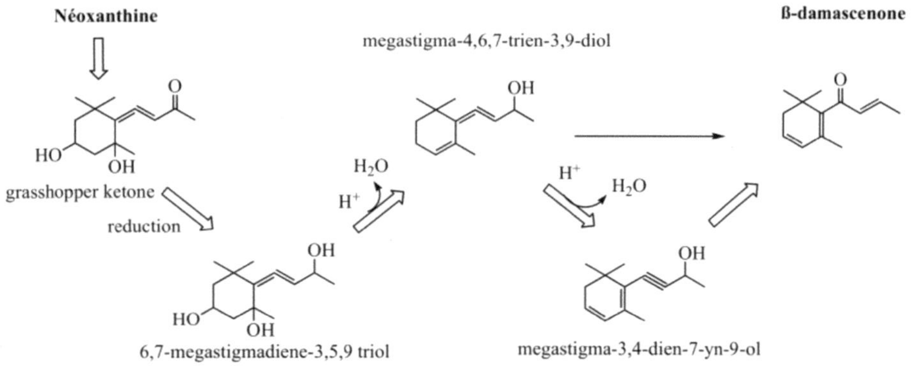

Figure 2: ß-damascenone precursors in grape and wine (adapted from [99, 100]).

IMPACT OF VOLATILE THIOLS

By definition, thiols are sulfur-containing compounds with a sulfhydryl group (-SH) attached to a carbon atom in their chemical structure. Adding a thiol group to a molecule with average sensory properties can transform it into a highly-potent flavour compound, decreasing the perception threshold by several orders of magnitude. Many sulfur compounds in the thiol family are held responsible for olfactory defects. However, it has clearly been demonstrated that they make a major contribution to the aromas of many fruits (blackcurrant, grapefruit

[105], passion fruit [105-107], guava or papaya [108-112], lychee, Durian fruit [113], *etc.*), plants (fringed rue, boxwood, broom, rhubarb, basil, tomato leaves, green tea, blackcurrant buds, *etc.* [114-118]) and foods (roasted coffee, popcorn, grilled meat, virgin olive oil, cheese, *etc.* [119-124]). Finally, the contribution of compounds in this family to the aroma of beer was also reported [125-127].

Table 5: Volatile thiols identified in *Vitis vinifera* wines.

Name	Structure	Description	Perception threshold [a], ng/L	Concentration in wines, ng/L
3-sulfanylhexanol **3SH**	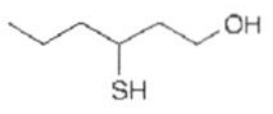	Passion fruit, grapefruit	60/17[b]	100-10 000
3-sulfanylhexyl acetate **3SHA**	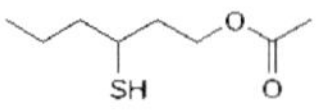	Boxwood, passion fruit	4/2	0-1 000
4-methyl-4-sulfanyl pentan-2-one **4MSP**	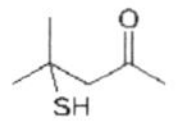	Boxwood, broom	0.8/0.1	0-120
4-methyl-4-sulfanyl pentan-2-ol **4MSPOH**	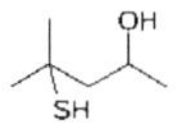	Citrus zest	55/20	0-150
Benzenemethanethiol **BM**	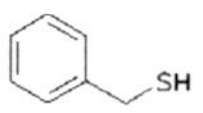	Gunflint, smoke	0.3/nd	1-40
3-methyl-3-sulfanyl butan-1-ol **3MSB**	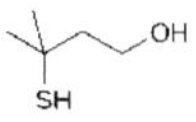	Cooked leeks	1500/1300	0-1 500
2-methyl-3-furanthiol **2M3F**	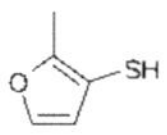	Meaty	4/1	0-300
2-furanmethanethiol **2FM**	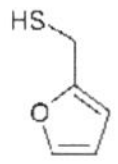	Coffee	0.4/nd	0-400
3-sulfanylpentan-1-ol **3SP**	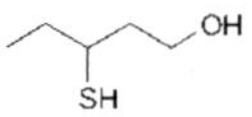	Citrus	900/600	0-400
3-sulfanylheptan-1-ol **3SHp**	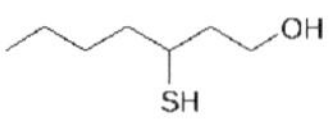	Grapefruit	35/10	0-80
2-methyl-3-sulfanyl butan-1-ol **2M3SB**	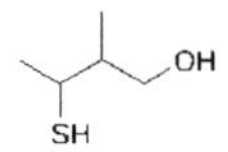	Raw onion	nd	10-70

[a] Determined at Faculté d'Oenologie, ISVV, University Bordeaux Segalen.

[b] Detection threshold in model solution and water, respectively.

nd: not determined.

Over the years, several of these compounds have been detected in wine and their positive contribution to wine flavour, particularly in the varietal aroma of Sauvignon Blanc wines and other white and red varieties, is now well documented. The three most important thiols in Sauvignon Blanc aroma are considered to be 3-sulfanylhexanol (3SH) with a grapefruit flavour, 3-sulfanylhexyl acetate (3SHA) and 4-methyl-4-sulfanylpentan-2-one (4MSP), having a box tree and broom flavour [128-130]. Descriptors such as box tree and broom for 4MSP and grapefruit/passion fruit for 3SH match the occurrence of these compounds in box tree, broom, grapefruit and yellow passion fruit, respectively. Several other odoriferous volatile thiols have also been identified in Sauvignon Blanc wine as 4-methyl-4-sulfanylpentan-2-ol with grapefruit zest flavour and 3-methyl-3-sulfanylbutan-1-ol with leek flavour [129, 131] (Table **5**). Although these varietal thiols were first identified in Sauvignon Blanc wine, they have also been found to contribute to the varietal aroma of wines made from other *Vitis vinifera* varieties, both red and white (Table **6**) [132-136], such as Gewürztraminer, Riesling, Semillon, Manseng [135], as well as Merlot and Cabernet Sauvignon [137].

Table 6: Volatile thiol concentrations (ng/L) in wines made from several *Vitis vinifera* grape varieties [132-136].

	4MSP	3SH	3SHA
Champagne wines	nd	250-640	
Colombard	nd	400-1100	20-60
Gewürztraminer	4-15	1000-3300	0-10
Macabeo	nd	nd	15-20
Merlot (rosé wines)	nd	0-7000	nd
Muscadet	nd	50-450	nd
Muscat	5-30	100-900	nd
Negrette	1-4	800-1500	8-22
Petit Manseng	nd	500-5000	50-150
Pinot Blanc	0-1	90-250	nd
Pinot Gris	0-2	310-1050	2-50
Riesling	2-10	400-1000	0-10
Sauvignon Blanc	5-60	250-15000	10-1000
Semillon	0-5	100-2000	10-100
Botrytised wines	0-100	1000-20000	nd
Sylvaner	0.2-0.5	60-150	nd
Verdejo	nd	nd	40-50

nd: not detected.

More recently, 3-sulfanylpentan-1-ol (3SP), 3-sulfanylheptan-1-ol (3SHp), 2-methyl-3-sulfanylbutan-1-ol (2M3SB) and 2-methyl-3-sulfanylpentan-1-ol (2M3SP) were identified in Sauternes wines [138] (Table **5**). These thiols, together with 3SH, significantly enhance the grapefruit flavour of botrytised sweet white wines, (Table **5**) with 3SP and 3SHp having a pronounced citrus zest aroma. Concentrations of 3SP in botrytised wines are always well below the perception threshold (900 ng/L), while 3SHp rarely exceeds its perception threshold (35 ng/L). However, an additive effect of these volatile thiols, combined with 3SH, has been demonstrated [138]. Concentrations of 3SH, 3SP and 3SHp in botrytised wines are strongly affected by the development of *Botrytis cinerea*. Wines made from healthy grapes contain 3SH but only traces of the other two thiols, while those in the *pourri plein* stage (entirely botrytised but not desiccated) of noble rot have much higher thiol concentrations [135, 139]. Moreover, other thiols contributing to wine flavour, not specifically to varietal flavour, have also been identified (Table **5**).

Due to their functional SH-group, thiol compounds sometimes occur in (*R*)- and (*S*)-enantiomer forms. The enantiomeric distribution of 3SH, which contains one chiral centre, was initially studied in passion fruit [140, 141] and investigated in wines made from Sauvignon Blanc and Semillon grapes by Tominaga *et al.* (2006) [139]. The (*R*)- and (*S*)-enantiomer ratios of these two thiols in dry white Sauvignon Blanc and Semillon wines are approximately 30:70 for 3SHA and 50:50 for 3-sulfanylhexanol. However, in sweet white wines made from botrytised grapes, the proportion of the *R* and *S* forms of 3SH is in the vicinity of 30:70. The aroma descriptors of the two enantiomers of 3SH and 3SHA are quite different, although their perception thresholds are similar. Therefore, the enantiomeric distribution of 3SH and 3SHA in wine may have an impact on the perception and complexity of dry and sweet white wine aromas.

NON-VOLATILE FORMS OF AROMA PRECURSORS

Many odourless and non-volatile compounds in grapes are the source of odoriferous compounds in wine. A distinction should be made between non-volatile compounds found in all grape varieties that could be called non-specific precursors of non-volatile/odourless forms of volatile compounds of grapes, and "linked" forms [142] found specifically, or more abundantly, in certain varieties, and whose release contributes to strengthen the typical varietal aromas of wine.

Non-specific Aroma Precursors

Unsaturated fatty acids with 18 carbon atoms such as linoleic acid and linolenic acid are converted during pre-fermentation operations into C6 aldehyde hexanal and 2 and 3-hexenal through grape lipoxygenase [143, 144], and then reduced to alcohol during fermentation. These C6 aldehydes and alcohols smell like cut grass, but considering their detection threshold in the mg/L range they rarely contribute to the herbaceous character of musts and wines.

Two phenolic acids (p-coumaric acid and ferulic acid) can be decarboxylated by the yeast *S. cerevisiae* hydroxycinnamate decarboxylase to produce volatile phenols such as vinyl-4-phenol and vinyl-4-guaiacol, the former having a smell of heavy gouache and the latter a spicy smell [145, 146]. The enzymatic decarboxylation reaction occurs only in white wine, seeing as the enzyme activity is inhibited by grape flavan-3-ols during the making of red wine [146]. Phenolic acids in grapes exist mainly in the form of tartaric acid esters whose concentration depends on the grape variety and ripening conditions [147]. The hydrolysis of esters has been shown to occur during ripening and especially during vinification due to cinnamate esterase, which up to the 1990s was present in a commercial fungal pectolytic preparation used for winemaking [148, 149]. The presence of certain red wine yeasts belonging to the genus *Brettanomyces sp.* can induce the same phenolic acids to form ethylphenols (ethyl-4-phenol and ethyl-4-gaiacol), whose leather and phenolic odours can affect red wine flavour during ageing [150].

Bound Forms of Varietal Grape Flavour

The fact that a fraction of typical grape aromas in a non-volatile and odourless form are released by a simple chemical or enzymatic reaction is a longstanding hypothesis [41] proved for the first time during the 1970s through the identification of monoterpenes bound to sugar in rose petals [151] and in grapes [142].

Glycosidic Forms

Several teams [152-156] have shown that the main grape terpenols, volatile terpene polyols and some C13 norisoprenoids are bound to sugars to form glycosides [157-159]. Sugars involved in glycosidic forms are ß-D-glucopyranose and disaccharides (α-L-arabinofuranosyl-β-D-glucopyranose; α-L-rhamnopyranosyl-β-D-glucopyranose or rutinose, β-D-apiofuranosyl-β-D-glucopyranose) (Fig. **3**). Glycosylated forms constitute a large portion of the total monoterpenes and C13-norisoprenoids found in grapes [156, 158]. The glycosilation process of sereval volatile monoterpenes and C13-norisoprenoids by *Vitis vinifera* L. cv. Gamay cell suspension cultures was evidenced [63].

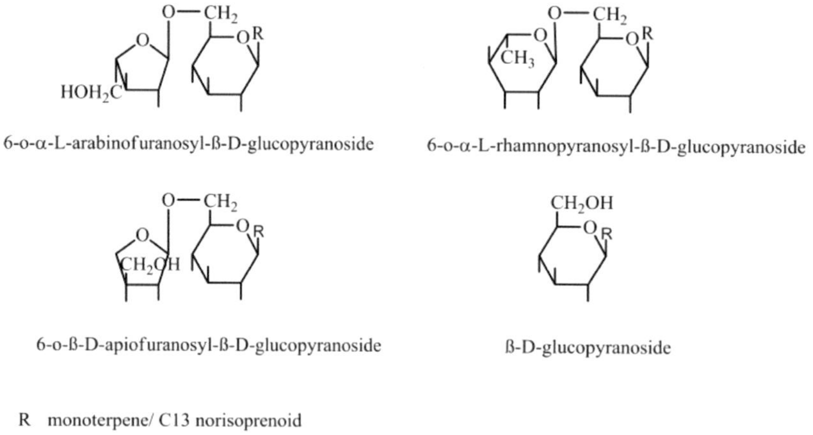

Figure 3: Glycosides identified in *Vitis vinifera* grapes [152-156].

These glycosylated aroma precursors can be released chemically through acid hydrolysis [160, 161] or in the presence of oxidase activities generally following a sequential hydrolysis phenomenon, which depends on the glycoside alpha-arabinosidase, alpha-rhamnosidase, apiosidase hydrolysis and ß-glucosidase hydrolysis [162]. Grape ß-glucosidase activity in grapes, as *S. cerevisiae* ß-glucosidase are not significantly active due to the pH of must [163, 164]. Even so, oxidases produced by *Aspergillus niger* are a way of liberating volatile monoterpenes, C13 norisoprenoids [165, 166], and commercial enzyme preparations derived from *A.niger* have been developed for winemaking since the 1990s [167].

Much more water-soluble than the aglycones, the monoterpene glycosides are considered as forms of transportation and accumulation of hydrophobic substances such as monoterpenes in plants [168]. Glycosides accumulate in the fruit during grape ripening starting from berry set and throughout ripening, in accordance with the proportions previously reported for both volatile and non-volatile forms *i.e.* mainly glycosylated forms, with far fewer volatile odoriferous forms [71, 72, 169]. Natural and viticultural parameters can influence their biogenesis.

Volatile Thiol Precursors: S-conjugates

Sauvignon Blanc musts, like those of many grape varieties with relatively simple aromas, are not highly odoriferous. The characteristic aroma of the grape variety is released during alcoholic fermentation. Volatile thiol formation has now been described in some detail. These compounds, almost totally absent from must, are principally formed during alcoholic fermentation by *Saccharomyces cerevisiae* wine yeast. 3-Sulfanylhexan-1-ol acetate (3SHA) is produced following the release of 3SH by alcohol acetyltransferase, encoded by the ATF1 gene [170]. The final concentration of 3SHA (and other fermentation esters) depends on the balance of alcohol acetyltransferase (promoting esterification of the corresponding alcohol) and esterase activities, encoded by the IAH1 gene (promoting their hydrolysis) in *S. cerevisiae*.

The precursor forms of Sauvignon Blanc aroma compounds were first identified at the Bordeaux Faculty of Enology in the 1990s. A β-lyase specific to *S*-cysteine conjugates was used to release 4MSP, 4MSPOH and 3SH from a non-volatile extract of Sauvignon Blanc aroma precursors, suggesting that these three thiols were present in grapes in cysteinylated form (Fig. **4**) [171, 172]. Tominaga and co-workers first identified the following *S*-cysteinyl conjugates: *S*-4-(4-methylpentan-2-one)-L-cysteine, *S*-4-(4-methylpentan-2-ol)-L-cysteine and *S*-3-(hexan-1-ol)-L-cysteine, as the precursors of 4MSP, 4MSPOH, and 3SH, respectively, *via* a *β*-elimination reaction catalysed by a carbon-sulfur *β*-lyase activity of *S. cerevisiae*. The final yields of these thiols are low compared to the precursor concentrations, with calculated yields ranging from <1% to about 5% [173, 174]. Recent investigations using genetic screens led to the identification of several yeast genes that influence volatile aroma release [175, 176]. In 2002, another form of 3SH precursor, *S*-3-(Hexan-1-ol)-glutathione (Pgsh-3SH), was identified in Sauvignon Blanc must [177]. More recently, *S*-4-(4-methylpentan-2-one)-L-glutathione (Pgsh-4MSP) was identified in Sauvignon Blanc must by high-performance liquid chromatography-tandem mass spectrometry [178]. The presence of these glutathionylated forms of aroma precursors in grape juice may be related to the biosynthesis of the relevant *S*-cysteine conjugates and the subsequent formation of volatile thiols in wine.

The development of several assay methods made it possible to determine the influence of ripening conditions on grape aroma potential and identify the location of cysteinylated thiol precursors in grapes. The changes in their levels during ripening were shown to be dependent on environmental conditions, soil and climate parameters and vineyard management techniques [179-181]. The *S*-cysteine conjugates are not distributed uniformly in Sauvignon Blanc berries. Distribution differs according to the type of precursor and is independent of the stage of ripeness of the grape (Fig. **5**). Pcys-4MSP is mainly localised in the pulp, with approximately 20% in the skin. In contrast, over 50% of Pcys-3SH occurs in the skin [182]. Similarly, a majority (60%) of Pcys-3SH is located in the skins of Cabernet Sauvignon and Merlot grapes [183]. These observations were recently confirmed and completed by studying the distribution of cysteinylated and glutathionylated precursors of 3SH and 4MSP. In Sauvignon Blanc varieties, glutathionylated forms are mainly present in grape skins [184]. This distribution of aroma precursors explains why skin contact enhances the aromatic potential of Sauvignon Blanc musts (and rosés made from Merlot and Cabernet

Sauvignon). During ripening, the aroma potential of grapes varies widely depending on vine water and nitrogen status. Peyrot des Gachons *et al.* [181] demonstrated that grape aroma potential was highest in vines with a moderate water deficit and non-excessive nitrogen supply.

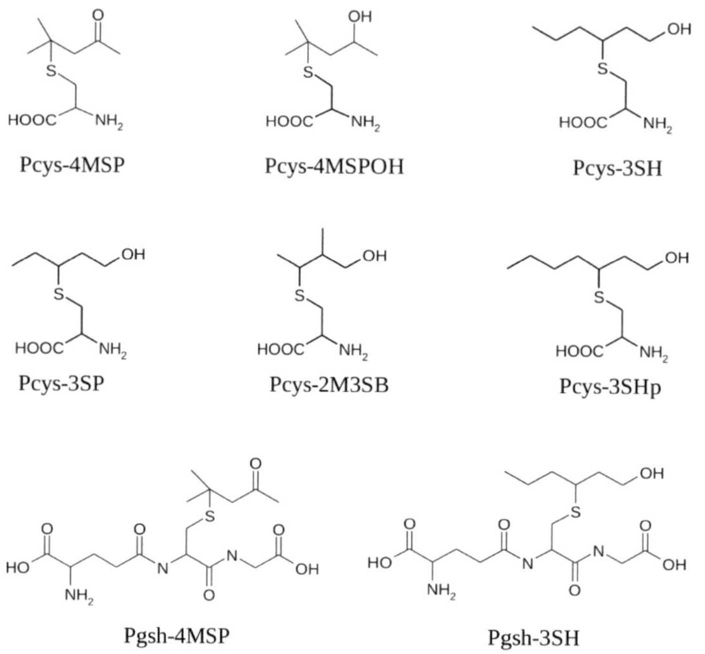

Figure 4: *S*-conjugate precursors of volatile thiols identified in grape must. *S*-4-(4-methylpentan-2-one)-L-cysteine (Pcys-4MSP), *S*-4-(4-methylpentan-2-ol)-L-cysteine (Pcys-4MSPOH), *S*-3-(hexan-1-ol)-L-cysteine (Pcys-3SH), *S*-3-(pentan-1-ol)-L-cysteine (Pcys-3SP), *S*-3-(2-methylbutan-1-ol)-L-cysteine (Pcys-2M3SB), and *S*-3-(heptan-1-ol)-L-cysteine (Pcys-3SHp), *S*-4-(4-methylpentan-2-one)-L-glutathione (Pgsh-4MSP), and *S*-3-(hexan-1-ol)-L-glutathione (Pgsh-3SH) [172, 177, 178].

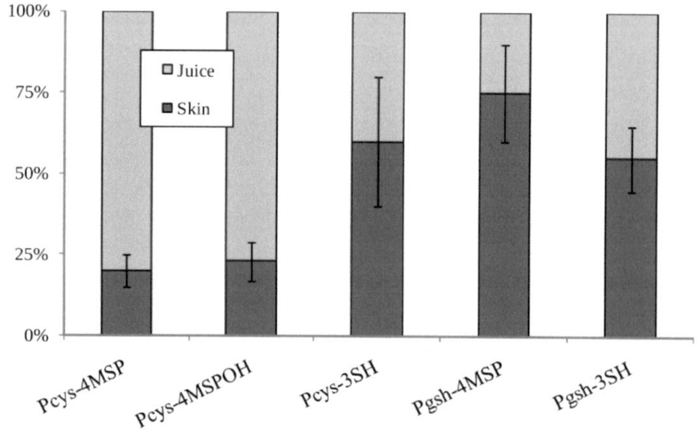

Figure 5: Distribution of cysteinylated and glutathionylated precursors in ripe Sauvignon Blanc berries [182, 184].

The origin of the cysteinylated precursor in grapevines has not been completely determined. The presence of *S*-3-(hexan-1-ol)-glutathione (Pgsh-3SH) in white grape juice may indicate that Pgsh-3SH is a ''pro-precursor'' of Pcys-3SH [177, 185, 186]. Glutathione *S*-conjugates are often involved in the detoxification systems of living organisms. First, toxic compounds are conjugated with glutathione by glutathione *S*-transferase [187, 188], and two enzymes, γ-glutamyltranspeptidase (to eliminate glutamic acid) and carboxypeptidase (to eliminate glycine), split these compounds to form a cysteine *S*-conjugate [189, 190]. Moreover, as *S*-cysteinyl conjugates are found alongside the metabolic degradation pathway of the corresponding *S*-glutathionyl conjugates, *S*-3-(hexan-1-ol)-glutathione has been proposed as the biogenetic precursor of *S*-3-(hexan-1-ol)-L-cysteine.

More recently, a publication by Subileau *et al.* (2008) ruled out *S*-3-(hexan-1-ol)-L-cysteine as the major 3SH precursor, accounting for only 3 - 7% of the total 3SH detected in Sauvignon Blanc wines [191]. Glutathionylated forms of both 4MSP [178] and 3SH [192, 193] were also subsequently found in juice. Glutathionylated 4MSP is converted to 4MSP by yeast with a similar efficiency to the cysteinylated precursor [194]. Glutathionylated 3SH is also converted to thiols by yeast [185], but at a lower efficiency than the cysteinylated version [186]. This conversion by yeast is likely to involve production of the cysteinylated form as an intermediate. Estimates of the ratio of glutathione to cysteine precursors of 3SH in juice vary by up to 100-folds [192, 193], so the relative contribution of the two forms of precursor to 3SH in wine is currently unclear.

An alternative pathway for the biogenesis of 3SH in wine, studied by Schneider *et al.* (2006) [195], started with (E)-*2*-hexenal added to the must at crushing due to the oxidative breakdown of unsaturated fatty acids. This pathway requires the addition of the sulfhydryl group from a sulfur compound, which has yet to be identified (hydrogen sulfide and L-cysteine have been proposed), to the conjugated carbonyl compound. However, the conversion rate of deuterated hexenal to 3SH was found to be very low, suggesting that this pathway only accounted for a small percentage of the total varietal thiol production.

Impact of Grape Botrytization on S-conjugate Biosynthesis

Sauternes, one of the most famous French dessert wines, is produced from *Vitis vinifera* L. cv. Sauvignon Blanc and Semillon grapes. Specific climatic conditions and, above all, the presence of *Botrytis cinerea* in its noble rot form promote a specific over-ripening process, essential for producing these excellent wines. Recently, it was observed that the development of *B. cinerea* on grapes induced significant changes in the volatile thiol content (*1-4*). Indeed, volatile thiol concentrations in botrytised wines are much higher than in dry white wine (over 30-folds for 3SH). Using a new method for Pcys-3SH assays in must, Thibon *et al.* [196, 197] showed that precursor levels in must from both varieties increased considerably as healthy grapes became botrytised (Table **7**). In molar terms, Pcys-3SH concentrations increased by 126- and 82-

Table 7: Quantitative assays of Pcys-3SH in must obtained from botrytised Sauvignon Blanc and Semillon grapes. Comparison of Pcys-3SH levels in relation to decrease in mean berry volume and increase in mean sugar concentrations [197].

Varieties	**Botrytisation stage**	**Variation in berry volume (%)**[a]	**Sugar concentration (g/L)**	**Pcys-3SH**	
				nmol/L	**pmol/berry**
Sauvignon B.	healthy	100	209	57[b] ± 34[c]	53 ± 31
	pourri plein	89	247	7176 ± 535	5876 ± 438
	pourri rôti	41	362	4685 ± 418	1776 ± 158
	late pourri rôti	33	436	5747 ± 325	1738 ± 98
Semillon	healthy	100	200	28 ± 3	29 ± 3
	pourri plein	82	259	2287 ± 335	1950 ± 286
	pourri rôti	55	382	2251 ± 23	1292 ± 46
	late pourri rôti	32	462	1932 ± 286	676 ± 100

[a] Based on the volume obtained by crushing 1000 berries.

[b] Mean value (n=4).

[c] Standard deviation *s* (n=4).

fold after one week when the berries became entirely botrytised but relatively little desiccated. At later stages, concentrations remained stable at about 5000 nmol/l and 2000 nmol/l for Sauvignon Blanc and Semillon, respectively. Analysis of the quantity of precursor per berry (pmol/berry) revealed the same increase factor for Pcys-3SH levels at the *pourri plein* stage, confirming that noble rot induced a significant Pcys-3SH gain in berries that was not simply explained by the desiccation of the grapes. Moreover, a new method for analysing Pcys-3SH highlighted an unusual diastereoisomeric distribution of this precursor in botrytised must, with a higher proportion of the *RS* form of Pcys-3SH (approximately *RR:RS*=30:70) as compared to healthy grape juice (around 50:50) [196]. These results demonstrate the predominant role of botrytisation in developing grape aroma potential. Previous studies showed that the concentrations of volatile thiols, particularly 3-sulfanylhexan-1-ol (3SH), formed during alcoholic fermentation were considerably higher when *Botrytis cinerea* had developed in the grapes [138]. Therefore, the increase in Pcys-3SH observed here explains the higher 3SH concentrations in sweet wines compared to dry wines made from the same grape varieties (30 to 60-fold).

The recent identification and quantification of three new cysteine-*S*-conjugate precursors in must made from botrytised grapes (*S*-3-(pentan-1-ol)-L-cysteine, *S*-3-(heptan-1-ol)-L-cysteine and *S*-3-(2-methylbutan-1-ol)-L-cysteine (Fig. **6**)) confirmed the predominant role of botrytisation in developing grape aroma potential [198]. Indeed, mean Pcys-3SP, Pcys-3SHp and Pcys-2M3SB levels in botrytised must were in the vicinity of 700, 50 and 500 nM, respectively, whereas concentrations in healthy must ranged from 0 to 50 nM.

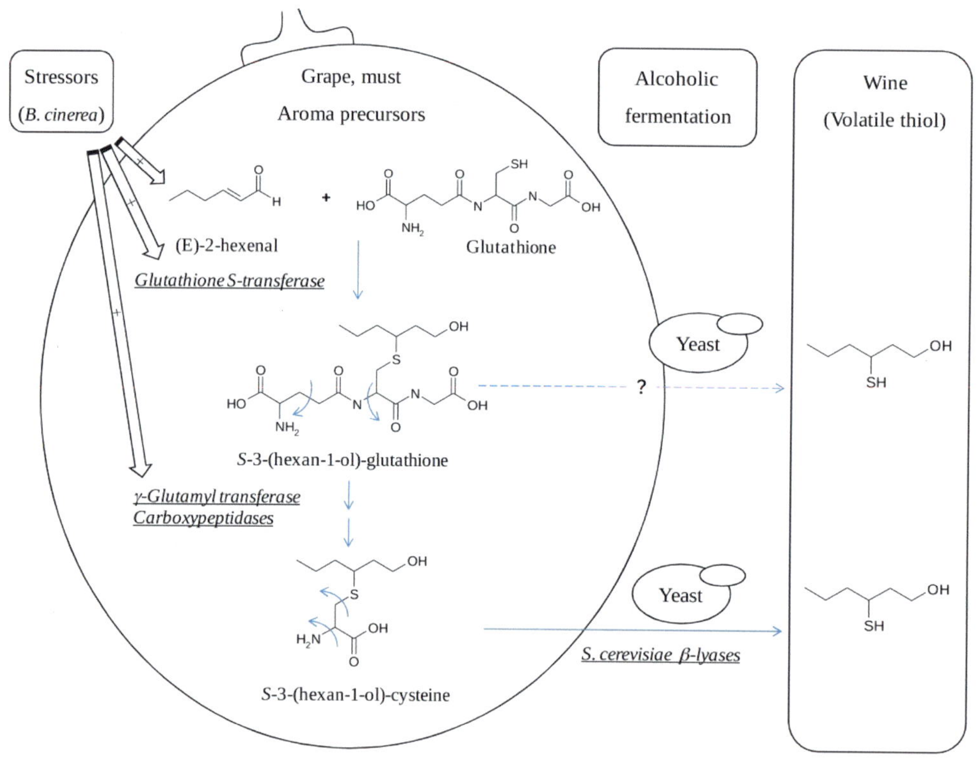

Figure 6: Hypothetical *S*-3-(hexan-1-ol)-glutathione (Pgsh-3SH) and *S*-3-(hexan-1-ol)-cysteine (Pcys-3SH) formation pathway in grapevines, induced by abiotic or biotic (*B. cinerea*) stressors. The production of 3-sulfanylhexanol (3SH) in wine occurred during alcoholic fermentation [205].

The role of *B. cinerea* in developing the aromatic potential of grape juice was clarified by monitoring the diastereoisomeric distribution of Pcys-3SH during several grape overripening processes. In order to determine whether *B. cinerea* enzymes were directly implicated in the formation of Pcys-3SH or whether fungal development stimulated the vine's metabolic pathway, the diastereoisomeric distribution of this precursor was studied during overripening with and without *B. cinerea*. The same Pcys-3SH distribution was observed in healthy overripe and botrytised grapes, with a prevalence of the *RS* form (approximately

RR:*RS*=30:70). Since all enzymes and metabolic pathways have a characteristic stereoselectivity, the Pcys-3SH formation pathways were apparently identical in overripe and botrytised grapes. Consequently, *B. cinerea* is not directly involved in precursor formation but probably stimulates the vine metabolic pathway implicated in Pcys-3SH formation.

The concomitance of cysteine and glutathione *S*-conjugates in must [177, 186, 192, 193, 199] indicates that volatile thiol precursors are synthesised by the pathways involved in vine detoxification processes. Detoxification pathways in plants are activated in response to a variety of stressors, including senescence as well as abiotic (oxidation, injury, *etc.*), and biotic (pathogen) stresses. *Botrytis cinerea* development in grapes stimulates this pathway: lipoxygenase VvLOXC and -O transcripts accumulate rapidly in *B. cinerea*-infected berries [200]. The lipoxygenase-hydroperoxide-lyase-enzyme pathway releases (*E*)-2-hexenal [201-203] and other reactive aldehydes [188]. (*E*)-2-hexenal induces glutathione *S*-transferases [188, 204], thus promoting the formation of *S*-glutathione conjugates, such as the Pgsh-3SH ''pro-precursor". Therefore, the cleavage of Pgsh-3SH by detoxification enzymes may be responsible for the rapid, large increase in Pcys-3SH levels in botrytised grapes. This stimulating effect of the fungus facilitated the determination of the origin of Pcys-3SH in grapes.

More recently, a new simple grapevine model was developed to clarify the role of *B. cinerea* in Pcys-3SH formation and determine the origins of *S*-3-(hexan-1-ol)-L-cysteine and *S*-3-(hexan-1-ol)-L-glutathione in grapes [205]. This model revealed that the plant bioconversion of Pgsh-3SH was considerably amplified when *B. cinerea* was cultured in grape cells. Like some cysteine *S*-conjugates, Pcys-3SH is produced by the breakdown of the corresponding glutathione *S*-conjugate (Pgsh-3SH). The addition of (*E*)-2-hexenal and glutathione to grapevine cells in the presence of *B. cinerea* also resulted in a considerable increase in Pcys-3SH biosynthesis. Therefore, (*E*)-2-hexenal makes a greater contribution than Pgsh-3SH to plant Pcys-3SH production. This result may be due to the fact that reactive aldehydes, including (*E*)-2-hexenal, induce glutathione *S*-transferase [188]. Overall, this study elucidated the origin of Pcys-3SH in grapes, where it derives from the cleavage of *S*-3-(hexan-1-ol)-L-glutathione, itself generated by the conjugation of glutathione on (*E*)-2-hexenal (Fig. **6**).

CONCLUSION

This chapter has summarized our knowledge on four groups of grape derived compounds in which so called "key odorants" are likely to contribute in wines to some of their aromatic characteristics. In general, our understanding of the biosynthesis of these compounds from grapes that contribute to wine flavour is limited. There is no doubt that the evolution of analytical equipment and knowledge of the grape genome will allow in the next ten years to grow in knowledge of the biosynthetic pathways and regulatory mechanisms for these compounds and their precursors. Indeed, the wine aroma been related to hundreds of volatile compounds, probably, the future understanding of the wine aroma will also be related in the understanding of the numerous perceptual interactions phenomena occurring at our brain level of the wine tasters between the numerous volatile compounds that make wine.

ABBREVIATIONS

2M3SB = 2-methyl-3-sulfanylbutan-1-ol

2M3SP = 2-methyl-3-sulfanylpentan-1-ol

3SH = 3-sulfanylhexanol

3SHA = 3-sulfanylhexyl acetate

3SP = 3-sulfanylpentan-1-ol

3SHp = 3-sulfanylheptan-1-ol

4MSP	=	4-methyl-4-sulfanylpentan-2-one
DMAPP	=	Dimethyl allyl pyrophosphate
DXP	=	1-deoxy-D-xylulose-5-phosphate
FLS	=	Flavonol synthase
IBMP	=	2-methoxy-3-isobutylpyrazine
IPMP	=	2-methoxy-3-isopropylpyrazine
IPP	=	Isopentenyl pyrophosphate
OMT	=	O-methyltransferase
Pcys-3SH	=	*S-3-(hexan-1-ol)-L-cysteine*
Pcys-4MSP	=	*S*-4-(4-methylpentan-2-one)-L-cysteine
Pcys-4MSPOH	=	*S*-4-(4-methylpentan-2-ol)-L-cysteine
Pgsh-3SH	=	*S*-3-(hexan-1-ol)-glutathione
Pgsh-4MSP	=	*S*-4-(4-methylpentan-2-one)-L-glutathione
TDN	=	1,1,6-trimethyl-1,2-dihydronaphthalene
TPB	=	1-(2,3,6-trimethylphenyl)buta-1,3-diene
Nd	=	Not detected

REFERENCES

[1] Shepherd GM. Smell images and the flavour system in the human brain. Nature *2*006; 444(7117), 316-21.

[2] Maga JA. Pyrazines in foods: an update. Crit Rev Food Sci Nutr 1982; 16: 1-48.

[3] Buttery RG, Seifert RM, Guadagni DG, Ling LC. Characterization of some volatile constituents of bell peppers. J Agric Food Chem 1969; 17: 1322-7.

[4] Murray KE, Whitfield FB. The occurrence of 3-alkyl-2-methoxypyrazines in raw vegetables. J Sci Food Agric 1975; 26: 973-86.

[5] Roujou de Boubée D. Recherche sur la 2-méthoxy-3-isobutylpyrazine dans les raisins et les vins. Approches analytique, biologique et agronomique 2000.

[6] Bayonove C, Cordonnier RE, Dubois P. Etude d'une fraction caractéristique de l'arôme du raisin de la variété Cabernet-Sauvignon : mise en évidence de la 2-méthoxy-3-isobutylpyrazine. CR Acad Sci Paris D 1975; 60: 1321-9.

[7] Harris RLN, Lacey MJ, Brown WV, Allen MS. Determination of 2-methoxy-3-alkylpyrazines in wine by gas chromatography- mass spectrometry. Vitis 1987; 26: 201-7.

[8] Lacey MJ, Allen MS, Harris RLN, Brown WV. Methoxypyrazines in Sauvignon blanc grapes and wines. Am J Enol Viticult 1991; 42: 103-8.

[9] Allen MS, Lacey MJ, Boyd S. Determination of methoxypyrazines in red wines by stable isotope dilution gas chromatography-mass spectrometry. J Agric Food Chem 1994; 42: 1734-8.

[10] Calo A, Di Stefano R, Costacurta A, Calo G. Caratterizzazione di Cabernet franc e Carmenère (*Vitis* sp.) echiarimenti sulla loro coltura in Italia. Riv Vitic Enol,1991; 44: 3-25.

[11] Belancic A, Agosin E. Methoxypyrazines in grapes and wines of *Vitis vinifera* cv. Carmenere. Am J Enol Viticult 2007; 58: 462-9.

[12] Culleré L, Escudero A, Campo E, Cacho J, Ferreira V. Multidimensional gas chromatography-mass spectrometry determination of 3-alkyl-2-methoxypyrazines in wine and must. A comparison of solid-phase extraction and headspace solid-phas extraction methods. J Chromatogr A 2009; 1216: 4040-5.

[13] Allen MS, Lacey MJ. Methoxypyrazine grape flavour : Influence of climate, cultivar and viticulture. Wein-Wiss 1993; 48: 211-3.

[14] Alberts P, Stander MA, Paul SO, de Villiers A. Survey of 3-alkyl-2-methoxypyrazine content in South African Sauvignon blanc wines using a novel LC-APCI-MS/MS method. J Agric Food Chem 2009; 57: 9347-55.

[15] Augustyn OPH, Rapp A, Van Wyk CJ. Some volatile aroma components of Vitis vinifera L. cv. Sauvignon blanc. South Afric J Enol Vitic 1982; 3(2): 53-60.

[16] Allen MS, Lacey MJ, Harris RLN, Brown WV. Methoxypyrazines of grapes and wines. In: Waterhouse AL, Ebeler SE, Eds. Chemistry of Wine Flavor. American Chemical Society: Washington, DC 1999; Vol. 714, pp. 31-8.

[17] Roujou de Boubée D, Van Leeuwen C, Dubourdieu D. Organoleptic impact of 2-methoxy-3- isobutylpyrazine on red Bordeaux and Loire wines. effect of environmental conditions on concentrations in grapes during ripening. J Agric Food Chem 2000; 48: 4830-4.

[18] Escudero A, Campo E, Farina L, Cacho J, Ferreira V. Analytical characterization of the aroma of five premium red wines. Insights into the role of odor families and the concept of fruitiness of wines. J Agric Food Chem 2007; 55: 4501-10.

[19] Allen MS, Lacey MJ, Harris RLN, Brown WV. Contribution of methoxypyrazines to Sauvignon blanc wine aroma. Am J Enol Viticult 1991; 42: 109-12.

[20] Parr WV, Green JA, White KG, Sherlock RR. The distinctive flavour of New Zealand Sauvignon blanc : sensory characterization by wine professionals. Food Qual Pref 2007; 18: 849-61.

[21] Escudero A, Gogorza B, Melús MA, Ortín N, Cacho J, Ferreira V. Characterization of the aroma of a wine from Maccabeo. Key role played by compounds with low odor activity values. J Agric Food Chem 2004; 52: 3516-24.

[22] Pineau B, Barbe JC, Van Leeuwen C, Dubourdieu D. Which impact for beta-damascenone on red wines aroma? J Agric Food Chem 2007; 55(10): 4103-8.

[23] Hashizume K, Umeda N. Methoxypyrazine content of Japanese red wines. Biosci Biotech Biochem 1996; 60: 802-5.

[24] Roujou de Boubée D, Cumsille AM, Pons M, Dubourdieu D. Location of 2-methoxy-3-isobutylpyrazine in Cabernet sauvignon grape bunches and its extractability during vinification. Am J Enol Viticult 2002; 53: 1-5.

[25] Maggu M, Winz R, Kilmartin PA, Trought MC, Nicolau L. Effect of skin contact and pressure on the composition of Sauvignon blanc must. J Agric Food Chem 2007; 55: 10281–8.

[26] Marais J. Effect of grape temperature, oxidation and skin contact on Sauvignon blanc juice and wine composition and wine quality. S Afr J Enol Vitic 1998; 19: 10-6.

[27] Gallois A, Kergomard A, Adda J. Study of the biosynthesis of 3-isopropyl-2-methoxypyrazine produced by *Pseudomonas taetrolens*. Food Chem 1988; 28: 299-309.

[28] Hashizume K, Samuta T. Grape maturity and light exposure affect berry methoxypyrazine concentration. Am J Enol Vitic 1999; 50: 194-8.

[29] Hashizume K, Tozawa K, Endo M, Aramaki I. S-Adenosyl-L-methionine-dependent O-methylation of 2-hydroxy-3-alkylpyrazine in wine grapes: A putative final step of methoxypyrazine biosynthesis. Biosci Biotech Biochem 2001; 65(10): 795-801.

[30] Hashizume K, Tozawa K, Hiraga Y, Aramaki I. Purification and characterization of a O-methyltransferase capable of methylating 2-hydroxy-3-alkylpyrazine from *Vitis vinifera* L. (cv. Cabernet Sauvignon). Biosci Biotech Biochem 2001; 65(10): 2213-9.

[31] Dunlevy JD, Soole KL, Perkins MV, *et al.* Two O-methyltransferases involved in the biosynthesis of methoxypyrazines: grape-derived aroma compounds important to wine flavour. Plant Mol Biol 2010; 74: 77-89.

[32] Ryona I, Pan BS, Intrigliolo DS, Lakso AN, Sacks GL. Effects of cluster light exposure on 3-isobutyl-2-methoxypyrazine accumulation and degradation patterns in red wine grapes (*Vitis vinifera* L. cv. Cabernet franc). J Agric Food Chem 2008; 56(22): 10838-46.

[33] Ryona I, Pan BS, Sacks GL. Rapid measurement of 3-Alkyl-2-methoxypyrazine content of winegrapes to predict levels in resultant wines. J Agric Food Chem 2009; 57: 8250-7.

[34] Falcao LD, de Revel G, Perello MC, Moutsiou A, Zanus MC, Bordignon-Luiz MT. A survey of seasonal temperatures and vineyard altitude influences on 2-methoxy-3-isobutylpyrazine, C13-norisoprenoids, and the sensory profile of Brazilian Cabernet Sauvignon wines. J Agric Food Chem 2007; 55: 3605-12.

[35] Heymann H, Noble AC, Boulton RB. Analysis of methoxypyrazines in wines. 1. Development of a quantitative procedure. J Agric Food Chem 1986; 34: 268-71.

[36] Maga JA. Sensory and stability properties of added methoxypyrazines to model and authentic wines. In: Charalambous G, Ed. Flavors and off-flavors. Elsevier. Amsterdam 1989; pp. 61-70.

[37] Sala C, Busto O, Guasch J, Zamora F. Influence of vine training and sunlight exposure on the 3-alkyl-2-methoxypyrazines content in musts and wines from the *Vitis vinifera* variety Cabernet Sauvignon. J Agric Food Chem 2004; 52: 3492-7.

[38] Scheiner JJ, Sacks GL, Pan B, *et al.* Impact of severity and timing of basal leaf removal on 3-isobutyl-2-methoxypyrazine concentrations in red winegrapes. Am J Enol Viticult 2010; 61: 358-64.

[39] Sala C, Busto O, Guasch J, Zamora F. Contents of 3-alkyl-2-methoxypyrazines in musts and wines from *Vitis vinifera* variety Cabernet Sauvignon: influence of irrigation and plantation density. J Sci Food Agric 2005; 85: 1131-6.

[40] Pickering G, Lin J, Riesen R, Reynolds A, Brindle I, Soleas G. Influence of *Harmonia axyridis* on the sensory properties of white and red wine. Am J Enol Viticult 2004; 55: 153-9.

[41] Cordonnier RE. Recherches sur l'aromatisation et le parfum des vins doux naturels et des vins de liqueur. Ann Technol Agric 1956; 5: 75-110.

[42] Ribéreau-Gayon P, Boidron JN, Terrier A. The aroma of Muscat grape variety. J Agric Food Chem 1975; 23: 1042-7.

[43] Marais J. Terpenes in the aroma of grapes: a review. S Afric J Enol Vit 1983; 1: 49-60.

[44] Strauss CR, Wilson B, Gooley PR, Williams PJ. Role of monoterpenes in grape and wine flavor. In: Parliament C, Ed. Biogeneration of aromas. American Chemical Society: Washington DC 1986; pp. 222-42.

[45] Rapp A. Studies on terpene compounds in wines. In: Charalambous G, Ed. Frontiers of Flavor. Elsevier, Amsterdam 1987; pp. 799-813.

[46] Bayonove C. L'arôme variétal: le potentiel aromatique du raisin. In: Flanzy C, Ed. Oenologie: Fondements scientifiques et technologiques. Lavoisier Paris 1998; pp. 165-77.

[47] Mateo JJ, Jimenez M. Monoterpenes in grape juice and wines. J Chromatogr 2000; 881(1-2): 557-67.

[48] Darriet P, Gheusi G, Mouret A, Ortega-Pérez I, Vincent JD, Lledo PM. A contribution of olfactory bulb maps to odor mixture perception. In Oenologie: Fondements scientifiques et technologiques. Bordeaux 2009.

[49] Crespan M. The parentage of Muscat of Hamburg. Vitis 2003; 42: 193-7.

[50] Lacombe T, Boursiquot JM, Laucou V, Dechesne F, Varès D, This P. Relationships and genetic diversity within the accessions related to Malvasia held in the Domaine de Vassal grape germplasm repository. Am J Enol Viticult 2007; 58: 124-31.

[51] Luan F, Mosandl A, Gubesch M, Wüst M. Enantioselective analysis of monoterpenes in different grape varieties during berry ripening using stir bar sorptive extraction- and solid phase extraction-enantioselective-multidimensional gas chromatography-mass spectrometry. J Chromatogr A 2006; 1112: 369-74.

[52] Boelens MH, Boelens H, van Gemert LJ. Sensory Properties of Optical Isomers. Perfum Flavor 1993; 18(6): 1-15.

[53] Guth H. Quantitation and sensory studies of character impact odorants of different white wine varieties. J Agric Food Chem 1997; 45: 3027-32.

[54] Koslitz S, Renaud L, Kohler M, Wüst M. Stereoselective formation of the varietal aroma compound rose oxide during alcoholic fermentation. J Agric Food Chem 2008; 56(7): 1371-5.

[55] Strauss CR, Wilson B, Williams PJ. Novel monoterpene diols and diol glycosides in *Vitis vinifera* grapes. J Agric Food Chem 1988; 36: 569-73.

[56] Versini G, Rapp A, Reniero F, Mandery H. Structural identification and presence of some p-menth-1-enediols in grape products. Vitis 1991; 30: 143-9.

[57] Dugelay I, Gunata Z, Sapis JC, Baumes R, Bayonove C. Etude de l'origine du citronellol dans les vins. J Int Sci Vigne Vin 1992; 26: 177-84.

[58] Rohmer M. The discovery of a mevalonate-independent pathway for isoprenoid biosynthesis in bacteria, algae and higher plants. Nat Prod Rep 1999; 16: 565-74.

[59] Luan F, Wust M. Differential incorporation of 1-deoxy-D-xylulose into (3S)-linalool and geraniol in grape berry exocarp and mesocarp. Phytochem 2002; 60: 451-9.

[60] Cseke L, Dudareva N, Pichersky E. Structure and evolution of linalool synthase. Mol Biol Evol 1998; 15: 1491-8.

[61] Yang T, Li J, Wang HX, Zeng Y. A geraniol-synthase gene from Cinnamonum tenuipilum. Phytochem 2005; 52: 2036-41.

[62] Luan F, Hampel D, Mosandl A, Wust M. Enantioselective analysis of free and glycosidically bound monoterpene polyols in *Vitis vinifera* L. cvs. Morio Muscat and Muscat Ottonel: Evidence for an oxidative monoterpene metabolism in grapes. J Agric Food Chem 2004; 56(7): 1371-5.

[63] Mathieu S, Wirth J, Sauvage FX, Lepoutre J-P, Baumes R, Gunata Z. Biotransformation of C13-norisoprenoids and monoterpenes by a cell suspension culture of cv. Gamay (*Vitis vinifera*). Plant Cell Tiss Organ Cult 2009; 97: 203-13.

[64] Jaillon O, Aury J-M, Noel B, *et al.* The grapevine genome sequence suggests ancestral hexaploidization in major angiosperm phyla. Nature 2007; 449: 463-7.

[65] Martin DM, Bohlmann J. Identification of *Vitis vinifera* (-)-alpha-terpineol synthase by *in silico* screening of full-length cDNA ESTs and functional characterization of recombinant terpene synthase. Phytochem 2004; 65: 1223-9.

[66] Duchêne E, Butterlin G, Claudel P, Dumas V, Jaegli N, Merdinoglu D. A grapevine (*Vitis vinifera* L.) deoxy-D-xylulose synthase gene colocates with a major quantitative trait loci for terpenol content. Theor Appl Genet 2009; 118: 541-52.

[67] Battilana J, Costantini L, Emanuelli F, *et al.* The 1-deoxy-D-xylulose 5-phosphate synthase gene co-localizes with a major QTL affecting monoterpene content in grapevine. Theor Appl Genet 2009; 118: 653-69.

[68] Wilson B, Strauss CR, Williams PJ. The distribution of free and glycosidically-bound monoterpenes among skin, juice, and pulp fractions of some white grape varieties. Am J Enol Vitic 1986; 37: 107-11.

[69] Gunata Z, Bayonove CL, Baumes RL, Cordonnier RE. The aroma of grapes. Localisation and evolution of free and bound fractions of some grape aroma components cv Muscat during first development and maturation. J Sci Food Agric 1985; 36: 857-62.

[70] Park S, Morrison JC, Adams D, Noble AC. Distribution of free and glycosidically bound monoterpenes in the skin of Muscat of Alexandrie grapes during development. J Agric Food Chem 1991; 39: 514-8.

[71] Wilson B, Strauss CR, Williams PJ. Changes in free and glycosidically bound monoterpenes in developing muscat grapes. J Agric Food Chem 1984; 32: 919-24.

[72] De Billerbeck GM, Ebang-Oke JP, Ambid C. La S-linalyl synthase chez *Vitis vinifera* L cv. Muscat de Frontignan - Approches biochimique et moléculaire In: Lonvaud A, de Revel G, Darriet Ph, Eds. Oenologie 2003. Tec&Doc Lavoisier, Paris 2003; pp. 89-93.

[73] Belancic A, Agosin E, Ibacache A, Bordeu E, Baumes R, Razungles A, Bayonove C. Influence of sun exposure on the aromatic composition of Chilean Muscat grape cultivars Moscatel de Alejandria and Moscatel rosada. Am J Enol Vitic 1997; 48: 181-5.

[74] Boidron JN. Relation entre les substances terpéniques et la qualité du vin (Role de *Botrytis cinerea*). Ann Technol Agric 1978; 27: 141-5.

[75] Mirata M-A, Wüst M, Mosandl A, Schrader J. Fungal biotransformation of (±)-linalool. J Agric Food Chem 2008; 56: 3287-96.

[76] Gramatica P, Manitto P, Maria Ranzi B, Delbianco A, Francavilla M. Stereospecific reduction of geraniol to (R)-(+)-citronellol by *Saccharomyces cerevisiae*. Experientia 1982; 38: 775-6.

[77] Garcia Moruno E, Ribaldone M, Di Stefano R, Conterno L, Gandini A. Study of five strains of *Saccharomyces cerevisiae* with regard to their metabolism towards geraniol. J Int Sci Vigne Vin 2004; 36: 221-5.

[78] Pedersen DS, Capone DL, Skouroumounis GK, Pollnitz AP, Sefton MA. Quantitative analysis of geraniol, nerol, linalool, and a-terpineol in wine. Anal. Bioanal Chem 2003; 375: 517-22.

[79] Marais J, Van Wyk C, Rapp A. Effect of storage time, temperature and region on the levels of 1,1,6-Trimethyl-1, 2-dihydronaphthalene and other volatiles, and on quality of Weisser Riesling wines. S Afr J Enol Vitic 1992; 13: 33-44.

[80] Luan F, Degenhardt A, Mosandl A, Wüst M. Mechanism of wine lactone formation: Demonstration of stereoselective cyclization and 1,3-hydride shift. J Agric Food Chem 2006; 54; 10245-52.

[81] Coelho E, Rocha SM, Delgadillo I, Coimbra M. A headspace-SPME applied to varietal volatile components evolution during *Vitis vinifera* L. cv. "Baga" ripening. Anal Chim Acta 2006; 563: 204-14.

[82] Siebert TE, Wood C, Elsey GM, Pollnitz AP. Determination of rotundone, the pepper aroma impact compound, in grapes and wine. J AgricFood Chem 2008; 56: 3745-48.

[83] Wood C, Siebert TE, Parker M, *et al.* From wine to pepper: rotundone, an obscure sesquiterpene, is a potent spicy aroma compound. J Agric Food Chem 2008; 56: 3738-44.

[84] Winterhalter P, Sefton MA, Williams PJ. Volatile C-13- Norisoprenoid compounds in Riesling wine are generated from multiple precursors. Am J Enol Vitic 1990; 41: 277-83.

[85] Mendes-Pinto MM. Carotenoid breakdown products the -norisoprenoids- in wine aroma. Arch Biochem Biophys 2009; 483: 236-45.

[86] Demole E, Enggist P, Sauberli U, Stoll M, Kováts E. Structure et synthèse de la damascénone, constituant odorant de l'essence de Rose Bulgare. Helv Chim Acta 1970; 541-51.

[87] Schreier P, Drawert F, Junker A. Identification of volatile constituents from grapes. J Agric Food Chem 1976; 24(2): 331-6.

[88] Kotseridis Y, Baumes RL, Bertrand A, Skouroumounis GK. Quantitative determination of free and hydrolytically liberated β-damascenone in red grapes and wines using a stable isotope dilution assay. J Chromatogr A 1999; 849: 245-54.

[89] Kotseridis Y, Baumes RL, Bertrand A, Skouroumounis GK. Quantitative determination of ß-ionone in red wines and grapes of Bordeaux using a stable isotope dilution assay. J Chromatogr A 1999; 848: 317-25.

[90] Janusz A, Capone DL, Puglisi CJ, Perkins MV, Elsey GM, Sefton MA. (E)-1-(2,3,6-Trimethylphenyl)buta-1,3-diene: A Potent Grape-Derived Odorant in wine. J Agric Food Chem 2003; 51: 7759-63.

[91] Cox A, Skouroumounis GK, Elsey GM, Perkins MV, Sefton MA. Generation of (E)-1-(2,3,6-trimethylphenyl)buta-1,3-diene from C 13-norisoprenoid precursors. J Agric Food Chem 2005; 53(17): 6777-83.

[92] Cox A, Capone DL, Elsey GM, Perkins MV, Sefton MA. Quantitative Analysis, Occurrence, and Stability of (E)-1-(2,3,6-Trimethylphenyl)buta-1,3-diene in wine. J Agric Food Chem 2005; 53: 3584-91.

[93] Simpson RF. 1,1,6-trimethyl-1,2-dihydronaphtalene: an important contributor to the bottle aged bouquet of wine. Chem Ind 1976; 37.

[94] Eggers NJ, Bohna K, Dooley B. Determination of vitispirane in wines by stable isotope dilution assay. Am J Enol Vitic 2006; 57: 226-32.

[95] Razungles A, Gunata Z, Pinatel S, Baumes R, Bayonove C. Etude quantitative de composés terpéniques, norisoprénoïdes et de leurs précurseurs dans diverses variétés de raisins. Sci. Alim 1993; 13: 59-72.

[96] Marais J, Hunter JJ, Haasbroek PD. Effect of canopy microclimate, season and region on Sauvignon blanc grape composition and wine quality. S Afr J Enol Vitic 1999; 20: 19-30.

[97] Baumes R, Wirth J, Bureau S, Gunata Y, Razungles A. Biogeneration of C13-norisoprenoid compounds: Experiments supportive for an apo-carotenoid pathway in grapevines. Anal Chim Acta 2002; 458(1): 3-14.

[98] Mathieu S, Terrier N, Procureur J, Bigey F, Günata Z. A Carotenoid cleavage dioxygenase from *Vitis vinifera* L: functional characterization and expression during grape berry development in relation to C-norisoprenoid accumulation . J Exp Bot 2005; 56: 2721-31.

[99] Puglisi CJ, Elsey GM, Prager RH, Skouroumounis GK, Sefton MA. Identification of a precursor to naturally occurring ß-damascenone. Tetrahedr Lett 2001; 42: 6937-9.

[100] Puglisi CJ, Daniel MA, Capone DL, Elsey GM, Prager RH, Sefton MA. Precursors to damascenone: synthesis and hydrolysis of isomeric 3,9-dihydroxymegastigma-4,6,7-trienes. J Agric Food Chem 2005; 53(12): 4895-900.

[101] Winterhalter P. 1,1,6-Trimethyl-1,2-dihydronaphthalene (TDN) formation in wine. 1. Studies on the hydrolysis of 2,6,10,10-tetramethyl-1-oxaspiro[4.5]dec-6-ene-2,8-diol rationalising the origin of TDN and related C13-norisoprenoids in Riesling wine. J Agric Food Chem 1991; 39: 1825-9.

[102] Daniel MA, Capone DL, Sefton MA, Elsey GM. Riesling acetal is a precursor to 1,1,6-trimethyl-1,2-dihydronaphthalene (TDN) in wine. Aust J Grape Wine Res 2009; 15: 93-6.

[103] Galy B. Incidence qualitative des matériels d'extraction des moûts dans la région de Cognac. In: 7ème colloque viticole et oenologique EUROVITI 1993.

[104] Linsenmeier AW, Löhnertz O. Changes in norisoprenoid levels with long-term nitrogen fertilisation in different vintages of *Vitis vinifera* var. Riesling wines. S Afr J Enol Vitic 2007; 28(1): 17-24.

[105] Buettner A, Schieberle P. Characterization of the most odor-active volatiles in fresh, hand-squeezed juice of grapefruit (*Citrus paradisi* Macfayden). J Agric Food Chem 1999; 47(12): 5189-93.

[106] Engel KH, Tressl R. Identification of new sulfur-containing volatiles in yellow passionfruit (*Passiflora edulis f. flavicarpa*). J Agric Food Chem 1991; 39: 2249-52.

[107] Tominaga T, Dubourdieu D. Identification of cysteinylated aroma precursors of certain volatile thiols in passion fruit juice. J Agric Food Chem 2000; 48(7): 2874-6.

[108] Clery RA, Hammond CJ. New sulfur components of pink guava fruit (Psidium guajava L.). J Essent Oil Res 2008; 20(4): 315-7.

[109] Konig T, Ruff C, Kleinschnitz M, Schreier P, Fischer N, Neugebauer W. Enantiomeric distribution of 2-pentanethiol in guava fruit (*Psidium guajava* L.) by multidimensional gas chromatography with sulfur chemiluminescence detection. HRC J High Res Chrom 1998; 21(6): 371-2.

[110] Vernin G, Vernin C, Pieribattesti JC, Roque C. Analysis of the volatile compounds of *Psidium cattleianum* Sabine fruit from Reunion Island. J Essent Oil Res 1998; 10(4): 353-62.

[111] Idstein H, Keller T, Schreier P. Volatile constituents of mountain papaya (*Carica candamarcensis*, syn. *C. pubescens* Lenne et Koch) fruit. J Agric Food Chem 1985; 33(4): 663-6.

[112] Idstein H, Schreier P. Volatile constituents from guava (*Psidium guajava*, L.) fruit. J Agric Food Chem 1985; 33(1): 138-43.

[113] Weenen H, Koolhaas WE, Apriyantono A. Sulfur-containing volatiles of Durian fruits (*Durio zibethinus* Murr.). J Agric Food Chem 1996; 44(10): 3291-3.

[114] Escher S, Niclass Y, van de Waal M, Starkenmann C. Combinatorial synthesis by nature: volatile organic sulfur-containing constituents of *Ruta chalepensis* L. Chem Biodivers 2006; 3(9): 943-57.

[115] Tominaga T, Dubourdieu D. Identification of 4-mercapto-4-methylpentan-2-one from the box tree (*Buxus sempervirens L.*) and broom (*Sarothamnus scoparius (L.) Koch.*). Flavour Frag J 1997; 12(6): 373-6.

[116] Patterson RL. Catty odours in food: confirmation of the identity of 4-mercapto-4-methylpentan-2-one by GC-MS. Chemistry and Industry (London): 1969; 2: 48-9.

[117] Kumazawa K, Masuda H. Identification of potent odorants in Japanese green tea (Sen-cha). J Agric Food Chem 1999; 47(12): 5169-72.

[118] Masuda H, Kumazawa K. The change in the flavor of green and black tea drinks by the retorting process. In: ACS Symposium Series 2000; Vol. 754, pp. 337-46.

[119] Gasser U, Grosch W. Identification of volatile flavour compounds with hight aroma values from cooked beef. Z Lebensm Unters Forsch 1988; 186: 489-94.

[120] Guth H, Grosch W. Identification of the character impact odorants of stewed beef juice by instrumental analysis and sensory studies. J Agric Food Chem, 1994; 42: 2862-6.

[121] Schieberle P. Primary odorants in popcorn. J Agric Food Chem 1991; 39: 1141-4.

[122] Tressl R, Silwar R. Investigation of sulfur-containing components in roasted coffee. J Agric Food Chem 1981; 29: 1078-1082.

[123] Reiners J, Grosch W. Concentration of 4-methoxy-2-methyl-2-butanethiol in Spanish virgin olive oils. Food Chem 1999; 64(1): 45-7.

[124] Sourabié AM, Spinnler HE, Bonnarme P, Saint-Eve A, Landaud S. Identification of a powerful aroma compound in Munster and Camembert cheeses: Ethyl 3-mercaptopropionate. J Agric Food Chem 2008; 56(12): 4674-80.

[125] Vermeulen C, Lejeune I, Tran TT, Collin S. Occurrence of polyfunctional thiols in fresh lager beers. J Agric Food Chem 2006; 54(14): 5061-8.

[126] Takoi K, Degueil M, Shinkaruk S, *et al.* Specific flavor compounds derived from Nelson Sauvin hop and synergy of these compounds. BrewingScience 2009; 62: 108.

[127] Takoi K, Degueil M, Shinkaruk S, *et al.* Identification and characteristics of new volatile thiols derived from the hop (*Humulus luplus* L.) cultivar nelson Sauvin. J Agric Food Chem 2009; 57(6): 2493-502.

[128] Tominaga T, Darriet P, Dubourdieu D. Identification of 3-mercaptohexyl acetate in Sauvignon wine, a powerful aromatic compound exhibiting box-tree odor. Vitis 1996; 35(4): 207-10.

[129] Tominaga T, Furrer A, Henry R, Dubourdieu D. Identification of new volatile thiols in the aroma of *Vitis vinifera* L. var. Sauvignon blanc wines. Flavour Frag J *1*998; 13(3): 159-62.

[130] Darriet P, Tominaga T, Lavigne V, Boidron JN, Dubourdieu D. Identification of a powerful aromatic component of *Vitis vinifera* L. var. Sauvignon wines: 4-mercapto-4-methylpentan-2-one. Flavour Frag J 1995; 10(6): 385-92.

[131] Tominaga T, Murat ML, Dubourdieu D. Development of a method for analyzing the volatile thiols involved in the characteristic aroma of wines made from *Vitis vinifera* L. cv. Sauvignon blanc. J Agric Food Chem 1998; 46: 1044-8.

[132] Fretz CB, Känel S, Luisier JL, Amado R. Analysis of volatile components of Petite Arvine wine. Eur Food Res Technol 2005; 221: 504-10.

[133] Fretz CB, Luisier JL, Tominaga T, Amado R. 3-Mercaptohexanol : An aroma impact compound of Petite Arvine wine. Am J Enol Vitic 2005; 56(4): 407-10.

[134] Mateo-Vivaracho L, Ferreira V, Cacho J. Automated analysis of 2-methyl-3-furanthiol and 3-mercaptohexyl acetate at ng L-1 level by headspace solid-phase microextracion with on-fibre derivatisation and gas chromatography-negative chemical ionization mass spectrometric determination. J Chromatogr A 2006; 1121(1): 1-9.

[135] Tominaga T, Baltenweck-Guyot R, Peyrot des Gachons C, Dubourdieu D. Contribution of volatile thiols to the aromas of white wines made from several *Vitis vinifera* grape varieties. Am J Enol Vitic *2*000; 51: 178-81.

[136] Tominaga T, Guimbertau G, Dubourdieu D. Role of certain volatile thiols in the bouquet of aged champagne wines. J Agric Food Chem 2003; 51(4): 1016-20.

[137] Bouchilloux P, Darriet P, Henry R, Lavigne-Cruege V, Dubourdieu D. Identification of volatile and powerful odorous thiols in Bordeaux red wine varieties. J Agric Food Chem 1998; 46(8): 3095-9.

[138] Sarrazin E, Shinkaruk S, Tominaga T, Bennetau B, Frerot E, Dubourdieu D. Odorous impact of volatile thiols on the aroma of young botrytized sweet wines: identification and quantification of new sulfanyl alcohols. J Agric Food Chem 2007; 55(4): 1437-44.

[139] Tominaga T, Niclass N, Frérot E, Dubourdieu D. Stereoisomeric distribution of 3-mercaptohexan-1-ol and 3-mercaptohexyl acetate in dry and sweet white wines made from *Vitis vinifera* (var. sauvignon blanc and Semillon). J Agric Food Chem 2006; 54(19): 7251-5.

[140] Weber B, Dietrich A, Maas B, Marx A, Olk J, Mosandl A. Stereoisomeric flavour compounds. LXVI. Enantiomeric distribution of the chiral sulphur-containing alcohols in yellow and purple passion fruits. Z Lebensm Unters Forsch 1994; 199: 48-50.

[141] Weber B, Maas B, Mosandl A. Stereoisomeric Flavor Compounds. 72. Stereoisomeric Distribution of Some Chiral Sulfur-Containing Trace Components of Yellow Passion Fruits. J Agric Food Chem 1995; 43(9): 2438-41.

[142] Cordonnier RE, Bayonove C. Mise en evidence dans la baie de raisin, variété Muscat d'Alexandrie, de monoterpènes liés, révélables par une ou plusieurs enzymes du fruit. CR Acad Sci Paris 1974; 278: 3387–90.

[143] Drawert F, Heimann W, Emberger R, Tressl R. Über die Biogenese von Aromastoffen bei Pflanzen und Früchten, II. Enzymatische Bildung von Hexen-(2)-al-(1), Hexanal und deren Vorstufen. Justus Liebigs Ann Chem 1966; 694: 200-8.

[144] Gardner HW. Recent investigations into the lipoxygenase pathway of plants. Biochem Biophys Acta 1991; 1084 221-39.

[145] Albagnac G. La décarboxylation des acides cinnamiques substitués par les levures. Ann Technol Agric 1975; 24: 133-41.

[146] Chatonnet P, Dubourdieu D, Boidron JN, Lavigne V. Synthesis of volatile phenols by S. cerevisiae in wines. J Sci Food Agric 1993; 62: 191-202.

[147] Romeyer F, Macheix JJ, Goiffon JP, Reminiac C, Sapis JC. Browning capacity of grapes. 3. Changes and importance of hydroxycinnamic acid-tartaric acid esters during development and maturation of the fruit. J Agric Food Chem 1983; 31: 346-9.

[148] Dugelay I, Gunata Z, Sapis J-C, Baumes R, Bayonove C. Role of cinnamoyl esterase activities from enzyme preparations on the formation of volatile phenols during winemaking. J Agric Food Chem 1993; 41(11): 2092-6.

[149] Barbe C, Dubourdieu D. Characterisation and purification of a cinnamate esterase from *Aspergillus niger* industrial pectinase preparation. J Sci Food Agric 1998; 78(4): 471-8.

[150] Chatonnet P, Dubourdieu D, Boidron JN, Pons M. The origin of ethylphenols in wines. J Sci Food Agric. 1992; 60: 165-78.

[151] Francis MJO, Allcock C. Geraniol ß-D-glucoside: occurrence and synthesis in rose flowers *Phytochem.*1969; 8: 1339-47.

[152] Williams PJ, Strauss CR, Wilson B. Novel monoterpene disaccharide glycoside of *Vitis vinifera* grapes and wines. Phytochem 1982; 21: 2013-20.

[153] Voirin S, Baumes R, Bitteur S, Günata Z, Bayonove C. Novel monoterpene disaccharide glycosides in *Vitis vinifera* grapes. J Agric Food Chem 1990; 38: 1373-8.

[154] Gunata Z, Bayonove CL, Baumes RL, Cordonnier RE. The aroma of grapes. I. Extraction and determination of free and glycosidically bound fraction of some grape aroma components. J Chromatogr A 1985; 331: 83-90.

[155] Strauss CR, Wilson B, Williams PJ. Novel monoterpene diols and diol glycosides in *Vitis vinifera* grapes. J Agric Food Chem 1988; 36: 569-73.

[156] Voirin S, Baumes R, Sapis JC, Bayonove C. Analytical methods for monoterpene glycosides in grape and wine. II Qualitative and quantitative determination of monoterpene glycosides in grape. J Chromatogr A 1992; 595: 269-81.

[157] Winterhalter P, Sefton MA, Williams PJ. Two-dimensional GC-DCCC analysis of the glycoconjugates of monoterpenes, norisoprenoids, and shikimate-derived metabolites from Riesling wine. J Agric Food Chem 1990; 38(4): 1041-8.

[158] Winterhalter P, Schreier P. C13-norisoprenoid glycosides in plant tissues: An overview on their occurrence, composition and role as flavour precursors. Flav Fragr J 1994; 9: 281-7.

[159] Baumes R, Aubert C, Gunata Y, De Moor W, Bayonove C. Structure of two C13 norisoprenoid glycosidic precursors of wine flavor. J Essent Oil Res 1994; 6: 587-99.

[160] Williams PJ, Strauss CR, Wilson B, Massy-Westropp RA. Studies on the hydrolysis of *Vitis vinifera* monoterpene precursor compounds and model ß-D-glucosides rationalizing the monoterpene composition of grape. J Agric Food Chem 1982; 30: 1219-23.

[161] Skouroumounis GK, Sefton MA. Acid-catalyzed hydrolysis of alcohols and their β-D-glucopyranosides. J Agric Food Chem 2000; 48: 2033-9

[162] Gunata Z, Bitteur S, Brillouet JM, Bayonove C, Cordonnier RE. Sequential enzymic hydrolysis of potentially aromatic glycosides from grape. Carbohydr Res 1988; 194: 139-49.

[163] Aryan AP, Wilson B, Strauss CR, Williams PJ. The properties of glycosidases of *Vitis vinifera* and comparison of their β-glucosidase activity with that of exogenous enzymes. An assessment of possible applications in enology. Am J Enol Viticult 1987; 38: 182–8.

[164] Delcroix A, Gunata Z, Sapis JC, Salmon JM, Bayonove C. Glycosidase activities of three enological yeast strains during wine making. Effect on the terpenol content of Muscat wine. Am J Enol Vitic 1994; 45: 291-6.

[165] Maicas S, Mateo JJ. Hydrolysis of terpenyl glycosides in grape juice and other fruit juices: a review. Appl Microbiol Biotechnol 2005; 67: 322-35.

[166] Gunata Z, Bayonove C, Tapiero C, Cordonnier R. Hydrolysis of grape monoterpenyl β-D-glucosides by various β-glucosidases. J Agric Food Chem 1990; 38: 1222-36.

[167] Cabaroglu T, Selli S, Canbas A, Lepoutre J-P, Günata Z. Wine flavor enhancement through the use of exogenous fungal glycosidases. Enz Microb Techn 2003; 33(5): 581-7.

[168] Stahl-Biskup E. Monoterpene glycosides, state-of-the-art. Flav Fragr J 1987; 2(2): 75-82.

[169] Razungles A, Günata Z, Pinatel S, Baumes R, Bayonove C. Etude quantitative de composés terpéniques, norisoprénoïdes et de leurs précurseurs dans diverses variétés de raisins. Sci Alim 1993; 13: 59-72.

[170] Swiegers JH, Willmott R, Hill-Ling A, *et al.* Modulation of volatile thiol and ester aromas by modified wine yeast. In: Developments in Food Science 2006; Vol. 43, pp. 113-6.

[171] Tominaga T, Masneuf L, Dubourdieu D. Mise en évidence d'un S-conjugué de la cystéine, précurseur d'arôme du Sauvignon. J Int Sci Vigne Vin 1995; 29(4): 227-32.

[172] Tominaga T, Peyrot Des Gachons C, Dubourdieu D. A new type of flavor precursors in *Vitis vinifera* L. cv. Sauvignon blanc: S-cysteine conjugates. J Agric Food Chem 1998; 46(12): 5215-9.

[173] Dubourdieu D, Tominaga T, Masneuf I, Peyrot Des Gachons C, Murat ML. The role of yeasts in grape flavor development during fermentation: The example of Sauvignon blanc. Am J Enol Vitic 2006; 57: 81-8.

[174] Murat M. L, Masneuf I, Darriet P, Lavigne V, Tominaga T, Dubourdieu D. Effect of *Saccharomyces cerevisiae* yeast strains on the liberation of volatile thiols in Sauvignon blanc. Am J Enol Vitic,. 2001; 52(2): 136-9.

[175] Howell KS, Klein M, Swiegers JH, *et al.* Genetic determinants of volatile-thiol release by *Saccharomyces cerevisiae* during wine fermentation. Appl Environ Microbiol 2005; 71(9): 5420-6.

[176] Thibon C, Marullo P, Claisse O, Cullin C, Dubourdieu D, Tominaga T. Nitrogen catabolic repression controls *Saccharomyces cerevisiae.* FEMS Yeast Res *2*008; in press.

[177] Peyrot des Gachons C, Tominaga T, Dubourdieu D. Sulfur aroma precursor present in *S*-glutathione conjugate form: identification of *S*-3-(hexan-1-ol)-glutathione in must from *Vitis vinifera* L. cv. Sauvignon blanc. J Agric Food Chem 2002; 50(14): 4076-9.

[178] Fedrizzi B, Pardon KH, Sefton MA, Elsey GM, Jeffery DW. First Identification of 4-S-glutathionyl-4-methylpentan-2-one, a potential precursor of 4-mercapto-4-methylpentan-2-one, in Sauvignon blanc juice. J Agric Food Chem 2009; 57(3): 991-5.

[179] Choné X, Lavigne-cruège V, Tominaga T, *et al.* Effect of vine nitrogen status on grape aromatic potential : Flavor precursors (S-cysteine conjugates), glutathione and phenolic content in *Vitis vinifera* L. cv. Sauvignon blanc. J Int Sci Vigne Vin 2006; 40(1): 1-6.

[180] Peyrot des Gachons C, Tominaga T, Dubourdieu D. Measuring the aromatic potential of *Vitis vinifera* L. Cv. Sauvignon blanc grapes by assaying *S*-cysteine conjugates, precursors of the volatile thiols responsible for their varietal aroma. J Agric Food Chem 2000; 48(8): 3387-91.

[181] Peyrot des Gachons C, Van Leewen C, Tominaga T, Soyer JP, Gaudillère JP, Dubourdieu D. Influence of water and nitrogen deficit on fruit ripening and aroma potential of *Vitis vinifera* L. cv. Sauvignon blanc in field conditions. J Sci Food Agric 2005; 85: 73-85.

[182] Peyrot des Gachons C, Tominaga T, Dubourdieu D. Localization of *S*-cysteine conjugates in the berry: Effect of skin contact on aromatic potential of *Vitis vinifera* L. cv. Sauvignon blanc must. Am J Enol Viticult 2002; 53(2): 144-6.

[183] Murat ML, Tominaga T, Dubourdieu D. Assessing the aromatic potential of Cabernet Sauvignon and Merlot musts used to produce rose wine by assaying the cysteinylated precursor of 3-mercaptohexan-1-ol. J Agric Food Chem 2001; 49(11): 5412-7.

[184] Roland A, Schneider R, Charrier F, Cavelier F, Rossignol M, Razungles A. Distribution of varietal thiol precursors in the skin and the pulp of Melon B. and Sauvignon Blanc grapes. Food Chem 2011; 125(1): 139-44.

[185] Grant-Preece PA, Pardon KH, Capone DL, *et al.* Synthesis of wine thiol conjugates and labeled analogues: Fermentation of the glutathione conjugate of 3-mercaptohexan-1-ol yields the corresponding cysteine conjugate and free thiol. J Agric Food Chem 2010; 58(3): 1383-9.

[186] Kobayashi H, Takase H, Kaneko K, Tanzawa F, Takata R, Suzuki S, Konno K. Analysis of S-3-(hexan-1-ol)-glutathione and S-3-(hexan-1-ol)-L-cysteine in *Vitis vinifera* L. cv. Koshu for aromatic wines. Am J Enol Vitic 2010; 61(2): 176-85.

[187] Kishimoto K, Matsui K, Ozawa R, Takabayashi J. Volatile C6-aldehydes and allo-ocimene activate defense genes and induce resistance against Botrytis cinerea in Arabidopsis thaliana. Plant Cell Physiol 2005; 46(7): 1093-102.

[188] Mueller MJ. Archetype signals in plants: the phytoprostanes. Curr Opin Plant Biol 2004; 7(4): 441-8.

[189] Jakoby WB, Stevens J, Duffel MW, Weisiger RA. The terminal enzymes of mercapturate formation and the thiomethyl shunt. Rev Biochem Tox 1984; 6: 97-115.

[190] Wolf AE, Dietz KJ, Schroder P. Degradation of glutathione S-conjugates by a carboxypeptidase in the plant vacuole. FEBS Lett 1996; 384(1): 31-4.

[191] Subileau M, Schneider R, Salmon JM, Degryse E. New insights on 3-mercaptohexanol (3MH) biogenesis in Sauvignon Blanc wines: Cys-3MH and (E)-hexen-2-al are not the major precursors. J Agric Food Chem, 2008; 56(19): 9230-5.

[192] Capone DL, Sefton MA, Hayasaka Y, Jeffery DW. Analysis of Precursors to Wine Odorant 3-Mercaptohexan-1-ol Using HPLC-MS/MS: Resolution and Quantitation of Diastereomers of 3-S-Cysteinylhexan- 1-ol and 3-S-Glutathionylhexan-1 -ol. J Agric Food Chem 2010; 58(3): 1390-5.

[193] Roland A, Schneider R, Guernevé CL, Razungles A, Cavelier F. Identification and quantification by LC-MS/MS of a new precursor of 3-mercaptohexan-1-ol (3MH) using stable isotope dilution assay: Elements for understanding the 3MH production in wine. Food Chem 2010; 121(3): 847-55.

[194] Roland Al, Schneider RM, Razungles A, Le Guernevé C, Cavelier F. Straightforward synthesis of deuterated precursors to demonstrate the biogenesis of aromatic thiols in wine. J Agric Food Chem 2010; 58(19): 10684-9.

[195] Schneider R, Charrier F, Razungles A, Baumes R. Evidence of an alternative biogenetic pathway leading to 3-mercaptohexanol and 4-mercapto-4-methylpentan-2-one in wines. Anal Chim Acta 2006; 563: 58-64.

[196] Thibon C, Shinkaruk S, Tominaga T, Bennetau B, Dubourdieu D. Analysis of the diastereoisomers of the cysteinylated aroma precursor of 3-sulfanylhexanol in *Vitis vinifera* grape must by gas chromatography coupled with ion trap tandem mass spectrometry. J Chromatogr A 2008; 1183(1-2): 150-7.

[197] Thibon C, Dubourdieu D, Darriet P, Tominaga T. Impact of noble rot on the aroma precursor of 3-sulfanylhexanol content in *Vitis vinifera* L. cv Sauvignon blanc and Semillon grape juice. Food Chem 2009; 114(4): 1359-64.

[198] Thibon C, Shinkaruk S, Jourdes M, Bennetau B, Dubourdieu D, Tominaga T. Aromatic potential of botrytized white wine grapes: Identification and quantification of new cysteine-S-conjugate flavor precursors. Anal Chim Acta 2010; 660(1-2): 190-6.

[199] Roland A, Vialaret J, Razungles A, Rigou P, Schneider R. Evolution of S-cysteinylated and S-glutathionylated thiol precursors during oxidation of Melon B. and Sauvignon blanc musts. J Agric Food Chem 2010; 58(7): 4406-13.

[200] Podolyan A, White J, Jordan B, Winefield C. Identification of the lipoxygenase gene family from *Vitis vinifera* and biochemical characterisation of two 13-lipoxygenases expressed in grape berries of Sauvignon Blanc. Funct Plant Biol 2010; 37(8): 767-84.

[201] Hatanaka A. The biogeneration of green odour by green leaves. Phytochem 1993; 34(5): 1201-18.

[202] Hamilton-Kemp TR, Archbold DD, Collins RW, Yu K. Emission patterns of wound volatile compounds following injury of ripe strawberry fruit. J Sci Food Agric 2003; 83(4): 283-8.

[203] Myung K, Hamilton-Kemp TR, Archbold DD. Biosynthesis of trans-2-hexenal in response to wounding in strawberry fruit. J Agric Food Chem 2006; 54(4): 1442-8.

[204] Thoma I, Loeffler C, Sinha AK, *et al.* Cyclopentenone isoprostanes induced by reactive oxygen species trigger defense gene activation and phytoalexin accumulation in plants. Plant J 2003; 34(3): 363-75.

[205] Thibon C, Cluzet S, Mérillon JM, Darriet P, Dubourdieu D. 3-Sulfanylhexanol precursor biogenesis in grapevine cells: the stimulating effect of *Botrytis cinerea.* J Agric Food Chem 2011; 59(4): 1344-51.

[206] Seifert RM, Buttery RG, Guadagni DG, Black DR, Harris JG. Synthesis of some 2-methoxy-3-alkylpyrazines with strong bell pepper-like odors. J Agric Food Chem 1970; 18: 246-9.

[207] Seifert RM, Buttery RG, Guadagni DG, Black DR, Harris JG. Synthesis and odor properties of some additional compounds relates to 2-isobutyl-3-methoxypyrazine. J Agric Food Chem 1972; 20; 135-7.

CHAPTER 7

Polyamines and Grape Berry Development

M. N. Panagiotis[1,*] and A. Aziz R. -A. A. Kalliopi[2]

[1]*Department of Biology, University of Crete, Vassilika Vouton, Heraklion, Crete, Greece and* [2]*URVVC-Stress & Environment EA 2069, PPDD, University of Reims, F-51687 Reims cedex 2, France*

Abstract: Polyamines have been correlated with numerous plant processes, and the elucidation of their physiological role(s) has been a state-of the-art research topic in the recent years. Both their anabolism and catabolism seem to possess a regulatory role in development, stressing the necessity of their fine-tuning. Amongst the most important is the control of fruit-set and development. In this Chapter, the role of polyamines in the different stages of the grape berry development is discussed. In addition, references to other plant species are also included to strengthen the pan-species' specific role of polyamines.

Keywords: Abscisic acid, Abscission, Hormone, Photosynthetic efficiency, Polyamine biosynthesis, Polyamines, Programmed cell death, Putrescine, Reactive oxygen species, Spermidine, Spermine.

INTRODUCTION

In winemaking, the quality of the vintage is directly correlated with optimal grape maturity. Site selection and grape growing practices significantly affect maturity. Thus, the physiological and biochemical characteristics of the grape berry are affected by the concerted action of the genome and the environment. A considerable amount of research has identified strategies that can be used to optimise grape maturity at harvest, such as irrigation [1-3], canopy management [4-9], and crop levels [10-16].

Grape berry is essentially an independent biochemical factory [17]. Grape berry growth is supported by imports of water, mineral nutrients and carbohydrates. However, the berry has the potential to synthesize all other berry components (amino acids, pigments, flavour and aroma compounds, and others). Moreover, the fruit is not hydraulically isolated from the parent plant by xylem occlusion, but rather it is "hydraulically buffered" by water delivered *via* the phloem [18].

The first phase of rapid growth that occurs after fruit-set is the result of massive cell divisions and cell expansion. Cell division in the pericarp is largely completed during the first weeks of development. This phase of rapid berry growth is followed by a lag phase, during which little or no growth occurs. Thereafter, a second growth phase follows, which coincides with the onset of ripening (veraison) (Fig. **1**). This growth phase results from the expansion of pericarp cells [19]. Grape berry growth and development is concomitant with an influx of water, carbon and mineral nutrients, as well as distinct biosynthetic activities of primary and secondary metabolites.

The beginning of veraison is characterised by softening and colouring of the berry. Grape berry ripening, like other developmental processes, clearly involves the coordination of a large number of events. The striking changes observed in gene expression during veraison [20-22] indicate that the transition into ripening is driven to a large extent by changes in gene transcription. Overall, the berry approximately doubles in size between the beginning of veraison and harvest. Many of the solutes that accumulate during the first period of development remain at harvest, yet due to the increase in berry volume, their concentration is reduced significantly. Some metabolic activities that occur prior to veraison, such as photosynthesis and organic acid accumulation, are either turned off or are at least down-regulated. On the other hand, processes such as synthesis and accumulation of anthocyanins in berry skins appear during veraison. The timing and extent of ripening is of considerable importance not only for scientists, but also for the grape industry.

***Address correspondence to M. N. Panagiotis:** Department of Biology, University of Crete, Vassilika Vouton, Heraklion, Crete, Greece; E-mail: panagis1982@yahoo.gr

Hernâni Gerós, M. Manuela Chaves and Serge Delrot (Eds)

In general, all fruits undergo similar changes as they ripen: they store energy-rich metabolites, they soften and they accumulate colour, flavour and aroma compounds. Traditionally, fruits have been classified as climacteric and non-climacteric [23]. The climacteric fruits undergo a peak in their respiratory activity during the ripening process along with an ethylene burst. The non-climacteric fruits tend to show a slow decline in respiratory activity during ripening with no major peak in ethylene levels [24]. Grapes are regarded as non-climacteric [25].

Hormones have long been implicated in grape berry developmental changes. Ethylene emission during veraison, if any, in grape berries is rather small when compared to climacteric fruits. While there are still debates regarding the changes in ethylene levels that occur around veraison, the situation for abscisic acid (ABA) is much clearer. In numerous studies an increase in free ABA levels concomitant with sugar accumulation and colour development has been reported [25-30].

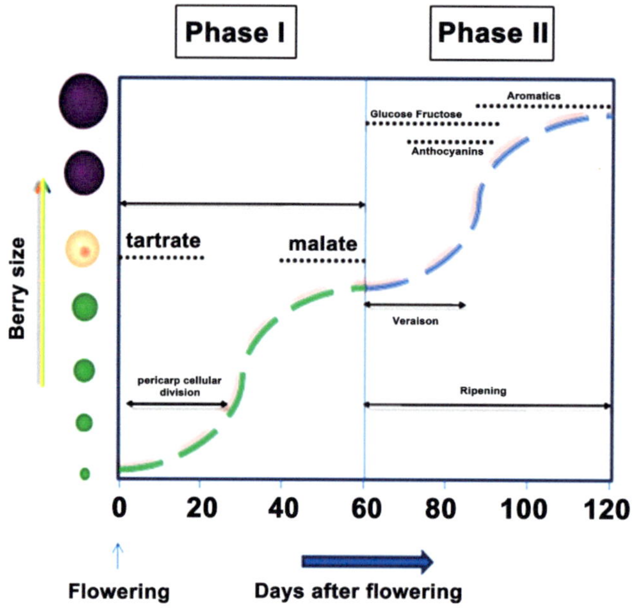

Figure 1: Double sigmoidal curve of grape berry growth and development. Grape berry development can be divided in two phases, namely phase I and phase II with the progression of berry size exhibiting a lag phase which is equally distributed between the two phases.

Early in development, ABA levels are high, decreasing prior to veraison and increasing again during veraison and reaching a peak two to three weeks later, after which they decline as the fruit approaches ripeness [31-35]. In 2006, Symons *et al.* [36] revealed the implication of Brassinosteroids (BRs) in the control of berry ripening. In this study brassinolide (BL) was shown to induce grape ripening. Moreover, changes in the levels of gibberellins (GAs) during development are consistent with an active role of GAs in the early stages of plant development. Auxin, cytokinins (CK) and salicylic acid (SA) are considered to be inhibitors of ripening. Auxin levels are high early in development, but decrease later at veraison [27, 31, 35, 37]. Cytokinins are involved in berry set and growth promotion. The content of zeatin and zeatin riboside are high early in berry development but decrease rapidly to become low at veraison [35]. Little is known about the role of SA, but it has been demonstrated that SA injected directly into the berry of the Syrah variety pre-veraison delayed skin colour development and perhaps delayed ripening in general [38]. For an extensive review, see Davies and Böttcher [39] and Böttcher and Davies, this edition.

Other molecules like biogenic amines may play a fundamental role in grape berry set and development, solely or in contribution with the aforementioned phytohormones. Polyamines (PAs) are the most important biogenic amines found so far in plants, as they have been associated with numerous developmental and stress-related processes. The most abundant and well-described PAs, putrescine (Put), spermidine (Spd) and spermine (Spm) (Fig. **2**) exist in a free form (soluble; S-) or conjugated to small (soluble hydrolysed; SH-;

such as hydrocinnamic molecules) or to large (pellet hydrolysed; PH-; such as proteins and high molecular weight compounds) molecules. PAs are involved in a broad spectrum of physiological processes in plant growth and development [40], and are also considered to delay senescence [41]. Increasing evidence supports that PA-derived H_2O_2 participates in signalling networks, which may affect developmental processes that range from cell division and morphogenesis to stress responses [42-47].

The primary step in PA biosynthesis is the formation of Put from Orn or Arg (which derive mainly from the urea cycle) *via* the rate limiting enzymes Arg decarboxylase (ADC; EC 4.1.1.19; two genes in Arabidopsis) or Orn decarboxylase (ODC; EC 4.1.1.17; absent from the Arabidopsis genome), respectively. Put is converted to Spd and Spm by Spd synthase (SPDS; EC 2.5.1.16; probably two genes in Arabidopsis) and Spm synthase (SPMS; EC 2.5.1.22; probably a single gene in Arabidopsis), respectively, which add aminopropyl groups derived from S-adenosyl-L-methionine (SAM) by the enzymatic activity of SAM decarboxylase (SAMDC, EC 4.1.1.50) (Fig. **2**) [48].

Diamines are catabolised by one or more diamine oxidases (DAO; EC 1.4.3.6). The Arabidopsis genome contains twelve putative genes that encode for DAOs. The DAOs characterised so far oxidise the primary amino groups of the diamine Put. The higher PAs are catabolised by PA oxidases (PAO; EC 1.5.3.11; five genes in Arabidospis), which oxidise the secondary amino groups of Spd and Spm (but are incapable to exert any effect upon Put), and yield, depending on the species, Δ^1-pyrroline and 1,5-diazabicyclononane (or 4-aminobutanal), respectively, along with 1,3-diaminopropane (Dap) and H_2O_2 (Fig. **2**) [49]. PAOs are mainly localised to the apoplast. Recently, it was shown that peroxisome-localised PAOs in Arabidopsis are responsible for the back-conversion of Spm to Spd and of Spd to Put [50, 51]. Interestingly, PAOs seem to negatively affect tolerance to abiotic stress and to enhance tolerance to biotic factors [51-56]. The mechanism involves exodus of PA to the apoplast and catabolism by PAOs producing H_2O_2 which, depending on its size, could further induce defensive responses or the programmed cell death (PCD) syndrome.

In this chapter we attempt to illustrate some correlative roles of PAs in grape berry development, summarising the current literature, and to give the state of the art concerning the ongoing research in this field.

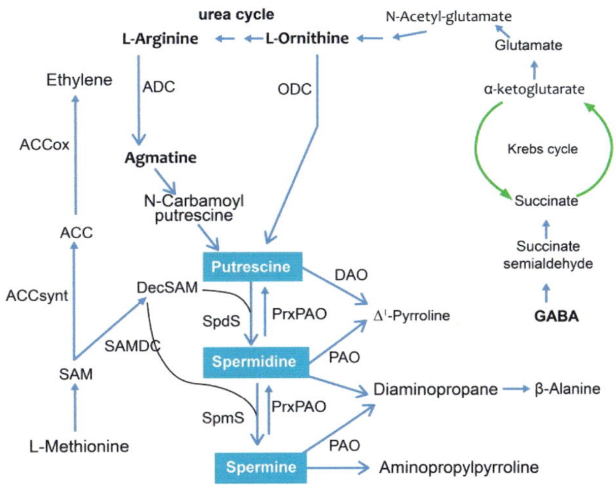

Figure 2: Polyamine biosynthesis and catabolism pathways in plants. The first step is the synthesis of Put exerted by the activity of ADC or ODC. Put is sequentially converted to Spd and Spm by SPDS and SPMS, respectively, which acquire a dcSAM molecule for its action as donor of the aminopropyl moiety. decSAM or dcSAM (SAM origin; prime precursor methionine) is produced by the activity of SAMDC. SAM is also the precursor of ethylene. At this node, the two pathways seem to be antagonising for the same precursor. PAs are terminally catabolised by the apoplastic PAO or backconverted by the peroxisomal PAO (PrxPAO). The existence of the peroxisomal pathway in the peroxisomes has not yet been shown in grapevine. ADC, Arginine decarboxylase; ODC, ornithine decarboxylase; DAO, diamine oxidase; PAO, polyamine oxidase; SAM, S-adenosylmethionine; SAMDC, SAM decarboxylase; Dec SAM, decarboxylated SAM; ACC, aminocycloparopane carboxylic acid; ACCsynt, ACC synthase; ACCox, ACC oxidase; GABA, γ-aminobutyric acid.

POLYAMINE TITRES IN VEGETATIVE AND REPRODUCTIVE TISSUES/ORGANS DURING DEVELOPMENT

As the development of an organ is progressing, a general rule of thumb is that PA catabolism substitutes PA anabolism [56]. The first solid evidence for this notion came from the work of Paschalidis and Roubelakis-Angelakis [44, 45] who showed an inverse correlation between ageing and PA titres in *Nicotiana tabacum* plants. The authors extended their work to grapevine, showing that this is plausible at least for Spd and Spm along the grape plant axis [57] (Fig. **3A**). More importantly, there is a strong correlation between the detailed expression patterns of the PA biosynthetic and catabolic genes and PA titres (Fig. **3B**). However, Orn transport from young to old tissues with parallel enhancement of the ODC pathway led to an inverse, to Spd and Spm, Put gradient with high levels of Put accumulating in the basal plant part and in the roots. Furthermore, PA catabolism by the enhanced expression of DAO and PAO generated H_2O_2, which correlated well with the levels of peroxidases (POXs) and phenolics during vascular differentiation [57].

More specifically, in the grapevine, total (S+SH+PH)-Put increased with explant age, whereas total Spd and Spm decreased. the total PA content in leaves (lamina and petiole) of the grapevine decreased from 878±74 to 613±56 nmol gFW^{-1} in the youngest (1st) to the oldest (25th) leaf, respectively [57]. The individual total-Put, total-Spd and total-Spm contents ranged from 190±13 to 464±14, from 538±10 to 195±8 and from 150±6 to 37±3 nmol gFW^{-1}, respectively, from the youngest to the oldest leaf. All fractions (S-, SH- and PH-) of Spd and Spm decreased with increasing ontogenetic stage and age of plant organs, and reached very low levels in the senescing (25th) ones. Spd decreased more than 2-fold from younger to older lamina tissues and even more in the vascular tissues (petioles and internodes); in the youngest internodes, Spd levels were almost 6-fold higher than in the oldest ones. Furthermore, very low levels of Spm were found in roots, while PH-Spm was below the detection limit in old leaves. On the contrary, all fractions of Put increased with age.

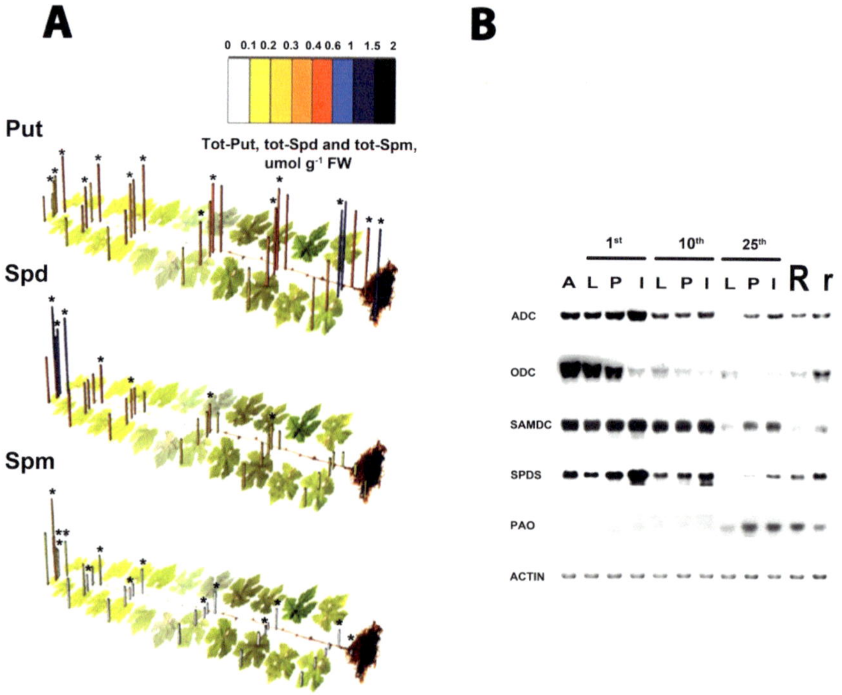

Figure 3: Polyamine titres and transcripts of PA biosynthetic/catabolic genes along the grapevine plant axis. A, Total Put, Spd and Spm titres along the grape plant axis analysed by High Performance Liquid Chromatography (HPLC) and shown as bars. B, RNA-gel blot analysis of the corresponding mRNA levels of the genes *ADC, ODC, SAMDC, and SPDS* (biosynthetic) and PAO (catabolic) normalised against the gene actin. A, apex; L, leaf; P, petiole; I, internode; R, primary and r, secondary root (modified from [57]).

Moreover, in leaves, all the biosynthetic enzyme activities (ADC, ODC, SAMDC, SPDS and SPMS) decreased with leaf age, as did also the transcripts of *ADC*, *ODC*, *SAMDC* and *SPDS* (Fig. **3B**). On the contrary, *PAO* transcripts and PAO activities increased with increasing age. In older vascular tissues, the synergistic action of PAO/DAO and POXs may play a role in the differentiation process. Thus, progress in plant development probably requires the induction of *DAO* and *PAO* expression, which results in an increase in H_2O_2 accumulation in the apoplast, further driving POX-catalysed cross-linking and lignification to complete cell wall stiffening and differentiation. Alternatively, H_2O_2 participates in the loosening of cell walls to promote cell expansion. Since POX was found to follow DAO and PAO activity along the grapevine plant axis, the view that PA oxidation is involved in the spatio-temporal supplement of site-specific ROS load on the cell wall is strengthened. Furthermore, DAO and PAO are fulfiling a rather developmental role in this occasion instead of solely controlling PA titres.

These results reveal a positive correlation between PA anabolic activities and PA titres in grapevine and an inverse relationship between PA catabolism and PA titres, with the exception of Put which seems not to comply with this rule in grapevine as it did in tobacco. Thus, Put may be a rather highly transportable nitrogenous molecule. At any case, the inverse gradient of Put could be the result of a basipetal Put transport, as shown by the early findings of Friedman *et al.* [58], who reported that xylem exudates and phloem sap of the grapevine contain high amounts of Put.

Whether in fact PAs, and especially Put, are transported from the vegetative tissues to grape berries still remains an open question, as is the capacity by which berries synthesise PAs. What is known is that during grape berry development PA titres profile are modified as is the case of vegetative tissues. In general, at flowering SH-PAs predominate because these compounds are associated with floral induction and floral development, probably due to the detoxification effect exerted by their conjugation with inhibitory molecules on flowering, such as hydrocinnamic acids [59]. At fruit-set, total PA levels, and especially PH- and SH-PAs, decrease [60, 61]. At stage I of development tot-PAs decrease with increasing age and at veraison, there is a second peak of PAs, as shown in Cabernet Sauvignon and Merlot [62], as well as Muscat Bailey [63] and Tempranillo [64]. Finally, at harvest the levels of PAs strongly decrease.

More specifically, at maturity there are 4 types of berries in the same grape cluster: normal red berries (8-15mm in diameter), very small berries (2-3mm in diameter), medium green berries (4-6mm in diameter) and medium red berries (5-7mm in diameter; [65] and references therein). Comparative studies and contents of PAs in these berry categories showed that shot berries have abnormal PA metabolism (Table **1**). Very small berries have cellular structure and levels of PAs typical of normal young berries at the time of fruit-set (Stage 22 of [66]; stage J of Baggiolini (Fig. **4**)). Cellular structure and PA contents of medium green berries are typical of normal grown berries (Stage 23 of [66]; stage K of Baggiolini), while cellular structure and PA contents of medium red berries are typical of berries at the time of veraison (Stage 23 of [66]; stage K of Baggiolini). Thus, as development is progressing during phase I, PA titres decrease with age.

Table 1: Polyamine content in Merlot berries during development (modified from [81]).

	Polyamine content ($\mu g\ gFw^{-1}$)	**Baggiolini stages**
Normal berries in normal grape	33	N
Small berries	635	J
Medium green berries	506	K
Medium red berries	122	M
Normal berries in shot grape	46	N

Abnormal berries contain hyper optimal titres of PAs and especially Dap (a good correlative index of PA catabolism [55]), whose drastic increase could result in growth arrest probably, at least partly, due to cytotoxic effects of H_2O_2 generated by PA catabolism. Thus, the progression of normal berry development requires PAs but within limits. Hydrogen peroxide, the product of PA catabolism which is produced in equimolar amounts with Dap, may enhance cellular development during the initial cell division at phase I, or modulate cell expansion at phase II by its cell wall loosening effect [67]. Moreover, the ratio of S- to bound (SH- and PH-) PAs increases during berry ripening [64]. It was reported that a relative increase in the S-PA pool occurs during active growth stages [64, 68] or, alternatively, the higher S-PAs titres could serve as substrates for PAO to produce the required H_2O_2 that could promote the ripening process. In addition, it was suggested that high levels of S-Put and S-Spd during early development may be associated with cell proliferation in the pericarp in phase I of berry growth [63]. In the same study, it was shown that the seed conjugation of PAs is increasing in parallel with the developmental stage, while S-PAs decrease.

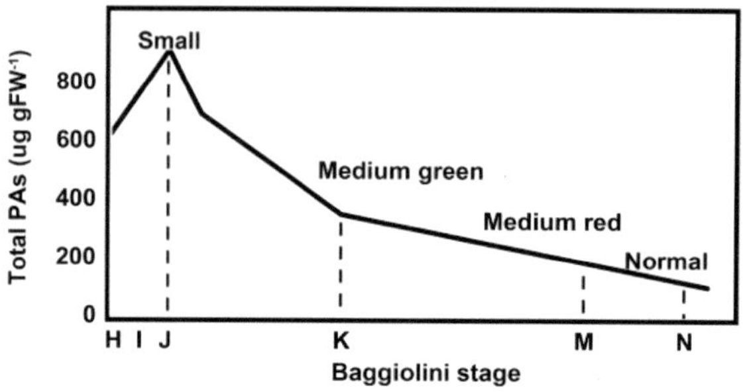

Figure 4: Polyamine titres in berries of Merlot during development (modified from [81]).

POLYAMINES DURING FRUIT-SET

The possible role of PAs in stimulating fruit-set and the initial phase of fruit development has been reported for several fruits, such as apple [69, 70], pear [68], pepper [71], olive [72], mango [73], tomato [74], citrus [75] and grapevine [76, 77]. Kushad and Orvos [78] found that a high percentage of total PA content was localised in the reproductive organs, and there were changes in conjugated and free PAs related to citrus flower growth. Pritsa and Voyiatzis [79] studied the fluctuations of free and conjugated Spd and Spm in different organs of two olive cultivars, showing a relationship between PA changes and developmental processes, such as floral differentiation, shoot growth, anthesis, fertilisation and fruit growth. A direct correlation has also been established between high levels of endogenous PAs or exogenous application of PAs and increased fruit-set in different grapevine cultivars [76, 77].

Lespy-Labaylette [80] and Geny [61] showed that the exogenous application of PAs at the 'end flowering' stage increased berry set (reviewed in [81]). Conversely, an opposite effect was observed when inhibitors of PA synthesis 'at button stage' were applied. These effects were reversed by the simultaneous administration of inhibitors and PAs. Interestingly, Dap levels (and more specifically its conjugated form) correlated well with berry set. Moreover, the effects of the exogenous application of Put and PA synthesis inhibitors at full bloom on the berry drop of Kyoho grapes were investigated by Shiozaki *et al.* [63]. They showed that inhibition of PA synthesis by using PA synthesis inhibitors accelerated berry drop. Thus, the rate of berry set was significantly lower in the grapes treated with PA synthesis inhibitors than in the corresponding controls. Neither Put nor the inhibitors affected seed number per berry. On the contrary, hyper optimal titres of PH-PAs (and Dap) have been correlated with malformation of grape berries [82].

In accordance with the above, much higher concentrations of Spd and Dap were found in floral organs and young berries of high fruit-bearing grapevine plants compared with other lower yielding plants [76]. Spd was also represented at the highest percentage among PAs covalently bound in the soluble fraction. Similar results have been reported in apricots [83], where fruit-set was associated with high PA contents in ovaries. At full flowering, free PAs were lower in grapevine cultivars exhibiting lower fruit-setting, like Merlot and Gewürztraminer. This difference was more evident in the inflorescences in comparison with the leaves. In the roots, free PA titres were very low. These results raise the possibility that fruit-set in grape could depend upon free PA level in the floral organs and a basipetal transport.

We also investigated the influence of the PA content (through the application of exogenous PAs or of their metabolic inhibitors) in fruit-set in the grapevine cultivars, Pinot Meunier, Pinot Noir, Merlot and Gewürztraminer, differing in flowering and fruit load, and their relation to fruit-set. The results strongly suggest some correlation between fruit-set and the endogenous PA level. At fruit-set stage, less than 35% of flowers in Merlot and Gewürztraminer plants have fruited against about 60% in Pinot Noir and Pinot Meunier. Free PA content was also influenced by organ and developmental stage for each cultivar. Moreover, exogenous application of 1 mM Spd before anthesis increased the percentage of fruit-set in all cultivars (Fig. **5**). A similar effect was found in plants treated with 1mM Dap. PAs could serve either as a nitrogenous source or as signal molecules regulating the reproductive development in the grapevine. This hypothesis is consistent with all the known facts on PA relation to reproductive activity [84].

We also showed that the percentage of fruit-set decreased in response to CHA (Fig. **5**), which also lowered free Spd, Spm and Dap in leaves and inflorescences of the grapevine (Fig. **6**). These effects revealed the contribution of the SPDS pathway in modulating Spd content in reproductive organs. SPDS activity was detected in pea ovaries and the gene encoding this enzyme was highly expressed in flowers at anthesis [85]. Other reports [86, 87] indicated a close connection between high levels of Spd and reproductive development in several lines of Arabidopsis. Wu *et al.* [88] showed that null mutants of the *AtPAO3* gene, a gene that encodes a peroxisomal PAO in Arabidopsis, have reduced seed set.

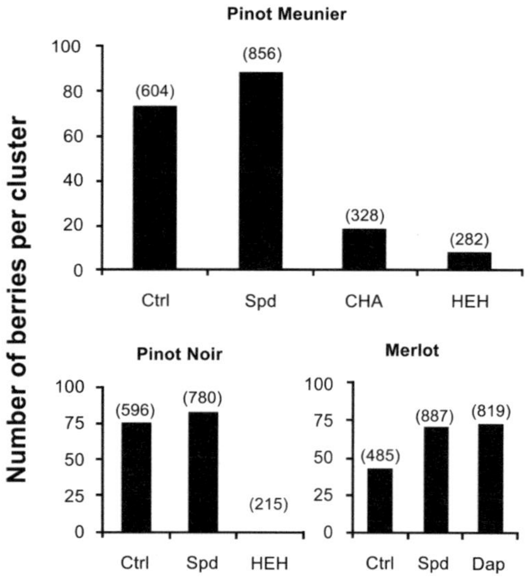

Figure 5: Effect of spermidine (Spd), its metabolic inhibitors and diaminopropane (Dap) on fruit-set in different grapevine cultivars. Seven-year-old grafted vines were sprayed on before anthesis with 1 mM Spd, Dap, cyclohexylamine (CHA) and β-hydroxyethylhydrazine (HEH), potent inhibitors of Spd synthase and PA-oxidases, respectively. Analysis was done after fruit-set (stage J of Baggiolini). Values in brackets on the histograms correspond to the total free polyamine content in young berries (nmol gFW^{-1}) at fruit-set stage.

Beta-hydroxyethylhydrazine (HEH), a potent inhibitor of PAOs, strongly reduced the fruit-set in selected grapevine cultivars (Fig. **5**). It has been reported that PAOs are involved in the adjustment of PA titres during sexual differentiation [89] and that PA oxidation ensures recycling of carbon and nitrogen to the Krebs cycle through the formation of Δ^1-pyrroline and GABA [90].

These observations are in agreement with the accumulating evidence implicating PAs in various plant growth and development processes, including floral development, fruit-set and ripening [91, 92]. Results also support that both Spd synthesis and catabolism are tightly coordinated and could be required for an optimal level of Spd, and subsequently for regular fruit-set. This is also in accordance with the findings of Burtin *et al.* [93], showing that inhibition of Spd synthesis in tobacco caused malformation of anthers and infertility. Furthermore, Applewhite *et al.* [86] suggested a role for Spd in the bolting and flowering in a delayed-flowering mutant of Arabidopsis.

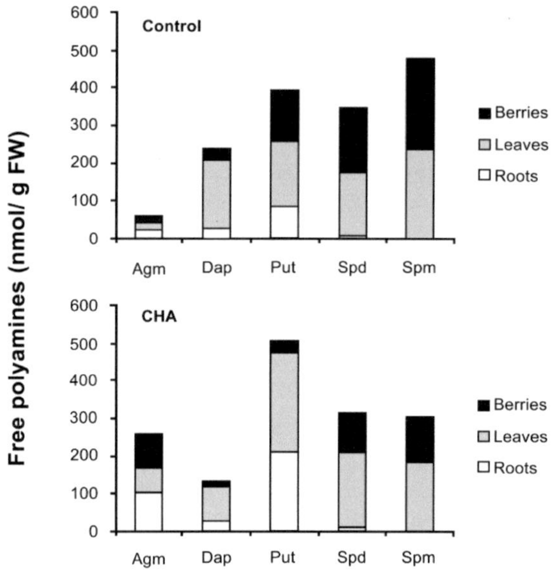

Figure 6: Free polyamine distribution in non-treated (control) and cyclohexylamine-treated (CHA) Pinot Meunier plants. CHA, inhibitor of Spd synthase, at 1 mM and water as control were sprayed before anthesis and polyamines were analysed after fruit-set (stage J of Baggiolini).

The balance between PA and ethylene synthesis may also be one of the major determinants regulating the fruit-set process, since SPDS requires an aminopropyl group donor (decarboxylated SAM) (Fig. **2**). During fruit development, rates of PAs and ethylene biosynthesis are contrasting. Early stages of fruit development are associated with increased titres of PAs [61, 76], while later in development PA levels decrease as ethylene levels increase [94]. These patterns are in agreement with the experimental evidence of the inhibitory effects of PAs on ethylene biosynthesis and *vice versa* [59, 94]. Moreover, the highest levels of ADC activity in grapevine were observed in pre-bloom flowers [95]. A predicted *ADC* (CTG1028599) was highly and uniquely expressed in pre-bloom flowers. The enzyme was previously cloned from *V. vinifera* cell suspension cultures [96].

Table 2: Effect of treatment of polyamine and polyamine biosynthetic inhibitors as well as Dap content (S-, SH- and PH-) on fruit-set at different developmental stages (end of flowering and floral button; modified from [81]).

Treatments (10^{-3}M)	Stage of treatment	%Fruit- set	DAP content (μg gFW^{-1})		
			S	SH	PH
Control		33	19	50	0
Put	End of flowering	38	10	50	45
Spd	End of flowering	50	12	49	37
(D-Arg + D-Orn)	Floral button	24	0	53	0
MGBG	Floral button	20	0	45	0
(D-Arg + D-Orn) + Put	Floral button	47	0	54	0
MGBG + Spd	Floral button	45	0	80	0

As a concluding remark, both Put and Spd are required for berry set. Put seems to be the first initiative signal, whereas Spd-derived H_2O_2 is the progressing signal required for pollen tube growth and fruit-set.

Exogenous applications of Spd prevented plants from bolting and flowering [86]. Similarly, Put applications overcame inhibition of short-day photoperiod, inducing Arabidopsis plants to flower [86]. Pre-bloom Put may be associated with the initiation of berry set in the grape, or as storage for Put supply of the later stages of berry development to produce Spd, which seems to be necessary for berry set progression. Recently, Wu *et al.* [88] showed that Spd oxidase-derived H_2O_2 regulates pollen plasma membrane Ca^{2+}-permeable channels and pollen tube growth in Arabidopsis. The previous results support that decreased PAs and/or conjugated-Dap result in reduced berry set and abnormal berry growth (see also Table **2**). In addition, PAOs may act indirectly by controlling intracellular PA levels. Thus, berry set seems to need the concerted action of PAs and PAOs to orchestrate berry set and berry development, and H_2O_2 produced from DAO and PAO may act as a regulator of this process [88, 97]. Moreover, the application of methylglyoxal-bis-guanylhydrazone (MGBG, a SAMDC inhibitor), which decreases Dap levels, strengthens the view that PAO is the main source of Dap from Spd oxidation (since MGBG inhibits Spd synthesis, which is the obligate precursor of Spm). The effect of MGBG is reversed by Spd application (Table **2**). Spd seems to be the most effective PA in increasing berry set, since it increases fruit-set by 17%.

POLYAMINE TITRES DURING VERAISON AND MATURATION

Before veraison, the majority of water reaches the berry *via* the xylem, whereas phloem transport predominates after veraison [98]. It has been suggested that the source of the increased PAs in the berry flesh at veraison could be the seeds (see above [63, 99]).

In 2008, Antolin *et al.* [99] showed that ABA and PAs show a kind of molecular cross-talk during berry development of Superior Seedless. The authors showed that PAs decrease during veraison; at the onset of veraison total PAs were 106.5 nmol gDW^{-1}, while in at the intermediate stage of veraison total PAs were 71.5 nmol gDW^{-1}, approximately 30% less. During harvesting PA titres in the berries further decreased, reaching 28.8 nmol gDW^{-1} (Fig. **7**). Moreover, these values were affected by the irrigation regime used, varying between Partial Root Zone (PRD) drying and Regulated Deficit Irrigation (RDI). The environmental factor was thus reflected on the final levels of PAs into berries (Fig. **8**). Moreover, Put seems to be the main PA during the onset under a normal irrigation regime, while Spd shows a slight increase later on [99]. Moreover, the profile of PAs is highly altered when different watering regimes are used, with Spd being the most prevalent PA.

POLYAMINES AND GRAPE BERRY ABSCISSION

In grapes, berry abscission occurs naturally just after full bloom and results in low fruit-set, which is known as 'coulure' [100] or 'Hana-burui' in Japanese. This shedding is induced by the dissolution of the cell wall between neighbour cells of the abscission zone of the pedicel. As in many fruit tree species, some grapevine cultivars exhibit massive fruitlet abscission even under favourable growing conditions.

This event is largely linked to metabolic and hormonal 'disorders' and repression of the nutrient supply to the inflorescence [101, 102]. Abscisic acid and ethylene are considered to be among the primary drivers of the procedure [103, 104], and PAs may act as regulators as well [99]. The translocation under photoperiodic flowering induction of free Spd to the inflorescence seems to be part of the complex mechanism which occurs during the transition of vegetative buds to flowers [105]. Moreover, Paschalidis and Roubelakis-Angelakis [81] suggested that limitation of S-Put and increases of S- and conjugated Spd production in the floral organs could positively influence the abscission process in the grapevine. It appears that the low titres of free PAs in the inflorescences are correlated with abscission sensitivity. At the fruit-set stage, about 65% of flowers and young fruits of Merlot (cultivar with high abscission rate) had abscissed *versus* 35% of Pinot Noir (cultivar with a low abscission rate), which exhibits a high level of free PAs (with Spd as the major one) in their inflorescences [77]. Thus, the high level of S-PA observed in inflorescences of the genotype with a low abscission rate could suggest that these compounds may have an important function in reproductive organ development and/or fertility as reported for other plants [76, 84, 106]. This is further reinforced by the findings of Wu *et al.* [88], as mentioned before.

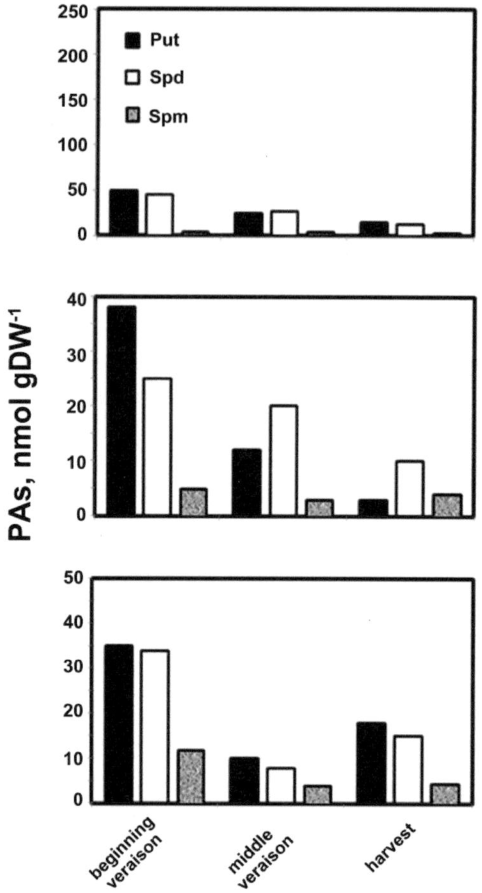

Figure 7: Polyamine titres expressed as S-, SH- and PH- during start (1) and middle (2) of veraison and harvesting (3) in Superior Seedless (modified from [99]).

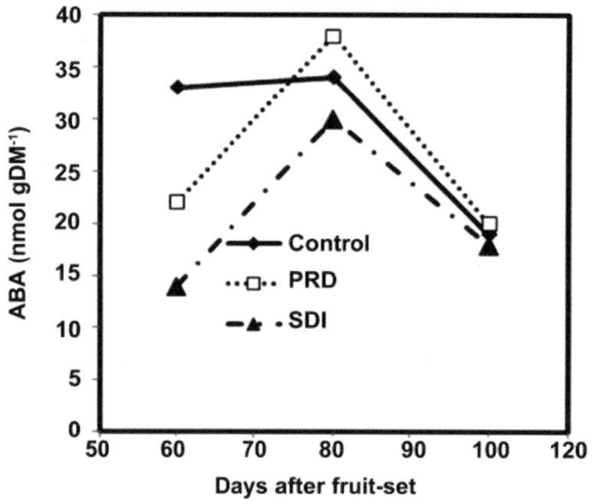

Figure 8: Effect of watering regime in the levels of abscisic acid. Abscisic acid levels in grape berries during development in control and in PRD and RDI water regimes of the Superior Seedless variety (modified from [99]).

Consistently with the above, inhibition of PA synthesis was shown to increase abscission. Thus, when the nutritive solution was supplemented with 1 mM DFMA (Difluoromethylarginine), a specific and irreversible inhibitor of ADC, the percentage of abscission increased by about 40 and 20% in Pinot and Merlot, respectively, compared to the corresponding controls (Fig. **9**). On the other hand, application of DFMO (Difluoromethylornithine), an ODC inhibitor, did not significantly affect the abscission levels. Thus, the PA pool produced *via* ADC may counteract floral organ abscission [77, 81].

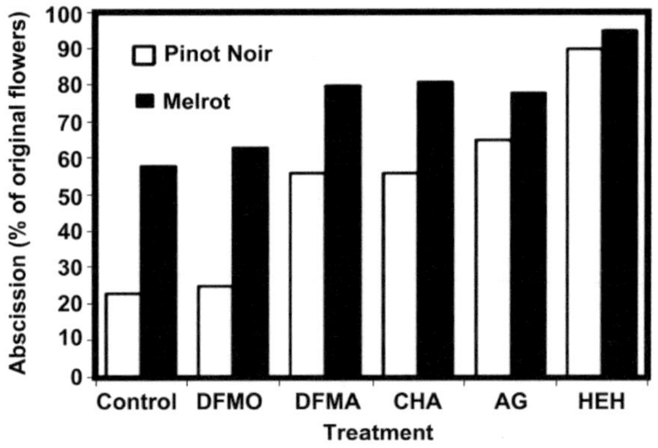

Figure 9: Effect of polyamine metabolic inhibitors on the abscission percentage in Pinot Noir (high abscission) and Merlot (low abscission rate) cultivars. DFMO, Difluoromethylornithine; DFMA, Difluoromethylarginine; CHA, cyclohexylamine; AG, aminoguanidine; HEH, ß-hydroxyethyl-hydrazine (modified from [81]).

Similarly, it was shown that the abscission percentage increased in response to CHA, added in the external medium (Fig. **9**). CHA also lowered Spd, Spm and Dap (primarily produced from Spd oxidation) in leaves and flowers of the grapevine [76]. These effects revealed the contribution of the SPDS pathway in modulating Spd concentration in inflorescences. SPDS activity was detected in pea ovaries and the gene encoding this enzyme was highly expressed in flowers during anthesis [85].

The metabolites resulting from the catabolism of Spd, such as H_2O_2 and Dap, may also be involved in the regulation of grapevine abscission, since HEH or AG, respectively, strongly increase the abscission of floral organs in both Merlot and Pinot Noir, noir (Fig. **9**). Total abscission was observed at the end of flowering in the presence of HEH (Fig. **9**). Martin-Tanguy [89] reported that in male-sterile flowers of tobacco, PAOs are involved in the adjustment of PA content during sexual differentiation. Furthermore, overexpression of PAO in tobacco plants significantly altered PA titres, reducing Spd and Spm, as well as Put, albeit slightly [55]. It has been reported that PA oxidation ensures recycling of the reduced carbon and nitrogen to Krebs cycle through the formation of Δ^1-pyrroline and GABA [53, 107], thus compensating for the energy loss which may result in abscission. Spd could also act by inhibiting the enzymatic activity of proteases involved in the abscission process [53, 108].

Competition for photoassimilates is responsible for fruit abscission [75, 109, 110]. In grapes, S-PA titres were strongly photo-dependent seeing as the darkening of Pinot Noir plants resulted in decreased S- and SH-PAs titres in both leaves and inflorescences [81]. 24 h of dark treatment were enough for a 55 and 45% reduction, respectively. Energy seems to be a significant factor preventing abscission in the grape berry.

Cultivars with low abscission rates (*e.g.* Merlot) under dark conditions mimic those exhibiting high abscission, due to decreased PAs and sugar contents and increased amino acid levels in inflorescences [76]. PA biosynthesis has been shown to be light-inducible and depends on diurnal rhythm [111]. Interestingly, PA catabolism is induced by light; darkness may therefore result in reduced PA catabolism [67]. Darkness had also a detrimental effect on growth, flowering and abscission (more than 85% of floral organs had abscissed) [76].

The fact that light regimes and hormones control fruit-set in grapevine was reported long ago [112]. This could be related to the localisation of ADC enzyme to chloroplasts [113]. The enzyme itself is photo-inducible [111, 114]. PAs are present in chloroplasts, thylakoids and PSII membranes and in the light-harvesting complex [115]. This led to the hypothesis that PAs could exert a functional role in the photosynthetic activity in the regulation of photoassimilate biosynthesis or in their partitioning between organs. Consequently, reduced PA titres due to reduced light could exhibit a negative effect on photosynthetic assimilates, which serve as energy substrates, to prevent abscission. In addition, the decrease of PA catabolism could result in higher energy deprivation due to the diminishing of carbon recycling in the Krebs cycle oriented by PA catabolism.

POLYAMINES AND ASSIMILATES IN THE GRAPE BERRY

In grapevines, inflorescences compete with the shoot apex for photosynthetic assimilates from the onset of their development to flowering and the early stage of berry development. Thus, a molecular mechanism must exist which determines the balance between productive and aborted fruit-set. This suggests that the allocation of assimilates into the inflorescence could be photo-regulated and implies communication, which could be mediated either directly *via* light (*e.g.* photoreceptors) [116, 117] or indirectly *via* photosynthesis and source/sink relationships [110]. Moreover, in tomato plants it was recently shown that overexpression of the *SAMDC* gene involved in the synthesis of Spd resulted in altered C:N signalling relationships [118]. Direct or indirect Spd effects might be mediated by the enhanced levels of Spd itself through its preferential accumulation in the inflorescences [76] or by the PAO-mediated generation of H_2O_2 [88], or even yet by a suspected control at the level of nutrient adjustment in sink organs.

Polyamines and Amino Acid Level in the Berry

PAs could also play a role in the modulation of reduced nitrogen in the cells and in the structure and functioning of the photosynthetic apparatus [115, 119] presumably due to their prosperity in amine groups and their biosynthesis through amino acid decarboxylases [120, 121]. It was shown that the total amino acid level declined during the post-anthesis stages, which coincides with sucrose accumulation in grapevine inflorescences [77]. Application of Spd by spraying it on grafted Pinot Meunier caused a marked decrease in total amino acid content in all plant organs and particularly in inflorescences and young fruits (Fig. **10A**). Put had no significant effect, while Dap increased the level of total amino acids in young fruits. Amino acid contents also increased in inflorescences and young berries of grafted plants when they were sprayed with CHA or HEH (Fig. **10A**). Proteolysis of proteins in different parts of the grapevine and translocation of free amino acids to reproductive organs may contribute to the increased amino acid levels. Furthermore, the target(s) of Spd could be located at the level of enzymes involved in glutamate metabolism [119, 122]. In response to Spd inhibitors, amino acid accumulation could reflect the extent of damage caused to protein synthesis, while the effects of Spd could be attributed to their anti-senescence properties, as reported in tomato [90] and rape leaf tissues [119] subjected to osmotic stress. Interestingly, pea shoots, oat leaves and soybean seedlings were shown to metabolise Put and Spd to GABA, glutamate, aspartate, sugars and organic acids [123]. In addition, although glutamate is a precursor of Put synthesis in tobacco plants, additional routes for Put synthesis could exist in tobacco plants (Velanis and Roubelakis-Angelakis, unpublished results).

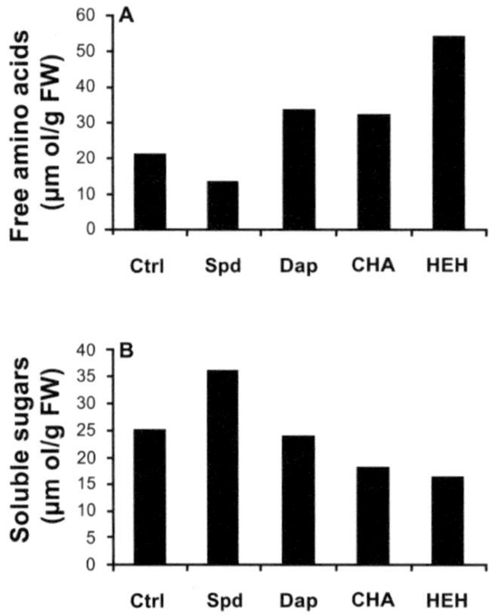

Figure 10: Effect of spermidine (Spd), diaminopropane (Dap) and inhibitors of Spd synthesis and oxidation on free amino acid. A, soluble sugar. B, levels in young berries of Pinot Meunier. Spd, Dap, cyclohexylamine (CHA) and β-hydroxyethylhydrazine (HEH), as potent inhibitors of Spd synthase and PA-oxidases, were sprayed at 1 mM on seven-year-old grafted vines before anthesis. Analysis was done 7 days after fruit-set (stage J of Baggiolini).

Furthermore, exogenous Spd enhanced the protein content in the inflorescences, in consistency with the findings that PAs inhibit protease activity [55, 108, 124]. Interestingly, Moschou *et al.* [55] showed that tobacco plants overexpressing the *MPAO* gene, which have reduced Spd titres (the main reduction observed), showed increased protease activity during induced oxidative stress by pro-oxidants. This suggests that the reduction of amino acid titres relies to some extent on the inhibition of protein degradation by Spd. In contrast, CHA might increase the release of amino acids from proteolysis.

Polyamines and Carbohydrate Status

Changes in carbon metabolism in plants during development are known to be associated with alterations in nitrogen metabolism [125, 126]. The high level of sucrose in grapevine inflorescences post-anthesis was associated with high fruit-set [77]. Indeed, sucrose content increased markedly after anthesis in the inflorescences of Pinot Noir and Meunier. This post-anthesis sucrose accumulation was accompanied by a slight increase of hexoses in inflorescences. Spraying Spd on grafted Pinot Meunier increased the total soluble sugar levels in all organs and particularly in young berries (Fig. **10B**), while Put and Dap had no effect.

Similar effects of Spd were observed in fruiting cuttings from other cultivars [77]. In the presence of Spd the content of soluble sugars increased 2-folds in inflorescences and leaves of Merlot. On the contrary, when plants were sprayed with CHA or HEH (Fig. **10B**), the soluble sugar content was decreased in the inflorescences and young fruits. With HEH, the level of sugars was also reduced in these organs, suggesting that Spd metabolism may influence either sugar synthesis or their accumulation in favour of the sink organs. Other studies showed that the reduced level of sucrose in the inflorescences was associated with inhibition of sucrose synthase activity, which seems to be a determinant factor for tomato fruit-set [138].

Polyamines and Photosynthetic Efficiency

Taking into account that the changes occurring in PA content precede those of sugars and amino acids, Aziz [77] explored the relationship between PAs and these compounds at full bloom. For instance, exogenous Spd induced an increase in soluble sugar content in both leaves and inflorescences, while CHA and HEH exhibited an opposite effect especially at the level of inflorescences. Moreover, Mattoo *et al.* [118] showed that PAs are a nitrogenous signal perceived during the sink/source or C:N cross-talk that enhances metabolism, which could be related to energy produced by PA catabolism and re-assimilation of the products within the Krebs cycle. Thus, the finding of Aziz could be related to this effect. Sucrose was the main sugar accumulated, which represents about 50% of the total soluble sugars. This result suggests that Spd metabolism may influence either sucrose synthesis and/or its accumulation in favour of the sink organs.

It is well-known that photosynthesis is one of the most sensitive processes affected by different stresses, and its tolerance is decisive to plant productivity. Previous studies have shown that Put, Spd and Spm are present in chloroplasts, thylakoid membranes, photosystem II membranes, and the light-harvesting complex [113, 115]. With the localisation of ADC in chloroplasts, Borrell *et al.* [113] concluded that the role of the PAs may be to maintain photosynthetic activity, preventing osmotic stress-induced senescence. Using PA inhibitors, a correlation between PA concentration, chlorophyll biosynthesis and photosynthetic rate has also been demonstrated [139]. During early stages of grape berry development, Rubisco was highly expressed and may be involved in light-mediated CO_2 assimilation and refixation of CO_2 released by respiration or other metabolic processes [140]. Expression patterns derived from EST frequencies are consistent with the abundance of Rubisco in the berry mesocarp up to the onset of ripening [141]. Other transcripts with predicted roles in photosynthesis are also highly expressed in flowers compared to berries. The expression patterns of a putative PSII type I chlorophyll *a/b*-binding protein is consistent with stages in flower-berry development where chlorophyll is abundant and flowers and berries fix CO_2 [142].

Kotzabasis *et al.* [115] reported that Put, Spd and Spm are associated with light harvesting complex (LCH) and PSII, suggesting that Pas, and especially the thylakoid-associated Pas, may play a decisive role in protecting the photosynthetic apparatus. In grapevine leaf discs exposed to high osmotic stress, an inhibition of photosynthetic capacity was observed (Fig. **11**). This decrease of Φ_{PSII} could be due essentially to the strong increase in the reduction state of the primary quinine acceptor, suggesting that a fraction of the PSII reaction centres was closed during osmotic stress. Treatment of leaf discs with DFMA or DFMO under non-stress conditions did not show any change in photosynthetic capacity. However, in stressed leaf discs both DFMA and DFMO inhibited Φ_{PSII} activity (Aziz *et al.* unpublished). Data suggest that under osmotic stress conditions both ADC and ODC pathways could be involved in plant cell homeostasis, probably through a regulatory effect on the photosynthetic apparatus. This could be due to the involvement of PAs in the regulation of the reaction centres at the PSII level and hence of the rate of electron transport, which seems to be reduced in stressed leaf discs. The use of aminoguanidine (AG), an inhibitor of DAO, under non-stressed conditions disrupts photosynthetic efficiency by a slight inhibition of Φ_{PSII}. However, the application of AG under osmotic stress led to a strong inhibition of Φ_{PSII} (Fig. **11**). The PSII reaction centres were apparently not damaged but they were not able to carry out electron transport. It is well-known that the catabolic degradation of Put by DAO is not only a means to regulate the cellular content of this diamine [136, 143]. Pyrroline, the reaction product of DAO, can be further catabolised to γ-aminobutyric acid (GABA) and subsequently transaminated and incorporated into the Krebs cycle (Fig. **2**). GABA has also been shown to play a key role in signal transduction pathways during stress response in many plants [144].

Overall, low levels of PAs and sugars are correlated with high amino acid content in the inflorescence and subsequent fruitlet abscission [77]. As a result, Spd and its metabolic pathways seem to be involved in the regulation of grapevine fruitlet abscission directly or indirectly through interaction with soluble sugar and amino acid accumulation in sink organs. This suggests that Spd could act as a component of the self-regulatory mechanism that adjusts fruitlet load with carbon and nitrogen compounds and, possibly, offer a physiological basis for the photoassimilate competition-induced abscission occurring under natural conditions.

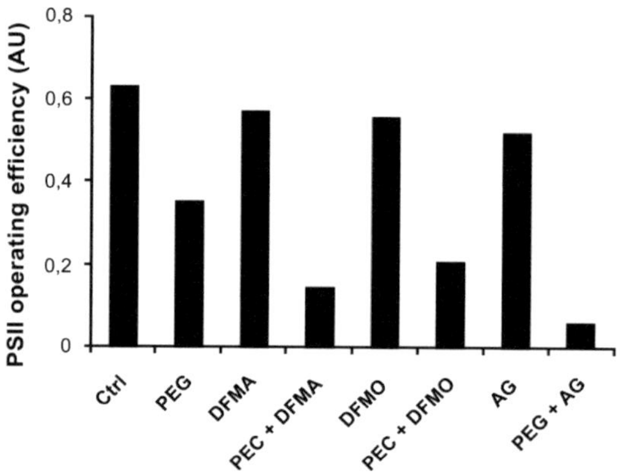

Figure 11: Effect of inhibitors of biosynthesis and catabolism of putrescine on photosynthetic efficiency in osmotic-stressed leaf discs of Chardonnay. Leaf discs were incubated with DFMA, DFMO and aminoguanidine (AG) as inhibitors of ADC, ODC and diamine-oxidase at 1 mM before they were submitted to osmotic stress (PEG, -1.5 MPa). PSII operating efficiency was measured by imaging PAM at 48h post-treatment .

INTERPLAY OF POLYAMINES AND HORMONES DURING GRAPE BERRY DEVELOPMENT

One of the changes observed during berry development are those of ABA titres. ABA is a central signal of plant responses to various environmental challenges including drought, salt and cold stresses (reviewed by [145]). In seeds, ABA is responsible for the biosynthesis of seed storage reserves, acquisition of desiccation tolerance and dormancy, and induction of stress tolerance [145]. In fleshy fruits such as the grape berry, ABA has been considered as a promoter of ripening as high levels of ABA were observed at the intermediate stage of veraison [127-129]. Also, ABA regulates various processes such as anthocyanin biosynthesis in coloured cultivars [130], assimilate uptake and sucrose metabolism of berries [131]. The source of ABA in berries is uncertain. Antolin *et al.* [128] suggested that under normal growth conditions the majority of ABA is transported from leaves to berries, where it accumulates. However, under water deficit, this is not clear [99]. Another possibility is that the source of ABA in the pericarp can be the pericarp itself [146]. Proteomic analysis following application of ABA to berries showed that ABA up-regulates a series of ripening-induced proteins [132].

Interestingly, Antolin *et al.* [99] showed that there is a strong relationship between the free to bound PA ratio and ABA concentration in berries (Fig. **12**). PAs and ABA may play a much more important role in a non-climacteric fruit, such as the grape berry, because of the lower importance of ethylene in their development and ripening [60]. Recently, it was reported that ABA could modulate PAs metabolism in response to water stress in Arabidopsis and grapes [133, 134], which suggests that, in many cases, ABA decreases PA levels by increasing or decreasing ADC activity and, in others, by increasing PAO activities [51-56, 99, 134]. A relationship between PAs and ABA was also observed in berries of Cabernet Sauvignon [135].

Moreover, there is an interconnection between PA metabolism and ethylene, since SPDS and SPMS require an aminopropyl group donor (dcSAM). It seems likely that the two biosynthetic pathways are struggling to steal from each other their precursor, SAM. Ruperti *et al.* [104] indicated that increased ethylene production during floral development was accompanied by an increased fruitlet abscission in peaches. In contrast, treatment with inhibitors of ethylene production stimulated the accumulation of free and conjugated PAs [44, 45, 136]. Ethephon, the ethylene-releasing chemical, aided grape berry removal. The percentage of berry recovery increased with increasing ethephon rates. Grape cultivars vary in their sensitivity to ethephon and require specific rates so as to induce an adequate degree of berry loosening without excessive leaf senescence [137]. Thus, it seems likely that abscission may depend on the balance between ethylene and PA synthesis.

With regard to GAs, it seems that GA berry treatment increases Put content, without further effect on other PAs [63]. Moreover, it was shown in the same study that Put (but not Spd and Spm) might affect the development of seedless grape berries regardless of the presence of GA_3. Consistently with the above, application of MGBG (*i.e.* inhibitor of Spd and Spm synthesis but not of Put) did not have any effect on berry development induced by GA_3. Simultaneous application of 500 ppm Put + 25 ppm GA_3 and 500 ppm Put increased to 111 and 112% the fresh weight of berries. This corroborates the hypothesis of the direct involvement of Put in the events signalled by GA_3.

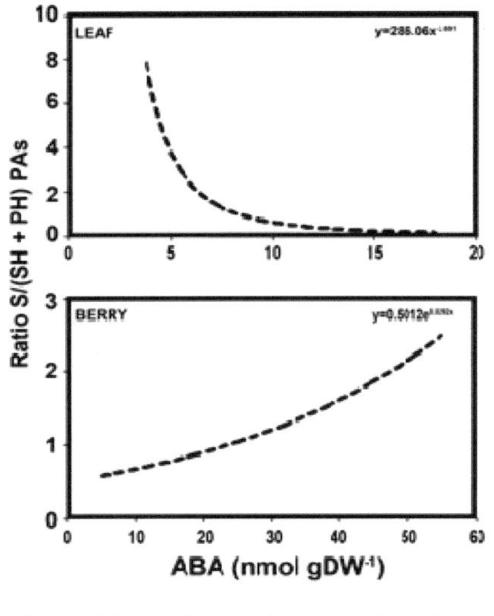

Figure 12: Relationships between abscisic acid and free to bound polyamine ratio in leaves and berries of Superior Seedless. The corresponding equations are depicted by the regression curves and the R values were 0.73 and 0.53, respectively.

Treatment of grape berries with the synthetic auxin-like compound benzothiazole-2-oxyacetic acid (BTOA) causes a delay in the onset of ripening of approximately 2 weeks [33]. BTOA treatment also delayed by 2 weeks the increase in ABA level that normally accompanies ripening and altered the expression of a number of developmentally regulated genes. These observations support the view that auxin (perhaps in conjunction with ABA) may have a role in the control of grape berry ripening by affecting the expression of genes involved in the ripening process. In addition, Paschalidis and Roubelakis-Angelakis [44, 45] showed that PAs are abundant in tissues with increased auxin, while auxin was shown to down-regulate the expression of the *PAO* gene in maize [67]. Thus, auxin may affect PA titres in the grape berry and may down-regulate *PAO* levels as well, increasing the content of S-PAs.

CONCLUDING REMARKS AND FUTURE PERSPECTIVES

Many aspects of PA metabolism in the grape berry need further clarification. One of the main aspects is the efficiency of the grape berry to synthesize and to catabolise PAs. Moreover, it seems very likely that transport of PAs between vegetative tissues/organs and the grape berry takes place during different developmental stages or under certain circumstances. Thus, the spatial and temporal PA accumulation, along with expression of biosynthetic and catabolic PA genes and enzymes in grapes, needs a more detailed study. The transport of PAs may have a signalling role since they are sensed as a nitrogenous signal that activates a cascade which finally reorients the C:N ratio. Therefore, PAs seem to have a regulatory role in grape berry development either directly, or *via* their catabolic products. The direct action remains largely unknown and transgenic grape plants overexpressing *PAO* or *SAMDC* genes may elucidate their potential roles. Tobacco transgenic plants overexpressing or down-regulating these genes, along with others of PA anabolism in tobacco led to the identification of the roles of PAs in development and under stress conditions [51-56, 88].

In addition, PAs seem to possess a key regulatory role in plant response(s) against abiotic and biotic challenge [52]. Paschalidis *et al.* [65] showed that DAO inhibition in grapes restricts accumulation of phytoalexins. Since the grape berry is constantly targeted by myriads of biotic challenges, especially fungi, DAOs and PAOs may compensate for these attack-related detrimental effects by producing H_2O_2, whose cytotoxic action could inhibit pathogen division-dispersal. Previously, Moschou *et al.* [52] showed that engineering PA catabolism leads to enhanced tolerance against oomycetes and bacteria in tobacco plants. The opposite effects were observed in tobacco plants which down-regulated the *PAO* gene when challenged with bacteria (but not withoomycetes). Paschalidis *et al.* [65] presented evidence for the direct involvement of PAs in the defence against *Botrytis cirenea* in grape vegetative tissues. More specifically, application of the PA metabolic inhibitor DFMO resulted in increased susceptibility to *B. cirenea*. Interestingly, the application of DFMA and DFMO resulted in lower PEG-induced phytoalexin accumulation in the grape, suggesting a role of PA anabolism in phytoalexin accumulation. Since DAO inhibition by AG application exerted the same effect, *i.e.* inhibition of phytoalexin accumulation, it seems that both PA anabolism and catabolism should be active for phytoalexin accumulation to occur.

PAs seem to act in parallel or antagonistically with hormones, promoting or delaying their effects. ABA seems to be a crucial hormone in the modulation of PA metabolism [134], while one paradigm of antagonistic action between a hormone and PAs is that of ethylene. The specification of the precise role of each PA in grape berry development will allow for the recognition of the molecular pathways affecting the cross-talk of PAs with hormones.

Finally, the development of transgenic grape plants with altered PA metabolism or even the isolation of specific mutants could allow in the future for the identification of the molecular mechanisms that dictate the mode of action of PAs in plants. Only recently some light was shed on the topic by discussing the use of transgenic plants in genetic engineering and plant regeneration.

ABBREVIATIONS

ABA	=	Abscisic acid
ACC	=	Aminocycloparopane carboxylic acid
ACCox	=	Aminocycloparopane carboxylic acid oxidase
ACCsynt	=	Aminocycloparopane carboxylic acid synthase

ADC	=	Arginine decarboxylase
AG	=	Aminoguanidine
Arg	=	Arginine
BTOA	=	Benzothiazole-2-oxyacetic acid
CHA	=	Cyclohexylamine
DAO	=	Diamine oxidase
Dap	=	1,3-diaminopropane
dcSAM/decSAM	=	Decarboxylated S-adenosylmethionine
DFMA	=	Difluoromethylarginine
DFMO	=	Difluoromethylornithine
EST	=	Expression sequence tag
GA	=	Gibberellic acid
GABA	=	γ-aminobutyric acid
HEH	=	Beta-hydroxyethylhydrazine
LCH	=	Light harvesting complex
MGBG	=	Methylglyoxal-bis-guanylhydrazone
ODC	=	Ornithine decarboxylase
Orn	=	Ornithine
PA/PAs	=	Polyamine(s)
PAO	=	Polyamine oxidase
PCD	=	Programmed cell death
PH-PAs	=	Pellet hydrolyzed polyamines
POXs	=	Peroxidases
PRD	=	Partial root zone drying
PSII	=	Photosystem II
Put	=	Putrescine
RDI	=	Regulated deficit irrigation

ROS	=	Reactive oxygen species
SAM	=	S-adenosylmethionine
SAMDC	=	S-adenosylmethionine decarboxylase
SH-PAs	=	Soluble hydrolysed polyamines
S-PAs	=	Soluble polyamines
SPDS	=	Spermidine synthase
SPMS	=	Spermine synthase
Φ_{PSII}	=	Quantum yield of photosystem II

ACKNOWLEDGEMENTS

This work was implemented in the frame of COST (European Cooperation in Science and Technology) FA0605 and 858 actions.

REFERENCES

[1] Bravdo B, Hepner Y. Water management and the effect on fruit quality in grapevines. Proceedings of the 6th Australian Wine Industry Technical Conference 1986; pp. 150-8.

[2] Hardie WJ, Considine J. Response of grapes to water-deficit stress in particular stages of development. Amer J Enol Vitic 1976; 27: 55-61.

[3] Matthews MA, Cheng G, Weinbaum SA. Changes in Water Potential and Dermal Extensibility during Grape Berry Development. J Am Soc Hort Sci 1987; 112(2): 314-9.

[4] Crippen DD, Morrison JC. The effects of sun exposure on the compositional development of cabernet-sauvignon berries. Am J Enol Vitic 1986; 37(4): 235-42.

[5] Dokoozlian NK, Kliewer WM. Influence of light on grape berry growth and composition varies during fruit development. J Am Soc Hort Sci 1996; 121(5): 869-74.

[6] Hunter JJ, Villiers J, Watts JE. The effect of partial defoliation on growth characteristics of Vitis vinifera L. cv. Cabernet Sauvignon II. Reproductive growth. South Afr J Enol Vitic 1990; 11: 26-32.

[7] Hunter JJ, Ruffner HP, Volschenk CG, Le Roux DJ. Partial defoliation of Vitis vinifera LCv. Cabernet-Sauvignon/99-Richter-Effect on root-growth, canopy efficiency, grape composition, and wine quality. Amer J Enol Vitic 1995; 46(3): 306-14.

[8] Rojas-Lara BA, Morisson JC. Differential effects of shading fruit or foliage on the development and composition of grape berries. Vitis 1989; 28: 199-208.

[9] Smart RE. Vine manipulation to improve wine grape quality. Davis grape and wine centenial symposium proceedings. University of California 1980; pp. 362-75.

[10] Eschenbruch R, Mora SJ de, Knowles SJ, Leonard WK, Forrester T, Spedding DJ. The formation of volatile sulfur-compounds in unclarified grape juice. Vitis 1986; 25(1): 53-7.

[11] Kliewer WM, Wolpert JA, Benz M. Trellis and vine spacing effects on growth, canopy microclimate, yield and fruit composition of Cabernet Sauvignon. Proceedings of the Fifth International Symposium on Grapevine Physiology 2000; 526, pp. 21-31.

[12] Intrieri C, Poni S. Physiological response of winegrape to management practices for successful mechanization of quality vineyards. Proceedings of the Fifth International Symposium on Grapevine Physiology 2000; 526, pp. 33-47

[13] Jackson DI, Lombard PB. Environmental and management-practices affecting grape composition and wine quality-a Review. Amer J Enol Vitic 1993; 44(4): 409-30.

[14] Marais J, van Wyc CJ. Effect of grape maturity and juice treatments on terpene concentrations and wine quality of Vitis vinfera L. cv. Weisser Riesling and Bukettraube. South Afr J Enol Vitic 1986; 7: 26-35.

[15] Marais J, van Wyc CJ, Rapp A. Carotenoid levels in maturing grapes as affected by climatic regions, sunlight and shade. South Afr J Enol Vitic 1996; 12: 64-9.

[16] Reynolds AG, Bouthillier MJ, Wardle DA, Denby LG. 'Sooke Seedless' table grape. Hortsci 1997; 32(4): 745-6.

[17] Gholami M, Hepner Y, Coombe BG, Jackson JF, Robinson SP, Williams PJ. Biosynthesis of flavour compounds in Muscat Gordo Blanco grape berries. Austr J Grape Wine Res 1995; 1, pp. 19-24.

[18] Choat B, Gambetta GA, Wada H, Shackel KA, Matthews MA. The effects of Pierce's disease on leaf and petiole hydraulic conductance in Vitis vinifera cv. Chardonnay. Physiol Plant 2009; 136(4): 384-94.

[19] Ojeda H, Deloire A, Carbonneau A. Influence of water deficits on grape berry growth. Vitis 2001; 40(3): 141-5.

[20] Davies C, Robinson SP. Sugar accumulation in grape berries-Cloning of two putative vacuolar invertase cDNAs and their expression in grapevine tissues. Plant Physiol 1996; 111(1): 275-83.

[21] Terrier N, Glissant D, Grimplet J, *et al.* Isogene specific oligo arrays reveal multifaceted changes in gene expression during grape berry (Vitis vinifera L.) development. Planta 2005; 222(5): 832-47.

[22] Deluc LG, Grimplet J, Wheatley MD, *et al.* Transcriptomic and metabolite analyses of Cabernet Sauvignon grape berry development. BMC Genomics 2007; 8(1): 429.

[23] Tucker GA. Introduction. In: Seymour GB, Taylor JE, Tucker GA, Eds. Biochemistry of Fruit Ripening. Chapman and Hall, London 1993.

[24] Seymour G. Banana. In: Seymour GB, Taylor JE, Tucker GA, Eds. Biochemistry of Fruit Ripening. Chapman and Hall, London 1993.

[25] Coombe BG, Hale CR. Hormone content of ripening grape berries and effects of growth substance treatments. Plant Physiol 1973; 51(4): 629-34.

[26] Hale CR, Buttrose MS. Effect of temperature on ontogeny of berries of Vitis-Vinifera L Cv Cabernet-Sauvignon. J Amer Soc Hort Sci 1974; 99(5): 390-4.

[27] Cawthon DL, Morris JR. Relationship of seed number and maturity to Berry development, fruit maturation, hormonal changes, and uneven ripening of concord (Vitis-Labrusca L) grapes. J Amer Soc Hort Sci 1982; 107(6): 1097-104.

[28] Kataoka I, Sugiura A, Utsunomiya N, Tomana T. Effect of abscisic-acid and defoliation on anthocyanin accumulation in Kyoho Grapes (Vitis-Vinifera L X V Labruscana Bailey). Vitis 1982; 21(4): 325-32.

[29] Okamoto G, Kuwamura T, Hirano K. Effects of water deficit stress on leaf and berry ABA and berry ripening in Chardonnay grapevines (Vitis vinifera). Vitis 2004; 43(1): 15-7.

[30] Deytieux-Belleau C, Gagne S, L'Hyvernay A, Doneche B, Geny L. Possible roles of both abscisic acid and indol-acetic acid in controlling grape berry ripening process. J Int Des Sci De La Vigne Et Du Vin 2007; 41(3): 141-8.

[31] Inaba A, Ishida M, Sobajima Y. Changes in endogenous hormone concentrations during berry development in relation to ripening of delaware grapes. J Jap Soc Hort Sci 1976; 45(3): 245-52.

[32] Scienza A, Miravalle R, Visai C, Fregoni M. Relationships between seed number, gibberellin and abscisic-acid levels and ripening in Cabernet Sauvignon grape berries. Vitis 1978; 17(4): 361-8.

[33] Davies C, Boss PK, Robinson SP. Treatment of grape berries, a nonclimacteric fruit with a synthetic auxin, retards ripening and alters the expression of developmentally regulated genes. Plant Physiol 1997; 115(3): 1155-61.

[34] Kondo S, Masuda E, Inoue K. Relation between ABA application and fruit quality of 'Pionnier' grape (Vitis sp.). Postharvest '96 - Proceedings of the International Postharvest Science Conference. 1998; 464, pp. 35-40.

[35] Zhang XR, GuoGuang L, RongHui W, Jing W, Himelrick DG. Growth and developmental responses of seeded and seedless grape berries to shoot girdling. J Amer Soc Hort Sci 2003; 128(3): 316-23.

[36] Symons GM, Davies C, Shavrukov Y, Dry IB, Reid BJ, Thomas RM. Grapes on steroids. Brassinosteroids are involved in grape berry ripening. Plant Physiol 2006; 140(1): 150-8.

[37] Baydar NG, Harmankaya N. Changes in endogenous hormone levels during the ripening of grapevine cultivars having different berry set mechanisms. Turk J Agr For 2005; 29: 205-10.

[38] Kraeva E, Andary C, Carbonneau A, Deloire A. Salicylic acid treatment of grape berries retards ripening. Vitis 1998; 37(3): 143-4.

[39] Davies C,Böttcher C. Hormonal control of grape berry ripening. In: Roubelakis-Angelakis KA, Ed. Grapevine Molecular Physiology and Biotechnology. Springer Dodrecht 2009; pp. 229-61.

[40] Kaur-Sawhney R, Tiburcio AT, Galston AW. Leaf area/crop weight ratios of grapevines: influence on fruit composition and wine quality. Proceedings of the American Society for Enology and Viticulture 50th Anniversary Anual Meeting. 2003; pp. 285-95.

[41] Bouchereau A, Aziz A, Larher F, Martin-Tanguy J. Polyamines and environmental challenges: recent development. Plant Sci 1999; 140(2): 103-25.

[42] Perez-Amador MA, Leon L, Green JP, Carbonell J. Induction of the arginine decarboxylase ADC2 gene provides evidence for the involvement of polyamines in the wound response in Arabidopsis. Plant Physiol 2002; 130(3): 1454-63.

[43] Vuosku J, Jokela A, Läärä E, *et al.* Consistency of polyamine profiles and expression of arginine decarboxylase in mitosis during zygotic embryogenesis of Scots pine. Plant Physiol 2006; 142(3): 1027-38.

[44] Paschalidis KA, Roubelakis-Angelakis KA. Spatial and temporal distribution of polyamine levels and polyamine anabolism in different organs/tissues of the tobacco plant. Correlations with age, cell division/expansion, and differentiation. Plant Physiol 2005a; 138(1): 142-52.

[45] Paschalidis KA, Roubelakis-Angelakis KA. Sites and regulation of polyamine catabolism in the tobacco plant. Correlations with cell division/expansion, cell cycle progression, and vascular development. Plant Physiol 2005b; 138(4): 2174-84.

[46] Kusano T, Yamaguchi K, Berberich T, Yamaguchi K, Berberich T, Takahashi Y. Advances in polyamine research in 2007. J Plant Res 2007; 120(3): 345-50.

[47] Cona A, Rea G, Angelini, R, Federico R, Tavladoraki P. Functions of amine oxidases in plant development and defence. Trends Plant Sci 2006; 11(2): 80-8.

[48] Tiburcio AF, Altabella T, Borrell A, Masgrau C. Polyamine metabolism and its regulation. Physiol Plant 1997; 100(3): 664-74.

[49] Cohen SS. A Guide to Polyamines. Oxford University Press, New York, Oxford 1998.

[50] Tavladoraki P, Rossi MR, Saccuti G, *et al.* Heterologous expression and biochemical characterization of a polyamine oxidase from Arabidopsis involved in polyamine back conversion. Plant Physiol 2006; 141(4): 1519-32.

[51] Moschou PN, Sanmartin M, Andriopoulou AH, Rojo E, Sanchez-Serrano JJ, Roubelakis-Angelakis KA. Bridging the gap between plant and mammalian polyamine catabolism: a novel peroxisomal polyamine oxidase responsible for a full back-conversion pathway in Arabidopsis. Plant Physiol 2008; 147(4): 1845-57.

[52] Moschou PN, Sarris PF, Skandalis N, *et al.* Engineered polyamine catabolism preinduces tolerance of tobacco to bacteria and oomycetes. Plant Physiol 2009; 149(4): 1970-81.

[53] Moschou PN, Paschalidis KA, Roubelakis-Angelakis KA. Plant polyamine catabolism: The state of the art. Plant Signal Behav 2008; 3(12): 1061-6.

[54] Moschou PN, Paschalidis KA, Delis ID, *et al.* Spermidine exodus and oxidation in the apoplast induced by abiotic stress is responsible for H_2O_2 signatures that direct tolerance responses in tobacco. Plant Cell 2008; 20(6): 1708-24.

[55] Moschou PN, Delis ID, Paschalidis KA, Roubelakis-Angelakis KA. Transgenic tobacco plants overexpressing polyamine oxidase are not able to cope with oxidative burst generated by abiotic factors. Physiol Plant 2008; 133(2): 140-56.

[56] Angelini R, Cona A, Federico R, Fincato P, Tavladoraki P, Tisi A. Plant amine oxidases "on the move": an update. Plant Physiol Biochem 2010; 48(7): 560-4.

[57] Paschalidis KA, Moschou PN, Toumi I, Roubelakis-Angelakis KA. Polyamine anabolic/catabolic regulation along the woody grapevine plant axis. J Plant Physiol 2009; 166(14): 1508-19.

[58] Friedman R, Levin N, Altman A. Presence and identification of polyamines in xylem and phloem exudates of plants. Plant Physiol 1986; 82(4): 1154-7.

[59] Martin-Tanguy J. Metabolism and function of polyamines in plants: recent development (new approaches). Plant Growth Reg 2001; 34(1): 135-48.

[60] Geny L, Deytieux C, Doneche B. Importance of hormonal profile on the onset of ripening in grape berries of Vitis vinifera L. Proceedings of the 5th International Postharvest Symposium, Vols 1-3, 2005; 682, pp. 99-105

[61] Geny L, Broquedis M, Martin-Tanguy J, Bouard J. Free, conjugated, and wall-bound polyamines in various organs of fruiting cuttings of Vitis vinifera L cv Cabernet Sauvignon. Amer J Enol Vitic 1997; 48(1): 80-4.

[62] Geny L, Colin L, Brezillon I, Broqueidis M. Hydroxycinnamic acid amines in flowers and berries of Vitis vinifera Vitis 1999; 38(4): 157-60.

[63] Shiozaki S, Ogata T, Horiuchi S. Endogenous polyamines in the pericarp and seed of the grape berry during development and ripening. Scientia Horticulturae 2000; 83(1): 33-41.

[64] Baigorri H, Antolin C, De Luis I, Geny L, Broquedis M, Aguirrezabal F, Sanchez-Diaz M. Influence of training system on the reproductive development and hormonal levels of Vitis vinifera L. cv. Tempranillo. Amer J Enol Vitic 2001; 52(4): 357-63.

[65] Paschalidis KA, Moschou PN, Aziz A, Toumi I, Roubelakis-Angelakis KA. Polyamines in grapevine: an update. In: Roubelakis-Angelakis KA, Ed. Grapevine Molecular Physiology and Biotechnology. Springer Dodrecht 2009; pp. 207-28.

[66] Fougere-Rifot M, Park HS, Bouard J. Berry pericarp ontogenesis from fertilization to maturity of Vitis vinifera L. var. Merlot. J Intern Sci Vigne Et Du Vin 1997; 31: 109-18.

[67] Cona A, Cenci F, Cervelli M, *et al.* Polyamine oxidase, a hydrogen peroxide-producing enzyme, is up-regulated by light and down-regulated by auxin in the outer tissues of the maize mesocotyl. Plant Physiol 2003; 131(2): 803-13.

[68] Crisosto CH, Lombard PB, Sugar D, Polito VS. Putrescine influences ovule senescence, fertilization time, and fruit-set in comice pear. J Amer Soc Hort Sci 1988; 113(5): 708-12.

[69] Costa G, Bagni N. Effects of polyamines on fruit-set of apple. Hortsci 1983; 18(1): 59-61.

[70] Biasi R, Costa G, Bagni N. Polyamine metabolism as related to fruit-set and growth. Plant Physiol Biochem 1991; 29(5): 497-506.

[71] Serrano M, Martinez-Madrid CM, Riquelme F, Romojaro F. Endogenous Levels of polyamines and abscisic-acid in pepper fruits during growth and ripening. Physiol Plant 1995; 95(1): 73-6.

[72] Rugini E, Mencuccini M. Increased yield in the olive with putrescine treatment. Hortsci 1985; 20(1): 102-3.

[73] Singh Z, Singh L. Increased fruit-set and retention in mango with exogenous application of polyamines. J Hortic Sci 1995; 70(2): 271-7.

[74] Antognoni F, Ghetti F, Mazzucato A, Francascetti M, Bagni N. Polyamine pattern during flower development in the parthenocarpic fruit (pat) mutant of tomato. Physiol Plant 2002; 116(4): 539-47.

[75] Gomez-Cadenas A, Mehouachi J, Tadeo FR, Primo-Millo E, Talon M. Hormonal regulation of fruitlet abscission induced by carbohydrate shortage in citrus. Planta 2000; 210(4): 636-43.

[76] Aziz A, Brun O, Audran JC. Involvement of polyamines in the control of fruitlet physiological abscission in grapevine (Vitis vinifera). Physiol Plant 2001; 113(1): 50-8.

[77] Aziz A. Spermidine and related-metabolic inhibitors modulate sugar and amino acid levels in Vitis vinifera L.: possible relationships with initial fruitlet abscission. J Exp Bot 2003; 54(381): 355-63.

[78] Kushad MM, Orvos AR, Yelenosky G. Relative Changes in Polyamines during Citrus Flower Development. Hortsci 1990; 25(8): 946-8.

[79] Pritsa TS, Voyiatzis DG. Seasonal changes in polyamine content of vegetative and reproductive olive organs in relation to floral initiation, anthesis, and fruit development. Austr J Agric Res 2004; 55(10): 1039-46.

[80] Lespy-Labaylette P. Recherches sur des polyamines des baies de raisin au cours de ler developpement et des pepins au cours de leur germination. University Thesis, Bordeaux 1994; p. 183.

[81] Paschalidis KA, Aziz A, Geny L, Primikirios NI, Roubelakis-Angelakis KA. Polyamines in grapevine. In: Roubelakis-Angelakis KA, Ed. Molecular Biology and Biotechnology of the Grapevine. Kluwer Academic Publishers, The Netherlands 2001; pp. 109-52.

[82] Colin L, Cholet C, Geny L. Relationships between endogenous polyamines, cellular structure and arrested growth of grape berries. Austr J Grape Wine Res 2008; 8: 101-8.

[83] Pritsa TS, Voyiatzis DG. Correlation of ovary and leaf spermidine and spermine content with the alternate bearing habit of olive. J Plant Physiol 2005; 162(11): 1284-91.

[84] Galston AW, Galston AW, Kaur-Sawhney R, Altabella T, Tiburcio AF. Plant polyamines in reproductive activity and response to abiotic stress. Botan Acta 1997; 110(3): 197-207.

[85] Alabadi D, Carbonell J. Differential expression of two spermidine synthase genes during early fruit development and in vegetative tissues of pea. Plant Mol Biol 1999; 39(5): 933-43.

[86] Applewhite PB, Kaur-Sawhney R, Galston AW. A role for spermidine in the bolting and flowering of Arabidopsis. Physiol Plant 2000; 108(3): 314-20.

[87] Tassoni A, van Buuren M, Franceschettia M, Fornalè S, Bagni N. Polyamine content and metabolism in Arabidopsis thaliana and effect of spermidine on plant development. Plant Physiol Biochem 2000; 38(5): 383-93.

[88] Wu J, Shang Z, Wu J, Jiang X, *et al.* Spermidine oxidase-derived H_2O_2 regulates pollen plasma membrane hyperpolarization-activated Ca^{2+}-permeable channels and pollen tube growth. Plant Journal 2010; 63(6):1042-53.

[89] Martin-Tanguy J. Conjugated polyamines and reproductive development: Biochemical, molecular and physiological approaches. Physiol Plant 1997; 100(3): 675-88.

[90] Aziz A, Martin-Tanguy J, Larher F. Stress-induced changes in polyamine and tyramine levels can regulate proline accumulation in tomato leaf discs treated with sodium chloride. Physiol Plant 1998; 104(2): 195-202.

[91] Kakkar RK, Sawhney R. Polyamine research in plants-a changing perspective. Physiol Plant 2002; 116(3): 281-92.

[92] Kakkar RK, Rai VK. Plant Polyamines in Flowering and Fruit Ripening. Phytochem 1993; 33(6): 1281-8.

[93] Burtin D, Martin-Tanguy J, Tepfer D. Alpha-Dl-Difluoromethylornithine, a specific, irreversible inhibitor of putrescine biosynthesis, induces a phenotype in tobacco similar to that ascribed to the root-inducing, left-hand transferred DNA of Agrobacterium rhizogenes. Plant Physiol 1991; 95(2): 461-8.

[94] Mehta RA, Cassol T, Li N, Ali N, Handa AK, Mattoo AK. Engineered polyamine accumulation in tomato enhances phytonutrient content, juice quality, and vine life. Nature Biotechnol 2002; 20(6): 613-8.

[95] da Silva FG, Iandolino A, Al-Kayal F, *et al.* Characterizing the grape transcriptome. Analysis of expressed sequence tags from multiple vitis species and development of a compendium of gene expression during berry development. Plant Physiol 2005; 139(2): 574-97.

[96] Primikirios NI, Roubelakis-Angelakis KA. Cloning and expression of an arginine decarboxylase cDNA from Vitis vinifera L-cell-suspension cultures. Planta 1999; 208(4): 574-82.

[97] Moschou PN, Paschalidis KA, Delis ID, *et al.* Spermidine exodus and oxidation in the apoplast induced by abiotic stress is responsible for H_2O_2 signatures that direct tolerance responses in tobacco. Plant Cell 2008; 20(6): 1708-24.

[98] Bondada BR, Matthews MA, Shackel KA. Functional xylem in the post-veraison grape berry. J Exp Bot 2005; 56(421): 2949-57.

[99] Antolin MC, Santesteban H, Santa María E, Aguirreolea J, Sánchez-Díaz M. Involvement of abscisic acid and polyamines in berry ripening of Vitis vinifera (L.) subjected to water deficit irrigation. Austr J Grape Wine Res 2008; 14(2): 123-33.

[100] Winkler R, Cook JA, Kliewer WK, Lider LA. General viticiculture. University of California Press Berkley 1962; p. 700.

[101] Kaska N. Bud, flower and fruit drop in citrus and other fruit trees. In: Osborne DJ, Jackson MD, Eds. Cell separation in plants. Berlin, Springer Verlag 1989; pp. 309-21.

[102] Gonzalez-Carranza ZH, Lozoya-Gloria E, Roberts JA. Recent developments in abscission: Shedding light on the shedding process. Trends in Plant Sci 1998; 3(1): 10-4.

[103] Sexton R, Roberts JA. Cell biology of abscission. Annual Rev Plant Physiol Plant Mol Biol 1982; 33: 133-62.

[104] Ruperti B, Bonghi C, Tonutti P, Ramina A. Ethylene biosynthesis in peach fruitlet abscission. Plant Cell Envir 1998; 21(7): 731-7.

[105] Caffaro SV, Antognoni F, Scaramagli S, Bagni N. Polyamine translocation following photoperiodic flowering induction in soybean. Physiol Plant 1994; 91(2): 251-6.

[106] Bagni N, Torrigiani P. Polyamines-a new class of growth-substances. Progr Plant Growth Reg 1992; 13, pp. 264-75.

[107] Aziz A, Larher F. Osmotic stress induced changes in lipid composition and peroxidation in leaf discs of Brassica napus L. J Plant Physiol 1998; 153(5-6): 754-62.

[108] Balestreri E, Cioni P, Romagnoli A, Bernini S, Fissi A, Felicioli R. Mechanism of polyamine inhibition of a leaf protease. Arch Biochem Biophys 1987; 255(2): 460-3.

[109] Moss GI, Steer BT, Kriedema PE. Regulatory Role of Inflorescence Leaves in Fruit-Setting by Sweet Orange (Citrus-Sinensis). Physiol Plant 1972; 27(3): 432-8.

[110] Goldschmidt EE, Koch KE. Citrus. In: Zamski E, Schaffer A, Eds. Photoassimilate Distribution in Plant and Crops. New York, Basel, Hong Kong, Dekker 1996; pp. 797-823.

[111] Gemperlova L, Novakova M, Vankova R, Eder J, Cvikrova M. Diurnal changes in polyamine content, arginine and ornithine decarboxylase, and diamine oxidase in tobacco leaves. J Exp Bot 2006; 57(6): 1413-21.

[112] Roubelakis KA, Kliewer WM. Influence of light-intensity and growth-regulators on fruit-set and ovule fertilization in grape cultivars under low-yemperature conditions. Amer J Enol Vitic 1976; 27(4): 163-7.

[113] Borrell A, Culianez- Macia FA, Altabella T, Besford RT, Flores D, Tiburcio AF. Arginine decarboxylase is localized in chloroplasts. Plant Physiol 1995; 109(3): 771-6.

[114] Gemperlova L, Eder J, Cvikrova M. Polyamine metabolism during the growth cycle of tobacco BY-2 cells. Plant Physiol Biochem 2005; 43(4): 375-81.

[115] Kotzabasis K, Fotinou C, Roubelakis-Angelakis KA, Ghanotakis D. Polyamines in the photosynthetic apparatus-photosystem-Ii highly resolved subcomplexes are enriched in spermine. Photos Res 1993; 38(1): 83-8.

[116] Galston AW. Polyamines as modulators of plant development. Biosci 1983; 33(6): 382-8.

[117] Yoshida I, Hirasawa E. Photoinduction of arginine decarboxylase activity in leaves of Pharbitis nil. Phytochem 1998; 49(8): 2255-9.

[118] Mattoo AK, Sobolev PA, Neelam A, Goyal KR, Handa KA, Segre AL. Nuclear magnetic resonance spectroscopy-based metabolite profiling of transgenic tomato fruit engineered to accumulate spermidine and spermine reveals enhanced anabolic and nitrogen-carbon interactions. Plant Physiol 2006; 142(4): 1759-70.

[119] Larher F, Aziz A, Deleu C, *et al.* Suppression of the osmoinduced proline response of rapeseed leaf discs by polyamines. Physiol Plant 1998; 102(1): 139-47.

[120] Evans PT, Malmberg RL. Do polyamines have roles in plant development. Annual Rev Plant Physiol Plant Mol Biol 1989; 40: 235-69.

[121] Walden R, Cordeiro A, Tiburcio AF. Polyamines: Small molecules triggering pathways in plant growth and development. Plant Physiol 1997; 113(4): 1009-13.
[122] Brugiere N, Dubois F, Limami MA, *et al.* Glutamine synthetase in the phloem plays a major role in controlling proline production. Plant Cell 1999; 11(10): 1995-2011.
[123] Rastogi R, Davies PJ. Effects of light and plant growth regulators on polyamine metabolism in higher plants. In: Slocum RD, Flores HE, Eds. Biochemistry and physiology of polyamines in plants. Boca Raton FL, CRC Press 1991; pp. 187-99.
[124] Navakoudis E, Vrentzou K, Kotzabasis K. A polyamine- and LHCII protease activity-based mechanism regulates the plasticity and adaptation status of the photosynthetic apparatus. BBA-Bioenergetics 2007; 1767(4): 261-271.
[125] Huppe HC, Turpin DH. Integration of Carbon and Nitrogen-Metabolism in Plant and Algal Cells. Annual Rev Plant Physiol Plant Mol Biol 1994; 45: 577-607.
[126] Skopelitis DS, Paranychianakis NV, Kouvarakis A, Spyros A, Stephanou EG, Roubelakis-Angelakis KA. The isoenzyme 7 of tobacco NAD(H)-dependent glutamate dehydrogenase exhibits high deaminating and low aminating activities *in vivo*. Plant Physiol 2007; 145(4): 1726-34.
[127] Antolin MC, Ayari M, Sanchez-Diaz M. Effects of partial rootzone drying on yield, ripening and berry ABA in potted Tempranillo grapevines with split roots. Austr J Grape Wine Res 2006; 12(1): 13-20.
[128] Antolin MC, Baigorri H, de Luis I, *et al.* ABA during reproductive development in non-irrigated grapevines (Vitis vinifera L. cv. Tempranillo). Austr J Grape Wine Res 2003; 9(3), pp. 169-76.
[129] Gagne S, Esteve K, Deytieux C, Sauvier C, Geny L. Influence of abscisic acid in triggering "veraison" in grape berry skins of Vitis vinifera L. cv. Cabernet-Sauvignon. J Int Sci Vigne Du Vin 2006; 40(1): 7-14.
[130] Jeong ST, Goto-Yamamoto M, ,c S, Esakac, S.; Esaka, M. Effects of plant hormones and shading on the accumulation of anthocyanins and the expression of anthocyanin biosynthetic genes in grape berry skins. Plant Sci 2004; 167(2): 247-52.
[131] Pan QH, Pan Q, Li M-J, *et al.* Abscisic acid activates acid invertases in developing grape berry. Physiol Plant 2005; 125(2): 157-70.
[132] Giribaldi M, Geny L, Delrot S, Schubert A. Proteomic analysis of the effects of ABA treatments on ripening Vitis vinifera berries. J Exp Bot 2010; 61(9): 2447-58.
[133] Alcazar R, Cuevas JC, Patron M, Altabella T, Tiburcio F. Abscisic acid modulates polyamine metabolism under water stress in Arabidopsis thaliana. Physiol Plant 2006; 128(3): 448-55.
[134] Toumi I, Moschou PN, Paschalidis KA, *et al.* Abscisic acid signals reorientation of polyamine metabolism to orchestrate stress responses *via* the polyamine exodus pathway in grapevine. J Plant Physiol 2010; 167(7): 519-25.
[135] Geny L, Deytieux C, Darrieumerlou A, Doneche B. Hormonal status in grape berry during ripening: Importance of calcium to polyamine and abscisic acid synthesis. Proceedings of the Seventh International Symposium on Grapevine Physiology and Biotechnology 2005; 689: pp. 243-50.
[136] Kumar A, Altabella T, Taylor MA, Tiburcio AF. Recent advances in polyamine research. Trends Plant Sci 1997; 2(4): 124-30.
[137] Hedberg PR, Goodwin PB. Factors Affecting Natural and Ethephon-Induced Grape Berry Abscission. Am J Enol Vitic 1980; 31(2): 109-13.
[138] D'Aoust MA, Yelle S, Nguyen-Quoc B. Antisense inhibition of tomato fruit sucrose synthase decreases fruit-setting and the sucrose unloading capacity of young fruit. Plant Cell 1999; 11: 2407-18.
[139] Beigbeder AR. Influence of polyamine inhibitors on light independent and light dependent chlorophyll biosynthesis and on the photosyntetic rate. J Photochem Photobiol 1995; 28: 235-42.
[140] Schwender J, Goffman F, Ohlrogge JB, Shachar-Hill Y. Rubisco without the Calvin cycle improves the carbon efficiency of developing green seeds. Nature 2004; 432: 779-82.
[141] Famiani F, Walker RP, Tecsi L, Chen Z-H, Proietti P, Leegood RC. An immunohistochemical study of the compartmentation of metabolism during the development of grape (Vitis vinifera L.) berries. J Exp Bot 2000; 51: 675-83.
[142] Blanke MM, Lenz F. Fruit photosynthesis. Plant Cell Environ 1989; 12: 31-46.
]143] Bagni N, Tassoni A. Biosynthesis, oxidation and conjugation of aliphatic polyamines in higher plants. Amino Acids 2001; 20: 301-17.
[144] Shelp BJ, Bown AW, McLean MD. Metabolism and functions of gamma- aminobutyric acid. Trends Plant Sci 1999; 4: 446-52.
[145] Leung J, Giraudat J. Abscisic acid signal transduction. Annual Rev. Plant Physiol Plant Mol Biol 1998; 49: 199-222.
[146] Koussa T, Broquedis M, Brouard J. Mise en evidence d'une relation entre les teneurs en acide abscissique des feuilles de vine et des baies de raisin à l'époque de la véraison. J Inter Sci Vigne Vin 1993; 27: 263-76.

CHAPTER 8

Grape Cell Vacuoles: Structure-Function and Solute Transport Across the Tonoplast

N. Fontes[1,2], S. Delrot[3] and H. Gerós[1,2,*]

[1]Centro de Investigação e de Tecnologias Agro-Ambientais e Biológicas (CITAB), Quinta de Prados, 5001-801 Vila Real, Portugal; [2]Departamento de Biologia, Universidade do Minho, Campus de Gualtar, 4710-057 Braga, Portugal and [3]UMR 1287 Ecophysiology and Grape Functional Genomics, University of Bordeaux, INRA, Institut des Sciences de la Vigne et du Vin, Domaine de la Grande Ferrade, 210 chemin de Leysotte, 33883 Villenave d'Ornon, France

Abstract: Grape berry cells have a complex vacuolar membranous system highly specialised in solute storage. Vacuoles are the main reservoirs of sugars, organic acids, aromas, flavours, ions and water. After veraison, when growth occurs exclusively by cell enlargement, the vacuole volume greatly increases due to the massive sugar and water uptake. A large number of tonoplast proteins, including pumps, carriers, ion channels and receptors support the numerous functions of the plant vacuole. Some of them have been well characterised in several plant models at the biochemical and molecular level, including the most abundant ones, V-ATPase, V-PPase, and water channels (aquaporins). The present chapter provides an overview on the diversity and storage role of the vacuole of grape cells, and the molecular mechanism involved in solute transport across the tonoplast is updated and discussed.

Keywords: ATP-binding cassette transporter, Fluorescein diacetate, Grape vacuoles, Solute storage, Tonoplast, Tonoplast intrinsic protein, Tonoplast transporters, Vacuolar ATPase, Vacuolar pyrophosphatase, Vacuole isolation.

INTRODUCTION

The vacuoles of plant cells are widely diverse in form, size, content, functional dynamics and play central roles in plant growth and development. They are involved in protein turnover, pH and ion homeostasis, turgor pressure maintenance, sequestration of toxic compounds, pigmentation, and solute and water accumulation [1, 2]. The central vacuole, which can occupy more than 80% of the total plant cell volume, is separated from the surrounding cytosol by the tonoplast which controls the passage of inorganic and organic solutes through a wide range of pumps, carriers, ion channels and receptors [3].

In the mesocarp of fleshy fruits, the large central vacuole plays a prominent role in cell expansion, fruit size and fruit quality. In grape cells, vacuoles are the main reservoir of sugars, organic acids, aromas, flavours, ions and water, which are unevenly distributed in berry tissues. Sugars (glucose + fructose) and organic acids (malic acid + tartaric acid) may accumulate in vacuoles at concentrations in the molar range [1, 4]. Thus, the vacuole content plays important roles in fruit and wine quality.

Several progresses have been achieved regarding the understanding of the physiological, biochemical, and molecular aspects of grape berry maturation, but still little is known regarding the identity and functioning of vacuolar transporters at the molecular level. This is true, for instance, for tonoplast sugar transporters, although sugar storage in grape cells has a major economic importance. These developments will be illustrated and discussed in this chapter. Approaches aiming at the isolation and purification of intact vacuoles from grape cells are of special interest to understand the role of this organelle in solute compartmentation during grape berry development and ripening. Independent proteomic studies focused on

***Address correspondence to H. Gerós:** Centro de Investigação e de Tecnologias Agro-Ambientais e Biológicas (CITAB), Quinta de Prados, 5001-801 Vila Real, Portugal and Departamento de Biologia, Universidade do Minho, Campus de Gualtar, 4710-057 Braga, Portugal E-mail: geros@bio.uminho.pt

Hernâni Gerós, M. Manuela Chaves and Serge Delrot (Eds)

intact vacuoles from Arabidopsis were already developed [3, 5, 6], but so far in grape they have just been started. The progresses achieved in our laboratories on the isolation and functional characterisation of intact vacuoles from grape berry cells will be also reviewed in the present chapter.

STRUCTURE, DIVERSITY AND STORAGE ROLE

A recent report focuses on the isolation of mesocarp cells from ripened berries of both wine and table varieties to study their cytoplasm organisation, function and viability, combining bright-field, fluorescence and confocal microscopy and flow cytometry [2]. Cell wall digestion was performed with cellulase Y-C and pectolyase Y-23 and protoplasts were purified in a discontinuous gradient of sucrose and sorbitol [2, 7, 8]. Protoplasting from grape berry mesocarp tissue yields a population of intact and viable cells. Berry-derived protoplasts were stained with the fluorescent membrane marker *N*-(3-triethylammoniumpropyl)-4-(4-(dibutylamino)styryl)pyridinium dibromide (FM1-43) and observed under the confocal microscope, which provided a beautiful picture of the vacuole organisation in the cytoplasm and clearly demonstrated the integrity of the biological membranes (Fig. **1**). Cell viability and membrane integrity was further assessed by fluorescein diacetate (FDA), a non-fluorescent compound that freely diffuses across the plasma membrane and is hydrolysed to fluorescein (non-permeant and brightly fluorescent under blue light) and acetate by non-specific esterases in the cytoplasm [9]. The acidotropic stain Neutral Red confirmed the diversity and the acidic character of most vacuoles of the isolated cells from ripe berry. The vacuoles of a mesocarp cell from ripe berry vary from large and small colourless to numerous small acidic vacuoles distributed throughout the cytoplasm, with a large central-vacuole also being observable in some cells [2, 7, 8] (Fig. **2**).

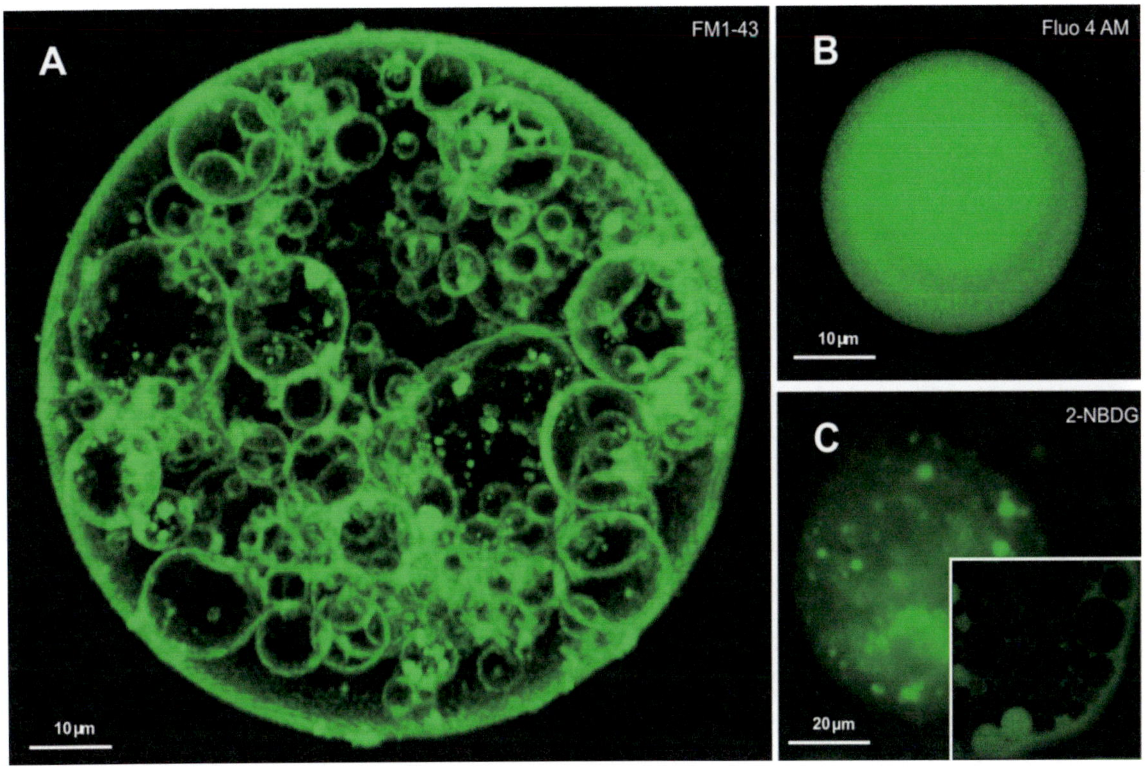

Figure 1: Ultrastructure of grape berry cells under the confocal microscope. Protoplasts were imaged by confocal laser scanning after being immersed overnight with the styryl dye FM1-43, and a maximum Z projection of 20 sections covering approximately 30 μm is represented (A). Intact vacuoles imaged after staining with the calcium fluorescent probe Fluo-4 AM (B). Plasma membrane integrity assessed under the fluorescence microscope with the fluorescent glucose analogue 2-NBDG (16 h incubation) (C). *Inset*: single section of protoplast loaded with 2-NBDG observed by confocal microscopy 16 h after incubation. Reproduced with permission from the American Journal of Enology and Viticulture [2].

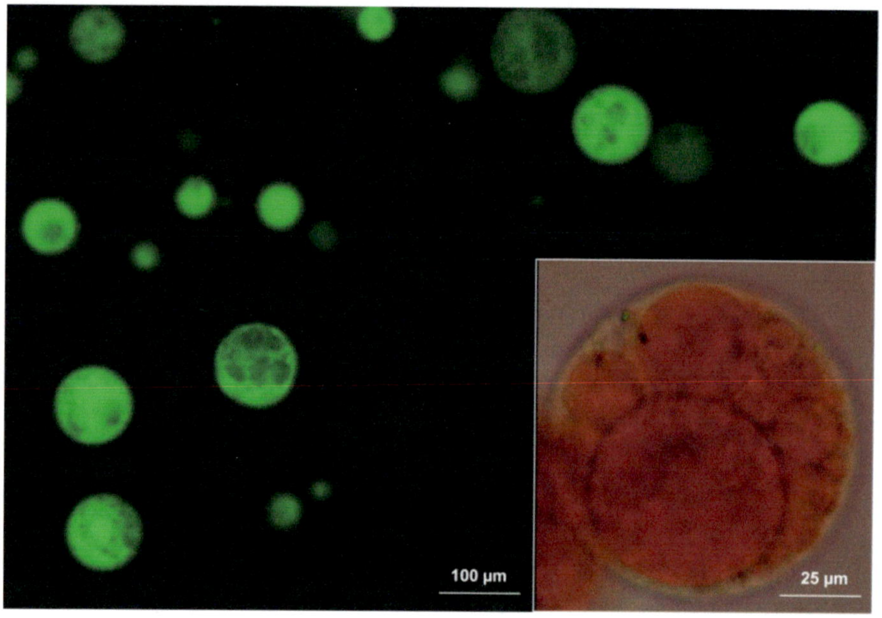

Figure 2: Protoplasts from the pulp of ripened grape berries observed under UV light (epifluorescence) after staining with FDA to measure viability and membrane integrity. *Inset*: a close-up view of an intact protoplast highlighting the integrity of both the plasma membrane and tonoplast. *Inset*: intact protoplast labelled with Neutral Red showing the acidic nature and integrity of the vacuolar apparatus.

The physiological/structural status of the soft ripe berry has been debated over the last decades. A loss of compartmentation of the berry mesocarp was initially suggested to be associated with fruit softening at the maturity [10, 11]. Recently, Krasnow and co-workers [12] concluded that there is a substantial maintenance of cell viability when sliced grapes were observed under the confocal microscope. These observations were confirmed when mesocarp cells were isolated and purified from ripened berries [2, 8], as reported above.

The diversity in size and lumen acidity of the vacuoles probably parallels its distinct biological roles and biogenesis. More than one kind of vacuole has been observed in cells undergoing maturation, where some function primarily as storage organelles and others as lytic compartments [1, 13-16]. Also, “tannin vacuole”, “mucilage vacuole”, “lipid vacuole” and “phenolic vacuole” have been reported (reviewed by Fontes *et al.* [17]). Grape cells may have different types of storage vacuoles according to its position in the berry tissue. Thus, the polyphenol-containing cells are widely found in the berry skin, but are also associated with the vascular network in the outer mesocarp where polyphenolics may shield this network from parasitic attack [18]. Anthocyanoplasts are found inside and outside the vacuoles and may be involved in anthocyanin synthesis [19].

Young fruit is unappealing due to the accumulation of organic acids, tannins and pyrazines in the vacuole. At anthesis, most mesocarp cells appear to be univacuolated but at veraison, the large vacuole splits into smaller vacuoles generating a complex internal membrane structure (Fig. **1**). The berry “fattening” that occurs at the ripening stage is mostly due to the massive accumulation of sugars, water and phenolics in the vacuole. Thus, changes in the vacuolation degree may be involved in the maintenance of turgor pressure or result from an increase of its storage function [20]. Parallel to cell expansion, other metabolic changes take place at maturity, namely a relative decrease in acid content, colour development and aroma and flavour compound synthesis and accumulation [21]. Overall, this results in an increase of the organoleptic properties of the fruit, the vacuole being the main reservoir of those appealing compounds.

TONOPLAST TRANSPORTERS AND SOLUTE STORAGE

The storage function of grape berry vacuoles has been recently reviewed [17]. Different compounds are imported and/or compartmented through the coordinated activity of plasma membrane and tonoplast transporters. The grape berry vacuoles maintain an acidic pH, ranging from pH 2.5 in the green stage to pH 3.5 during ripening. In fruit cells, the maintenance of a low vacuolar pH is guaranteed by two different processes, namely proton pumping across the tonoplast, and synthesis and accumulation of organic acids in the vacuolar sap [22]. Two distinct primary proton pumps, the vacuolar ATPase (V-ATPase) and the vacuolar inorganic pyrophosphatase (V-PPase), generate a proton electromotive force, which, in turn, energises the secondary active transport of sugars, organic acids and inorganic ions. The activity of V-PPase predominates in tonoplast vesicles from mature grape berries [23] and intact vacuoles from CSB (Cabernet Sauvignon Berry) suspension cultured cells [24].

The pH of the grape berry and wine relies mostly on tartaric and malic acid content. The biosynthesis and metabolism of both compounds in the berry tissues is extensively reviewed in **Chapter 4**. In Arabidopsis, malate exchange between the vacuole and the cytosol is mediated by the tonoplast malate transporter AttDT [25] and by the tonoplast malate channel AtALMT9 [26]. In grape berry, four good malate transporter candidates (VvALMT9:1, 2, 3, 4) have been identified by blast analysis of *Vitis vinifera* genome with the AtALMT9 protein sequence [27]. Despite the progresses on the elucidation of malate transport mechanisms across the vacuolar membrane, nothing is yet known about tartaric acid transport (reviewed by Fontes *et al.* [17]).

The accumulation of high sugar concentrations in the vacuoles of plant cells has important economical repercussions. Industrial production of sugars is still mainly based on the extraction of sugar cane stalks and sugar beet roots, where glucose and fructose (sugar cane) or sucrose (sugar beet) are accumulated in the vacuoles [28-31]. Grapevine is another crop where sugar accumulation in the vacuole plays an important role, determining the sweetness of table grapes and the alcohol content of wine. Thus, photoassimilate transport in grapevine has been a matter of intense research over the past decade. The literature is very rich on this research topic, and extensive reviews have been published [17, 32, 33, chapter by Davies *et al.* in this edition]. Sugar transport and accumulation into the vacuole of the grape berry mesocarp cells begins at the veraison stage and this massive sugar import and accumulation (up to 1 M glucose and fructose) continues until harvest [21, 32, 33]. In spite of its importance for crop quality, the route(s) of sugar import and storage in the vacuole and their regulations are not yet well understood. Sugar accumulation in the vacuole of storage cells implies sugar uptake into the cell across the plasma membrane and compartmentation into the vacuole. Given the high vacuolar sugar concentration, this compartmentation is an energy-dependent process.

In Arabidopsis three AtTMT (*A. thaliana* Tonoplast Monosaccharide Transporters) isoforms were localised at the tonoplast by GFP fusion [34] and the tonoplast glucose/H^+ antiporter AtVGT1 (At3g03090) was identified [35]. The disaccharide transporters, HvSUT2 from *Hordeum vulgare* and AtSUT4 (AtSUC4) (*A. thaliana* Sucrose Transporter) were identified as tonoplast-localised sucrose transporters [6]. This suggests that the *V. vinifera* disaccharide transporter VvSUC11 is located at the tonoplast due to its similarity to HvSUT2 and AtSUC4. In grapevine, the monosaccharide transporters VvHT2 and VvHT6 (*Vitis vinifera* Hexose Transporter 2 and 6) are hypothetically targeted to the tonoplast. The sequence of VvHT6 is similar to that of three Tonoplast Monosaccharide Transporters (TMTs) in Arabidopsis [36] and its pattern of expression is consistent with VvHT6 having a role in post-veraison import of hexoses into the vacuole (see the chapter by Davies *et al.*, this edition). Our groups have recently identified distinct mechanisms that may account for sugar compartmentation into grape vacuoles. The observation, made by confocal microscopy, that isolated mesocarp cells are able to incorporate in vesicles the fluorescent glucose analogue 2-[N-(7-nitrobenz-2-oxa-1,3-diazol-4-yl amino]-2-deoxy-D-glucose (2-NBDG) (Fig. **1**) strongly supports that endocytosis is involved in the transport and intracellular compartmentation of apoplastic sugars [2]. Also, the involvement of a sucrose facilitated diffusion and an H^+-dependent monosaccharide transport system was suggested when purified tonoplast vesicles and intact vacuoles were used in transport experiments with radioactive sugars and fluorescent markers (N. Fontes, H. Gerós and S. Delrot, unpublished).

Sugar and water fluxes during the unloading and accumulation of sugars in the vacuoles of the flesh cells may be functionally linked as suggested by the co-expression of some aquaporins and sugar transporters [37]. Indeed, there is a fast accumulation of water and sugar when phloem unloading in the berry changes from symplastic to apoplastic, at the time, or just before veraison. At this time, the observed loss of turgor in companion and mesocarp cells due to the high apoplast osmolarities may contribute to fasten water uptake [38]. A paradoxical finding in *Olea europaea* cultured cells has suggested that proteinaceous channels are involved in glucose transport [39], implying that aquaporins, or aquaporin-like proteins may be involved in both water transport and organic nutrition [40]. The transport of water and sugar through such hydrophilic channel at the cell membrane and tonoplast would enable rapid cellular uptake or exit of sugar with minimal osmotic perturbation. This is a particularly important mechanism during fruit ripening when high amounts of sugars and water are accumulated [39].

Aquaporins belong to the major intrinsic protein (MIP) family and are mostly implicated in water transport across biological membranes [40] (see the chapter by Tyerman *et al.*, this edition). Those proteins, very abundant in the tonoplast of plant cells, are named TIP (from Tonoplast Intrinsic Protein) and play important roles in water accumulation, vacuole expansion and cell growth. In the grapevine, the tonoplast aquaporin VvTIP2;1 and the plasma membrane aquaporin VvPIP2;1 are highly expressed in dividing and elongating cells and in cells involved in water and solute transport [41].

The grape berry vacuoles accumulate several inorganic cations and anions that have a pivotal role during fruit development and ripening as well as in wine quality. This topic is explored in detail in the chapter by Martins *et al.* (this edition). For instance, calcium contributes to the maintenance of fruit integrity because it plays structural roles in the cell wall and membranes and regulates the ripening timing due to its well-known role in cell signaling. Potassium is the major cation in grape berry being involved in the regulation of phloem transport rates and assimilation partitioning patterns. Thus, it has a tremendous effect on wine organoleptic properties. Tonoplast transporters of mineral nutrients are also much less known than their plasma membrane counterparts. Regarding Ca^{2+} uptake by vacuoles, the activity of primary active transporters like Ca^{2+}-ATPase and ATP-binding cassette (ABC) transporters was already reported in plant cells [42]. Grape vacuoles also accumulate Ca^{2+} through secondary active transporters dependent on the H^+ gradient dissipation (see below). Regarding K^+ transport, the gene *VvNHX1* that encodes a vacuolar cation/H^+ antiporter from *V. vinifera* was cloned and characterised by Hanana *et al.* [43]. This transporter displays higher affinity for K^+ (12.8 mM) than for Na^+ (40.2 mM), which is striking due to its similarty to the Arabidopsis vacuolar Na^+/H^+ antiporter AtNHX1 (*Arabidopsis thaliana* Sodium Proton Exchanger) that plays an important role in Na^+ sequestration in the vacuole during salt stress [44].

Grape berry vacuoles also play an important role in the sequestration of phenolic compounds and aromas in the grape berry. They are frequently accumulated as glycosides, which increases solubility, transport and storage processes [17, 45]. The biochemical and molecular mechanisms involved in the transport of these compounds across the tonoplast is still rudimentary, although their elucidation is of the utmost biochemical and biotechnological importance due to the effect of these compounds on the colour and organoleptic properties of fruit and wine. Non-flavonoid phenolics accumulate primarily in the vacuoles of mesocarp cells, and flavonoids accumulate in the dermal cells of the skin tissue. Tannins are accumulated in "tannin vacuoles". Regarding pigments, whereas flavonoid pigments are accumulated in cell vacuoles, carotenoids are predominantly stored in plastids [46]. Anthocyanins are synthesised at the endoplasmic reticulum and transported to the vacuole after being glycosilated [47]. Anthocyanic vacuolar inclusions (AVIs), loosely termed as anthocyanoplasts, are believed to sequester anthocyanins primarily to increase their stability, but also to reduce inhibition of certain vacuolar enzymes [48, 49]. The grapevine proteins anthoMATE1 (AM1) and anthoMATE3 (AM3) that belong to the Multidrug and Toxic Extrusion (MATE) family transport acylated anthocyanins across the tonoplast in an H^+ dependent manner [50]. Glycosylated flavonoids conjugated with glutathione may be incorporated in to the vacuole through ABC transporters (see the chapter by Castellarin *et al.*, this edition). Furthermore, aroma compounds are frequently accumulated as glycosides [45, 51, 52], and a number of glycosides and glucosides also exist in the vacuolar sap. Terpenes, norisoprenoids and thiols conjugated with sugars or amino acids are accumulated in the vacuoles of exocarp cells [53].

EXPERIMENTAL APPROACHES TO STUDY TONOPLAST TRANSPORT AND VACUOLE FUNCTION

Vacuole Isolation and Characterisation

Despite the importance and uniqueness of fleshy fruit vacuoles, several vacuolar functions and tonoplast transporters are still poorly studied, especially those from grape cells. The major bottleneck for such studies is the obtention of highly purified vacuoles with a good yield. Also, the increasing impact of proteomic studies in plant biology has generated an unexpected interest in the purification of this extremely fragile organelle and led to independent proteomic studies focused on intact vacuoles from Arabidopsis [3, 5, 6].

An efficient way to isolate intact vacuoles is based on osmotic lysis of protoplasts at a relatively high temperature followed by a Ficoll step gradient centrifugation [24] (Fig. **3**). Grapevine protoplasts have been isolated from leaves, stems, roots, callus, embryogenic tissue and mesocarp tissue of the fruit [8, 54, 55]. Highly pure, intact and functional vacuole populations from cultured cells-derived protoplasts are routinely obtained at the top of the Ficoll gradient. The cytosolic glucose-6-phosphatase may be used as a marker enzyme to monitor vacuole purification. Usually, less than 2% of the marker is recovered in the vacuolar fraction, indicating that this sample is strongly depleted in protoplasts and cytosolic contaminations. This conclusion may be further supported by microscopy observation, flow cytometry analysis and functional studies.

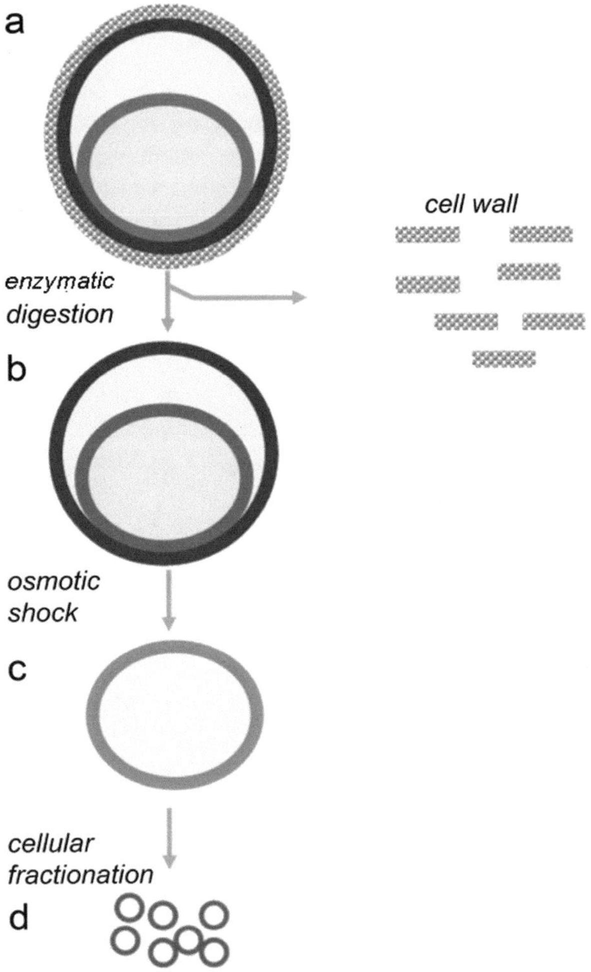

Figure 3: Cell fractionation procedure to obtain protoplasts (b), intact vacuoles (c) and tonoplast vesicles (d) from grape cells (a).

The styryl dye FM1-43 is very appropriate to visualise the vacuoles with the fluorescence microscope. This fluorescent probe exhibits weak fluorescence in aqueous medium, but shines brightly when inserted into membranes. Incubation results in a strong staining of the tonoplast, demonstrating the intactness of these extremely fragile organelles. Other membrane surrounded organelles or membrane vesicles are rarely observed (Fig. **4A**). Most of the intact vacuoles, ranging in size from 10 to 50 µm, are able to maintain an internal acidic pH as they stain with the lipophilic phenazine dye Neutral Red (not shown). The incubation with the fluorescent probe Fluo 4AM indicated their ability to accumulate calcium [24] (Fig. **4B**).

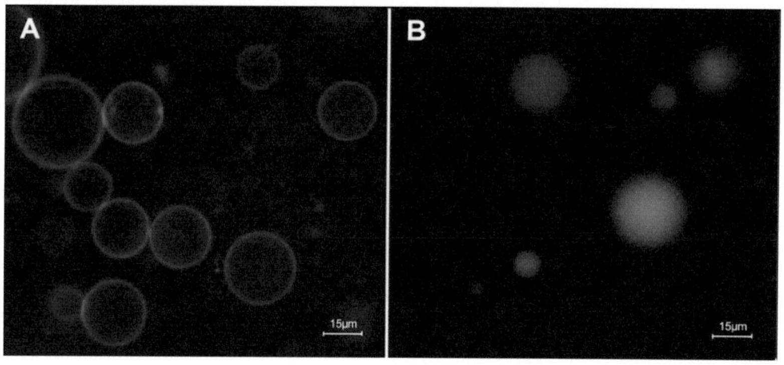

Figure 4: Fluorescence microscopy of intact vacuoles labeled with the fluorescent membrane markers FM1-43 (to stain the tonoplast, A) and Fluo4 AM (to stain calcium, B) Adapted from [24].

Study of Proton Pumps and Secondary Active Transporters

Monitoring the transmembrane proton gradient in intact vacuoles is a proper approach to study the mechanisms of vacuolar acidification, because intact vacuoles are physiologically closer to the *in vivo* plant system than tonoplast vesicles. ACMA (9-amino-6-chloro-2-methoxyacridine) is a highly sensitive pH-dependent fluorescent dye that may be used for proton-pumping measurements by spectrofluorimetry. The PPi-dependent and ATP-dependent H^+ pumping activities as measured by the fluorescence quenching of ACMA are depicted in Fig. **5A** and **B**. In intact vacuoles from cultured cells the V-PPase seems to be the main tonoplast proton pump [24].

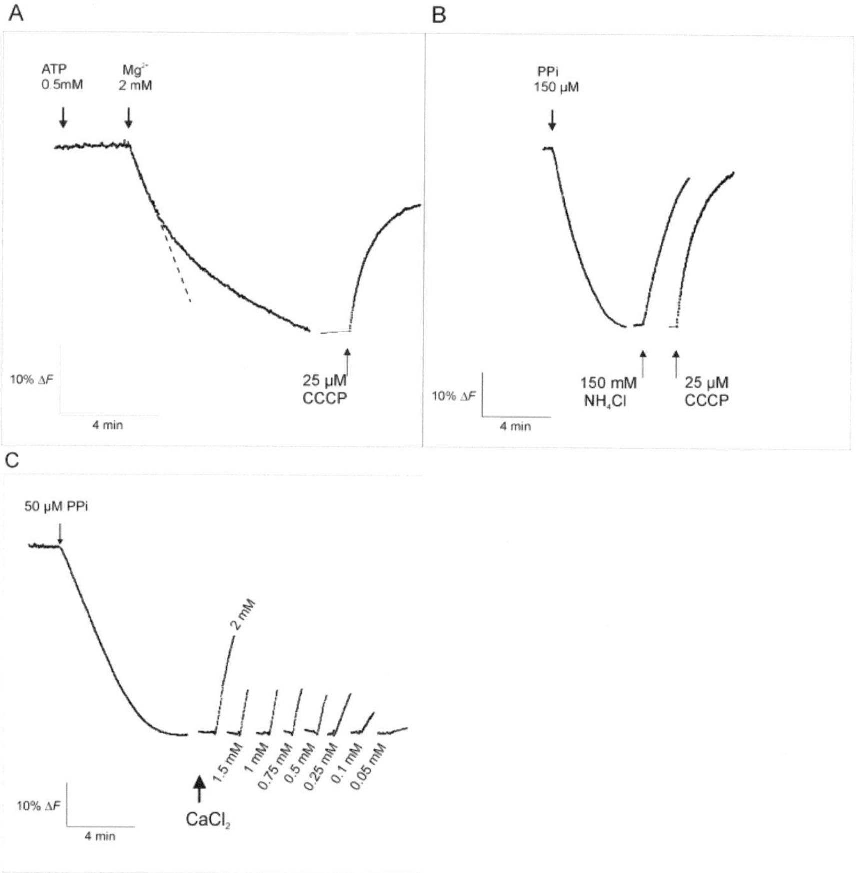

Figure 5: Proton pumping activity of V-ATPase (A) and V-PPase (B) and H^+-dependent Ca^{2+} uptake (C) in intact vacuoles purified from grape protoplasts. The accumulation/efflux of protons was determined by spectrofluorimetry with the fluorescence probe ACMA. Adapted from [24].

In this intact organelle system the capacity of a selected substrate to dissipate the pre-established ΔpH gradient across the tonoplast may suggest the involvement of an H^+-dependent transport system for that substrate. This is the case of Ca^{2+} because the addition of $CaCl_2$ to energised intact vacuoles through the activation of V-PPase resulted in an immediate dissipation of the proton gradient (Fig. **5C**). The first plant Ca^{2+}/H^+ antiporter identified was named CAX1 [56]. Recently, similar proteins were found in both Arabidopsis [57] and rice [58]. Overall intact vacuoles are good experimental models to monitor the mechanisms of vacuolar acidification and solute uptake [24].

Isolated Vacuoles for Proteomic Approaches

Proteomic methodologies can provide important insights into proteins' potential functions based on their subcellular localisations or changes in their expression level in response to a stimulus. Additionally, functional analysis, together with molecular characterisation of tonoplast transporters, should advance our understanding of vacuole function and may promote the genetic engineering of fruits and vegetables [5, 42]. As referred to above, two independent proteomic studies were developed in Arabidopsis vacuoles. In grape berry, several proteomic studies have been performed including the mesorcarp proteome [59], berry proteome [60], and the plasma membrane proteome from CSB (Cabernet Sauvignon Berry) suspension cultured cells [61], but not vacuole proteome.

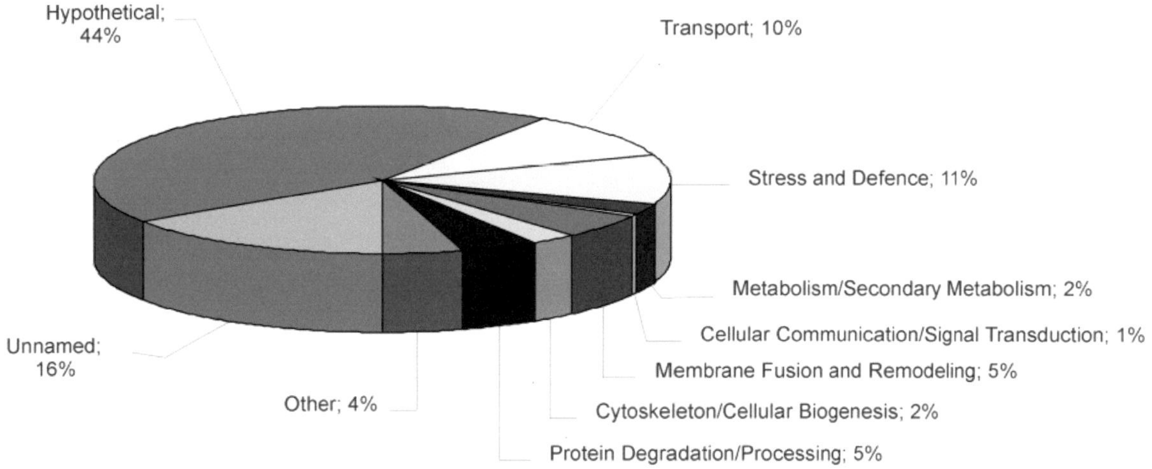

Figure 6: Functional characterisation of the proteins from intact vacuoles derived from grape cells based on the entries in the GeneBank dbEST (NCBI) (Fontes and co-workers, unpublished).

As an earlier attempt to characterise the grape vacuolar proteome, vacuoles were isolated from protoplasts derived from CSB (Cabernet Sauvignon Berry) cultured cells. Total proteins were extracted from purified vacuolar preparations and were subject to one-dimensional SDS-PAGE coupled with liquid chromatography/tandem mass spectrometry (LC MS/MS). The sensitivity of our analysis is indicated by the high coverage of known tonoplast proteins: V-ATPase, V-PPase and TIPs. A total of 671 proteins were identified (Fig. **6**), including 252 vacuolar and putative vacuolar proteins (35% of the total), which were classified and categorised into seven major groups. About 51% of the identified vacuolar proteins are unnamed and of unknown function. Known non-vacuolar proteins represented 14% of the total proteins identified and include plasma membrane, cytoplasm, mitochondrial, ER and chloroplast contaminants. Since the vacuolar membrane contains less than 1% of the cellular protein and this membrane is less protein dense compared to most other cellular membranes, small contaminations may result in the detection of a large number of nonvacuolar proteins. Consistently, other vacuolar proteomic studies have shown similar proportions of contamination [3, 5, 6, 62]. However, it must be stressed that the presence of these proteins, or part of them, in the vacuole samples may not be simply regarded as contamination. Other proteomic studies [5] have in fact reported that the presence of these proteins in the vacuolar sap may be explained by the role of the vacuole in protein and organelle recycling. At least seven subunits of the well-characterised V-type H^+-ATPase were detected, as well as the V-type H^+-. 90% of the identified peptides of all ATPases belong to V-ATPase, which suggests a good purification of our sample in tonoplast membranes. Five TIPs and one vacuolar invertase were also identified. As for transporters, putative potassium transporters, sugar transporters, ABC transporters, calcium/H^+ antiporters, among others, were identified. The results of this study are still preliminary but provide testable hypotheses for determining the molecular components of many processes, including tonoplast fusion, vacuole biogenesis, cytoskeletal attachments, protein degradation, solute storage, among others (Fontes and co-workers, unpublished).

CONCLUSIONS AND FUTURE PERSPECTIVES

Recent research has further demonstrated that the vacuole is an important organelle for the overall plant physiology. They share some basic properties with the vacuoles of algae and yeast and the lysosomes of animal cells. The cytosol of a ripened berry cell possesses a complex vacuole system with recognised function in solute storage. Although our understanding of plant vacuoles remains rudimentary, rapid progress is being made in the areas of tonoplast transport and the regulation and import of vacuolar proteins. However, the same cannot be said for other, equally fundamental aspects of vacuole biology. The mechanisms that control vacuole identity are unknown, so too are the means by which one vacuole fuses with another or fragments into many. Highly pure, intact and functional vacuole populations may be obtained from grape cells. The capacity of the vacuole population to sequester protons and to accumulate Ca^{2+} strongly suggests the intactness and physiological integrity of these extremely fragile organelles. Intact vacuoles may be used as models for both basic research such as solute uptake and compartmentation and proteomics. The reported proteomic analysis is a first step towards understanding the function of this organelle in grapevine. Additional bioinformatic analysis for protein identification and characterisation and for pattern searching is being performed, together with the search for potential contaminants by immunoblotting staining of specific marker enzymes to improve the complete characterisation of the isolated vacuole fractions. Moreover, GFP fusion assays will validate the reliability of protein identification and localisation.

ABBREVIATIONS

2-NBDG	=	2-[N-(7-nitrobenz-2-oxa-1,3-diazol-4-yl amino]-2-deoxy-D-glucose
ABC transporter	=	ATP-binding cassette transporter
ACMA	=	9-amino-6-chloro-2-methoxyacridine
FDA	=	Fluorescein diacetate
FM1-43	=	*N*-(3-triethylammoniumpropyl)-4-(4-(dibutylamino)styryl)pyridinium dibromide
MIP family	=	Major intrinsic protein family
TIP	=	Tonoplast intrinsic protein
V-ATPase	=	Vacuolar ATPase
V-PPase	=	Vacuolar pyrophosphatase

ACKNOWLEDGEMENTS

This work was supported in part by the Portuguese Foundation for Science and Technology (research project no. PTDC/AGR-ALI/100636/2008; grant no. SFRH/BD/23169/2005 to N.F).

REFERENCES

[1] Martinoia E, Maeshima M, Neuhaus HE. Vacuolar transporters and their essential role in plant metabolism. J Exp Bot 2007; 58: 83-102.

[2] Fontes N, Côrte-Real M, Gerós H. New observations on the integrity, structure and physiology of flesh cells from fully ripened grape berry. Am J Enol Vitic 2011. In press (doi: 10.5344/ajev.2011.10126)

[3] Shimaoka T, Ohnishi M, Sazuka T, *et al.* Isolation of intact vacuoles and proteomic analysis of tonoplast from suspension-cultured cells of *Arabidopsis thaliana.* Plant Cell Physiol 2004; 45: 672-83.

[4] Coombe BG. Distribution of solutes within the developing grape berry in relation to its morphology. Am J Enol Vitic 1987; 38: 120-7.

[5] Carter C, Pan S, Zouhar J, Ávila EL, Girke T, Raikhel N. The vegetative vacuole proteome of Arabidopsis thaliana reveals predicted and unexpected proteins. Plant Cell, 2004; 16: 3285-03.

[6] Endler A, Meyer S, Schelbert S, *et al.* (2006) Identification of a vacuolar sucrose transporter in Barley and Arabidopsis mesophyll cells by a tonoplast proteomic approach. Plant Physiol 2006; 141: 196-207.

[7] Fontes N, Delrot S, Gerós H. PT103851. 2009. (http://www.inpi.pt/servicos/patentes_online.html).

[8] Fontes N, Delrot S, Gerós H. A method for the isolation of protoplasts from grape berry mesocarp tissue. Rec Pat Biotechnol 2010; 4: 125-9.

[9] Jones KH, Senft JA. An improved method to determine cell viability by simultaneous staining with fluorescein diacetate–propidium iodide. J Histochem Cytochem 1985; 33: 77-9.

[10] Lang A, Thorpe MR. Xylem, phloem and transpiration flows in a grape: application of a technique for measuring the volume of attached fruits to high-resolution using Archimedes Principle. J Exp Bot 1989; 40: 1069-78.

[11] Lang A, During H. Partitioning control by water potential gradient: evidence for compartmentation breakdown in grape berries. J Exp Bot 1991; 42: 1117-22.

[12] Krasnow M, Matthews M, Shackle K. Evidence for substantial maintenance of membrane integrity and cell viability in normally developing grape (*Vitis vinifera* L.) berries throughout development. J Exp Bot 2008; 1-11.

[13] Marty F. Plant vacuoles. Plant Cell 1999; 11: 587-99.

[14] Paris N, Stanley CM, Jones RL, Rogers JC. Plant cells contain two functionally distinct vacuolar compartments. Cell, 1996; 85: 563-72.

[15] Bethke PC, Jones RL. Vacuoles and prevacuolar compartments. Curr Opin Plant Biol 2000; 3: 469-75.

[16] Jiang L, Phillips TE, Hamm CA, *et al.* The protein storage vacuole: a unique compound organelle. J Cell Biol 2001; 155: 991-1002.

[17] Fontes N, Gerós H, Delrot S. Grape berry vacuole: a complex and heterogeneous membrane system specialized in the accumulation of solutes. Am J Enol Vitic 2011. In press (doi: 10.5344/ajev.2011.10125).

[18] Hardie WJ, O'Brien TP, Jaudzems VG. Morphology, anatomy and development of the pericarp after anthesis in grape, *Vitis vinifera* L. Aust J Grape and Wine Res 1996; 2: 97-142.

[19] Pecket CR, Small CJ. Occurrence, location and development of anthocyanoplasts. Phytochem 1980; 19: 2571-6.

[20] Klymchuk DO, Kordyum EL, Vorobyova TV, Chapman DK, Brown CS. Changes in vacuolation in the root apex cells of soybean seedlings in microgravity. Adv Space Res 2003; 31: 2283-8.

[21] Conde C, Silva P, Fontes N, Dias ACP, *et al.* Biochemical Changes throughout grape berry development and fruit and wine quality. Food 2007; 1: 1-22.

[22] Shiratake K, Martinoia E. Transporters in fruit vacuoles. Plant Biotechnol 2007; 24: 127-33.

[23] Terrier N, Deguilloux C, Sauvage FX, Martinoia E, Romieu C. Proton pumps and anion transport in *Vitis vinifera*: The inorganic pyrophosphatase plays a predominant role in the energization of the tonoplast. Plant Physiol Biochem 1998; 36: 367-77.

[24] Fontes N, Silva R, Vignault C, Lecourieux F, Gerós H, Delrot S. Purification and functional characterization of protoplasts and intact vacuoles from grape cells. BMC Res Notes 2010; 3:19.

[25] Emmerlich V, Linka N, Reinhold T, *et al.* The plant homolog to the human sodium/dicarboxylic cotransporter is the vacuolar malate carrier. Proc Natl Acad Sci USA 2003; 100: 11122-6.

[26] Kovermann P, Meyers S, Hortensteiner S, *et al.* The Arabidopsis vacuolar malate channel is a member of the ALMT family. The Plant J 2007; 52: 1169-80.

[27] Rongala J. Identification and localization of vacuolar organic aid carriers in grapevine berries. Ph.D. thesis, University of Adelaide, Faculty of Science, School of Agriculture, Food and Wine, Waite Campus, 2008.

[28] Thom M, Komor E. Role of the ATPase of sugar-cane vacuoles in energization of the tonoplast. Eur J Biochem 1984; 138(1), 93-9.

[29] Preisser J, Sprügel H, Komor E. Solute distribution between vacuole and cytosol of sugarcane suspension cells: Sucrose is not accumulated in the vacuole. Planta 1992; 186(2), 203-11.

[30] Getz HP. Accumulation of sucrose in vacuoles released from isolated beet root protoplasts by both direct sucrose uptake and UDP-glucose dependent group translocation. Plant Physiol Biochem 1987; 25(5), 573-9.

[31] Getz HP, Klein M. Characteristics of sucrose transport and sucrose-induced H^+ transport on the tonoplast of Red Beet (*Beta vulgaris* L.) storage tissue. Plant Physiol 1995; 107: 459-67.

[32] Agasse A, Vignault C, Glissant D, Conde C, Gerós H, Delrot S. Molecular aspects of sugar transport and its regulation during grape berry ripenning. In: Jeandet P, Clément C, Conreux A, Eds. Macromolecules and Secondary Metabolites of Grapevine and Wine. Intercept, Lavoisier, France 2007.

[33] Agasse A, Vignault C, Kappel C, Conde C, Gerós H, Delrot S. Sugar transport and sugar sensing in grape. In: Roubelakis-Angelakis KA, Ed. Molecular Biology & Biotechnology of the Grapevine (2nd edition). Springer, Netherlands 2009.

[34] Neuhaus HE. Transport of primary metabolites across the plant vacuolar membrane. FEBS Lett 2007; 581: 2223-6.

[35] Aluri S, Büttner M. Identification and functional expression of the Arabidopsis thaliana vacuolar glucose transporter 1 and its role in seed germination and flowering. Proc Natl Acad Sci USA 2007; 104: 2537-42.

[36] Büttner M. The monosaccharide transporter(-like) gene family in Arabidopsis. FEBS Lett 2007; 581: 2318-24.

[37] Delrot S, Picaud S, Gaudillère JP. Water transport and aquaporins in grapevine. In: Roubelakis-Angelakis KA, Ed. Molecular Biology and Biotechnology of the Grapevine. Kluwer Academic Publishers, Dordrecht, The Netherlands 2001; pp. 241-62.

[38] Tilbrook J, Tyerman SD. Hydraulic connection of grape berries to the vine: varietal differences in water conductance into and out of berries, and potential for backflow. Funct Plant Biol 2009; 36: 541-50.

[39] Conde C, Silva P, Agasse A, Tavares RM, Delrot S, Gerós H. An Hg-sensitive channel mediates the diffusional component of glucose transport in olive cells. Biochim Biophys Acta 2007; 1768: 2801-11.

[40] Gomes D, Agasse A, Thiébaud P, Delrot S, Gerós H, Chamount F. Aquaporins are multifunctional water and solute transporters highly divergent in living organisms. Bioch Biophy Acta 2009; 1213-28.

[41] Fouquet R, Léon C, Ollat N, Barrieu F. Identification of grapevine aquaporins and expression analysis in developing berries. Plant Cel Rep 2008; 27: 1541-50.

[42] Maeshima M. Tonoplast transporters: organization and function. Annu Rev Plant Physiol Plant Mol Biol 2001; 52: 469-97.

[43] Hanana M, Cagnac O, Yamaguchi T, Hamdi S, Ghorbel A, Blumwald E. A grape berry (*Vitis vinifera* L.) cation/proton antiporter is associated with berry ripening. Plant Cell Physiol 2007; 48: 804-11.

[44] Silva P, Façanha AR, Tavares RM, Gerós H. Role of tonoplast proton pumps and Na^+/H^+ antiport system in salt tolerance of *Populus euphratica* Oliv. J Plant Growth Regul 2009; 29: 23-34.

[45] De DN. Plant cell vacuoles: an introduction. CSIRO, Australia, 2000; pp. 38-248.

[46] Jackson RS. Grapevine Structure and Function (Chapter 3). In: Wine Science Principles and Applications (3rd edition). Elsevier Inc. San Diego, California, USA 2008; pp. 50-106.

[47] Grotewold E. The challenges of moving chemicals within and out of cells: insights into the transport of plant natural products Planta 2004; 219: 906-9.

[48] Conn S, Zhang W, Franco C. Anthocyanic vacuolar inclusions (AVIs) selectively bind acylated anthocyanins in *Vitis vinifera* L. (grapevine) suspension culture. Biotechnol Lett 2003; 25: 835-9.

[49] Conn S, Franco C, Zhang W. Characterization of anthocyanic vacuolar inclusions in Vitis vinifera L. cell suspension cultures. Planta 2010; 231(6), 1343-60.

[50] Gomez C, Terrier N, Torregrosa L, *et al.* Grapevine MATE-type proteins act as vacuolar H^+-dependent acylated anthocyanin transporters. Plant Physiol 2009; 150: 402-15.

[51] Terrier N, Ollé D, Verriès C, Cheynier V. Biochemical and molecular aspects of flavan-3-ol synthesis during berry development In: Roubelakis-Angelakis KS, Ed. Grapevine Molecular Physiology and Biotechnology (2nd edition). Springer Academic Publishers, Netherlands 2009; pp. 429-60.

[52] Dunlevy JD, Kalua CM, Keyzers RA, Boss PK. The production of flavour and aroma compounds in grape berries. In: Roubelakis-Angelakis KA, Ed. Grapevine Molecular Physiology and Biotechnology (2nd edition). Springer Academic Publishers, Netherlands, 2009; pp. 429-60.

[53] Lund ST, Bohlmann J. The molecular basis for wine grape quality - a volatile subject. Science 2006; 311: 804-5.

[54] Papadakis AK, Reustle GM, Roubelakis-Angelakis KA. Protoplast technology in grapevine In: Roubelakis-Angelakis KA, Ed. Molecular Biology and Biotechnology of the Grapevine (1st edition). Kluwer Academic Publishers, Netherlands 2001.

[55] Papadakis AK, Fontes N, Gerós H, Roubelakis-Angelakis KA. Protoplast technology in Vitis spp. In: Roubelakis-Angelakis KA, Ed. Molecular Biology and Biotechnology of the Grapevine (2nd edition). Springer, Netherlands, 2009; pp. 429-60.

[56] Hirschi K, Zhen R, Cunningham KW, Rea PA, Fink GR. CAX1, a H+/Ca2+ antiporter from Arabidopsis. Proc Natl Acad Sci USA 1996; 93: 8782-6.

[57] Cai X, Lytton J The cation/Ca^{2+} exchanger superfamily: Phylogenetic analysis and structural implications. Mol Biol Evol 2004; 21: 1692-703.

[58] Kamiya T, Maeshima M. Residues in internal repeats of the rice cation/H^+ exchanger are involved in the transport and selection of cations. J Biol Chem 2004; 279: 812-9.

[59] Sarry JE, Sommerer N, Sauvage FX, *et al.* Grape berry biochemistry revisited upon proteomic analysis of the mesocarp. Proteomics 2004; 4: 201-15.

[60] Giribaldi M, Perugini L, Sauvage FX, Schubert A. Analysis of protein changes during grape berry ripening by 2-DE and MALDI-TOF. Proteomics 2007; 7: 3154-70.

[61] Zhang J, Ma H, Feng J, Zeng L, Wang Z, Chen S. Grape berry plasma membrane proteome analysis and its differential expression during ripening. J Exp Bot 2008; 59: 2979-90.

[62] Schmidt UG, Endler A, Schelbert S, *et al.* Novel tonoplast transporters identified using a proteomic approach with vacuoles isolated from Cauliflower buds. Plant Physiol 2007; 145: 216-29.

CHAPTER 9

Tackling the Cell Wall of the Grape Berry

L. F. Goulao[1], J. C. Fernandes[2], P. Lopes[2] and S. Amâncio[2,*]

[1]*Centro de Ecofisiologia, Bioquímica e Biotecnologia Vegetal, Instituto de Investigação Científica Tropical, IP, Quinta do Marquês, Av. da República 2784-505 Oeiras, Portugal and* [2]*Centro de Botânica Aplicada à Agricultura (CBAA)/ DRAT, Instituto Superior de Agronomia, UTL, Tapada da Ajuda 1349-017 Lisboa, Portugal*

Abstract: The cell wall (CW) is the dynamic border of plant cells. In grape berries, the CW decisively accounts for the difference between the pulp and skin cells, with direct consequences on the grape characteristics, wine quality and wine-making methods. The softening of mature berries results from the de-polymerisation and solubilisation of CW polymers. Modifications of grape pulp and skin CW provide the flexibility for cell expansion during fruit growth and to modulate the final texture. Wine making and berry processing methods are directly related with the absence, in white wines, or the presence, in red wines, of skin CW in the fermenting must. Anthocyanin extraction depends directly on skin yielding of the pigment upon CW degradation. During fruit growth and ripening, the cooperative action between different enzyme families is capital in CW metabolism. The sequencing and public availability of the *Vitis* genome allowed us to focus on individual pathways, to profile the expression pattern of isoforms associated with each tissue, developmental phase or stress response, anticipating the effects on berry (and wine) production and quality. Retrieving the sequences of genomic coding regions and the predicted enzymes that act on the *Vitis,* CW allows us for the first time to tackle the grape berry Cell Wall*ome*.

Keywords: Cell wall enzymes, Cellulose, Glycoproteins, Hemicelluloses, Lignin, Microfibrils, Pectins, Phenolic compounds, Polysaccharides, Primary cell wall, Secondary cell wall, Wallome, Xyloglucans.

INTRODUCTION

The plant cell wall (CW) is a complex macromolecular structure that surrounds and protects the cell. Functions of the primary wall include plant structural and mechanical support, determination and maintenance of cell shape, resistance to internal turgor pressure of the cell, control over growth at a precise rate and direction, regulation of diffusion through the apoplast, and protection against pathogens, dehydration and environmental factors [1]. Thus, the CW is an important source of biologically active signalling molecules, regulating cell-to-cell interactions and also a carbohydrate storage reserve. Remodelling of the fruit CW is mandatory to provide the flexibility required for cell expansion during fruit growth and to modulate final texture attributes which, together with flavour and aroma, render the fruit attractive to a variety of seed-dispersing organisms [2]. Therefore modifications of the wall polymers must be fine-tuned to regulate the CW dynamics needed to accommodate growth and ripening.

The nutraceutical effect of wine, grape and grape derivatives is commonly associated with the antioxidant properties of the phenolic species they contain [3, 4]. The colour, astringency and antioxidant properties of wines, in particular of red wines, can be assigned to phenolic acids, to simple flavonoids like anthocyanins or to condensed flavonoids as proanthocyanidins (PA) and tannins [3, 5]. These phenolic compounds can be solubilised into the vacuole or linked to the CW polysaccharides. Hence, the CW of grape berry skin cells is also of main relevance to wine- making and other grapevine processing methods, since it forms a hydrophobic barrier to the diffusion of phenols, holding the main control of extractability [6].

The release of the *Vitis* genome [7, 8] hastens *omics*-related research. Profiling the expression patterns of genes associated with *Vitis* berry CW during growth, development and in response to abiotic and biotic stresses provides the understanding of CW impact on grape and wine production and quality.

***Address correspondence to S. Amâncio:** Centro de Botânica Aplicada à Agricultura (CBAA)/ DRAT, Instituto Superior de Agronomia, UTL, Tapada da Ajuda 1349-017 Lisboa, Portugal; E-mial: author: samport@isa.utl.pt

Hernâni Gerós, M. Manuela Chaves and Serge Delrot (Eds)

THE PLANT CELL WALL STRUCTURE AND COMPOSITION

Primary Cell Wall

The primary CW of dicotyledonous and non-commelinoid monocot species (Type-I cell walls, according to Carpita and Gibeaut [9]) is composed of approximately 90% polysaccharides [10] from three major classes that form its structural elements: cellulose, matrix cross-linking glycans (henceforth referred to as hemicelluloses) and pectic polysaccharides, which, in fruits, represent about 35%, 15% and 40% of the CW mass, respectively [1]. Structural glycoproteins, phenolic esters, minerals, and enzymes are also present, directing modifications on its physical and chemical properties.

Cellulose is a linear polysaccharide consisting of long unbranched β-1,4-linked cellobiose chains. It forms a crystalline or semi-crystalline microfiber phase, *via* extensive hydrogen bonding between individual strands (microfibrils) that, winded together, provide most of the tensile strength to the plant cell matrix and forms the framework around which the other components are positioned. Cellulose microfibrils are embedded in a matrix phase consisting of hemicelluloses and pectic polysaccharides.

Hemicelluloses are cross-linking glycans that can interact non-covalently trough hydrogen-bonds to cellulose microfibrils, having the capacity to coat and tether them together to form an extensive framework. Hemicelluloses consist of polysaccharides with a backbone of 1,4- linking β-D-pyranosyl residues in which *O*-4 is in the equatorial orientation. They differ from cellulose due to its substitution with other sugars, which results in considerable variation in their composition and structure.

Xyloglucans (XGs) are the predominant hemicelluloses in the dicot primary CW, representing 15-25% [9]. Non-Solanaceae Type-I CW XGs are composed of repeating heptasaccharide units to which variable amounts of sugar residues are added during synthesis up to about 75% of the β-1,4-D-Glc*p* backbone residues [11], resulting in a family with large heterogeneity. Short side chains holding xylose-containing mono- (xylose), di- (xylose-galactose) or tri- (xylose-galactose-fucose) saccharides are linked by α-1,6 bonds at regular sites to the *O*-6 position of the glucose units of the linear backbone of XG. XGs occur at distinct locations in the wall, either binding tightly to portions of exposed faces of glucan chains in the cellulose microfibrils, or spanning the distance between adjacent microfibrils to lock them into place. Recently, XGs and xylans have been localised to cell junctions in ripening fruits, suggesting a role of hemicelluloses in cell adhesion [12], which was previously attributed to pectic homogalacturonans (see below).

Other hemicelluloses include mannans (a β-1,4-mannose backbone, with or without galactose linked by an α-1,6 bond), including glucomannans, galactomannans and galactoglucomannans, and xylans (a backbone of β-1,4-linked xylosyl residues, substituted by β-linked 4-*O*-methylglucuronic acid and by acetyl esters on C2, and α-linked arabinose on C2 or C3) of some xylosyl residues, forming arabinoxylans, glucuronoxylans and glucuronoarabinoxylans.

Pectins are embedded within the cellulose/hemicellulose network, forming hydrophilic gels that impose mechanical features to the wall, such as regulation of the hydration status and ion transport, definition of the porosity and stiffness which, in this way, determines the water holding capacity, controls the permeability of the wall for enzymes and provides additional strength to the matrix. Molar mass, neutral sugar content, proportions of smooth and hairy regions, ferulic acid substitution, amounts of methoxyl and acetyl esters and distribution of ester groups on the polymer characteristically define its fine structure which, in turn, determines functional properties of micro-domains, such as surface charge, pH and ion balance and establishes the biological roles within the CW. Pectins are complex, structurally heterogeneous acidic polysaccharides composed of a range of 1,4-linked α-D-galactosyluronic acid (Gal*p*A) residue-containing linear chains, assembled with a range of modifications and substitutions with variable degrees of ramifications by single sugars or complex side chains [13]. Structural classes of pectins include homogalacturonan (HG), rhamnogalacturonan-I (RG-I), rhamnogalacturonan-II (RG-II) and, at a lower extent, arabinan, arabinogalactan-I (AG-I) and arabinogalactan-II (AG-II), as well as substituted galacturonans like apiogalacturonan (AGA) and xylogalacturonan (XGA).

HGs are polymers formed by α-1,4-linked linear chains of more than 72-100 Gal*p*A residues [14], and can account for more than 60-65% of the total plant pectins. The walls of fruits such as tomato and mango have up to 35% and 52% of uronic acid, respectively [15, 16]. HG Gal*p*A residues may be methyl-esterified at the C-6 carboxyl and/or acetylated at the *O*-2 or *O*-3 position. Methyl-esterification is tightly regulated in a developmental and tissue-specific way. Methyl-esterified regions have neutral charge, but the unmethylated Gal*p*A residues are negatively charged and may be ionically cross-linked with Ca^{2+} to form stable gels with other pectic molecules, when stretches of 10 or more consecutive un-methyl-esterified residues occur. The hypothesised *in vivo* structure of the HG-calcium complex is referred to as the "egg-box" [17] and describes the close packing of HG that occurs upon Ca^{2+}-induced gelling in the CW of plants. Methyl-esterification neutralises the charge on Gal*p*A residues and thereby abolishes their ability to cross-link calcium ions. The occurrence of micro-domains inside the pectic polysaccharides means the localisation of precise areas with distinct properties as the result, to some extent, of a different, localised demethylation mechanism which may lead to stiffening or loosening of the wall (reviewed by Goulao [18]).

RG-I is the major branched, heterogeneous and hydrated component of the middle lamella and primary CWs. It consists of a backbone holding a variable number of α-1,4-linked Gal*p*A and α-1,2-linked rhamnose repeats, and three types of neutral sugar side groups attached to the 4-position of approximately 20-80% of the rhamnose backbone units, depending on the source of the polysaccharide [19]. These side-chains can derive from single or polymeric substitutions and are mainly composed of α-1,5-L-arabinans, β-1,4-D-galactans and arabinogalactans, where arabinose is usually terminal and galactose links can be connected through C-4, C-3 or C-6. Its abundance is developmentally and differentially regulated [20].

RG-II molecules are stretches of HG backbone approximately 7-9 α-1,4-D-Gal*p*A residues long, substituted with clusters of four highly complex and well-defined conserved side chains that contain 12 different types of sugars, in more than 20 different linkages [21]. Its structure consists of self-associated dimers cross-linked by single borate diesters [22, 23] and stabilised by the presence of calcium [24].

The three main pectin domains, HG, RG-I and RG-II, are described as being covalently linked to form the pectic matrix, envisioned as a unique and complex macromolecule [25-27], although the nature of their covalent arrangements is still unclear. A representation of the pectin network was proposed by Vincken *et al.* [26] and afterwards supported by Coenen *et al.* [27], in which RG-I supplies the main backbone to which HG, RG-II and the other less abundant pectic domains are covalently cross-linked to form side-chains of the same molecule.

In addition to the polysaccharides, primary CW contains about 10% structural proteins, and protein rods act as supporting brackets to the long polysaccharide chains [28]. Five classes of structural apoplastic proteins have been described: extensins, glycine-rich proteins (GRPs), proline-rich proteins (PRPs), arabinogalactan proteins (AGPs), and solanaceous lectins [29]. Extensins are rich in hydroxyproline amino acid residues that may covalently cross-link polysaccharides to form an interlocking framework where the ends of the protein rods are wrapped around the cellulose microfibrils [30]. AGPs are proteoglycans that have been mainly implicated in cell adhesion [31].

Secondary Cell Wall

When the cell stops dividing and expanding, in some tissues lignin is deposited within the cellulose microfibrils and matrix carbohydrates, establishing chemical bonds with non-cellulosic carbohydrates, forming a thick secondary CW. According to the chemical groups that stabilise polysaccharide-phenol complexes, two types of bonds are identified: hydrogen bonds between the hydroxyl groups of phenols and the oxygen atoms of CW polysaccharides sugar moieties or hydrophobic interactions with secondary structures of some polysaccharides [3]. Generally, secondary CWs consist of three layers: outer (S1), middle (S2), and inner (S3) [3, 32]. The formation of secondary walls occurs mainly in xylem vessels, structural fibers, seed pods and seed integument, as in grapevine berry seeds [33, 34]. The process starts in the middle lamella and the primary wall (initiation of S1 formation). When the polysaccharide matrix of the S2 layer is completed, lignification proceeds through the secondary wall [35], in particular at the final stage of xylem differentiation [36]. Lignin deposition is then developmentally programmed assuring structural integrity and waterproof of the CW and enabling the transport of water and solutes through the vascular system, although its biosynthesis can also be induced by biotic and abiotic stress conditions [35, 37].

Lignin, the second most abundant plant organic compound, is a branched heteropolymer of phenylpropanoids synthesised from the polymerisation of the three most abundant *p*-hydroxycinnamyl monolignols, *p*-coumaryl, *p*-coniferyl and *p*-sinapyl alcohols, which, once incorporated into the polymer, are referred to as *p*-hydroxyphenyl (H), guaiacyl (G) and syringil (S) units, respectively [34, 38, 39]. The relative amount of each unit varies between species, tissues and environmental conditions [34]. Dehydrogenated monolignols can form dimers through covalent bonds between the central carbon of the monolignol tail β-β type [34], or between the β carbon and C atoms of the aromatic ring, *e.g.* β-*O*-4 or β-5. After a new dehydrogenation of the dimer, another covalent bond can be established by a polymerisation process of one unit at a time. Molecular species other than the canonical monolignols can be integrated in the lignin polymer, which explains the plasticity of the polymerisation process and the variability of the final polymer [34, 36]. The lignin of angiosperms, as stands for the grapevine, is almost exclusively composed by G and S subunits. In poplar, a woody plant, the linear lignin length is between 13 and 20 units [40], but no reports are available for the length of the grapevine lignin chain.

MODELS OF SUPRA-MOLECULAR ARCHITECTURE OF THE PRIMARY CELL WALL

The CW is represented as a three-dimensional network containing interconnected fluid-filled pores that form pathways for solutes through the walls. Although the complexity of the primary CW supraorganisation and architecture is under continuous debate, a model of supramolecular organisation of the dicot primary CW based on the "tethered network" model [11, 41] has been the most consensual in the last years. In this model, XG is proposed to form hydrogen bonds with cellulose microfibrils, acting as a load bearing tether between the microfibrils, which reinforces the CW. Its location both in the inner and outer surfaces of microfibrils allows for the binding of adjacent microfibrils, while preventing hydrogen bonding between cellulose microfibrils, and thus facilitating each microfibril to slide during cell expansion. Yet, only about ca 8% [42] of the cellulose microfibril surfaces are covered with XGs and not all of the XG is adsorbed to cellulose [42, 43]. Moreover, XGs bind to the surface of cellulose microfibrils making CW-XG a composite structure in which cellulose crystallites are embedded in a matrix of XG with a semi-rigid (straightened backbone) conformation, that is, a matrix that is partly ordered rather than amorphous [44].

This XG-cellulose network is considered to be organised independently and embedded in a second network formed by an amorphous pectin matrix, which acts as a cement (reviewed by Cosgrove [45]) where the negatively charged chains of polygalacturonic acid provide the capacity of interacting and binding with positively charged molecules such as polyamines, cations and positive charges of proteins. However, *in muro* covalent linkages between RG-I-arabinan side chains and cellulose microfibrils [46-51], and anionic complexes derived from covalent linkages between XG and pectins have been reported [52-57]. RG-I was found to be very firmly integrated into the wall [58, 59], providing structural links between the two major CW networks, which are expected to have a role in maintaining the structure of the wall. Pectic polymers operating in cell adhesion are possibly tethered into CW structures by links through XG located in CW regions that are important for maintaining cell adhesion [12]. Moreover, as pectic chains are much more flexible than hemicellulose molecules [60], the alignment of the rod-like chain segments with the microfibrillar surface is less likely, and it seems possible that the hydrogen-bonded interface is relatively disordered [61].

Finally, a third network of structural proteins covalently bound to each other and to other cell components is also often considered [62]. Several models have been proposed to explain the CW architecture (reviewed in [18, 45, 63]) but, to date, there seems to be no definitive evidence favouring a given model over the others. Realistic wall models should consider a highly cross-linked wall wherein pectin-pectin, pectin-XG, pectin-cellulose, pectin-phenolics, pectin-protein and XG-cellulose provide a cohesive network.

COMPOSITION OF THE GRAPE BERRY CELL WALL

The mesocarp of mature grapes follows the typical Type-I CW model, consisting of approximately 90% by weight of polysaccharides and less than 10% of a protein fraction rich in arginine and hydroxyproline residues. Cellulose and polygalacturonans are the major constituents, each accounting for ca 30-40% by weight of the polysaccharide component of the walls [64, 65]. They display, however, significant varietal differences in the relative abundance of the two polysaccharides. While the mesocarp cells of Traminer and Sauvignon Blanc berries have thin CWs, skins, on the other hand, consist of thick-walled epidermal and hypodermal cells [66]. In the exocarp, polysaccharides account for 50% of the CW material [67], with a glycosyl-residue composition similar to mesocarp walls [64, 68]. Neutral polysaccharides (cellulose, XG, arabinan, galactan, xylan and mannan) account for 30%, while acidic pectin substances (of which ca 62% are methylesterified) account for 20%. The remaining part is composed of 15% insoluble proanthocyanidins, <5% structural proteins [67] and lignin [69].

The pectic fraction is composed of 65% HG, 10% RG-I, 2% RG-II and 23% neutral side chains [64, 65, 68]. Arabinans and AGI contribute with 4-6% by weight to the pectic polysaccharides [64]. Noticeably, all berries contain overall high amounts of HG in comparison with other fruits, and grapes have higher proportions of RG-I compared to other berries [70]. Differences in pectin composition have been observed between the pulp and the skin [65]. 75% of the grape berry walls (by weight) originate from the skin, representing 25% of the total fresh berry weight [65, 71]. The relative molar distribution (mol %) of the different polysaccharides in the red wine grape skins was estimated to be 57-62 HG, 6-14 cellulose, 10-11 XG, 7 mol %, 4.5-5 RGI, 3.5-4 RGII, 3 mol % AG, and 0.5-1 mannans [72]. Also, the relative abundance of grape skin CW polymers differs from the pulp. A 3-folds higher content of pectic polysaccharide fractions was detected in the skin as compared to flesh tissues [65]. On the other hand, pulp tissues contain a 2-fold more buffer-soluble AGPs and pectins than skins [65, 68].

Hemicellulosic polysaccharides consisting mainly of XGs, comprise approximately 8-12% of the total wall polysaccharide fraction, both in the pulp and skin [64, 71], and the remainder is made up of smaller amounts of mannans, heteroxylans, arabinans, galactans and arabinogalactans [64, 67, 71]. Although grape XG and XGs isolated from the walls of other dicot plants have similar structures, the amount is lower than the typical XG content of dicot walls [71]. XGs isolated from mesocarp and exocarp CWs of grape berries are composed of eight types of oligosaccharides (XXXG, XLXG, XXLG, XLLG, XXFG, XLFG, XFFG and XXG; see Fry *et al.* [73] for nomenclature) in similar proportions in the skin and pulp, except for XXFG, which is more abundant in the pulp, and XLFG, which is more abundant in the skin [71].

Concerning secondary compounds, it is assumed that most phenolic compounds are nearly absent from the grape berry flesh, mainly embodied in the skin and seeds. The CW of grape skins includes ca 15% tannins with an average degree of polymerisation of 28 [3]. Recently, Bidon *et al.* [69] have reported a PA skin fraction accounting for 54% of the total extractable PA. Interestingly, a different interaction pattern occurs between flesh and skin CW material and PA, with possible effects on PA extraction and winemaking.

CELL WALL MODIFICATIONS DURING BERRY GROWTH AND RIPENING

Remodelling of the fruit CW is mandatory to provide the flexibility required for cell expansion during fruit growth and to modulate final texture attributes which, together with flavour and aroma, render the fruit attractive to a variety of seed-dispersing organisms. In fact, fruit softening during ripening is one of the developmental events whereby most changes occur in the CWs, which explains why most research on fruit CW metabolism has focused primarily on the ripening phase of development.

Grapes develop according to a double sigmoidal curve. The first growth phase (phase I) is due to cell division and subsequently to cell enlargement followed by the lag phase (phase II), which is characterised by the lack of changes in berry weight and volume. The end of this phase coincides with the onset of ripening (veraison, V) and takes place ca 5-8 weeks before maturity. Following veraison, the second growth phase (phase III) occurs with increasing size of the central mesocarp cells, entirely due to cell expansion within the berry. Grape ripening represents the last third of berry development, so grapes soften at the same time as they expand, during this second growth stage. In grapes, after the onset of ripening, in addition to sugar accumulation and water influx, growth results from the synthesis of new CW material [74].

Changes in Pulp Cell Walls

During ripening, the fruit CW experiences a general increase in pectin solubility, losses of non-glucosyl neutral sugars from pectin side-chains, and loosening of the xyloglucan-cellulose network [75-77]. These events are common to every fruit species and may or not be accompanied by a decrease in the molecular mass of matrix polysaccharides [76, 77]. Depolymerisation varies among fruit species and, even within a species, different timing and extent of modifications may occur according to specific genotypes [77].

In grape berries, dissolution of the pulp CW during ripening is observed [78], but apparently no dramatic changes in wall polysaccharide composition seem to occur [74]. Instead, more subtle structural modifications of specific constituent components may contribute to softening.

The total amount of uronic acids (UA) per berry increases rapidly during phase I due to marked increases of newly synthesised highly methyl-esterified HGs [79]. However, this new synthesis of HG occurs more slowly during phases II and III, concomitantly with slower cell enlargement and thinning of walls to accommodate fruit expansion [80], and it is not sufficient to offset thickening and enlargement of walls during these stages [81], as grape pericarp CWs do not thicken appreciably during ripening [66]. However, as a consequence of the increase in cell volume without concomitant wall synthesis, the CW becomes in fact thinner at the end of maturation [74, 82] both in mesocarp and exocarp [83], which can explain the lower amount of isolated CW material as ripening progresses, particularly during the last weeks of grape development [82].

The most significant change in the composition and chemical properties of the wall of berry mesocarp cells during the onset of ripening is the decrease in its galactose/galactan content, particularly the β-1,4-linked Gal*p*A residues [74] corresponding to a significant loss of AG-I from side chains of pectic polysaccharides, from before veraison (BV) to ripe berries [74]. This event has been reported as a crucial step associated with the initiation of softening [82].

Total pectins from Muscat Gordo grapes decline from 58%, two weeks BV, to 47%, four weeks post-veraison (PV) [74]. The type of pectin found in grape berries also changes during ripening, as CW-bound pectins decrease together with an increase in the water-soluble fraction [84]. In fact, at PV, the decrease in wall-bound pectins is accompanied by a two-fold increase in water-soluble polysaccharides [74, 82], disclosing solubilisation mainly of galacturonan as ripening progresses and the grapes soften [74, 84]. A higher solubility of polygalacturonic acid (PGA), AG-II and arabinan is also noticed [74]. However, during veraison, galactose and arabinose content in the water-soluble fraction did not specifically change [85], suggesting that the increase in UA content of the water-soluble fraction is due to degradation or demethylation of pectin, more easily extracted in hot water. It has been suggested that de-esterification increases pectin solubilisation by creating electronic repulsion between negatively-charged molecules that could result in the loosening of weakly attached pectins [18, 86]. The presence of high molecular mass pectic polysaccharides at BV probably reflects differences in the degree of esterification from initial to later stages. Changes in the degree of pectin methylation seem to be cultivar-specific. While in Gordo an initial decline in the degree of esterification is observed during ripening, in Ugni Blanc, it decreases from 68% in green stages to less than 20% as ripening progresses [74, 81].

A decrease of the molecular masses of both pectic and hemicellulosic polysaccharides and a reduction in the cellulose and total hemicellulose content are observed at V and proceed throughout subsequent phases [85]. It should be noted that, according to Nunan *et al.* [74] at PV cellulose and XG levels decrease on a fresh weight basis but not on a mol% basis.

In pectin fraction, total amount and the neutral and acidic sugars of the water soluble fraction, temporally increase from BV to V stages, decreasing rapidly PV [84, 85]. In contrast, neutral sugar of the hemicellulosic fraction decrease from BV and throughout V [85]. This discloses a pattern of temporal modifications in grape berry softening in which CW polysaccharides of the mesocarp are promptly modified from BV to V [85] (Fig. **1**).

Another noteworthy wall modification in grape ripening is an important increase in the content of proteins at PV [74]. From phase II throughout final maturity, the protein content of the mesocarp CW increase by more than 50% [74]. Amino acid analyses reveal that such increase is largely due to hydroxyproline-rich proteins, including extensins [74, 87] (Fig. **1**).

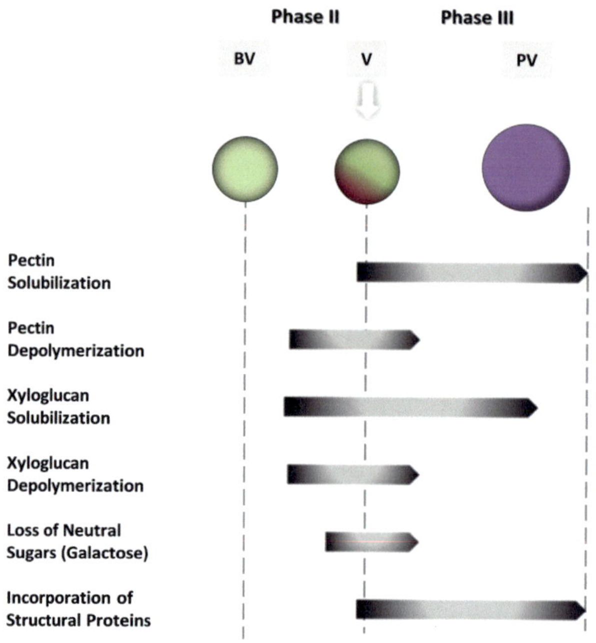

Figure 1: Cell wall modifications during berry growth and ripening. Most significant changes in the composition and chemical properties of the wall of berry cells take place at veraison and post veraison stages, at phases II and III of growth and ripening.

Changes in Skin Cell Walls

In all development phases the active metabolism of the skin severely influences the final characteristics of the grape berry. Berry size does not change from BV to V, even with the advent of softening, most likely due to the unchanged stiffness of the outer skin [85]. Besides differences in composition (see previous sub-chapter), there are also differences in the skin CW dynamics when compared to pulp tissues. In fact, the massive dissolution of CW found in the pulp PV [78] was not observed in the skin [66]. However, skin loosens continuously PV. Ultra-structural changes of the loosening skin include wall swelling in epidermis and sub-epidermis cells and degradation of the middle lamella in the hypodermis cells. Additionally, the wall surfaces become ''wavy'' as the ionic calcium bridge among pectin molecules is broken [88].

Similarly to the pulp, changes in the constituents of the grape skin CW are well related with the degree of ripening, namely the decrease in the amount of CW material and galactose [83] in solubilisation of AGP-I, and in the degree of pectin methyl- and acetyl- esterification (most varieties) [89]. Changes in the degree of pectin methylation seems to be cultivar specific since it decreases as ripening progresses in some cultivars like Cabernet Sauvignon, Merlot and Monastrell, while it hardly changes in Syrah [83]. These changes are accompanied by the accumulation of glucose, while other neutral sugars showed no significant variations. However, the amount of skin CW polysaccharides was estimated to correspond to ca 4.2 mg/berry in Shiraz grapes, remaining constant during ripening [89].

During ripening, a more than 2-folds increase in water-soluble fraction occurs both in the pulp and the skin. However, while it stands for 10 to 23% of the mesocarp CW material [74], it represents a small fraction (3 to 8%) in the skin [89]. Even considering that in grapes the pulp to skin weight ratio is about 1:5, a smaller proportion of water-soluble polysaccharides is present in the skin [89], which corresponds to increases of 450-920 μg/berry in the pulp of Gordo Muscat compared to a 220-270 μg/berry increase for the Shiraz skin tissue [74, 89]. Likewise, in Grenache Blanc grapes, the water-soluble fraction accounts for 30% and 13% of the pulp and skin, respectively [65]. Partial loss of wall structural polysaccharides is compensated by the incorporation of structural proteins and formation of phenolic cross-linkages that happen at the end of the maturation period especially in the walls of epidermis and sub-epidermis cells [88]. Since proteins are confined to the outmost four layers of skin cells, they might contribute to the necessary strength of the tissue to maintain berry integrity, acting as a protective tissue [83, 88]. This aspect may be an important difference in CW modification between the skin and the pulp [88].

Therefore, changes in the skin CWs are continuous but restrained in comparison to other fruits or other grapevine tissues, particularly the small proportion of water-soluble polysaccharides [89], making skin more resistant to solubilisation and maintaining berry integrity.

Modulation of Berry Growth by Mesocarp and Exocarp Cell Wall Modifications

During phase I, cellular expansion proceeds throughout all tissues, while during the transition between phases II and III only the exocarp cells expand [82], and during phase III (berry growth and ripening) only the expansion of the mesocarp cells occurs. Therefore, grape berry skin controls PV growth by remodelling its CW [88, 90-92]. The loosening of mesocarp CW allows for the accumulation of soluble sugars and takes place prior to the loosening of exocarp CW [91]. Since veraison is observed at the end of the slow growth phase between the first and the second growth stages, little differences occur in berry size during this period, suggesting that rapid structural modifications of CW polysaccharides verified during veraison without any large change in berry size are due to the unchanged stiffness of the outer skin [85]. This evidence is further supported by differential transcription of genes encoding CW-modifying enzymes (as discussed below). Taken as a whole, changes in CW components lead to skin loosening and consequently berry enlargement at early PV, and loosening of pulp tissue CW contributes to berry softening [88, 91].

Secondary Cell Wall

Intense research has focused on the phenylpropanoid metabolism but little attention was given to the characterization of its role in fruit development, particularly in grape berries [38]. In the grape berry and other fleshy fruits, both xylem and phloem vessels deliver water, depending mainly on the developmental stage of the fruit. As a fleshy fruit matures, there is a clear reduction in the proportion of water entering the fruit *via* the xylem. It was accepted that lignified xylem cells are unable to increase in length during the second stage of berry growth, occurring then a disruption in xylem continuity. However, as under the influence of a hydrostatic gradient, the water movement was recovered, and it is now accepted that the second growth stage is apparently not related to a lack of functionality of xylem vasculature but to an increase in phloem transport [93]. Clues for monolignol polymerisation were reported in Gamay rouge berries as being distributed in the whole fruit but specifically localised to berry xylem vessels at veraison [94].

In mature grape seeds, lignin accounts for 44% of the CW [95]. In grape seeds, a histochemical study showed that lignin is present in cells of medium integument in increasing amounts as the seed matures. BV lignin is slightly detected in the thin walls of large cells, while at V and until harvest the staining with specific dyes for lignin was visible in thick walls of increasingly smaller sized cells [33]. The deposition of seed phenolics is associated with berry development and maturation: changes in seed coat colour are related to berry anthocyanins and total skin phenolics, so the colour of the seed coat can be used as an indicator of ripeness [96].

CONNECTION OF THE CELL WALL CHARACTERISTICS TO BERRY TECHNOLOGICAL APPLICATIONS

Differences between Cultivars

Contrasting softening behaviours during ripening between fruit genotypes and cultivars from the same species are associated with distinct CW composition and related enzymatic metabolism [77]. In grapes, the CW composition differs sufficiently between varieties to allow for their discrimination, even with respect to technological differences [83]. The most striking difference between the mesocarp walls of the firm table grape cultivar Ohanez and the softer, multipurpose cultivar Muscat Gordo Blanco lies in the relative proportions of cellulose and pectic polysaccharides, and in the hydroxyproline composition of wall-associated structural proteins [64], with Ohanez having significantly higher cellulose and hydroxyproline contents than Muscat Gordo Blanco [64]. In fact, among the most abundant polysaccharides of the grape berry walls, the measured cellulose content is 39% and 31% by weight in the Ohanez and Muscat Gordo Blanco cultivars, respectively. Additionally, in Ohanez walls the extensin network is more abundant, which is consistent with its firmer texture [64]. In contrast, galacturonans account for 29% to 41% in the same varieties. Therefore, Gordo walls appear to require a larger pectic matrix phase than the firmer Ohanez grapes. Similarly, XG which are probably closely associated with cellulose microfibrils [9], account for 8% by weight of the walls in Gordo and 12% in walls from Ohanez [64].

Implications of the Cell Wall in Winemaking

Different wine processing methods are directly related with the CW. White wines are made by fermenting grape juice, which contains little amounts of skin CW, in contrast to the process of fermenting whole berries in red wines. A first consequence is the amount of RG-II, which is a major polysaccharide component of red wine. One litre of red wine may contain between 100 to 150 mg of RG-II while white wine typically contains 20 to 30 mg of RG-II per litre.

In addition, the CW of grape berry skin cells is of main relevance to wine making, since it forms a hydrophobic barrier to the diffusion of phenols, holding the main control of extractability [6, 97]. Pinelo *et al.* [3] propose that phenolic substances, including tannins, can be deposited into the lignin-polysaccharide matrix of lignified secondary CW or bind to macromolecules, including CW polysaccharides. It is assumed that most phenolic compounds are nearly absent from the grape berry flesh, mainly embodied in the skin and seeds, and can be released during the wine making process [3]. However, normally there is no reference to lignin, although this phenylpropanoid polymer ornates the secondary walls of xylem vascular bundles in a network that crosses the whole volume of the flesh layer, the main contributor to wine volume [98]. On a per berry basis, tannins accumulate during the first growth period and decline during the second growth stage (phase III) [99]. Then, the degradation of CW polysaccharides is crucial for the yield of phenolic compounds from grape skin cells [69]. The retention of phenols by the CW depends on the composition, structure and molecular weight of the phenol molecule and of CW physical traits [69, 97]. Porosity, structure and chemical composition can influence the aggregation between conformational CW polysaccharides and phenolic substances. The curious concept of *phenolic ripeness* [100] is associated with anthocyanin content and extractability. Anthocyanin extraction from the grape skin and diffusion into must and wine depends on anthocyanin content but also on the capacity of the berry skin to yield up the pigment as a consequence of CW degradation. When phenolic ripeness is attained, the pectin-rich middle lamella between cells is degraded, and the CWs are perforated and allow for extraction and diffusion [6].

In berry seeds, tannins have the same constitutive units as the skin tannins but a lower degree of polymerisation [3]. Along with seed growth and development, tannins accumulate when the seed acquires a green colour (phase I); they reach a maximum accompanied by tannin oxidation when seeds show a yellow colour (phase II); and they decrease as the seed dries and matures, taking a brown colour (phase III). The decrease in tannin level is probably related to aggregation of the oxidised forms of the seed coat.

THE *Vitis* CELL WALL*OME*

Vitis Cell Wall Key Enzymes

The covalent modifications of CW polysaccharides during fruit growth and ripening result largely from the concerted activity of a set of hydrolases and transglycosylases [2, 77, 101]. The cooperative action between members of several different enzyme families, including expansins, endo-β-1,4-glucanases (EGase), xyloglucan endotransglycosylases/hydrolases (XTH), β-xylanases (Xyn), endomannanases, polygalacturonases (PG) or pectate lyases (PL) are of primary interest in CW metabolism during fruit development. On the other hand, esterases like pectin methylesterases (PME) and pectin acetylesterases (PAE), exo-acting hydrolases and other glycosidases such as β-galactosidases (β-gal), α-Larabinofuranosidases (AFase) or xylosidases (Xyl) are also involved through cooperative action with hydrolases in the pectic or hemicellulosic polymer metabolism [101, 102]. Removal of side chains containing neutral sugar may be necessary to expose the polysaccharide's backbone for cleavage, thus facilitating its solubilisation [103] and promoting a decrease in the degree of polymerisation which, in turn, can modify the binding between polymers [104]. Cooperative events can result from the disassembly of the hemicellulosic network proved necessary for modifications in the pectic network due to the physical accessibility of pectolytic enzymes to pectin substrates. The panorama is even more complex because new components are synthesised and integrated into the CW, even during ripening [105, 106]. Nonetheless, the real contribution of the referred enzymes still remains to be fully elucidated.

As far as the secondary CW is concerned, after the deamination of phenylalanine by phenylalanine ammonia lyase (PAL) to form cinnamic acid, the lignin synthesis pathway includes hydroxylations of the aromatic ring, methylation of one or two hydroxyl groups and two reductions of the carboxylic to the monolignol alcohol side chain [34]. The crucial enzymes and respective coding genes are cinnamate 4-hydroxylase (C4H), 4-coumarate:CoA ligase (4CL), cynnamoyl CoA reductase (CCR), β-hydroxycinnamoyl-CoA:quinate/shikimate/*p*-hydroxy-cinnamoyl transferase (HCT), cinnamyl alcohol dehydrogenase (CAD) and ferulate 5-hydroxylase (F5H). The end products are the monolignols *p*-coumaroyl, *p*-coniferyl and *p*-sinapyl which differ in their methylation degree [3, 34]. Polymerisation occurs after monolignol dehydrogenation by large families of apoplastic peroxidases and laccases which can vary in specificity according to each type of monolignol [34, 36].

The *Vitis* Cell Wall-Related Genome

The sequencing and public availability of the *Vitis* genome [7, 8] makes it possible to focus on individual pathways, to profile the expression pattern of isoforms associated with each tissue, developmental phase and response to the different stresses affecting grape berries during their development, and anticipating the effects on wine production and quality. For that reason, retrieving the sequences of coding regions and predicted amino acid primary structure of enzymes known to act on the *Vitis* CW is mandatory to support *omics*-related research. Table **1** provides a list of the number of genes related with primary CW biosynthesis and modification and with secondary CW biosynthesis present in the genome of higher plant sequenced species *Vitis vinifera*, *Oryza sativa*, *Populus trichocarpa* and *Arabidopsis thaliana*.

In silico analysis shows that, in *Vitis*, most of the primary CW related gene families present a number of members similar to *Arabidopsis*, rice and *Populus*, suggesting the conservation of mechanisms associated with CW biosynthesis and modification along plant evolution (Table **1**). As expected, a higher number of genes involved in the biosynthesis of lignin were retrieved in the woody species *Vitis* and *Populus* as compared with *Arabidopsis* and *Oryza*, (Table **1**) but it is worthy to note that *Vitis* genome holds more genes than *Populus* [107].

CesA (Cellulose synthase), XTH (Xyloglucan endotransglucosylase/hydrolase), EGase (β-1,4-endoglucanase), PME (Pectin methylesterase), PMEI (Pectin methylesterase/nvertase inhibitor), PAE (Pectin acetylesterase), PG (Polygalacturonase), PL (Pectate lyase), C4H (Cinnamate 4-hydroxylase), HCT (Hydroxycinnamoyl-CoA shikimate/quinate hydroxycinnamoyl transferase), C3H (Coumarate 3-hydroxylase), CCR (Cinnamoyl-CoA reductase), F5H (Ferulate 5-hydroxylase), CAD (cinnamyl alcohol dehydrogenase). The protein sequences were extracted from public databases: Arabidopsis sequences from TAIR (http://www.arabidopsis.org/ index.jsp) and Cell Wall Navigator (http://bioweb.ucr.edu/Cellwall/); rice sequences from orygenesdb (http://orygenesdb.cirad.fr/cgi-bin/gbrowse/odb_japonica/?name=Os_1:1..10000) and Cell Wall Navigator; Populus sequences from JGI (http://genome.jgi-psf.org/Poptr1/poptr1.home.html) and *Vitis* sequences from Genoscope genomic database 8X (http://www.genoscope.cns.fr/externe/GenomeBrowser/Vitis/). Databases last access September 2010.

Amino acid sequence similarity analyses reveal that *Vitis* CW-related sequences cluster with orthologs from monocot, dicot and woody model species in most families (Fig. **2A**, **B**). Noticeably, in families such as XTHs some clusters are enriched with *Vitis* sequences, suggesting specification with respect to evolution or a possible correlation with substrate specificity (Fig. **2C**). This aspect may be important to unravel CW dynamics in this species and deserves further investigation.

Table 1: Number of genes related to primary cell wall biosynthesis and modification and to secondary cell wall biosynthesis retrieved *in silico* from the higher plant sequenced species *Vitis vinifera*, *Oryza sativa*, *Populus trichocarpa* and *Arabidopsis thaliana* genomes.

	Vitis vinifera	*Oryza sativa*	*Populus trichocarpa*	*Arabidopsis thaliana*
Primary Cell Wall				
CesA	11	10	18	10
Expansin	30	56	36	36
XTH	33	29	24	33
β-1,3-glucanase	43	65	73	49
EGase	21	24	31	25
PME	36	37	84	66
PMEI	11	22	48	33
PAE	7	10	11	9
PG	60	48	45	66
PL	16	10	27	27
Secondary Cell Wall				
C4H	5	3	2	1
HCT	16	9	11	2
C3H	3	1	2	1
CCR	20	12	18	7
F5H	3	2	3	2
CAD	8	12	12	9

The Grape Berry Cell Wall Transcriptome

One way of tackling the pathways of a given physiological event is to understand the regulation of the transcription of related genes. Initial studies based on "candidate gene" approaches in grapes associated β-gal, α-gal, PME, PL enzyme activity and gene expression with changes in mesocarp CW pectin composition, during berry ripening [81, 82, 87, 109]. These events were accompanied by the up-regulation of genes involved in the cellulose:hemicellulose network, such as expansins and XTH, at veraison, coincidently with the depolymerization of XG [87, 109, 110]. In grape berries, the expression of EGase genes seems to be confined to the initial growth stages and was not detected during ripening [87], which contrasts with the large majority of fruit species, except for apple [111, 112]. Therefore these genes are pointed out as strong candidates to be involved in the metabolism of the grape berry CW.

In the skin, as stated above, the remodelling of the CW apparently exerts a marked influence on the control of berry growth during phases I and III [92]. The transcription profiles of candidate genes for CW-modifying enzymes support this assumption. A bimodal trend is observed, with high levels of expression coincident with periods of rapid berry growth as well as cellular expansion and low expression levels during growth arrest [92]. Despite major differences in grape cell morphology between the exocarp and mesocarp, most genes for CW metabolic processes follow similar expression profiles in both tissues throughout berry development [92]. However, different patterns in genes for some CW-modifying families, including β-1,3-endoglucanase, PME, PL, and two genes from the expansin family (*EXP3* and *EXPL*) were observed [92]. The differences are more evident during the rapid growth phase associated with the beginning of phase III, with an up-regulation of β-1,3-endoglucanase, *EXP3* and *EXL* gene expression in the exocarp accompanied by a down-regulation in the mesocarp, and a lag in the up-regulated expression of exocarp PL, PME and EXPL during phase II [92]. Noticeably, the up-regulation of exocarp transcripts during transition phases II and III is accompanied by a tissue-specific expansion of both epidermal and hypodermal cells. This exocarp up-regulation followed by the down-regulation of β-1,3-endoglucanase and expansin-like genes discloses patterns of gene expression concurrent with changes in the epidermis and hypodermis CW thickness, indicating a role in CW loosening to accommodate expansion of the mesocarp tissues [92]. Moreover, exocarp tissues exhibit a larger increase in transcript level of *PME* and *EGase*, in contrast to the increased transcription of *EXPL* in the mesocarp [92]. It should be emphasised that this research followed a "candidate gene" approach, so the possible involvement of other isoforms and gene families is expected.

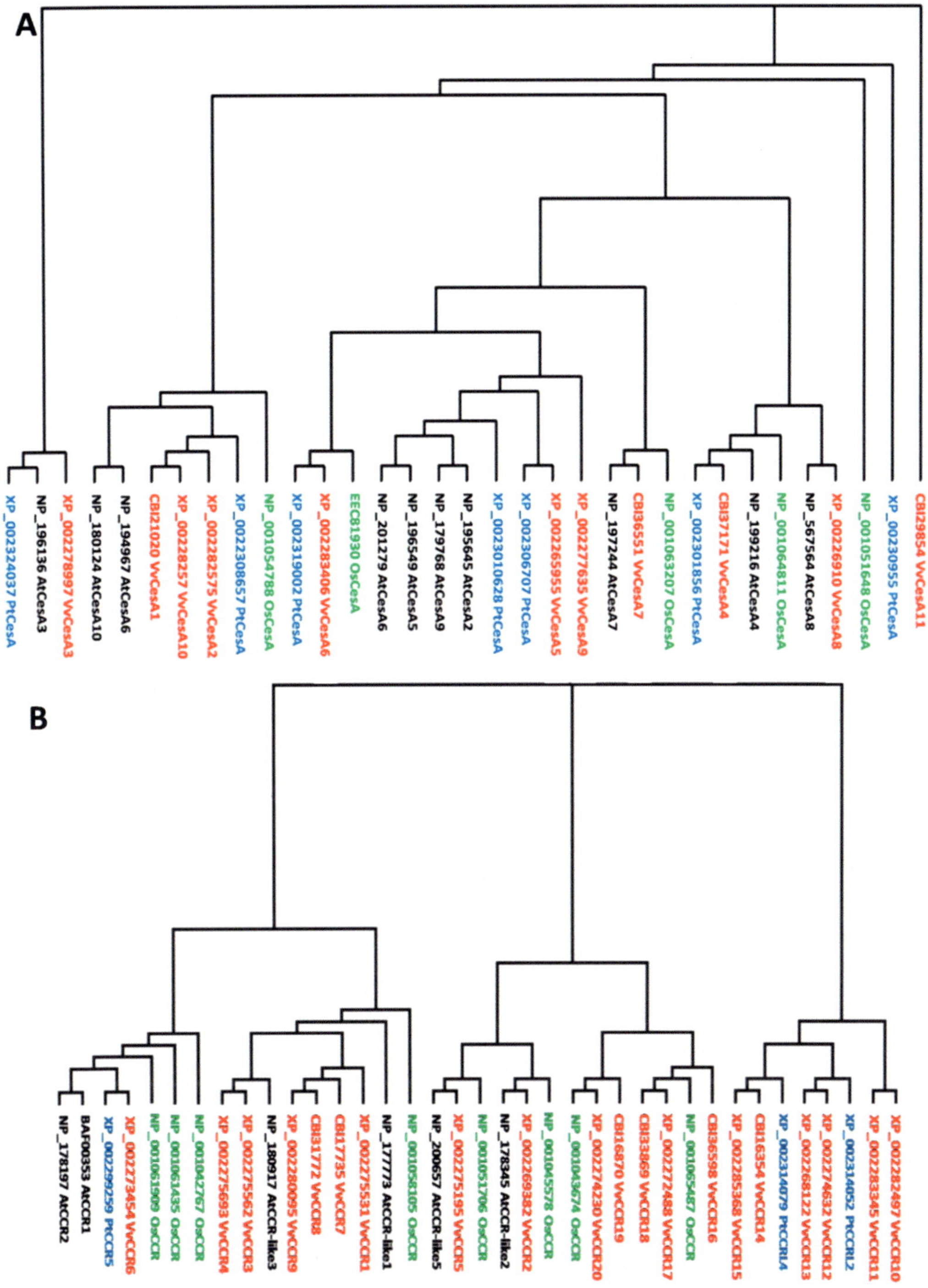

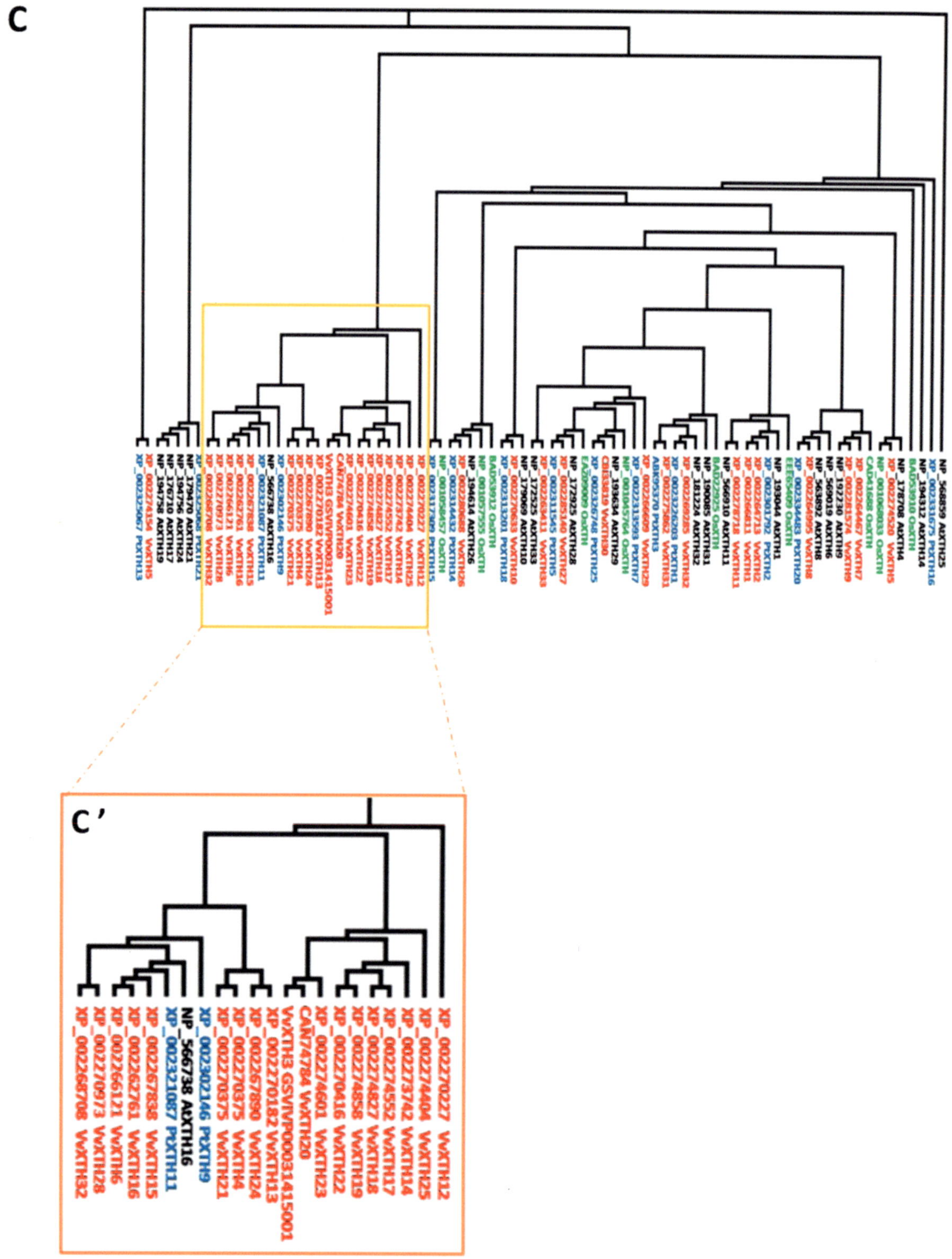

Figure 2: Phylogenetic dendrogram of cellulose synthases (CesA, A), cynnamoyl CoA reductases (CCR, B) and xyloglucan endotransglycosylases/hydrolases (XTH, C; magnification of cluster enriched in *Vitis* sequences, C'). Dendrograms were generated using ClustalX and TreeView software [108] based on mature protein sequences extracted as in Table **1** of all *Vitis* (red) family members annotated so far and representative members of Arabidopsis (black), Oryza (green) and Populus (blue) protein families.

With large Expressed Sequence Tags (EST)-based microarrays, new insights about the regulation of CW in *Vitis* berries was achieved, taking advantage of sequence annotation and high throughput gene expression. The public release of the *Vitis* genome and the annotation of *Vitis* genes allowed for the in-depth study of the functional genomics of berry CW, including the genes coding for enzymes associated with CW biosynthesis, modification during cell growth and fruit ripening as well as the deposition of secondary wall polymers, providing a better picture of the associated pathways. In general, large scale high-throughput transcriptomic microarray data indicate several categories of genes that are significantly differentially expressed during berry development (Table **2**). An incidence of genes related to CW organisation and biogenesis are included in clusters showing overexpression during development, thus disclosing the prevalence of metabolic processes involved in CW synthesis and loosening PV [113].

Table 2: Summary of representative primary cell wall biosynthesis (CesA) and modification gene families (cellulose-hemicellulose network: XTH and expansin; pectin network: PG, PME, PMEI) strongly differentially expressed in grape berry tissues during developmental phases or imposed treatments.

Gene family	Method	Condition	Tissue	References
CesA	Affymetrix *Vitis* GeneChip® Microarray	Berry development	Deseeded berry	[113]
			Whole berry	[114]
	Array-Ready Oligo Set™ for *Vitis*	C_2H_4 (ethylene) treatment	Whole berry [a]	[115]
XTH	Affymetrix *Vitis* GeneChip® Microarray	Berry development	Deseeded berry	[113]
			Whole berry	[114]
		Water deficit	Pulp, skin, seed	[116]
	Array-Ready Oligo Set™ for *Vitis*	C_2H_4 (ethylene) treatment	Whole berry [a]	[115]
	Oligo Array	Berry development	Whole berry	[117]
Expansin	Affymetrix *Vitis* GeneChip® Microarray	Berry development	Deseeded berry	[113]
			Whole berry	[114]
		Water deficit	Pulp, skin, seed	[116]
	Array-Ready Oligo Set™ for *Vitis*	C_2H_4 (ethylene) treatment	Whole berry [a]	[115]
	Oligo Array	Berry development	Whole berry	[117]
	EST sequencing	Berry development	Whole berry	[118, 119]
PG	Affymetrix *Vitis* GeneChip® Microarray	Water deficit	Pulp, skin, seed	[116]
		ABA treatment	Skin	[120]
	Array-Ready Oligo Set™ for *Vitis*	C_2H_4 (ethylene) treatment	Whole berry [a]	[115]
PME	Affymetrix *Vitis* GeneChip® Microarray	Berry development	Deseeded berry	[113]
			Whole berry	[114]
		Water deficit	Pulp, skin, seed	[116]
		ABA treatment	Skin	[120]
	Array-Ready Oligo Set™ for *Vitis*	C_2H_4 (ethylene) treatment	Whole berry [a]	[115]
	Oligo Array	Berry development	Whole berry	[117]

	EST sequencing	Berry development	Whole berry	[119, 121]
PMEI	Affymetrix *Vitis* GeneChip® Microarray	Berry development	Deseeded berry	[113]
			Whole berry	[114]
		Water deficit	Pulp, skin, seed	[116]
	EST sequencing	Berry development	Whole berry	[118]

[a] Gene expression was further investigated individually in the pulp, skin and seed *via* quantitative real-time RT-PCT.

The previously described involvement of some gene families in berry growth and softening was confirmed with large-scale transcriptomics. Among these are included pectin modifying enzymes (PME, PG and PL) and cellulose-hemicellulosic ones (expansins and XTH). However, in moving from "candidate gene" approaches to "large scale high throughput" transcriptomics, it was possible to identify in the berry members of previously CW-related overlooked families such as pectin methylesterase inhibitors (PMEIs) or cellulose synthases (CesA) [114], as well as additional members of previously studied families associated with CW modifications. For instance, by using the Affymetrix *Vitis* GeneChip genome array, 10 members of the XTH family were modulated during development, four of them strongly up-regulated during ripening [113] compared to the two "candidate genes" previously investigated both in berry skin and flesh [87, 109]. Moreover, comprehensive comparisons of gene expression between pulp, skin and seed tissues are now facilitated [116].

Functional genomics of lignin biosynthesis is established for *Arabidopsis* with key genes identified by screening of mutant and transgenic plants, gene silencing, overexpression and other reverse-genetics approaches [34, 39, 122]. In the general lignin biosynthetic pathway, C4H, a cytochrome P450-dependent mono-oxygenase, converts cinnamic acid, the product of phenylalanine deamination, into *p*-coumaric acid, which is esterified with CoA to *p*-coumaroyl-CoA by 4CL. Then CCR, the first enzyme of the monolignol specific biosynthetic pathway converts the lateral chain of coumaroyl-CoA ester to its respective aldehyde. A further reduction step catalysed by CAD forms *p*-coumaryl alcohol - the monomer for H lignin. At *p*-coumaroyl- CoA level the pathway can give rise to a distinct branch: HCT converts *p*-coumaroyl-CoA to *p*-coumaroyl shikimic acid, which is converted to caffeoyl shikimic acid by C3H (also a cytochrome P450-dependent mono-oxygenase), then converted to caffeoyl-CoA by HCT, which is methylated to feruoyl-CoA and finally converted to coniferyl aldehyde by CCR. From this metabolite two new branches can proceed: one catalysed by CAD giving rise to coniferyl alcohol, the monomer for G lignin; a second one with F5H, the third cytochrome P450-dependent mono-oxygenase of the pathway, catalysing the hydroxylation of coniferyl aldehyde to hydroxyconiferyl aldehyde, which is methylated to sinapyl aldehyde and finally dehydrogenated by CAD to sinapyl alcohol, the monomer for S lignin.

The expression of genes for most of the enzymes catalysing specific steps of lignin biosynthesis is suggested in the grape berry [114], although no experimental evidence is available at cellular or biochemical levels to support the above hypothesis. Furthermore, no reports are available on the crosslinks between lignin and polysaccharides in the grape seed secondary CW. In the grape berry pulp vascular tissue, immunogold labelling localised PAL to primary and secondary CW and 4CL to secondary thick walls [38]. The authors refer to a previous localisation of PAL to the cytoplasm and organelles, and C4H to the endoplasmic reticulum. The physiological significance of the compartmentalisation of phenylpropanoid pathway enzymes is therefore not totally clarified.

So far, only a few transcriptomic analyses of grape berry through Affymetrix *Vitis* GeneChip genome array refer the expression of genes for lignin biosynthesis enzymes. One transcript subset includes transcripts assigned to "phenolic acid" function. Quoting the NCBI GeneBank and Genoscope annotations, it was possible to establish the correspondence with specific genes coding for lignin biosynthesis enzymes (Table **3**). The CCR and the first CAD listed express in ripening grapes eventually due to seed lignification at the maturation stage; whereas the second listed CAD, expressed in berries at the green stage, could be associated with xylem vasculature.

Table 3: Correspondence between transcripts related to the grape berry "Phenolic acid metabolism" [114] pool, Unique Gene in Genoscope and ESTs and Proteins from NCBI GeneBank grape berry libraries.

Families of genes for lignin biosynthesis	GeneBank annotation [114]	Genoscope unique gene	GeneBank EST	GeneBank protein
CCR	CF517687	GSVIVP00033763001	CF415449; CABSAU36	XP_002273454
CAD	CF512464	GSVIVP00008719001	CV179328; CABSAU36	XP_002279832
CAD	CF517155	GSVIVP00024587001	BQ798918	XP_002285368

When grape berry mRNA expression profiles were analysed in the skin, pulp and seed tissues [123], two CCRs showed seed specific expression, one showed skin specific expression and two others were expressed in the pulp and skin or in the skin and seed. Four CAD isoforms showed preferential accumulation in the skin or in the skin and pulp, certainly in relation to vascular bundle formation. In a study oriented to identify genes specifically involved in berry ripening [113], transcripts associated with primary CW were mostly repressed before veraison and induced onwards, while a number of secondary metabolism genes were repressed BV but a higher number was induced PV. This category included genes of the phenylpropanoid pathway. The authors interpret the pattern of expression of one 4CL isoform, negatively modulated throughout the ripening process, as being involved in lignin biosynthesis, while a second 4CL isoform, positively modulated throughout the whole period of berry development and ripening, would be involved in the anthocyanin pathway.

ABBREVIATIONS

4CL = 4-coumarate:CoA ligase

AFase = α-L-arabinofuranosidase

AGA = Apiogalacturonan

AG-I = Arabinogalactan-I

AG-II = Arabinogalactan-II

AGP = Arabinogalactan protein

BV = Before veraison

C3H = Coumarate 3-hydroxylase

C4H = Cinnamate 4-hydroxylase

CAD = Cinnamyl alcohol dehydrogenase

CCR = Cinnamoyl-CoA reductase

CesA = Cellulose synthase

CW = Cell wall

EGase = β-1,4-endoglucanase

EST	=	Expressed Sequence Tags
F5H	=	Ferulate 5-hydroxylase
Gal*pa*	=	α-D-galactosyluronic acid
GRP	=	Glycine-rich proteins
HCT	=	Hydroxycinnamoyl-CoA shikimate/quinate hydroxycinnamoyl transferase
HG	=	Homogalacturonan
PA	=	proanthocyanidin
PAE	=	Pectin acetylesterase
PAL	=	Phenylalanine ammonia lyase
PG	=	Polygalacturonase
PGA	=	Polygalacturonic acid
PL	=	Pectate lyase
PME	=	Pectin methylesterase
PMEI	=	Pectin methylesterase/nvertase inhibitor
PRP	=	Proline-rich proteins
PV	=	Post-Veraison
RG-I	=	Rhamnogalacturonan-I
RG-II	=	Rhamnogalacturonan-II
UA	=	Uronic acid
V	=	Veraison
XG	=	Xyloglucan
XGA	=	Xylogalacturonan
XTH	=	Xyloglucan endotransglucosylase/hydrolase
Xyl	=	Xylosidase
Xyn	=	Xylanase
β -Gal	=	β-galactosidase

ACKNOWLEDGEMENTS

JCF and PL are awarded respectively the PhD grant SFRH/BD/64047/2009 and the annual grant for integration in research FCT/CBAA/BII-7-2009/2010. IICT is funded by MCTES and CBAA by the FCT "Plurianual" Programme.

REFERENCES

[1] Brett C, Waldron K. Physiology and Biochemistry of Plant Cell Walls. Chapman & Hall, London 1996.

[2] Goulao LF, Almeida DPF, Oliveira CM. Effect of Enzymatic Reactions on Texture of Fruits and Vegetables. In: Bayindirli A, Ed. Enzymes in Fruit and Vegetable Processing: Chemistry and Engineering Applications. CRC Press (Taylor & Francis), New York 2010; pp. 72-122.

[3] Pinelo M, Arnous A, Meyer AS. Upgrading of grape skins: Significance of plant cell-wall structural components and extraction techniques for phenol release. Trends Food Sci Technol 2006; 17(11): 579-90.

[4] Rapport L, Lockwood B. Nutraceuticals (6) Protoanthocyanidins and grape seed extract. Pharm J 2001; 17: 581-4.

[5] Bogs J, Jaffé F W, Takos AM, Walker AR, Robinson SP. The grapevine transcription factor VvMYBPA1 regulates proanthocyanidin synthesis during fruit development. Plant Physiol 2007; 143(3): 1347-61.

[6] Ortega-Regules A, Romero-Cascales I, Ros-García JM, López-Roca JM, Gómez-Plaza E. A first approach towards the relationship between grape skin cell-wall composition and anthocyanin extractability. Anal Chimica Acta 2006; 563(1-2): 26-32.

[7] Jaillon O, Aury J-M, Noel B, *et al.* The grapevine genome sequence suggests ancestral hexaploidization in major angiosperm phyla. Nature 2007; 449(7161): 463-7.

[8] Velasco R, Zharkikh A, Troggio M, *et al.* A high quality draft consensus sequence of the genome of a heterozygous grapevine variety. PLoS ONE 2007; 2(12): e1326.

[9] Carpita NC, Gibeaut DM. Structural models of primary cell walls in flowering plants: consistency of molecular structure with the physical properties of the walls during growth. Plant J 1993; 3(1): 1-30.

[10] McNeil M, Darvill AG, Fry SC, PA. Structure and function of the primary cell walls of plants. Annu Rev Biochem 1984; 53: 652-83.

[11] Hayashi T. Xyloglucans in the primary cell wall. Annu Rev Plant Physiol Plant Mol Biol 1989; 40(1): 139-68.

[12] Ordaz-Ortiz JJ, Marcus SE, Knox JP. Cell wall microstructure analysis implicates hemicellulose polysaccharides in cell adhesion in tomato fruit pericarp parenchyma. Mol Plant 2009; 2(5): 910-21.

[13] Voragen AGJ, Pilnik W, Thibault J-F, Axelos MAV, Renard CMGC. Pectins. In: Stephen AM, Ed. Food polysaccharides and their applications. Marcel Dekker, New York 1995; pp. 287-339.

[14] Thibault J-F, Renard CMGC, Axelos MAV, Roger P, Crépeau M-J. Studies of the length of homogalacturonic regions in pectins by acid hydrolysis. Carbohydrate Res 1993; 238: 271-86.

[15] Seymour GB, Colquhoun IJ, Dupont MS, Parsley KR, Selvendran R. Composition and structural features of cell wall polysaccharides from tomato fruits. Phytochemistry 1990; 29(3): 725-31.

[16] Muda P, Seymour GB, Errington N, Tucker GA. Compositional changes in cell wall polymers during mango fruit ripening. Carbohydrate Polym 1995; 26(4): 255-60.

[17] Liners F, Letesson JJ, Didembourg C, Van Cutsem P. Monoclonal antibodies against pectin: Recognition of a conformation induced by calcium. Plant Physiol 1989; 91(4): 1419-24.

[18] Goulao LF. Pectin de-esterification and fruit softening: revisiting a classical hypothesis. Stewart Postharvest Rev [Online] 2010; 6: 1-12.

[19] Albersheim P, Darvill AG, O'Neill MA, Schols HA, Voragen AGJ. In Pectins and pectinases, Visser J, Voragen AGJ, Eds. Elsevier Science B.V., Amsterdam, 1996; pp. 47-56.

[20] Caffall KH, Mohnen D. The structure, function, and biosynthesis of plant cell wall pectic polysaccharides. Carbohydrate Res 2009; 344(14): 1879-900.

[21] Voragen A, Coenen G-J, Verhoef R, Schols H. Pectin, a versatile polysaccharide present in plant cell walls. Struct Chem 2009; 20(2): 263-75.

[22] Kobayashi M, Nakagawa H, Asaka T, Matoh T. Borate-rhamnogalacturonan II bonding reinforced by Ca2+ retains pectic polysaccharides in higher-plant cell walls. Plant Physiol 1999; 119(1): 199-204.

[23] Ishii T, Matsunaga T, Pellerin P, O'Neill MA, Darvill A, Albersheim P. The plant cell wall polysaccharide rhamnogalacturonan II self-assembles into a covalently cross-linked dimer. J Biol Chem 1999; 274(19): 13098-104.

[24] O'Neill MA, Ishii T, Albersheim P, Darvill AG. Rhamnogalacturonan II: structure and function of a borate cross-linked cell wall pectic polysaccharide. Annu Rev Plant Biol 2004; 55: 109-39.

[25] Ridley BL, O'Neill MA, Mohnen D. Pectins: structure, biosynthesis, and oligogalacturonide-related. signaling Phytochemistry 2001; 57(6): 929-67.

[26] Vincken J, Schols HA, Oomen RJ, *et al.* If homogalacturonan were a side chain of rhamnogalacturonan I. implications for cell wall architecture. Plant Physiol 2003; 132(4): 1781-9.

[27] Coenen GJ, Bakx EJ, Verhoef RP, Schols HA, Voragen AGJ. Identification of the connecting linkage between homo- or xylogalacturonan and rhamnogalacturonan type I. Carbohydrate Polym 2007; 70(2): 224-35.

[28] O'Neill MA, York WS. The composition and structure of plant primary cell walls. In The Plant Cell Wall, Rose JKC, Ed. Blackwell Publishing, New York 2003; pp. 1-54.

[29] Showalter AM. Structure and function of plant cell wall proteins. Plant Cell 1993; 5(1): 9-23.

[30] Mort AJ, Lamport DTA. Anhydrous hydrogen-fluoride deglycosylates glycoproteins. Anal Biochem 1977; 82(2): 289-309.

[31] Majewska-Sawka A, Nothnagel EA. The multiple roles of arabinogalactan proteins in plant development. Plant Physiol 2000; 122(1): 3-10.

[32] Donaldson LA. Lignification and lignin topochemistry - an ultrastructural view. Phytochemistry 2001; 57(6): 859-73.

[33] Cadot Y, Miñana-Castelló MT, Chevalier M. Anatomical, histological, and histochemical changes in grape seeds from *Vitis* vinifera L. cv Cabernet franc during fruit development. J Agric Food Chem 2006; 54(24): 9206-15.

[34] Bonawitz ND, Chapple C. The genetics of lignin biosynthesis: connecting genotype to phenotype. Annu Rev Genet 2010; 44: 337-63.

[35] Boerjan W, Ralph J, Baucher M. Lignin biosynthesis. Annu Rev Plant Biol 2003; 54: 519-46.

[36] Vanholme R, Morreel K, Ralph J, Boerjan W. Lignin engineering. Curr Op Plant Biol 2008; 11(3): 278-85.

[37] Vanholme R, Demedts B, Morreel K, Ralph J, Boerjan W. Lignin biosynthesis and structure. Plant Physiol 2010; 153(3): 895-905.

[38] Chen J-Y, Wen P-F, Kong W-F, Pan Q-H, Wan S-B, Huang W-D. Changes and subcellular localizations of the enzymes involved in phenylpropanoid metabolism during grape berry development. J Plant Physiol 2006; 163(2): 115-27.

[39] Goujon T, Sibout R, Eudes A, MacKay J, Jouanin L. Genes involved in the biosynthesis of lignin precursors in Arabidopsis thaliana. Plant Physiol Biochem 2003; 41(8): 677-87.

[40] Stewart JJ, Akiyama T, Chapple C, Ralph J, Mansfield SD. The effects on lignin structure of overexpression of ferulate 5-hydroxylase in hybrid poplar. Plant Physiol 2009; 150(2): 621-35.

[41] Fry SC. The structure and functions of xyloglucan. J Exp Bot 1989; 40: 1-11.

[42] Bootten TJ, Harris PJ, Melton LD, Newman RH. Solid-state ^{13}C-NMR spectroscopy shows that the xyloglucans in the primary cell walls of mung bean (*Vigna radiata* L.) occur in different domains: a new model for xyloglucan-cellulose interactions in the cell wall. J Exp Bot 2004; 55: 571-83.

[43] Hanus J, Mazeau K. The xyloglucan-cellulose assembly at the atomic scale. Biopolymers 2006; 82(1): 59-73.

[44] Bootten TJ, Harris PJ, Melton LD, Newman RH. Solid-state ^{13}C NMR study of a composite of tobacco xyloglucan and *Gluconacetobacter xylinus* cellulose: molecular interactions between the component polysaccharides. Biomacromolec 2009; 10(11): 2961-7.

[45] Cosgrove DJ. Wall structure and wall loosening. A look backwards and forwards. Plant Physiol 2001; 125(1): 131-4.

[46] Oechslin R, Lutz MV, Amadò R. Pectic substances isolated from apple cellulosic residue: structural characterisation of a new type of rhamnogalacturonan I. Carbohydrate Polym 2003; 51(3): 301-10.

[47] Vignon MR, Heux L, Malainine ME, Mahrouz M. Arabinan-cellulose composite in *Opuntia ficus*-indica prickly pear spines. Carbohydrate Res 2004; 339(1): 123-31.

[48] Zykwinska AW, Ralet MCJ, Garnier CD, Thibault JFJ. Evidence for *in vitro* binding of pectin side chains to cellulose. Plant Physiol 2005; 139(1): 397-407.

[49] Zykwinska A, Rondeau-Mouro C, Garnier C, Thibault J-F, Ralet M-C. Alkaline extractability of pectic arabinan and galactan and their mobility in sugar beet and potato cell walls. Carbohydrate Polym 2006; 65(4): 510-20.

[50] Zykwinska A, Gaillard C, Buléon A, Pontoire B, Garnier C, Thibault J-F, Ralet M-C. Assessment of *in vitro* binding of isolated pectic domains to cellulose by adsorption isotherms, electron microscopy, and X-ray diffraction methods. Biomacromolec 2006; 8(1): 223-32.

[51] Zykwinska A, Thibault JF, Ralet MC. Organization of pectic arabinan and galactan side chains in association with cellulose microfibrils in primary cell walls and related models envisaged. J Exp Bot 2007; 58: 1795-802.

[52] Femenia A, Rigby NM, Selvendran RR, Waldron KW. Investigation of the occurrence of pectic-xylan-xyloglucan complexes in the cell walls of cauliflower stem tissues. Carbohydrate Polym 1999; 39(2): 151-64.

[53] Abdel-Massih R, Baydoun E, Brett C. *In vitro* biosynthesis of 1,4-ß-galactan attached to a pectin-xyloglucan complex in pea. Planta 2003; 216(3): 502-11.

[54] Cumming C, Rizkallah H, McKendrick K, Abdel-Massih R, Baydoun E, Brett C. Biosynthesis and cell-wall deposition of a pectin-xyloglucan complex in pea. Planta 2005; 222(3): 546-55.

[55] Popper ZA, Fry SC. Widespread occurrence of a covalent linkage between xyloglucan and acidic polysaccharides in suspension-cultured angiosperm cells. Ann Bot 2005; 96(1): 91-9.

[56] Popper Z, Fry S. Xyloglucan–pectin linkages are formed intra-protoplasmically, contribute to wall-assembly, and remain stable in the cell wall. Planta 2008; 227(4): 781-94.

[57] Marcus SE, Verhertbruggen Y, Hervé C, *et al.* Pectic homogalacturonan masks abundant sets of xyloglucan epitopes in plant cell walls. BMC Plant Biol 2008; 8: 60.

[58] Vissenberg K, Martinez-Vilchez IM, Verbelen JP, Miller JG, Fry SC. *In vivo* colocalization of xyloglucan endotransglycosylase activity and its donor substrate in the elongationzone of Arabidopsis roots. Plant Cell 2000; 12(7): 1229-38.

[59] McCartney L, Steele-King CG, Jordan E, Knox JP. Cell wall pectic (1→4)-β-D-galactan marks the acceleration of cell elongation in the Arabidopsis seedling root meristem. Plant J 2003; 33(3): 447-54.

[60] Ha M-A, Viëtor RJ, Jardine GD, Apperley D C, Jarvis MC. Conformation and mobility of the arabinan and galactan side-chains of pectin. Phytochemistry 2005; 66(15): 1817-24.

[61] Jarvis MC. Plant cell walls: Supramolecular assemblies. Food Hydrocol 2011; 25(2): 257-62.

[62] Gorshkova T, Mikshina P, Gurjanov O, Chemikosova S. Formation of plant cell wall supramolecular structure. Biochemistry (Moscow) 2010; 75(2): 159-72.

[63] Scheller HV, Ulvskov P. Hemicelluloses. Annu Rev Plant Biol 2010; 61: 263-89.

[64] Nunan KJ, Sims IM, Bacic A, Robinson SP, Fincher GB. Isolation and characterization of cell walls from the mesocarp of mature grape berries (*Vitis vinifera*). Planta 1997; 203(1): 93-100.

[65] Vidal S, Williams P, O'Neill MA, Pellerin P. Polysaccharides from grape berry cell walls. Part I: tissue distribution and structural characterization of the pectic polysaccharides. Carbohydrate Polym 2001; 45(4): 315-23.

[66] Hardie WJ, O'Brien TP, Jaudzems VG. Morphology, anatomy and development of the pericarp after anthesis in grape, *Vitis vinifera* L. Aust J Grape Wine Res 1996; 2(2): 97-142.

[67] Lecas M, Brillouet J-M. Cell wall composition of grape berry skins. Phytochemistry 1994; 35(5): 1241-3.

[68] Saulnier L, Thibault J-F. Enzymic degradation of isolated pectic substances and cell wall from pulp of grape berries. Carbohydrate Polym 1987; 7(5): 345-60.

[69] Bindon KA, Smith PA, Holt H, Kennedy JA. Interaction between grape-derived proanthocyanidins and cell wall material. 2. Implications for vinification J Agri Food Chem 2010; 58(19): 10736-46.

[70] Hilz H. Characterisation of cell wall polysaccharides in bilberries and black currants. PhD Thesis, U. Wageningeng: Wageningen, NL, 2007; p. 158.

[71] Doco T, Williams P, Pauly M, O'Neill MA, Pellerin P. Polysaccharides from grape berry cell walls. Part II. Structural characterization of the xyloglucan polysaccharides. Carbohydrate Polym 2003; 53(3): 253-261.

[72] Arnous A, Meyer A. Grape skins (*Vitis vinifera* L.) catalyze the *in vitro* enzymatic hydroxylation of *p*-coumaric acid to caffeic acid. Biotech Lett 2009; 31(12): 1953-60.

[73] Fry SC, York WS, Albersheim P, *et al.* An unambiguous nomenclature for xyloglucan-derived oligosaccharides. Physiol Plant 1993; 89(1): 1-3.

[74] Nunan KJ, Sims IM, Bacic A, Robinson SP, Fincher GB. Changes in cell wall composition during ripening of grape berries. Plant Physiol 1998; 118(3): 783-92.

[75] Gross KC, Sams CE. Changes in cell wall neutral sugar composition during fruit ripening: a species survey. Phytochemistry 1984; 23(11): 2457-61.

[76] Brummell DA. Cell wall disassembly in ripening fruit. Funct Plant Biol 2006; 33: 103-19.

[77] Goulao LF, Oliveira CM. Cell wall modifications during fruit ripening: when a fruit is not the fruit. Trends Food Sci Technol 2008; 19(1): 4-25.

[78] Zhang DP, Li M, Wang Y. Ultrastructural changes in the mesocarp cells of grape berry during its development. Acta Botanica Sinica 1997; 39: 389-96.

[79] Liners F, Van Cutsem P. Distribution of pectic polysaccharides throughout walls of suspension-cultured carrot cells. Protoplasma 1992; 170(1): 10-21.

[80] Fougère-Rifot M, Cholet C, Bouard J. Evolution des parois des cellules de l'hypoderme de la baie de raisin lors de leur transformation en cellules de pulpe. J Int Sci Vigne Vin 1996; 30: 47-51.

[81] Barnavon L, Doco T, Terrier N, Ageorges A, Romieu C, Pellerin P. Involvement of pectin methyl-esterase during the ripening of grape berries: partial cDNA isolation, transcript expression and changes in the degree of methyl-esterification of cell wall pectins. Phytochemistry 2001; 58(5): 693-701.
[82] Barnavon L, Doco T, Terrier N, Ageorges A, Romieu C, Pellerin P. Analysis of cell wall neutral sugar composition, beta-galactosidase activity and a related cDNA clone throughout the development of *Vitis* vinifera grape berries. Plant Physiol Biochem 2000; 38: 289-300.
[83] Ortega-Regules A, Ros-García JM, Bautista-Ortín AB, López-Roca JM, Gómez-Plaza E. Changes in skin cell wall composition during the maturation of four premium wine grape varieties. J Sci Food Agric 2008; 88: 420-8.
[84] Silacci MC, Morrison JC. Changes in pectin content of Cabernet Saubignon grape berries during maturation. Am J Enol Vitic 1990; 41: 111-5.
[85] Yakushiji H, Sakurai N, Morinaga K. Changes in cell-wall polysaccharides from the mesocarp of grape berries during veraison. Physiol Plantarum 2001; 111(2): 188-95.
[86] Grignon C, Sentenac H. pH and ionic conditions in the apoplast. Annu Rev Plant Physiol Plant Mol Biol 1991; 42: 103-28.
[87] Nunan K, Davies C, Robinson S, Fincher G. Expression patterns of cell wall-modifying enzymes during grape berry development. Planta 2001; 214(2): 257-64.
[88] Huang X-M, Huang H-B, Wang H-C. Cell walls of loosening skin in post-veraison grape berries lose structural polysaccharides and calcium while accumulate structural proteins. Sci Hortic 2005; 104(3): 249-63.
[89] Vicens A, Fournand D, Williams P, Sidhoum L, Moutounet M, Doco T. Changes in polysaccharide and protein composition of cell walls in grape berry skin (cv. Shiraz) during ripening and over-ripening J Agric Food Chem 2009; 57(7): 2955-60.
[90] Matthews MA, Cheng G, Weinbaum SA. Changes in water potential and dermal extensibility during grape berry development. J Am Soc Hortic Sci 1987; 112: 314-9.
[91] Huang X-M, Huang H-B. Early post-veraison growth in grapes: evidence for a two-step mode of berry enlargement. Aust J Grape Wine Res 2001; 7(3): 132-6.
[92] Schlosser J, Olsson N, Weis M, *et al.* Cellular expansion and gene expression in the developing grape (Vitis vinifera L.). Protoplasma 2008; 232(3): 255-65.
[93] Bondada BR, Matthews MA, Shackel KA. Functional xylem in the post-veraison grape berry. J Exp Bot 2005; 56(421): 2949-57.
[94] Morales M, Pedreño MA, Muñoz R, Barceló AR, Calderón AA. Purification and kinetic characterization of a basic peroxidase isoenzyme responsible for lignification in Gamay rouge grape (Vitis vinifera) berries. Food Chem 1993; 48(4): 391-4.
[95] Moldes D, Gallego PP, Rodríguez Couto S, Sanromán A. Grape seeds: the best lignocellulosic waste to produce laccase by solid state cultures of Trametes hirsuta. Biotechnol Lett 2003; 25(6): 491-5.
[96] Ristic R, Iland PG. Relationships between seed and berry development of *Vitis* Vinifera L. cv Shiraz: Developmental changes in seed morphology and phenolic composition. Aust J Grape Wine Res 2005; 11(1): 43-58.
[97] Hanlin RL, Hrmova M, Harbertson JF, Downey MO. Condensed tannin and grape cell wall interactions and their impact on tannin extractability into wine. Aust J Grape Wine Res 2010; 16(1): 173-88.
[98] Kennedy J. Understanding grape berry development. Practical Winery and Vineyard [Online], 2002.
[99] Kennedy JA, Hayasaka Y, Vidal S, Waters EJ, Jones GP. Composition of grape skin proanthocyanidins at different stages of berry development. J Agric Food Chem 2001; 49(11): 5348-55.
[100] Río Segade S, Rolle L, Gerbi V, Orriols I. Phenolic ripeness assessment of grape skin by texture analysis. J Food Compos Anal 2008; 21(8): 644-9.
[101] Fischer RL, Bennett AB. Role of cell-wall hydrolases in fruit ripening. Annu Rev Plant Physiol Plant Mol Biol 1991; 42: 675-703.
[102] Hadfield KA, Bennett AB. Polygalacturonases: many genes in search of a function. Plant Physiol 1998; 117(2): 337-43.
[103] Seymour GB, Gross KC. Cell wall disassembly and fruit softening. Postharvest News and Information 1996; 7: 45N-52N.
[104] Hayashi T, Takeda T, Ogawa K, Mitsuishi Y. Effects of the degree of polymerization on the binding of xyloglucans to cellulose. Plant Cell Physiol 1994; 35(6): 893-9.
[105] Mitcham EJ, Gross KC, Ng TJ. Tomato fruit cell wall synthesis during development and senescence 1: *In vivo* radiolabeling of wall fractions using [14C] sucrose. Plant Physiol 1989; 89(2): 477-81.

[106] Greve LC, Labavitch JM. Cell wall metabolism in ripening fruit 1: V. Analysis of cell wall synthesis in ripening tomato pericarp tissue using a d-[U-13C]glucose tracer and gas chromatography-mass spectrometry. Plant Physiol 1991; 97(4): 1456-61.

[107] Jansson S, Douglas CJ. Populus: a model system for plant biology. Annu Rev Plant Biol 2007; 58: 435-58.

[108] Page RD. TreeView: an application to display phylogenetic trees on personal computers. Comput Appl Biosci 1996; 12(4): 357-8.

[109] Ishimaru M, Kobayashi S. Expression of a xyloglucan endo-transglycosylase gene is closely related to grape berry softening. Plant Sci 2002; 162(4): 621-8.

[110] Ishimaru M, Smith DL, Gross KC, Kobayashi S. Expression of three expansin genes during development and maturation of Kyoho grape berries. J Plant Physiol 2007; 164(12): 1675-82.

[111] Goulao LF, Santos J, de Sousa I, Oliveira CM. Patterns of enzymatic activity of cell wall-modifying enzymes during growth and ripening of apples. Postharvest Biol Technol 2007; 43(3): 307-18.

[112] Goulao LF, Cosgrove DJ, Oliveira CM. Cloning, characterisation and expression analyses of cDNA clones encoding cell wall-modifying enzymes isolated from ripe apples. Postharvest Biol Technol 2008; 48(1): 37-51.

[113] Pilati S, Perazzolli M, Malossini A, *et al.* Genome-wide transcriptional analysis of grapevine berry ripening reveals a set of genes similarly modulated during three seasons and the occurrence of an oxidative burst at vèraison. BMC Genom 2007; 8: 428.

[114] Deluc LG, Grimplet J, Wheatley MD, *et al.* Transcriptomic and metabolite analyses of Cabernet Sauvignon grape berry development. BMC Genom 2007; 8: 429.

[115] Chervin C, Tira-umphon A, Terrier N, Zouine M, Severac D, Roustan J-P. Stimulation of the grape berry expansion by ethylene and effects on related gene transcripts, over the ripening phase. Physiol Plantarum 2008; 134(3): 534-46.

[116] Grimplet J, Deluc L, Tillett R, *et al.* Tissue-specific mRNA expression profiling in grape berry tissues. BMC Genom 2007; 8(1): 187.

[117] Terrier N, Glissant D, Grimplet J, *et al.* Isogene specific oligo arrays reveal multifaceted changes in gene expression during grape berry (Vitis vinifera L.) development. Planta 2005; 222(5): 832-47.

[118] Silva F, Iandolino A, Al-Kayal F, *et al.* Characterizing the grape transcriptome. Analysis of Expressed Sequence Tags from multiple *Vitis* species and development of a compendium of gene expression during berry development. Plant Physiol 2005; 139(2): 574-97.

[119] Terrier N, Ageorges A, Abbal P, Romieu C. Generation of ESTs from grape berry at various developmental stages. J Plant Physiol 2001; 158(12): 1575-83.

[120] Koyama K, Sadamatsu K, Goto-Yamamoto N. Abscisic acid stimulatedripening and gene expression in berry skins of the Cabernet Sauvignon grape. Funct Integr Genom 2010; 10(3): 367-281.

[121] Peng FY, Reid KE, Liao N, *et al.* Generation of ESTs in *Vitis* vinifera wine grape (Cabernet Sauvignon) and table grape (Muscat Hamburg) and discovery of new candidate genes with potential roles in berry development. Gene 2007; 402(1-2): 40-50.

[122] Raes J, Rohde A, Christensen JH, Van de Peer Y, Boerjan W. Genome-wide characterization of the lignification toolbox in Arabidopsis. Plant Physiol 2003; 133(3): 1051-71.

[123] Deluc L, Barrieu F, Marchive C, *et al.* Characterization of a grapevine R2R3-MYB transcription factor that regulates the phenylpropanoid pathway. Plant Physiol 2006; 140(2): 499-511.

CHAPTER 10

Hormonal Control of Grape Berry Development and Ripening

C. Böttcher and C. Davies*

Commonwealth Scientific and Industrial Research Organisation, Plant Industry, P.O. Box 350, Glen Osmond, South Australia 5064, Australia

Abstract: This chapter describes the role of plant hormones during the development of the non-climacteric grape berry. Advances in analytical methods have enabled a better understanding of hormone accumulation although the role of some hormones, for example ethylene, during berry development, remains unclear. Even though the biosynthesis pathways for a number of hormones have been elucidated in grapevine, our understanding of some, for example, the auxin biosynthesis pathway, is still embryonic in grapes. Recent work regarding the breakdown of ABA and auxins has demonstrated the importance of catabolism in determining the final hormone profile during development. Much attention has been focused on the role of hormones during ripening due to the importance of the ripening process to the relevant grape industries. Brassinosteroids have now been added to the list of hormones that can advance ripening. High throughput techniques for analysing changes in gene expression, such as microarray analysis and deep sequencing, are now being used to assess the role of hormones during berry development. These techniques are ideally suited to analyse the effects of exogenous hormones on gene transcript levels, and the information gained will provide a global view of hormonal control at the transcriptome level. It is much more difficult to obtain an overall picture of the changes in protein levels as a response to hormone treatment. However, proteomic studies have demonstrated changes in the levels of a number of proteins, and some of these changes reflect changes in gene transcript levels. This chapter is a review of the current state of knowledge regarding the hormonal control of berry development (excluding polyamines which are described in **Chapter 7**). A portion of this work has been referred to in a previous review [1].

Keywords: Abscisic Acid, Auxins, Brassinosteroids, Cytokinins, Ethylene, Gibberellins, Grape berry ripening, Growth regulators, Hormone, Jasmonates, Receptors, Salicylic acid.

INTRODUCTION

Grape berry development, from fruit set to full maturation, is a complex series of processes requiring the coordination of a large number of events involving substantial and rapid changes in a number of tissues. Due to its importance to the final end product as used by the wine, table grape, juice and dried-fruit industries, ripening has been more intensively studied than the earlier events such as fruit set and the early, major phase of cell division. In a broad sense, these early events appear to be similar in many other fruit (and other tissues) but this may be, at least in part, due to a dearth of detailed information of the hormonal effects at this stage. In contrast, the relative novelty of the mode of berry ripening has contributed to it being more extensively studied. For these reasons this review will focus more on the events surrounding ripening than the pre-ripening period despite its obvious importance in berry development.

Grape berry development has been divided into three phases [2]. The first phase includes fruit set, rapid cell division, cell expansion and metabolic events such as the accumulation of organic acids. The second phase, the lag phase, is somewhat problematic and in some varieties, as well as in some seasons, this superficially quiescent period is poorly defined. What is readily observed is the transition into ripening, veraison, which is marked by events such as skin colouration, berry softening and sugar accumulation. In this chapter, we will use the last time-point before an accumulation of sugars is recorded, as the working definition of veraison. Where possible, the data from papers has been reinterpreted to align with this definition.

***Address correspondence to C. Davies:** Commonwealth Scientific and Industrial Research Organisation, Plant Industry, P.O. Box 350, Glen Osmond, South Australia 5064, Australia; E-mail: christopher.davies@csiro.au

Hernâni Gerós, M. Manuela Chaves and Serge Delrot (Eds)

Although the changes that occur as the berry begins to ripen have received considerable attention both at the physical and biochemical level [3, 4], their overall control and coordination remains poorly understood.

Coordination is a crucial issue in berry development as the changes that occur at both the physical and biochemical levels are considerable and rapid, occurring over only a few weeks and involving a range of tissues and cell types. At the biochemical level, coordinated control is exerted over changes in primary metabolism and over a diverse and biosynthetically complex set of secondary metabolite pathways that are vital to the flavour and aroma of wine. This coordination is also flexible enough to allow responses to external influences such as viticultural practices, pests and disease, and importantly, changes in climatic conditions.

The complexity of the molecular control over berry ripening has been starkly exemplified by recent developments in the analysis of changes in gene transcript levels during berry development. For example, differential screening, cDNA and oligonucleotide microarray analysis have shown that the expression of thousands of genes, including large numbers of transcription factors, do actually change during grape berry ripening [5-11]. 2-D gel analysis has shown that there are changes in the levels of numerous proteins during development [12-14]. These numerous and complex changes are a reflection of the complexity of hormone biosynthesis, perception and signalling pathways, and the interactions between them, in plants [15, 16].

From an evolutionary point of view all fruit have the same end purpose, to assist in the reproductive process. However, despite some obvious similarities in growth and ripening between the wide diversity of fruit, there are also many differences. Perhaps some of these differences relate to the ontogeny of the fruit as described by Coombe [17]. One clear difference appears to be the manner in which fruit ripen, something that has received considerable attention. Plant physiologists have categorized different fruit species as being either climacteric or non-climacteric, based on physiological differences in their ripening [18]. Much of the research on the development and ripening of fruit has been conducted using climacteric fruit such as tomato and banana, which undergo a peak in respiratory activity during the ripening process [18, 19]. However, the significance of the respiratory peak in climacteric fruit ripening has been questioned because of its apparent absence in fruit attached to the plant and it has been suggested that a change in ethylene level is the critical marker that characterises climacteric fruit [20]. In many climacteric fruit there is an ethylene burst which is coincidental with, or lags behind, the respiratory peak.

In contrast, non-climacteric fruit tend to show a slow decline in respiratory activity during development, with no major peak in ethylene levels at around the time when ripening commences [18]. However, as will be discussed, the classification of climacteric and non-climacteric fruit on the basis of ethylene production at the time of ripening is also problematic. Grapes are regarded as non-climacteric [21] along with fruits such as capsicum, olive and strawberry. While ethylene seems to provide a pivotal role in climacteric fruit ripening the hormonal control of grape berry ripening remains the subject of debate.

HORMONES DURING BERRY DEVELOPMENT–BIOSYNTHESIS, LEVELS, SEQUESTRATION, MODIFICATION AND LOCALISATION

Ethylene

Ethylene Biosynthesis/Accumulation

The ethylene biosynthesis pathway is relatively simple and has been studied in a wide range of plants. S-adenosylmethionine (SAM) is produced from L-methionine by SAM synthase. S-adenosylmethionine is converted to 1-aminocyclopropane-1-carboxylic acid (ACC) by ACC synthase (ACS) and then to ethylene by ACC oxidase (ACO). The activity of ACS and ACO is thought to be more important in controlling ethylene levels than SAM synthase (as it participates in a wide range of other biosynthetic reactions) and most of the gene expression studies regarding ethylene synthesis relate to their mRNA levels (see below).

As suggested above, the role of ethylene in non-climacteric fruit development, and in particular ripening, is the subject of ongoing debate. This is partly due to the comparatively low levels of ethylene produced by non-climacteric fruit and the inherent difficulties in quantifying low levels of a gas. A number of different techniques have been used to collect ethylene released during berry development from a range of different cultivars. Not surprisingly, this has led to differing outcomes. Some procedures rely on capturing evolved ethylene, others involve vacuum treatments to extract 'internal' ethylene. Removal of berries from the vine and vacuum treatments may induce the production of wound response ethylene that further clouds the interpretation of results. Many published studies concentrated solely on measuring ethylene levels at around the time of veraison. Early studies, using Doradillo fruit removed from the vine and vacuum treated, did not detect an increase in ethylene at the onset of ripening [21]. Inaba *et al.* [22] measured ethylene production in bagged bunches of seeded and unseeded Delaware grapes on vines from flowering until after veraison. The trend for both varieties was to have a maximal ethylene level at flowering after which the levels decreased to be low at one week after flowering. They reported no peak in ethylene production at, or just before, veraison. Ethylene levels (µl/kg fresh weight) did not increase at veraison in Thompson Seedless and Carignane bunches removed from the vine and placed in jars without vacuum [23]. These authors also reported that ethylene levels were highest at flowering but were low by fruit set. Ethylene has been associated with flowering in a number of plant species [24], although there is no further information on grape regarding its role at these stages.

In contrast to the above results, some reports have indicated changes in ethylene production during, or before, ripening, but do not contain data regarding levels at earlier stages of development. Alleweldt and Koch [25] measured ethylene, essentially as described by Coombe and Hale [21], in B-6-18, Bacchus and Optima berries removed from the vine. They saw a 15-fold increase in ethylene evolution at around veraison, just prior to an increase in abscisic acid (ABA) levels, but concluded that grape berries should be considered to be non-climacteric fruit as the levels were still relatively low and the observed increase in respiration was small. A similar result for B-6-18 was reported by Düring *et al.* [26].

As none of the measurements in the above papers was replicated, evaluation of the data is difficult. Chervin *et al.* [27] determined ethylene levels in field grown Cabernet Sauvignon berries with replication. Ethylene was extracted from detached bunches at different stages of development by vacuum treatment. The levels were low but a small peak in ethylene evolution occurred at seven weeks post-flowering, at the time of berry colouration. In contrast to this, a recent study by Zhang *et al.* [28] did not detect a peak in ethylene evolution at around the time of veraison in fruit that had been removed from the vine, but not vacuum treated.

The above conflicting and somewhat confusing reports may be due to the manner in which the measurements were conducted. In other plants it has been shown that the detachment of fruit from the plant can affect ripening. This, along with vacuum treatment, may influence the experimental outcome in grapes. For example, in melons, ethylene evolution was generally much lower in detached fruit [20]. Conversely, detachment triggered autocatalytic ethylene production in apples [29]. In attached tomatoes and melons the expected climacteric rise was not detectable [20, 30].

Modest increases in ethylene levels have been reported in other non-climacteric fruit in addition to grapes. Recently, laser photoacoustic spectroscopy was used to measure small changes in ethylene and CO_2 evolution during strawberry ripening while attached to the plant [31]. Despite being characterised as non-climacteric, strawberries exhibited several peaks in ethylene levels at different stages of development. A linear increase in ethylene levels when strawberries ripened (fruit going from pink to red) was accompanied by an increase in CO_2 production. Rises in ethylene levels and/or in the transcript levels of ethylene biosynthetic genes, have also been reported in the non-climacteric fruit pineapple and litchi [32, 33].

Changes in the Levels of Ethylene Biosynthesis Gene Transcripts and Proteins

Gene transcript and protein levels of biosynthesis enzymes can also be used to investigate ethylene production. Chervin *et al.* [27] detected peaks in *ACO* transcript level and enzyme activity at seven weeks post-flowering, at the time of berry colouration and consistent with the small increase in ethylene evolution that they observed. The analysis of the expression of ESTs was used by da Silva *et al.* [34] to estimate gene expression in a range of grape tissues and berry developmental stages. They found an *ACO* homolog that was specifically expressed at flowering which correlates with the elevated ethylene evolution previously reported (see above). Pilati *et al.* [9] used microarrays to analyse changes in transcript levels in field grown Pinot Noir. The profiles were limited to three time-points (one pre-veraison sample at maximum acidity, one two weeks later at about the time of veraison and one three weeks after that, when ripening was well underway). They showed that *ACO* transcript levels were elevated at around veraison and that *ACS* was expressed up to veraison, after which time its transcript level declined. A different pattern of *ACO* expression was observed in an analysis of transcript changes during Cabernet Sauvignon development by Deluc *et al.* [8], who looked at seven different developmental stages. Again, no data is available for very early in berry development since the first tissue examined in this series was at stage 31 (just prior to three weeks after flowering) according to the modified E-L system of Coombe [35]. Deluc *et al.* [8] found no peak in transcript levels of either *ACS* or *ACO* at around the time of veraison.

As useful as knowledge of transcript levels is, protein levels provide an essential further layer of information. Unfortunately, due to the higher degree of difficulty in identifying and quantifying proteins there is less information available. Two-D gel analysis of protein levels during the development of Nebbiolo Lampia berries showed that the levels of ACS protein declined after veraison, but there was no information on very early development (before 30 days after flowering) [13]. On balance, it seems that both *ACS* and *ACO* transcript and protein levels decrease after veraison, although the presence of small peaks in levels before veraison is still unconfirmed and needs further investigation. Some, albeit limited, information is available regarding the localisation of ethylene biosynthesis gene expression in berries. Grimplet *et al.* [36] used arrays to compare gene expression between the seeds, flesh and skin of mature Cabernet Sauvignon berries. In well-watered vines two *ACO* genes were more highly expressed in the skin than in the seed or pulp. This presumably leads to a localised distribution of ethylene biosynthesis enzymes and hence ethylene concentration within the berry.

On the basis of the low levels of ethylene and the low, and fairly stable, respiration rates in berries, grapes can be justifiably referred to as non-climacteric fruit using the accepted definition. However, studies in a range of fruit show that the role of ethylene in non-climacteric fruit is still not well defined. Examples have been given above which illustrate this point. In summary, there is evidence that ethylene levels and transcript levels of biosynthesis genes might increase in grape berries at around the time of veraison. More work needs to be done to establish the role and mechanism of action of ethylene during grape berry ripening, and how the ethylene signalling pathway interacts with other hormones implicated in the control of ripening. Much of the evidence supporting an involvement of ethylene in ripening is correlative and the difficulties in comparing studies conducted at different developmental stages in different varieties only complicate assessment. Proof of the role it plays will require the analysis of transgenic plants with modified ethylene accumulation and/or perception.

Abscisic Acid

Abscisic Acid Biosynthesis/Accumulation

The ABA biosynthesis pathway is considerably more complex than the pathway for ethylene. Not all of the genes associated with this pathway [37] have been unequivocally confirmed in grapevine, although two genes considered important in controlling ABA formation have been identified with reasonable certainty on the basis of amino acid sequence homology [28, 38]. These are two plastidial enzymes, 9-*cis*-epoxy-carotenoid dioxygenase (NCED) and zeaxanthin epoxidase (ZEP), which are involved in crucial steps in ABA synthesis. NCED, which catalyses the first committed step in ABA biosynthesis by producing xanthoxin, is a member of a larger group of carotenoid cleaving dioxygenases. It is usually thought of as the key enzyme in the pathway [39].

Unlike the situation with ethylene, the pattern of ABA accumulation during berry development seems to be well defined. This may be due to both its relative ease of measurement and its relatively high levels in developing fruit. A reproducible pattern of ABA accumulation has been reported by a number of groups, who have used different systems for measurement in a wide range of cultivars and, in some cases, interspecific crosses. Free ABA levels were high in the flesh of young berries and then decreased rapidly to be low during the remainder of the pre-veraison phase [40-42]. All reports of changes in free ABA levels at around the time of veraison, showed that there was an increase in free ABA levels concomitant with sugar accumulation and colour development [11, 21, 28, 40-49]. After the peak in levels following veraison, ABA levels declined as the fruit approached ripeness [11, 22, 40-42, 50-52]. This pattern of accumulation provides correlative evidence for a role of ABA in ripening and perhaps in its initiation (see below).

Expression of Abscisic Acid Biosynthesis Genes

cDNAs encoding two putative *NCED* genes, based on phylogenetic analysis, have been cloned from grapevine [38]. A cDNA corresponding to a single putative *ZEP* was also isolated and Southern blot analysis suggested that there are likely to be only two *NCED* and one *ZEP* genes in grapevine [38].

Differing patterns of expression during berry development have been reported for *NCED* and *ZEP* genes. There appears to be considerable season to season variation in the patterns of transcript accumulation [8, 42, 53]. In general *ZEP* transcript levels increased until around veraison after which they decreased to be low late in berry development [8, 11, 42]. *NCED1* transcript levels tended to be higher during the lag phase and early ripening, after which they decreased to be low at berry maturity [8, 11, 42, 53]. *NCED2* was generally more highly expressed pre-veraison, but a great deal of season to season variation was observed [8, 42, 53]. The reason for the observed variable expression patterns could be related to environmental variables as both *ZEP* and *NCED* were responsive to water deficit [36, 54].

Differences in transcript profiles between the skin, flesh and seed of ripe berries have also been detected by microarray analysis. *NCED1* was more highly expressed in pulp than in either skin or seed, and its transcript levels were increased under water stress [36]. The higher *NCED1* transcript levels in pulp as compared to skin and seed tissue is in contrast to the distribution of ABA reported above.

Abscisic Acid Catabolites and Conjugates

In addition to the free form, ABA can also occur in conjugated (bound) forms, the most common of which is ABA glycosyl ester (ABA-GE) [37]. In Arabidopsis there is evidence suggesting that ABA-GE might act as a storage form from which free ABA can be released by hydrolysis [55]. However, there is no direct evidence for this occurring in grapes. The level of conjugated ABA has been measured by alkaline hydrolysis of berry extract to gain a value for total ABA from which the levels of free ABA are subtracted so as to gain a value for the conjugated form. Kondo and Kawai [51] used this technique to show that the levels of conjugated ABA in the seeds of Pione grapes increased steadily from 15 days post-flowering (dpf) to 80 dpf. During the same period, the levels in the skin of these berries also increased, but fluctuated wildly. Koussa *et al.* [56] found that the levels of putative ABA-GE in Cabernet Sauvignon were highest in flowers and then decreased to be low by bunch closure. No later samples were analysed. Sequestration through conjugation may play a role in the decrease in free ABA that occurs midway through the ripening phase. Wheeler *et al.* [42] showed that roughly a third of the decrease in free ABA levels observed midway through ripening could be accounted for by an increase in the levels of conjugated ABA. Twenty percent of the decrease in concentration was due to dilution as a result of berry size increase. The remainder was thought to be the result of breakdown *via* enzymic catabolism, or due to export from the berry (as described in other plants [57]). Detailed analyses of ABA, ABA-GE and various catabolites (*e.g.* dihydrophaseic acid) have recently been carried out. ABA-GE levels in skin and pulp increased from three weeks post-flowering (wpf) to their highest levels at veraison, after which they decreased [11, 41]. Differences in observed patterns compared to the earlier work could have arisen from the different approaches to conjugate measurement. In seeds ABA-GE levels appear to increase steadily from three wpf [41]. The high levels of dihydrophaseic acid early in berry development indicated that the oxidation pathway is active at this stage thus suggesting developmental differences in the way free ABA levels are managed [11, 41]. A similar pattern of dihydrophaseic acid accumulation was observed in seeds [41].

Distribution of Abscisic Acid within the Berry

The distribution of free ABA within grape berry tissues is not even, and this may give us clues as to its role during berry development. The levels in skin and seeds have been reported to be higher than in the flesh

[21, 45, 52], but the placenta may contain the highest levels, especially around the peripheral and central phloem in ripe berries [58]. The localization of ABA to the phloem is consistent with a role in assimilate unloading and uptake. All tissues showed a sharp increase in ABA levels at veraison followed by a decrease [21, 45, 58]. In very young berries, ABA was detected by immunogold labelling in phenol-rich cells, mainly in the nucleus, but also in the cytoplasm and vacuole [59]. High affinity, highly specific ABA-binding proteins have been localized to the tonoplast and may be involved in ripening as their affinity for ABA is highest at veraison [59].

Brassinosteroids

Until recently, brassinosteroids (BRs) have been associated with plant growth and stress response [60], but now there also appears to be a role for BRs in fruit development, in particular during ripening. Symons *et al.* [61] studied the levels of the brassinosteroid castasterone (CS) and its precursor 6-deoxoCS during berry development. As with some other hormones, the levels of both were elevated in flowers and young berries and then decreased prior to veraison. CS levels were highest during the pre-veraison period at two wpf. This may indicate a role for CS during the period of cell division and expansion that occurs early in berry development. The levels of 6-deoxoCS and CS increased sharply at about the time of veraison (CS was at its highest concentration in berries at 10 wpf), but brassinolide (BL), which is the most active BR [62], was not detected at any stage [61]. After peaking immediately after veraison, the levels of both 6-deoxoCS and CS declined. This pattern of accumulation suggested that CS may play a role in ripening-associated processes, for example, the post-veraison phase of berry growth, or perhaps it is part of a response to the stress resulting from the massive sugar influx that occurs after veraison.

The transcriptional control of CS synthesis seems quite complex. *DWF1*, responsible for an early step in the BR biosynthesis pathway, was expressed most highly in young berries. The levels declined until around the time of veraison, when they again increased and declined somewhat during the later stages of ripening [53, 61]. This pattern mimics that for CS levels. The BR6OX enzyme (encoded by *BR6OX1)* is responsible for converting 6-deoxoCS to CS and is thought to be the controlling step in CS synthesis. *BR6OX1* transcript levels were low in flowers, but increased steadily to reach a maximum at veraison after which they declined sharply [61]. Castellarin *et al.* [53] also studied the accumulation of *BR6OX* transcripts. They found that the patterns of accumulation varied considerably over two seasons, although the transcript levels were low during the ripening phase.

The patterns of mRNA accumulation are in contrast with the accumulation pattern of 6-deoxoCS and CS, which are at low levels pre-veraison and increase rapidly at veraison. The negative correlation between *BR6OX* transcript levels and the amount of the corresponding enzyme substrate and product has been reported in other species [63-65] and indicates post-transcriptional control.

Gibberellins

Gibberellins (GAs) are regulators of many processes during plant development involving cell division and expansion [66]. The consensus of a number of studies is that GA levels in the flesh of seeded berries were high at around flowering and early in berry development after which they decreased steadily [47, 52, 61, 67]. The observed pattern of GA accumulation is consistent with the proposed role for GAs in cell division and expansion early in berry development. In seeds, GA levels peaked at a time similar to that described for cytokinins, *i.e.* during the lag phase of berry development when seed maturation is occurring [52, 68]. It is possible that this pattern of accumulation may indicate a role for GAs in delaying ripening. However, any delaying of ripening by GA treatments may be attributable to changes in sink source relationships rather than a direct effect on the control of ripening as such (see later sections).

The expression of *GA20-ox* (which encodes a GA 20-oxidase involved in GA biosynthesis) and *GA2-ox* (which encodes an oxidase involved in the deactivation of GAs) reflects the accumulation pattern of GAs in grape berries. *GA20-ox* was most highly expressed in flower and young berries after which its transcript levels declined [67]. *GA2-ox* was most highly expressed a little later in development than *GA20-ox*, in agreement with the subsequent decline in GA levels.

Auxins

Auxin Biosynthesis/Accumulation

The biosynthetic pathway leading to the formation of auxins, in particular indole-3-acetic acid (IAA), in grape berries is unknown, but there are a number of reports indicating that IAA levels are high early in development, after which they decline steadily to be very low at veraison [22, 44, 47, 52, 69]. The high levels early in berry development are in agreement with the proposed role for auxin in cell division and expansion. The decline in auxin levels towards veraison was not observed by Symons *et al.* [61] using GC-MS technology, although the reason for this conflicting observation is unknown. As we shall see below, studies using exogenous auxins applied prior to veraison do actually provide some evidence in support of a decrease in auxin levels as veraison approaches.

Auxin Conjugates

A large proportion of IAA in plants does not occur in its free form, but is ester-linked to sugars or amide-linked to amino acids, peptides and proteins. These conjugates are generally thought to be biologically inactive and are associated with storage, transport and IAA degradation, thereby contributing to the control of auxin homeostasis [70, 71]. In the context of fruit development, the accumulation of IAA-amide conjugates at, and after, the onset of ripening has been reported for the climacteric fruit banana [72] and muskmelon [73], as well as the non-climacteric strawberry [74]. Since the formation of these conjugates was accompanied by low IAA levels, it could also provide a possible explanation for the low concentration of free auxins in ripening grape berries (see above). In support of this, a ripening-associated accumulation of an IAA-amino acid conjugate, IAA-aspartic acid (IAA-Asp), in Cabernet Sauvignon berries has recently been reported [69]. The concentration of the IAA-Asp conjugate was high in flowers, very low in berries up to the stage of véraison, and then increased to a maximum 12 wpf, after which it slowly declined. The formation of this conjugate in berries was further linked to the expression and activity of an indole-3-acetic acid-amido synthetase (GH3-1). Since a comparable accumulation of IAA-Asp was detected in ripening tomatoes, it might represent a mechanism for the maintenance of low auxin levels in ripening fruit.

Cytokinins

Like other plant hormones cytokinins are involved in a diverse range of processes [75]. The levels of two cytokinins, zeatin and zeatin riboside, were high in one week old berry flesh, but decreased rapidly to be low by the time of veraison [52]. This pattern of accumulation is in agreement with the proposed roles for cytokinins in flower development and fruit set [75]. The pattern is also consistent with the ability of cytokinins to delay berry development (see below). In seeds, zeatin and zeatin riboside levels reached a peak roughly in the middle of the lag phase (the peak in zeatin riboside levels was six days later than that for zeatin). The levels of both were declining or low by veraison [52]. This pattern suggests a role for cytokinins in grape seed development and maturation.

There is little molecular information about the metabolism of cytokinins apart from the observation regarding a putative cytokinin catabolic enzyme, cytokinin oxidase. Cytokinin oxidase transcript levels are high early in berry development, but steadily decrease to be at low levels when berry colour is developing [8]. Cytokinin oxidase could therefore be involved in the decrease in cytokinin levels approaching veraison.

Salicylic Acid

Salicylic acid (SA) and methyl-SA are involved in signalling in plants, particularly in the induction of defence and stress responses [76]. The levels of SA and methyl-SA during grape berry development have not been reported, and therefore it is difficult to propose a developmental role for salicylates at this stage.

Jasmonates

The levels of jasmonic acid (JA) and methyl-jasmonate (MeJA) have been reported in seeded and unseeded, GA-treated, grapes [77]. The first stage sampled was at 20 days after flowering and in seeded berries the levels of both JA and MeJA were elevated. This elevated level is coincident with the early stage of cell division and expansion, and jasmonates have been associated with a range of developmental processes including many aspects of reproduction as well as plant responses to stress [78]. By 30 days post-flowering, the levels had decreased and remained low throughout ripening. Thus, there was no veraison-associated increase in jasmonate levels that could be suggestive of a role in ripening. In grape seeds the levels of MeJA remained fairly constant, but JA levels were high at 45 days post-flowering (the first time-point sampled), decreased to be low at about the time of veraison, and then increased again to higher levels in seeds from ripe fruit [77]. This pattern was also found in apple seeds and could be associated with seed maturation. Jasmonate levels in seedless, GA-treated fruit remained low throughout ripening.

In summary, the accumulation of most hormones during berry development is fairly well understood. The possible exception to this is the accumulation of ethylene that still seems to be the subject of some debate. All seven of the hormones discussed above are accumulated to significant levels in flowers and early in berry development (Fig. **1**). Presumably, this is due to the extensive requirements for hormonal control during the phase of rapid cell division and expansion. The decrease in the levels of all of the hormones approaching veraison suggests that this requirement decreases, although there is evidence, at least for IAA and cytokinins, that the reduction in levels may be a prerequisite for ripening initiation (see below). Only two hormones seem to increase in concentration at veraison, *i.e.* ABA and BRs (Fig. **1**). This pattern, and other evidence (see below), indicates a role for these hormones in the initiation and progression of ripening. Note that polyamines are discussed in the chapter by Panagiotis *et al.* (**Chapter 7**).

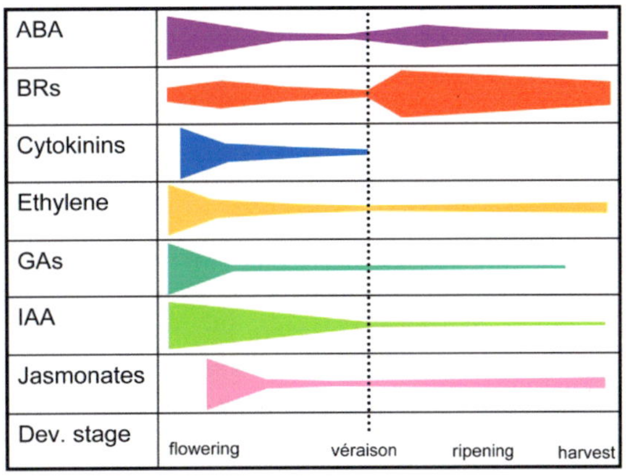

Figure 1: Stylised patterns of hormone accumulation (on a concentration basis) in deseeded berries during development (except for jasmonates where data was only available for skin tissue). The data for ABA were taken from Wheeler *et al.* [42] (and unpublished data). The BR data corresponds to castasterone levels [61]. The data for cytokinins refers to zeatin. Zeatin riboside levels were lower, but followed the same pattern as zeatin; the measurements commenced one week after flowering [52]. The ethylene data is a composite from two sources [22, 28]. Note that a peak in ethylene levels at around the time of veraison has also been reported and should also be considered [27]. The GA data refers to the levels reported for GA_1 [61]. The profile of IAA accumulation is taken from the paper by Böttcher *et al.* [69]. The jasmonate profile represents the pattern of jasmonic acid accumulation. Methyl jasmonate has a similar but lower profile [77]. The levels are relative only within the developmental series of each hormone, not between hormone profiles.

PERCEPTION – RECEPTORS

So far there is only information regarding putative ethylene and brassinosteroid receptors in grape.

Ethylene

Although the levels of ethylene evolved during grape berry development are low and difficult to accurately measure, this does not preclude a role for ethylene during grape berry development. It is possible that the response to ethylene may be modulated by changes in the sensitivity of perception. This may be the reason why the effect of applied hormones can vary considerably depending on the developmental stage of the fruit (see below), as it suggests that the ability to perceive, transduce and act upon hormone signals must vary during berry growth. There is evidence that the transcript levels of some grape ethylene receptors change during berry development. One receptor (*ERS1*) was expressed most highly at the pea size stage (stage 31, modified E-L system [35]), whereas the transcript levels decreased at bunch closure (stage 32) and then remained fairly stable until ripening [8]. The expression of another receptor (*EIN4/ETR5*) increased at the commencement of sugar accumulation and was most highly expressed late in development. As ethylene receptors are negative regulators of the ethylene response, any increase in receptor mRNA levels could mean the berries are less sensitive to ethylene at veraison. However, it has been shown in tomato that transcript levels do not always correlate with receptor protein levels [79]. A more detailed study in strawberries [80] showed that the transcript levels of three putative ethylene receptors changed during fruit development. The authors suggest that the change in the levels of the more sensitive type-II group receptor, *ETR2*, could make even small changes in ethylene concentration physiologically relevant. The situation in grapes will require more study before a conclusion can be reached.

Brassinosteroids

The putative BR receptor gene, *BRI1*, was expressed in all berry samples taken from flowering to harvest [61]. There was a decrease in transcript levels from flowering to young berries, but this was followed by an increase in the sample taken two weeks before veraison. Castellarin *et al.* [53] also detected *BRI1* transcripts at all stage of berry development sampled, but found considerable variation in profiles between successive seasons. The pre-veraison increase pre-empts the increase in *BR6OX* gene expression and CS levels by two weeks. The levels of *BRI1* transcripts reached a maximum at 10 wpf (two weeks after veraison), and then declined. There is no data regarding the levels of the receptor protein around the time of veraison, although the change in gene transcription may reflect an increase in sensitivity to BRs at this time.

CHANGES IN THE EXPRESSION OF GENES, AND THEIR PROTEINS, INVOLVED IN HORMONE SIGNALLING AND ACTION DURING BERRY DEVELOPMENT

A repeated theme throughout this review has been the fact that hormones have the ability, and indeed the purpose, to coordinate changes in a large number of genes in response to both developmental and external cues. Recent developments in technology now permit changes in the expression of genes during development, in response to hormone treatments and in hormone mutants, to be detected on a genome wide scale. Thus, the techniques of microarray analysis and high throughput sequencing are ideally suited to studying hormone biosynthesis, signalling and response. Even at this early stage in the use of these techniques, there is insufficient space here to properly outline all the data available. Data specific to hormone accumulation and perception have been detailed above. The value of using high throughput gene expression assay techniques to study the changes in hormone-related gene expression during development has been shown in a number of recent publications [8, 9, 36, 81-84]. Some of these data are discussed under other headings in this review, but it is not possible to summarise all the data and readers are referred to the papers for more detailed information. However, it is clear from the various studies that the patterns of putative hormone-associated gene expression are widely divergent even when looking at those relating to a single hormone. This serves to show the involvement of hormones throughout berry development in an extensive range of processes and hints at the complexity of hormone signalling and action.

High throughput techniques can also be used to study the response of the transcriptome to the application of hormones. So far these approaches have only been applied to the investigation of the effects of ABA and ethylene [84-86], and some of this work is discussed below.

Due to technical limitations proteomic studies are unable to provide the same comprehensive overview of changes in protein levels that is possible when studying changes in transcript level using 'deep sequencing' or microarray analysis. However, a number of studies have been completed in grape berries using 2D electrophoresis (2-DE) that have provided valuable information regarding changes in proteins involved in a wide range of processes, including some associated with the response to hormones (see below). Again, it is difficult to make a meaningful general summary of so much data, and readers are referred to the papers for further information [12, 13, 87].

EFFECTS OF EXOGENOUS PLANT GROWTH REGULATORS ON GENE EXPRESSION AND BERRY DEVELOPMENT

Ethylene, Ethylene Releasing Compounds and Inhibitors of Ethylene Perception

Ethylene and the more convenient ethylene producing compound 2-chloroethylphosphonic acid (CEPA) have a profound effect on ripening when applied to climacteric fruit, and CEPA is used commercially to hasten the ripening of some fruit. The effects of ethylene and CEPA application to non-climacteric fruit such as grape berries appear to be less dramatic than those described for climacteric fruit. The response to CEPA application is obviously complex and influenced by numerous factors. A review by Szyjewicz *et al.* [88] summarises many of the experiments that have been done using CEPA on a wide variety of grapevine cultivars. This review details the wide diversity of effects that are possible. The reason for these diverse effects is likely to be due to factors such as the developmental stage of the fruit at application, varietal differences in uptake and response, the method of application (including concentration, use of surfactants, *etc.*) and growth conditions. The most commonly reported 'positive' effect on grape berry ripening is an increase in the skin colour of red grapes [88].

CEPA treatment induces an increase in the transcript levels and enzyme activities of a number of genes and corresponding proteins involved in anthocyanin synthesis. For example, phenylalanine ammonia-lyase (PAL) activity was increased in the skin of isolated berries treated with light and sucrose, and this effect was enhanced by the addition of CEPA [89]. CEPA sprayed on Cabernet Sauvignon berries at a stage when 50% of the berries were coloured increased skin colour [90]. Northern blot analysis showed that chalcone synthase (*CHS*), flavanone-3-hydroxylase (*F3H*), leucoanthocyanidin dioxygenase (*LDOX*), UDP-glucose flavonoid 3-*O*-glucosyl transferase (*UFGT*), but not dihydroflavonol 4-reductase (*DFR*) transcript levels were elevated by the treatment. UFGT protein levels were also induced by CEPA application. The mechanism of ethylene action on UFGT transcription was further studied by Chervin *et al.* [91]. A promoter fragment from the grapevine *UFGT* gene, containing putative ethylene-responsive, ABA responsive and sugar responsive elements, was used to drive GFP expression in biolistic experiments using cultured cells. Sucrose, ABA and ethylene treatments all increased GFP expression to about the same level.

CEPA treatment has also been reported to increase total soluble solids (TSS) and hasten the timing of the initiation of ripening in *Vitis* species [88]. Interestingly, an opposite effect has also been demonstrated. Hale *et al.* [92] reported that grape berries exhibit a biphasic response to CEPA application. CEPA applied to Shiraz berries at 4, 5, 6, or 7 wpf delayed colour development while treatments at 8 and 9 wpf, *i.e.* just before veraison, appeared to hasten colouration. Measurement of the sugar to acid ratio suggested that ripening in general was delayed by the earlier treatments and advanced by treatments nearer to the time of veraison in control fruit. Therefore, polar effects can occur in response to CEPA (and by inference ethylene) depending on the developmental stage of the berries at the time of application. Changes in sensitivity of plants during development have been reported in other fruit species. For example, a change in the response to ethylene during development has been described in citrus, another non-climacteric species [93].

Despite the difficulties in application some experiments using ethylene have been conducted with grapes. In general, the results appear to agree with those obtained using CEPA. Hale *et al.* [92] treated Doradillo grapes in polythene bags with ethylene for 10 days from midway through the lag phase and measured berry compressibility. Both the timing and rate of berry softening were advanced, the commencement of softening was accelerated by about six days. The experiments of Tira-Umphon *et al.* [94] indicate that the mechanism of the induction of changes by exogenous hormones may not be achieved through the same mechanism as that occurring during normal berry development. When Tira-Umphon *et al.* [94] treated field grown Cabernet Sauvignon grapes with ethylene, or Gamay suspension cells with CEPA, they observed an increase in *UFGT* transcript levels, but saw no increase in the expression levels of the transcription factor *MybA* which controls anthocyanin accumulation during 'normal' ripening. Chervin *et al.* [85] used microarray analysis to study the effects of ethylene gas on Cabernet Sauvignon berry gene expression at veraison. They suggested that the small increase in berry diameter caused by the ethylene treatments may be related to the changes in expression they observed for genes putatively involved in water transport and cell wall modification.

1-methylcyclopropene (1-MCP) is an inhibitor of ethylene perception that works through irreversible binding to ethylene receptors [95]. Chervin *et al.* [27] treated Cabernet Sauvignon berries for 24 h periods with 1-MCP at weekly intervals (5, 6, 7, 8 and 9 wpf). Treatments at 6, 7, and 8 wpf inhibited colour development and berry size increase and resulted in less organic acid loss. Ripening in the control berries began sometime between 7 and 8 wpf. The 1-MCP treatments were most effective at the time of the observed peak in ethylene evolution (7 wpf). Therefore, it seems that 1-MCP was able to delay ripening when applied around the time of veraison. In a later experiment 1-MCP application to fruit on the vine at 10 wpf reduced *UFGT* transcript and protein accumulation [91]. This effect supports the other evidence (see above) for an involvement of ethylene in anthocyanin production.

The possible influence of ethylene on sugar accumulation has also been studied using 1-MCP. The expression of two sucrose transporter genes, *SUC11* and *SUC12,* normally increases at veraison [96, 97]. Cabernet Sauvignon bunches (at a stage with 50% coloured berries) were treated with 1-MCP for 24 h and sampled at 5 and 19 days after treatment [98]. Not only were sucrose levels decreased by 1-MCP, but the transcript levels of *SUC11* and *SUC12* also decreased.

Like the *UFGT* promoter, the promoter region of *ADH2*, the predominant grape alcohol dehydrogenase gene expressed during berry ripening, was found to contain sequences related to ethylene responsive elements [99]. The transcript levels of *ADH2* and to a lesser extent ADH enzyme activity in berries at veraison were reduced by 1-MCP treatment [100], again suggesting that ethylene may play some role in the ripening process, even if it might not be the major initiator.

Abscisic Acid

ABA, like ethylene, can promote grape berry ripening, or at least some aspects of it. As for ethylene, the timing of application seems crucial to the effects observed. In general, ABA treatment one or two weeks prior to the expected time of veraison advances the initiation of ripening, measured as earlier increases in sugar and changes in skin colour (Fig. **2**) [42, 43, 101]. There are a number of instances where ABA treatment had little or no effect on sugar accumulation, although increased anthocyanin accumulation was observed [45, 102-111]. The reason for this difference in effect may be the timing of ABA application. In many cases ABA application was at, or after, véraison, and in contrast to anthocyanin accumulation, ABA application may not be able to further enhance sugar accumulation once ripening has commenced.

Figure 2: Advancement of Shiraz berry ripening by application of ABA before veraison. Control on left, right panel showing fruit treated with 400 mg/L (+)-ABA.

The enhancement of anthocyanin accumulation by ABA appears to result from increased transcription of anthocyanin biosynthesis pathway genes [86, 112-115]. In some cases, the transcript levels of many of the anthocyanin biosynthesis pathway genes have been up-regulated by ABA. Ban *et al.* [114] showed that

ABA application to Kyoho grapes (*V. vinifera* L. x *V. labrusca* L.) at the onset of berry softening temporarily increased *PAL*, *CHS*, *CHI*, *DFR* and *UFGT* transcript levels and increased anthocyanin accumulation throughout development. Similar experiments in Cabernet Sauvignon showed that ABA application at veraison temporarily enhanced the expression of *CHS*, *CHI*, *F3H*, *DFR*, *LDOX* and *UFGT* genes, as well as that of a transcription factor, *MYBA1*, involved in the control of the pathway. As a consequence anthocyanin accumulation was also increased [115]. These results suggest that ABA application stimulates the activation of the entire anthocyanin biosynthesis pathway.

ABA application also affected the synthesis of flavonoids other than anthocyanins. ABA applied to berries at veraison increased transcript levels of two putative flavonol synthase genes in Merlot and increased the concentration in the skin tissue of quercetin [116]. Lacampagne *et al.* [49] showed that ABA application to green berries had modest effects on tannin accumulation. The activity of two enzymes involved in tannin synthesis *i.e.* leucoanthocyanin reductase (LAR) and anthocyanidin reductase (ANR) was generally reduced by ABA treatment, as were the transcript levels of the corresponding biosynthesis genes. The expression of two *MYB* genes involved in the control of tannin synthesis was also altered.

The effect of ABA on sugar accumulation may be in part due to its effects on invertases. Acid invertases are localized in the apoplast, cytoplasm and vacuole and are important in the uptake and accumulation of sugars. *cis*-(+)-ABA applied to Kyoho berry discs and intact berries increased the activity of both soluble and cell wall acid invertases [117]. The change in activity was reflected in corresponding changes in the amounts of enzyme present. The addition of a transcriptional inhibitor nullified the ABA-induced changes suggesting that changes in gene transcription were essential. Activation *via* phosphorylation was suggested as the protein kinase inhibitor quercetin and the addition of acid phosphatase enzyme suppressed the ABA induction of both cell wall and soluble invertase activity in berry discs without substantially altering the amount of enzyme. These results provide evidence of a link between ABA, invertase activity and sugar metabolism which could be an important part of the initiation of grape berry ripening.

In addition to affecting the expression of genes and enzymes involved in ripening, ABA can also affect its own metabolism. The transcript levels of *NCED* genes were auto-induced by ABA application in both berry skin [42, 86] and berry flesh [42]. Interestingly, *ZEP* expression was down-regulated in both skin and flesh by ABA treatment [42, 86]. The auto-induction of *NCEDs* could induce higher levels of ABA biosynthesis which may be an important factor in the alteration of berry development by ABA application, although this may also be part of the process that usually occurs during ripening.

A more global analysis of ABA-induced change in berry gene expression was undertaken by Koyama *et al.* [86] using the Affymetrix *V. vinifera* GeneChip. Many of the genes that were up-regulated (*e.g.* chitinase, thaumatin, genes encoding various cell wall proteins, sugar metabolism and transport *etc.*) are normally upregulated during the ripening phase, thus supporting the idea that ABA may be involved in the control of berry ripening. The expression of a number of structural genes in the phenylpropanoid and flavonoid pathways, and their corresponding Myb transcription factors, was upregulated (as was observed above). In contrast, the expression of most of the genes putatively involved in photosynthesis and associated processes were down-regulated by ABA. This also agrees with the decrease in photosynthesis and chlorophyll content as berries progress towards ripening. ABA application was also observed to affect the expression of genes involved in ethylene synthesis and response (mainly up-regulated) and auxin response (mainly down-regulated). These effects seem to agree with the notion that ABA is involved in the control of grape berry ripening. A similar conclusion was reached in another study using quantitative PCR analysis to study the expression of grape orthologs of various genes thought to be involved in ABA-signalling pathways in model species [84]. Studies showed that the expression of some genes putatively associated with ABA signalling (*e.g.* HB transcription factors and PP2C protein phosphatases) were induced during ripening in cultured and field-grown fruit and by the addition of sucrose and ABA to cultured berries.

Confirmation that ABA treatments not only alter transcript levels but also alter protein levels comes from a study using 2-DE [14]. ABA was applied to fruit before and at veraison, and changes in protein levels between treated and untreated berries were determined. Results showed that the levels of many proteins in skin and flesh were affected by ABA treatment and that the time of treatment affected which proteins were detected as differentially expressed. For example, the levels of some ripening-induced proteins involved in a range of processes (*e.g.* chitinase, xyloglucan endotransglycosylase, CHI, DFR) corresponding to genes whose expression was shown previously to be upregulated by ABA were also found to be increased. These results add further support to the view that ABA is involved in the control of ripening-associated genes, and therefore, ripening-associated processes.

Brassinosteroids, Brassinazole

The observed increase in CS at veraison, described above, is suggestive of a role in ripening. Further evidence in support of a role for BRs in berry ripening comes from the application of epi-BL, a synthetic brassinolide. Epi-BL was applied at four time-points commencing from approximately 5 wpf, well before veraison [61]. Ripening, as measured by the percentage of coloured berries, was advanced by these treatments. Conversely, brassinazole, an inhibitor of BR biosynthesis, applied at the same time, slowed berry colour development.

Gibberellins

Exogenous GAs have been widely used in the table grape industry to control bunch size, bunch openness, fruit set, berry size and seed formation. The effects vary considerably depending on the cultivar and the concentration, timing and frequency of application. For example, GA_3 sprayed onto Kyoho grapes at 10 days after bloom resulted in increased cluster length, berry weight and size, although total soluble solids and titratable acidity were unaffected [118]. Sato *et al.* [119] also found that GA applications at full bloom and 10-16 days thereafter had no effect on total soluble solids levels or titratable acidity, but they did suggest a negative effect on flesh firmness. In contrast, Teszlák *et al.* [120] stated that GA treatment of two red-skinned *V. vinifera* cultivars at flowering generally delayed ripening. In a three-year study where three sprays were applied (at 20% flowering, 14 dpf and 28 dpf) berry weight increased, although TSS accumulation was inhibited. Effects on other parameters varied depending on the year and the GA concentration [121]. The above GA treatments were applied early in development at a time of rapid cell expansion and division. Given this and the fact that in a number of cases berry size was increased, it is difficult to suggest that this represents support for a direct effect of GAs on berry ripening. GA-induced berry size increase may delay ripening simply because a greater amount of sugar needs to be accumulated to reach ripeness rather than GAs affecting ripening directly.

The effects of GA application on the level of invertase, an enzyme thought to be involved in hexose accumulation have been reported. Treatments of Sultana flowers/berries early in the development with GA_3 increased invertase activity and hexose accumulation on a per berry basis [122, 123], but not on a concentration basis [122]. Dreier *et al.* [122] found no direct relationship between invertase activity and the rate of assimilate import. Conversely, Pérez and Gómez [123] suggested a correlation between berry size, invertase activity and hexose content. The above effects refer to the treatment of unseeded berries with GA.

Auxin and Auxin Transport Inhibitors

A diverse array of functions in a range of plant developmental processes has been described for auxins [71, 124-127]. In both climacteric fruit (*e.g.* tomato, [128]) and non-climacteric fruit (*e.g.* strawberry, [129]) auxin application has been shown to delay ripening. Similarly, in grapes, there are numerous reports where the application of auxins to berries before veraison has delayed ripening or some part of the ripening process, such as the accumulation of sugars and anthocyanins, as well as the reduction of acidity and chlorophyll levels. A range of auxins and auxin-like compounds with varying degrees of effectiveness have been used. These include indole-3-acetic acid (IAA), indole-3-butyric acid (IBA), α-naphthaleneacetic acid (NAA), benzothiazole-2-oxyacetic acid (BTOA) and 2,4-dichlorophenoxyacetic acid (2,4-D) [21, 40, 43, 48, 92, 103, 114-116, 130-132].

NAA treatment reduced anthocyanin accumulation in Kyoho berries, when applied at veraison, possibly as a result of a reduced induction of PAL activity [112]. The increase in ABA in berry skins that usually occurs at veraison was also delayed. Ban *et al.* [114] applied 2,4-D to Kyoho grapes at veraison and showed that anthocyanin accumulation was much decreased. The transcript levels of *PAL*, *CHS*, *CHI*, *F3H*, *DFR*, *LDOX* and *UFGT* were all reduced by the 2,4-D treatment. In another study, NAA application at veraison inhibited anthocyanin accumulation in Cabernet Sauvignon berries [115]. Real time Q-PCR was used to show that the transcript levels of many of the anthocyanin biosynthesis pathway genes were lower in NAA-treated fruit. The accumulation of flavonols, which arise from dihydroflavonols and are precursors common to the anthocyanin pathway branch, was also reduced by auxin treatment. When applied at veraison, NAA caused a reduction in quercetin levels in Cabernet Sauvignon and Merlot berries [116]. The reduction in quercetin accumulation may well be due to a lack of substrate for flavonol synthase as the expression of genes in the flavonoid and general phenylpropanoid pathways were all downregulated by NAA.

BTOA application to berries prior to veraison delayed ripening and altered gene expression (Fig. **3**) [40]. The overall effect was to maintain the berry in the pre-veraison state. Berry softening, anthocyanin accumulation and sugar accumulation were delayed by approximately two weeks. The mRNA levels of a vacuolar invertase gene expressed during the pre-veraison period, whose expression normally decreased at, or shortly after, veraison was maintained. In contrast, the expression of genes associated with ripening (*CHS*, *UFGT GRIP4* and *CHI4*) was delayed as was the increase in ABA levels at veraison. The maintenance of the pre-veraison state by auxins is also shown by the delayed breakdown of chlorophylls. They are usually broken down as berries progress through the early stages of ripening [17], although IAA treatment of fruit at veraison delayed this process [48]. These fruit were also delayed in the expected increase in soluble solids and pH. Presumably, berries only commence ripening when free auxins are metabolised, thus releasing the berries from inhibition.

Figure 3: The delaying of ripening by BTOA treatment before veraison. The lower half of the bunch was dipped in a solution of BTOA (20 ppm) 6 and 8 weeks post-flowering.

Further support for a role for auxins in delaying ripening comes from the use of the auxin efflux inhibitor 2,3,5-triiodobenzoic acid (TIBA). When applied to berries at around the time of veraison, TSS and anthocyanin levels were increased [103]. Unfortunately, data showing the progress of berry development and the method used to determine veraison were not given in this paper, even though this approach deserves further attention.

Cytokinins

Synthetic cytokinins such as N-(2-Chloro-4-pyridinyl)-N'-phenylurea (CPPU) and thidiazuron, have been used to manipulate berry development. Various factors appear to affect the outcome of such applications, including the timing of application in regard to berry development, the type and concentration of cytokinin and the genetic background of the test vine. However, it seems that treatments close to the time of flowering increase the number of berries per bunch, but reduce berry size [133]. In general treatments after fruit set, when the fruit were approximately 5 mm in diameter, resulted in larger berries delayed in ripening. The delay in ripening was apparent due to lower TSS levels, increased acidity, reduced anthocyanin levels in red grapes and increased green colour compared to control fruit in white grapes [109, 118, 132-137]. Increased berry firmness was also recorded in response to CPPU treatment [109]. Two studies suggested that the response to CPPU was less evident in seeded fruit compared to unseeded fruit [136, 137]. However, it cannot be said whether this was due to the presence or absence of seeds or to other genetic differences between the cultivars used.

SA

SA does not appear to play developmental roles in other plants, but is involved in defence and stress responses. In grapes, SA delayed skin colour development and may also have delayed ripening in general, when injected into Shiraz berries two to three weeks before veraison [138]. In contrast, SA induced an increase in *PAL* mRNA levels and PAL enzyme and activity levels when applied to Cabernet Sauvignon berry slices (from fruit at veraison) [139]. While these observations may be interesting, a significant role for salicylates in the control of ripening remains unproven.

Jasmonates

MeJA application has been shown to induce *trans*-resveratrol accumulation in grape berries at 15 days post-veraison [140]. Interestingly, at 30 days post-veraison not only was *trans*-resveratrol accumulation not induced by MeJA treatment, but it was not even detectable. This is further evidence that the response of berries varies considerably during development and means that the timing of PGR applications is crucial to their effect. Although not directly relatable to whole berries, the response of grape cell cultures to jasmonates does tell us something about the control of some secondary pathways. MeJA has been shown to induce stilbene accumulation and the expression of genes involved in their biosynthesis [141-143]. Both JA and MeJA were shown to induce the production of numerous sesquiterpene-like compounds [143]. The induction of sesquiterpenes by MeJA was suppressed by the addition of SA. The expression of some of the genes from both the mevalonate and methylerythritol 4-phosphate pathways was induced. MeJA stimulated the accumulation of proanthocyanins with a concomitant induction of the expression of all the genes in the pathway. MeJA also induced the synthesis of peroxidase, chitinase and beta-1,3-glucanase proteins which are thought to be involved in plant defence responses [144].

INTERACTIONS BETWEEN DIFFERENT HORMONES/SUGARS

There is considerable evidence demonstrating that complex interactions occur between the various hormone signalling pathways in plants [15, 16]. However, due to various technical challenges the evidence for such interactions in grapes is fragmentary. A number of reports show that the application of one hormone altered the accumulation of another or the expression of genes or proteins involved in the metabolism, transport or signalling of another hormone (for example, [14, 21, 28, 40, 43, 85, 86, 145]). The effects on berry development of the application of combinations of plant growth regulators and hormones have, in some cases, indicated that interactions are possible (for example, [104, 135]).

Sugar signalling occurs in plants [146] and of particular interest is the relationship between ABA and sugar signalling where parts of the signalling pathways are common [147]. In grape berries a synergistic effect has been observed between sucrose and ABA resulting in enhanced anthocyanin accumulation in isolated tissue [102]. A study of the expression of grape orthologs putatively associated with sugar and ABA signalling has also indicated that both may be important in controlling berry ripening [84]. The work on a grape *ASR*-like gene indicates a relationship between glucose and ABA ([148], discussed in another chapter).

CONCLUDING REMARKS

Despite the expansion of our knowledge of the hormonal control of berry development there are still many questions to be answered. For instance, although the biosynthesis pathway for some hormones, *e.g.*

ethylene, is fairly well understood, we still have a limited understanding of others, *e.g.* IAA. Equally, although the receptors and parts of their signalling pathways are known at some level for some hormones, there is only limited knowledge of these aspects in respect to other hormones. In addition, little is understood of the interactions between the different signalling pathways, and how these affect berry development.

In regard to the control of berry ripening, the available evidence suggests that hormones can be divided into two groups: those involved in the pre-veraison cell division and berry expansion phase which may inhibit ripening (*e.g.* auxins and cytokinins), and those which have some role in promoting ripening (ABA, ethylene, BRs) (Fig. **4**). Although some interactions between hormones have been indicated (*e.g.* auxins and ABA), and between some others seem likely (*e.g.* auxins and BRs), there is still much to be learnt. Sugars are also likely to be involved in signalling during ripening and some links between sugars and ABA have been suggested. No doubt other relationships will be uncovered.

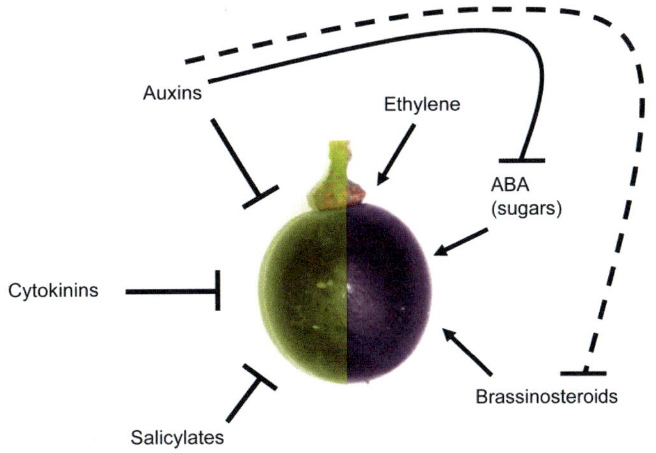

Figure 4: Schematic representation of the proposed influence of various hormones on grape berry ripening. The arrows represent a positive effect on ripening and the barred lines indicate an inhibitory effect. The dotted line indicates a speculative interaction between auxins and BRs.

It is anticipated that microarray analysis and 'deep sequencing' techniques will be useful tools in expanding our knowledge of the global effects of hormones on the transcriptome and the utility of these approaches is now being demonstrated. Changes in the proteome are more difficult to detect, but progress is also being made in this area.

The use of genetic approaches to study plant development has been successful in a number of plant species. These approaches are more difficult and time consuming in grapevines due to the time, cost and difficulty of grapevine transformation and a current lack of mutant populations coupled with the relatively long breeding cycle. The development of genetic tools for dissecting berry development is essential to the continuing effort to further refine our understanding of berry development and developing methods to manipulate it.

ABBREVIATIONS

ABA = Abscisic acid

ABE-GE = ABA glycosyl ester

ACC = 1-aminocyclopropane-1-carboxylic acid

ACO = 1-aminocyclopropane-1-carboxylic acid oxidase

ACS	=	1-aminocyclopropane-1-carboxylic acid synthase
ADH	=	Alcohol dehydrogenase
ANR	=	Anthocyanidin reductase
BL	=	Brassinolide
BR	=	Brassinosteroid
BR6OX	=	Brassinosteroid-6-oxidase
BTOA	=	Benzothiazole-2-oxyacetic acid
CEPA	=	2-chloroethylphosphonic acid
CHS	=	Chalcone synthase
CPPU	=	N-(2-Chloro-4-pyridinyl)-N'-phenylurea
CS	=	Castasterone
2,4-D	=	2,4-dichlorophenoxyacetic acid
DFR	=	Dihydroflavonol 4-reductase
Dpf	=	Days post-flowering
EST	=	Expressed sequence tag
F3H	=	Flavanone-3-hydroxylase
GA	=	Gibberellin
GA20-ox	=	Gibberellin20-oxidase
IAA	=	Indole-3-acetic acid
IBA	=	Indole-3-butyric acid
JA	=	Jasmonic acid
LAR	=	Leucoanthocyanin reductase
LDOX	=	Leucoanthocyanidin dioxygenase
1-MCP	=	1-methylcyclopropene
MeJA	=	Methyl jasmonate
NAA	=	α-naphthaleneacetic acid
NCED	=	9-*cis*-epoxy-carotenoid dioxygenase

PAL = Phenylalanine ammonia-lyase

SA = Salicylic acid

SAM = S-adenosylmethionine

TIBA = 2,3,5-triiodobenzoic acid

TSS = Total soluble solids

UFGT = UDP-glucose flavonoid 3-*O*-glucosyl transferase

Wpf = Weeks post-flowering

ZEP = Zeaxanthin epoxidase

ACKNOWLEDGEMENTS

We would like to acknowledge the support given to our work on grapes by the Grape and Wine Research and Development Corporation.

REFERENCES

[1] Davies C, Böttcher C. Hormonal Control of Grape Berry Ripening. In: Roubelakis-Angelakis KA. Ed. Grapevine Molecular Physiology and Biotechnology, Dordrecht, Springer, 2009; pp. 229-261.

[2] Coombe BG. Research on development and ripening of the grape berry. Am J Enol Vitic 1992; 43: 101-110.

[3] Ollat N, Diakou-Verdin P, Carde JP, Barrieu F, Gaudillère JP, Moing A. Grape berry development: A review. J Int Sci Vigne Vin 2002; 36: 109-131.

[4] Conde C, Silva P, Fontes N, *et al.* Biochemical changes throughout grape berry development and fruit and wine quality. Food 2007; 1: 1-22.

[5] Davies C, Robinson SP. Differential screening indicates a dramatic change in mRNA profiles during grape berry ripening. Cloning and characterization of cDNAs encoding putative cell wall and stress response proteins. Plant Physiol 2000; 122: 803-812.

[6] Terrier N, Glissant D, Grimplet J, *et al.* Isogene specific oligo arrays reveal multifaceted changes in gene expression during grape berry (*Vitis vinifera* L.) development. Planta 2005; 222: 832-847.

[7] Waters DLE, Holton TA, Ablett EM, Lee LS, Henry RJ. cDNA microarray analysis of developing grape (*Vitis vinifera* cv. Shiraz) berry skin. Funct Integr Genomics 2005; 5: 40-58.

[8] Deluc LG, Grimplet J, Wheatley MD, *et al.* Transcriptomic and metabolite analyses of Cabernet Sauvignon grape berry development. BMC Genomics [serial on the Internet]. 2007 Nov 22. Available from: http://www.biomedcentral.com/1471-2164/8/429

[9] Pilati S, Perazzolli M, Malossini A, *et al.* Genome-wide transcriptional analysis of grapevine berry ripening reveals a set of genes similarly modulated during three seasons and the occurrence of an oxidative burst at vèraison. BMC Genomics [serial on the Internet]. 2007 Nov 22. Available from: http://www.biomedcentral.com/1471-2164/8/428

[10] Lund ST, Peng FY, Nayar T, Reid KE, Schlosser J. Gene expression analyses in individual grape (*Vitis vinifera* L.) berries during ripening initiation reveal that pigmentation intensity is a valid indicator of developmental staging within a cluster. Plant Mol Biol 2008; 68: 301-315.

[11] Deluc LG, Quilici DR, Decendit A, *et al.* Water deficit alters differentially metabolic pathways affecting important flavour and quality traits in grape berries of Cabernet Sauvignon and Chardonnay BMC Genomics [serial on the Internet]. 2009 May 8. Available from: http://www.biomedcentral.com/1471-2164/10/212

[12] Deytieux C, Geny L, Lapaillerie D, Claverol S, Bonneu M, Donèche B. Proteome analysis of grape skins during ripening. J Exp Bot 2007; 58: 1851-1862.

[13] Giribaldi M, Perugini L, Sauvage FX, Schubert A. Analysis of protein changes during grape berry ripening by 2-DE and MALDI-TOF. Proteomics 2007; 7: 3154-3170.

[14] Giribaldi M, Gény L, Delrot S, Schubert A. Proteomic analysis of the effects of ABA treatments on ripening *Vitis vinifera* berries. J Exp Bot 2010; 61: 2447-2458.
[15] Gazzarrini S, McCourt P. Cross-talk in plant hormone signalling: What Arabidopsis mutants are telling us. Ann Bot 2003; 91: 605-612.
[16] Vandenbussche F, Van Der Straeten D. One for all and all for one: Cross-talk of multiple signals controlling the plant phenotype. J Plant Growth Regul 2007; 26: 178-187.
[17] Coombe BG. Development of fleshy fruits. Ann Rev Plant Physiol Plant Mol Biol 1976; 27: 207-228.
[18] Tucker GA. Introduction. In: Seymour GB, Taylor, JE, Tucker GA, Eds. Biochemistry of fruit ripening. Chapman and Hall, London 1993; pp. 1-51.
[19] Seymour GB. Banana. In: Seymour GB, Taylor, JE, Tucker GA, Eds. Biochemistry of fruit ripening. Chapman and Hall, London 1993; pp. 83-106.
[20] Bower J, Holford P, Latché A, Pech JC. Culture conditions and detachment of the fruit influence the effect of ethylene on the climacteric respiration of melon. Postharvest Biol Technol 2002; 26: 135-146.
[21] Coombe BG, Hale CR. Hormone content of ripening grape berries and effects of growth substance treatments. Plant Physiol 1973; 51: 629-634.
[22] Inaba A, Ishida M, Sobajima Y. Changes in endogenous hormone concentrations during berry development in relation to ripening of Delaware grapes. J Japan Soc Hort Sci 1976; 45: 245-252.
[23] Weaver RJ, Singh IS. Occurrence of endogenous ethylene and effect of plant growth regulators on ethylene production in grapevine. Am J Enol Vitic 1978; 29: 282-285.
[24] Lin Z, Zhong S, Grierson D. Recent advances in ethylene research. J Exp Bot 2009; 60: 3311-3336.
[25] Alleweldt G, Koch R. Ethylene content in ripening grape berries. Vitis 1977; 16: 263-271.
[26] Düring H, Alleweldt G, Koch R. Studies on hormonal control of ripening in berries and grape vines. Acta Hort 1978; 80: 397-405.
[27] Chervin C, El-Kereamy A, Roustan JP, Latché A, Lamon J, Bouzayen M. Ethylene seems required for the berry development and ripening in grape, a non-climacteric fruit. Plant Sci 2004; 167: 1301-1305.
[28] Zhang M, Leng P, Zhang G, Li X. Cloning and functional analysis of 9-*cis*-epoxycarotenoid dioxygenase (NCED) genes encoding a key enzyme during abscisic acid biosynthesis from peach and grape fruits. J Plant Physiol 2009; 166: 1241-1252.
[29] Lin SF, Walsh CS. Studies of the "tree factor" and its role in the maturation and ripening of 'Gala' and 'Fuji' apples. Postharvest Biol Technol 2008; 48: 99-106.
[30] Saltveit ME. Internal carbon-dioxide and ethylene levels in ripening tomato fruit attached to or detached from the plant. Physiol Plant 1993; 89: 204-210.
[31] Iannetta PPM, Laarhoven LJ, Medina-Escobar N, *et al.* Ethylene and carbon dioxide production by developing strawberries show a correlative pattern that is indicative of ripening climacteric fruit. Physiol Plant 2006; 127: 247-259.
[32] Cazzonelli CI, Cavallaro AS, Botella JR. Cloning and characterisation of ripening-induced ethylene biosynthetic genes from non-climacteric pineapple (*Ananas comosus*) fruits. Aust J Plant Physiol 1998; 25: 513-518.
[33] Wang HC, Huang HB, Huang XM. Differential effects of abscisic acid and ethylene on the fruit maturation of *Litchi chinensis* Sonn. Plant Growth Regul 2007; 52: 189-198.
[34] Da Silva FG, Iandolino A, Al-Kayal F, *et al.* Analysis of expressed sequence tags from multiple Vitis species and development of a compendium of gene expression during berry development. Plant Physiol 2005; 139: 574-597.
[35] Coombe BG. Adoption of a system for identifying grapevine growth stages. Aust J Grape Wine Res 1995; 1: 100-110.
[36] Grimplet J, Deluc LG, Tillett RL, *et al.* Tissue-specific mRNA expression profiling in grape berry tissues. BMC Genomics [serial on the Internet]. 2007 Jun 21 Available from: http://www.biomedcentral.com/1471-2164/8/187
[37] Nambara E, Marion-Poll A. Abscisic acid biosynthesis and catabolism. Annu Rev Plant Biol 2005; 56: 165-185.
[38] Soar CJ, Speirs J, Maffei SM, Loveys BR. Gradients in stomatal conductance, xylem sap ABA and bulk leaf ABA along canes of *Vitis vinifera* cv. Shiraz: molecular and physiological studies investigating their source. Funct Plant Biol 2004; 31: 659-669.
[39] Finkelstein RR, Rock CD. Abscisic Acid Biosynthesis and Response. In: Sommerville CR, Meyerowitz EM. Eds. The Arabidopsis Book, American Society of Plant Biologists, Rockville, MD 2002; pp. 1-48.
[40] Davies C, Boss PK, Robinson SP. Treatment of grape berries, a nonclimacteric fruit with a synthetic auxin, retards ripening and alters the expression of developmentally regulated genes. Plant Physiol 1997; 115: 1155-1161.

[41] Owen SJ, Lafond MD, Bowen P, Bogdanoff C, Usher K, Abrams SR. Profiles of abscisic acid and its catabolites in developing merlot grape (*Vitis vinfera*) berries. Am J Enol Vitic 2009; 60: 227-284.

[42] Wheeler S, Loveys B, Ford C, Davies C. The relationship between the expression of ABA biosynthesis genes, accumulation of ABA and the promotion of *Vitis vinifera* L. berry ripening by ABA. Aust J Grape Wine Res 2009; 15: 195-204.

[43] Hale CR, Coombe BG. Abscisic acid: an effect on the onset of ripening of grapes (*Vitis vinifera* L.). Royal Soc N Z Bull 1974; 12: 831-836.

[44] Cawthon DL, Morris JR. Relationship of seed number and maturity to berry development, fruit maturation, hormonal changes, and uneven ripening of Concord (*Vitis-labrusca* L.) grapes. J Am Soc Hort Sci 1982; 107: 1097-1104.

[45] Kataoka I, Sugiura A, Utsunomiya N, Tomana T. Effect of abscisic-acid and defoliation on anthocyanin accumulation in Kyoho grapes (*Vitis-vinifera* L. X *labruscana* BAILEY). Vitis 1982; 21: 325-332.

[46] Okamoto G, Kuwamura T, Hirano K. Effects of water deficit stress on leaf and berry ABA and berry ripening in Chardonnay grapevines (*Vitis vinifera*). Vitis 2004; 43: 15-17.

[47] Baydar NG, Harmankaya N. Changes in endogenous hormone levels during the ripening of grape cultivars having different berry set mechanisms. Turk J Agric For 2005; 29: 205-210.

[48] Deytieux-Belleau C, Gagne S, L'Hyvernay A, Donèche B, Geny L. Possible roles of both abscisic acid and indol-acetic acid in controlling grape berry ripening process. J Int Sci Vigne Vin 2007; 41: 141-148.

[49] Lacampagne S, Gagné S, Gény L. Involvement of abscisic acid in controlling the proanthocyanin biosynthesis pathway in grape skin: new elements regarding the regulation of tannin composition and leucoanthocyanidin reductase (LAR) and anthocyanidin reductase (NAR) activities and expression. J Plant Growth Regul 2010; 29: 81-90.

[50] Scienza A, Miravalle R, Visai C, Fregoni M. Relationships between seed number, gibberellin and abscisic-acid levels and ripening in Cabernet Sauvignon grape berries. Vitis 1978; 17: 361-368.

[51] Kondo S, Kawai M. Relationship between free and conjugated ABA levels in seeded and gibberellin-treated seedless, maturing 'Pione' grape berries. J Am Soc Hort Sci 1998; 123: 750-754.

[52] Zhang XR, Luo GG, Wang RH, Wang J, Himelrick DG. Growth and developmental responses of seeded and seedless grape berries to shoot girdling. J Am Soc Hort Sci 2003; 128: 316-323.

[53] Castellarin SD, Pfeiffer A, Sivilotti P, Degan M, Peterlunger E, Di Gaspero G. Transcriptional regulation of anthocyanin biosynthesis in ripening fruits of grapevine under seasonal water deficit. Plant Cell Environ 2007; 30: 1381-1399.

[54] Cramer GR, Ergül A, Grimplet J, *et al.* Water and salinity stress in grapevines: early and late changes in transcript and metabolite levels. Funct Integr Genomics 2007; 7: 111-134.

[55] Lee KH, Piao HL, Kim HY, *et al.* Activation of glucosidase *via* stress-induced polymerization rapidly increases active pools of abscisic acid. Cell 2006; 126: 1109-1120.

[56] Koussa T, Colin L, Broquedis M. Endogenous levels of abscisic acid in various organs of *Vitis vinifera* L. (cv. Cabernet Sauvignon) between flower buds separated and grape closed stages. J Int Sci Vigne Vin 2004; 38: 141-146.

[57] Cutler AJ, Krochko JE. Formation and breakdown of ABA. Trends Plant Sci 1999; 4: 472-478.

[58] Shiozaki S, Kamata Y, Ogata T, Horiuchi S, Kawase K. Localization of abscisic acid in grape berry by immunohistochemical techniques. J Japan Soc Hort Sci 1999; 68: 1-9.

[59] Zhang DP, Zhang ZL, Chen J, Jia WS. Specific abscisic acid-binding sites in mesocarp of grape berry: properties and subcellular localization. J Plant Physiol 1999; 155: 324-331.

[60] Haubrick LL, Assmann SM. Brassinosteroids and plant function: some clues, more puzzles. Plant Cell Environ 2006; 29: 446-457.

[61] Symons GM, Davies C, Shavrukov Y, Dry IB, Reid JB, Thomas MR. Grapes on steroids. Brassinosteroids are involved in grape berry ripening. Plant Physiol 2006; 140: 150-158.

[62] Wang ZY, Seto H, Fujioka S, Yoshida S, Chory J. BRI1 is a critical component of a plasma-membrane receptor for plant steroids. Nature 2001; 410: 380-383.

[63] Nomura T, Sato T, Bishop GJ, Kamiya Y, Takatsuto S, Yokota T. Accumulation of 6-deoxocathasterone and 6-deoxocastasterone in Arabidopsis, pea and tomato is suggestive of common rate-limiting steps in brassinosteroid biosynthesis. Phytochemistry 2001; 57: 171-178.

[64] Bancoş S, Nomura T, Sato T, *et al.* Regulation of transcript levels of the Arabidopsis cytochrome P450 genes involved in brassinosteroid biosynthesis. Plant Physiol 2002; 130: 504-513.

[65] Goda H, Shimada Y, Asami T, Fujioka S, Yoshida S. Microarray analysis of brassinosteroid-regulated genes in Arabidopsis. Plant Physiol 2002; 130: 1319-1334.

[66] Olszewski N, Sun T, Gubler F. Gibberellin signalling: biosynthesis, catabolism, and response pathways. Plant Cell 2002; 14: S61-S80.

[67] Pérez R, Espinoza A, Aquirre C, *et al.* Functional genomic analysis of table grape (*Vitis vinifera* L.) associated to gibberellic acid treatment. Acta Hort 2009; 827: 363-368.

[68] Pérez FJ, Viani C, Retamales J. Bioactive gibberellins in seeded and seedless grapes: identification and changes in content during berry development. Am J Enol Vitic 2000; 51: 315-318.

[69] Böttcher C, Keyzers RA, Boss PK, Davies C. Sequestration of auxin by the indole-3-acetic acid-amido synthetase GH3-1 in grape berry (*Vitis vinifera* L.) and the proposed role of auxin conjugation during ripening. J Exp Bot 2010; 61: 3615-3625.

[70] Woodward AW, Bartel B. Auxin: regulation, action, and interaction. Ann Bot 2005; 95: 707-735.

[71] Delker C, Raschke A, Quint M. Auxin dynamics: the dazzling complexity of a small molecule's message. Planta 2008; 227: 929-941.

[72] Purgatto E, Nascimento JRO, Lajolo FM, Cordenunsi BR. The onset of starch degradation during banana ripening is concomitant to changes in the content of free and conjugated forms of indole-3-acetic acid. J Plant Physiol 2002; 159: 1105-1111.

[73] Dunlap JR, Slovin JP, Cohen JD. Indole-3-acetic acid, ethylene, and abscisic acid metabolism in developing muskmelon (*Cucumis melo* L.) fruit. Plant Growth Regul 1996; 19: 45-54.

[74] Archbold DD, Dennis FG. Quantification of free and conjugated IAA in strawberry achene and receptacle tissue during fruit development. J Am Soc Hortic Sci 1984; 109: 330-335.

[75] Werner T, Schmülling T. Cytokinin action in plant development. Curr Opin Plant Biol 2009; 12: 527-538.

[76] Bari B, Jones JDG. Role of plant hormones in plant defence responses. Plant Mol Biol 2009; 69: 473-488.

[77] Kondo S, Fukuda K. Changes in jasmonates in grape berries and their possible role in fruit development. Sci Horti 2001; 91: 275-288.

[78] Avanci NC, Luche DD, Goldman GH, Goldman MHS. Jasmonates are phytohormones with multiple functions, including plant defense and reproduction. Genet Mol Res 2010; 9: 484-505.

[79] Kevany BM, Tieman DM, Taylor MG, Dal Cin V, Klee H. Ethylene receptor degradation controls the timing of ripening in tomato fruit. Plant J 2007; 51; 458-467.

[80] Trainotti L, Pavanello A, Casadoro G. Different ethylene receptors show an increased expression during the ripening of strawberries: does such an increment imply a role for ethylene in the ripening of these non-climacteric fruits? J Exp Bot 2005; 56: 2037-2046.

[81] Fernandez L, Torregrosa L, Terrier N, *et al.* Identification of genes associated with flesh morphogenesis during grapevine fruit development. Plant Mol Biol 2007; 63, 307-323.

[82] Tattersall EAR, Grimplet J, Deluc L, *et al.* Transcript abundance profiles reveal larger and more complex responses of grapevine to chilling compared to osmotic and salinity stress. Funct Integr Genomics 2007; 7: 317-333.

[83] Bellin D, Ferrarini A, Chimento A, *et al.* Combining next-generation pyrosequencing with microarray for large scale expression analysis in non-model species. BMC Genomics [serial on the Internet]. 2009 Nov 24. Available from: http://www.biomedcentral.com/1471-2164/10/555

[84] Gambetta GA, Matthews MA, Shaghasi TH, McElrone AJ, Castellarin SD. Sugar and abscisic acid signalling orthologs are activated at the onset of ripening in grape. Planta 2010; 232: 219-234.

[85] Chervin C, Tira-umphon A, Terrier N, Zouine M, Severac D, Roustan J-P. Stimulation of the grape berry expansion by ethylene and effects on related gene transcripts, over the ripening phase. Physiol Plant 2008; 134: 534-546.

[86] Koyama K, Sadamatsu K, Goto-Yamamaoto N. Abscisic acid stimulated ripening and gene expression in berry skins of the Cabernet Sauvignon grape. Funct Integr Genomics 2009; in press, doi:10.1007/s10142-009-0145-8.

[87] Negri AS, Prinsi B, Rossoni M, *et al.* Proteome changes in the skin of the grape cultivar Barbera among different stages of ripening. BMC Genomics [serial on the Internet]. 2008 Aug 8. Available from: http://www.biomedcentral.com/1471-2164/9/378

[88] Szyjewicz E, Rosner N, Kliewer WM. Ethephon ((2-Chloroethyl)phosphonic acid, Ethrel, CEPA) in viticulture-a review. Am J Enol Vitic 1984; 35: 117-123.

[89] Roubelakis-Angelakis KA, Kliewer WM. Effects of exogenous factors on phenylalanine ammonia-lyase activity and accumulation of anthocyanins and total phenolics in grape berries. Am J Enol Vitic 1986; 37: 275-280.

[90] El-Kereamy A, Chervin C, Roustan JP, *et al.* Exogenous ethylene stimulates the long-term expression of genes related to anthocyanin biosynthesis in grape berries. Physiol Plant 2003; 119: 175-182.

[91] Chervin C, Tira-Umphon A, Chatelet P, Jauneau A, Boss PK, Tesniere C. Ethylene and other stimuli affect expression of the UDP glucose-flavonoid 3-O-glycosyltransferase in a non-climacteric fruit. Vitis 2009; 48: 11-16.

[92] Hale CR, Coombe BG, Hawker JS. Effects of ethylene and 2-Chloroethylphosphonic acid on ripening of grapes. Plant Physiol. 1970; 45: 620-623.

[93] Katz E, Lagunes PM, Riov J, Weiss D, Goldschmidt EE. Molecular and physiological evidence suggests the existence of a system II-like pathway of ethylene production in non-climacteric Citrus fruit. Planta 2004; 219: 243-252.

[94] Tira-Umphon A, Roustan JP, Chervin C. The stimulation by ethylene of the UDP glucose-flavonoid 3-*O*-glucosyltransferase (UFGT) in grape tissues is independent from the MybA transcription factors. Vitis 2007; 46: 210-211.

[95] Huber DJ. Suppression of ethylene responses through application of 1-methylcyclopropene: a powerful tool for elucidating ripening and senescence mechanisms in climacteric and nonclimacteric fruits and vegetables. HortScience 2008; 43: 106-111.

[96] Davies C, Wolf T, Robinson SP. Three putative sucrose transporters are differentially expressed in grapevine tissues. Plant Sci 1999; 147: 93-100.

[97] Manning K, Davies C, Bowen HC, White PJ. Functional characterization of two ripening-related sucrose transporters from grape berries. Ann Bot 2001; 87: 125-129.

[98] Chervin C, Terrier N, Ageorges A, Ribes F, Kuapunyakoon T. Influence of ethylene on sucrose accumulation in grape berry. Am J Enol Vitic 2006; 57: 511-513.

[99] Verries C, Pradal M, Chatelet P, Torregrosa L, Tesniere C. Isolation and analysis of the promoter of VvAdh2, a grapevine (*Vitis vinifera* L.) ripening-related gene. Plant Sci 2004; 167: 1067-1074.

[100] Tesniere C, Pradal M, El-Kereamy A, *et al.* Involvement of ethylene signalling in a non-climacteric fruit: new elements regarding the regulation of ADH expression in grapevine. J Exp Bot. 2004; 55: 2235-2240.

[101] Matsushima J, Hiratsuka S, Taniguchi N, Wada R, Suzaki N. Anthocyanin accumulation and sugar content in the skin of grape cultivar Olympia treated with ABA. J Japan Soc Hort Sci 1989; 58: 551-555.

[102] Pirie A, Mullins MG. Changes in anthocyanin and phenolics content of grapevine leaf and fruit tissues treated with sucrose, nitrate, and abscisic acid. Plant Physiol 1976; 58: 468-472.

[103] Yakushiji H, Morinaga K, Kobayashi S. Promotion of berry ripening by 2,3,5-triiodobenzoic acid in 'Kyoho' grapes. J Japan Soc Hort Sci 2001; 70: 185-190.

[104] Delgado R, Martin P, del Álamo M, González MR. Changes in the phenolic composition of grape berries during ripening in relation to vineyard nitrogen and potassium fertilisation rates. J Sci Food Agric 2004; 84: 623-630.

[105] Mori K, Sugaya S, Gemma H. Decreased anthocyanin biosynthesis in grape berries grown under elevated night temperature condition. Sci Hortic 2005; 105: 319-330.

[106] Peppi MC, Fidelibus MW, Dokoozlian N. Abscisic acid application timing and concentration affect firmness, pigmentation, and color of 'flame seedless' grapes. HortScience 2006; 41: 1440-1445.

[107] Cantín CM, Fidelibus MW, Crisosto CH. Application of abscisic acid (ABA) at veraison advanced red color development and maintained postharvest quality of 'Crimson Seedless' grapes. Postharvest Biol Technol 2007; 46: 237-241.

[108] Peppi MC, Fidelibus MW, Dokoozlian NK. Application timing and concentration of abscisic acid affect the quality of 'Redglobe' grapes. J. Hortic. Sci. Biotechnol 2007; 82: 304-310.

[109] Peppi MC, Fidelibus MW. Effects of forchlorfenuron and abscisic acid on the quality of 'Flame Seedless' grapes. HortScience 2008; 43: 173-176.

[110] Amiri ME, Fallahi E, Mirjalili M. Effects of abscisic acid or ethephon at veraison on the maturity and quality of 'Beidaneh Ghermez' grapes. J Hortic Sci Biotechnol 2009; 84: 660-664.

[111] Laurie S, Ovadia R, Nissim-Levi A, *et al.* Abscisic acid improves colour development in 'Crimson Seedless" grapes in the vineyard and on detached berries. J Hortic Sci Biotechnol 2009; 84: 639-644.

[112] Kataoka I, Kubo Y, Tomana T. Effects of temperature, cluster shading and some growth regulators on L-phenylalanine ammonia-lyase activity and anthocyanin accumulation in black grapes. Memoirs of the College of Agriculture, Kyoto University 1984; 124: 35-44.

[113] Hiratsuka S, Onodera H, Kawai Y, Tatsuya K, Itoh H, Wada R. Enzyme activity changes during anthocyanin synthesis in 'Olympia' grape berries. Sci Hortic 2001; 90: 255–264.

[114] Ban T, Ishimaru M, Kobayashi S, Shiozaki S, Goto-Yamamoto N, Horiuchi S. Abscisic acid and 2,4-dichlorophenoxyacetic acid affect the expression of anthocyanin biosynthetic pathway genes in 'Kyoho' grape berries. J Hortic Sci Biotechnol 2003; 78: 586-589.

[115] Jeong ST, Goto-Yamamoto N, Kobayashi S, Esaka A. Effects of plant hormones and shading on the accumulation of anthocyanins and the expression of anthocyanin biosynthetic genes in grape berry skins. Plant Sci 2004; 167: 247-252.

[116] Fujita A, Goto-Yamamoto N, Aramaki I, Hashizume K. Organ-specific transcription of putative flavonol synthase genes of grapevine and effects of plant hormones and shading on flavonol biosynthesis in grape berry skins. Biosci Biotechnol Biochem 2006; 70: 632-638.

[117] Pan QH, Li MJ, Peng CC, *et al.* Abscisic acid activates acid invertases in developing grape berry. Physiol Plant 2005; 125: 157-170.

[118] Han DH, Lee CH. The effects of GA_3, CPPU and ABA application on the quality of Kyoho (*Vitis vinifera* L. x *V. labrusca* L.) grape. Acta Hort 2004; 640: 193–197.

[119] Sato A, Yamada H, Iwanami H, Mitani N. Quantitative and instrumental measurements of grape flesh texture as affected by gibberellic acid application. J Jap Soc Hort Sci 2004; 73: 7-11.

[120] Teszlák P, Gaál K, Shahin M, Nikfardjam MSP. Influence of grapevine flower treatment with gibberellic acid (GA_3) on polyphenol content of *Vitis vinifera* L wine. Anal Chim Acta 2005; 543: 275-281.

[121] Reynolds AG, Roller JN, Forgione A, De Savigny C. Gibberellic acid and basal leaf removal: implications for fruit maturity, vestigial seed development, and sensory attributes of Sovereign Coronation table grapes. Am J Enol Vitic 2006; 57: 41-53.

[122] Dreier LP, Hunter JL, Ruffner HP. Invertase activity, grape berry development and cell compartmentation. Plant Physiol Biochem 1998; 36: 865-872.

[123] Pérez J, Gómez M. Possible role of soluble invertase in the gibberellic acid berry-sizing effect in Sultana grape. Plant Growth Regul 2000; 30: 111-116.

[124] Boutté Y, Ikeda Y, Grebe M. Mechanisms of auxin-dependent cell and tissue polarity. Curr Opin Plant Biol 2007; 10: 616-623.

[125] Cheng YF, Dai XH, Zhao YD. Auxin synthesized by the YUCCA flavin monooxygenases is essential for embryogenesis and leaf formation in Arabidopsis. Plant Cell 2007; 19: 2430-2439.

[126] De Smet I, Jurgens G. Patterning the axis in plants-auxin in control. Curr Opin Genet Dev 2007; 17: 337-343.

[127] Zhao Y. Auxin biosynthesis and its role in plant development. Annu Rev Plant Biol 2010; 61: 19.1-19.6.

[128] Cohen JD. *In vitro* tomato fruit cultures demonstrate a role for indole-3-acetic acid in regulating fruit ripening. J Am Soc Hort Sci 1996; 121, 520-524.

[129] Given NK, Venis MA, Grierson D. Hormonal regulation of ripening in the strawberry, a non-climacteric fruit. Planta 1988; 174: 402-406.

[130] Weaver RJ. The effect of benzo-thiazole-2-oxyacetic acid on maturation of seeded varieties of grape. Am J Enol Vitic 1962; 29: 282-285.

[131] Hale CR. Growth and senescence of grape berry. Aust J Agric Res 1968; 19: 939-945.

[132] Kim CC. Effects of plant growth regulators on the morphological change of grapevine and maturation of grape berries. J Kor Soc Hort Sci 1991; 32: 199-205.

[133] Kim IL, Piao Y-L, Hwang YS, Lee JC. Effects of synthetic cytokinin, thidiazuron on berry size and quality of 'Campbell Early" (*Vitis labruscana*) grapes. J Kor Soc Hort Sci 2002; 43: 457-462.

[134] Reynolds AG, Wardle DA, Zurowski C, Looney NE. Phenylureas CPPU and thidiazuron affect yield components, fruit composition, and storage potential of 4 seedless grape selections. J Am Soc Hort Sci 1992; 117: 85-89.

[135] Retamales J, Bangerth F, Cooper T, Callejas R. Effects of CPPU and GA_3 on fruit quality of Sultanina table grape. Acta Hort 1995; 394: 149-157.

[136] Ezzahouani A. Effects of forchlorfenuron (CPPU) and girdling on table grape cultivars 'Perlette' and 'Italia'. J Int Sci Vigne Vin 2000; 34: 57-60.

[137] Zabadal TJ, Bukovac MJ. Effect of CPPU on fruit development of selected seedless and seeded grape cultivars. HortScience 2006; 41: 154-157.

[138] Kraeva E, Andary C, Carbonneau A, Deloire A. Salicylic acid treatment of grape berries retards ripening. Vitis 1998; 37: 143-144.

[139] Wen PF, Chen JY, Kong WF, Pan QH, Wan SB, Huang WD. Salicylic acid induced the expression of phenylalanine ammonia-lyase gene in grape berry. Plant Sci 2005; 169: 928-934.

[140] Larronde F, Gaudillère JO, Krisa S, Decendit A, Deffieux D, Mérillon JM. Airborne methyl jasmonate induces stilbene accumulation in leaves and berries of grapevine plants. Am J Enol Vitic 2003; 54: 63-66.

[141] Krisa S, Larronde F, Budzinski H, Decendit A, Deffieux G, Mérillon J-M. Stilbene production by Vitis vinifera cell suspension cultures: methyl jasmonate production and ^{13}C biolabeling. J Nat Prod 1999; 62: 1688-1690.

[142] Belhadj A, Telef N, Saigne C, *et al.* Effect of methyl jasmonate in combination with carbohydrates on gene expression of PR proteins, stilbene and anthocyanin accumulation in grapevine cell cultures Plant Physiol Biochem 2008; 46: 493-499.

[143] D'Onofrio C, Cox A, Davies C, Boss PK. Induction of secondary metabolism in grape cell cultures by jasmonates and the identification of induced genes using microarrays. Aust J Grape Wine Res 2009; 36: 323-338.

[144] Martinez-Esteso MJ, Sellés-Marchart S, Vera-Urbina JC, Pedreño MA, Bru-Martinez R. Changes of defense proteins in the extracellular proteome of grapevine (*Vitis vinifera* cv. Gamay) cell cultures in response to elicitors. J. Proteomics 2009; 73: 331-341.

[145] Toumi I, Moschou PN, Paschalidis KA, Bouamama B, *et al.* Abscisic acid signals reorientation of polyamine metabolism to orchestrate stress responses *via* the polyamine exodus pathway in grapevine. J. Plant Physiol 2009; 167: 519-525.

[146] Rolland F, Moore B, Sheen J. Sugar sensing and signaling in plants. Plant Cell 2002; 14: S185-S205.

[147] Rook F, Hadingham SA, Li Y, Bevan MW. Sugar and ABA response pathways and the control of gene expression. Plant Cell Environ. 2006; 29: 426-434.

[148] Çakir B, Agasse A, Gaillard C, Saumonneau A, Delrot S, Atanassova R. A grape ASR protein involved in sugar and abscisic acid signaling. Plant Cell 2003; 15: 2165-2180.

CHAPTER 11

Transcriptomics and Metabolomics for the Analysis of Grape Berry Development

G. B. Tornielli, A. Zamboni, S. Zenoni, M. Delledonne and M. Pezzotti*

Department of Biotechnology, University of Verona, 37134 Verona, Italy

Abstract: The investigation of key molecular events underlying grape berry development can provide useful data for the improvement of mature berry quality traits, a goal in this commercially important fruit crop. Berry development has been investigated at the transcriptional level to study global expression profiles during formation and ripening, while a smaller number of large-scale metabolite studies have been carried out till now. Here we present a meta-analysis of the "transcriptional story" of grape berry development by analysing the data from all transcriptional studies published to date using different analysis methods in diverse cultivars.

Keywords: Berry development, Cell wall metabolism, EST analysis, Expression profiles, Gene expression, Hormone biosynthesis, Metabolome, Microarray analysis, Primary metabolism, Secondary metabolism, Transcriptome.

INTRODUCTION

Grapevine is a commercially important fruit crop whose value depends to a large extent on the quality of its berries. Qualitative differences in berry characteristics impart unique organoleptic properties to the major grapevine products: fresh and dried fruit, juice and wine. Grape berries undergo a long developmental process to reach their ripened state, which involves a series of dramatic physical and biochemical changes. Much remains to be learned about the mechanisms underlying these profound events, but with the advent of molecular analysis tools there has been significant progress over the last two decades.

The ability to analyse transcriptional changes in berries has yielded large amounts of data, beginning with the detection and quantification of single transcripts but now encompasses techniques that allow thousands of transcripts to be monitored simultaneously. Furthermore, the availability of the complete grapevine genome sequence [1, 2] allows for comprehensive transcriptomic experiments. Therefore, it is now much easier to identify candidate genes governing the key processes underlying berry development, particularly those relating to quality traits. Such candidate genes represent a valuable pool of material for functional analyses, and painstaking research is still required to confirm genomic annotations and avoid annotation errors that would otherwise result in the misinterpretation of transcriptomic data.

Large-scale methods for the analysis of gene expression evolved from genomic sequencing by the Sanger method, and involved the census sequencing of cDNA libraries to generate collections of expressed sequence tags (ESTs) or short markers that were joined together in techniques such as serial analysis of gene expression (SAGE). For a period of about 10 years, sequence-based methods were largely replaced with microarrays, which allowed large numbers of transcripts to be quantified by hybridisation. In the last few years, the ultra-high-throughput potential of next-generation sequencing methods has allowed massive cDNA census sequencing technologies (collectively RNAseq) to become more and more prevalent, and as the cost of sequencing continues to fall it is likely to become the dominant approach once more.

The analysis of grape berry metabolism has undergone a similar evolutionary process, although the technical challenges have taken longer to overcome. From early studies based on chromatography and the

***Address correspondence to M. Pezzotti:** Department of Biotechnology, University of Verona, 37134 Verona, Italy; E-mail: mario.pezzotti@univr.it

Hernâni Gerós, M. Manuela Chaves and Serge Delrot (Eds)

detection of one or more target metabolites, it is now possible with mass spectrometry to analyse and quantify thousands of metabolites simultaneously. The large-scale analysis of metabolites (metabolomics) allows for the investigation of changing metabolic profiles during berry development and also allows the quality traits of different cultivars to be explained in terms of the organic molecules they produce. Although this approach lags somewhat behind the more established transcriptomic platforms, the first detailed metabolomic studies of berry development have recently been published.

We will gain the most complete picture of berry development by combining the data emerging from these novel high-throughput studies and extracting the common principles, as well as identifying cultivar-specific differences.

ANALYSIS OF BERRY DEVELOPMENT DURING THE TECHNOLOGICAL EVOLUTION OF TRANSCRIPTOPMICS AND METABOLOMICS

The transcriptome is the complete set of RNA molecules produced by a cell, tissue or organism. It includes mRNA, rRNA, tRNA and other non-coding RNAs, although in many cases the mRNA profile is the most sought after because it corresponds to the expression of protein-encoding genes. The transcriptome depends on gene expression and therefore changes qualitatively and quantitatively according to cell type, developmental stage and in response to external conditions or physiological states. The dynamic nature of the transcriptome is highly informative, and in the case of berry development it would indicate which genes are expressed during early development, veraison and ripening, and even during post-harvesting processes such as withering.

Transcriptomic methods are based on one of two principles, both of which allow for a highly parallel analysis of thousands of transcripts:

1) The digital analysis of sequence data by counting the number of times a particular sequence is represented in a sample (also known as census sequencing);

2) The quantitative analysis of hybridisation data.

Regardless of approach, the first and principal limiting step in grapevine transcriptomics has been RNA extraction from tissues often rich in phenolic compounds and polysaccharides. Initially this required time-consuming RNA extraction protocols based on sodium perchlorate [3, 4], guanidine slats [5, 6] or the hot borate method [7], but these have been largely supplanted by the availability of commercial kits such as RNeasy Plant MidiKit (Qiagen) and Spectrum^{TM} Plant Total RNA kit (Sigma) that allow for the rapid preparation of high-quality total RNA.

The first studies of gene expression in the grapevine were performed using low-throughput Sanger sequencing approaches (Table **1**), which were expensive and often did not provide quantitative data [8]. For example, Davies and Robinson [9] carried out the differential screening of a Shiraz post-veraison berry cDNA library using probes produced from pre-veraison and post-veraison RNA samples, and Ablet *et al.* [10] analysed 2,479 ESTs from a pre-veraison Chardonnay cDNA library. Terrier *et al.* [11] analysed another Shiraz library comprising 275 ESTs produced from three developmental stages (green, softening and ripening berries). A collection of 1,743 ESTs Pinot Noir berries from the veraison phase was screened by Moser *et al.* [12].

During the same period, a gel-based cDNA-AFLP (amplified fragment length polymorphism) analysis method was developed to characterise gene expression in the absence of prior sequence information, and this was used to profile transcripts in Chardonnay berries at six developmental stages [13]. The same technique was used to find differences between immature and mature Clairette Blanche and Cabernet Sauvignon berries [14] and to look at gene expression during Corvina berry development and withering on and off the vine [15].

As the quantity and availability of grapevine sequence information increased, it became possible to carry out gene expression analyses *in silico* [16, 17]. Furthermore, it became possible to produce cDNA-based microarrays and oligonucleotide arrays based on these sequences, leading to hybridisation-based gene expression profiling in grapevine berry development.

The first large-scale hybridisation experiments used microarrays prepared from the sequences in cDNA libraries. Terrier *et al.* [18] analysed changes in gene expression in Shiraz berries during nine developmental stages using an array of 50-mers representing 3,175 grapevine unigenes. The same array was later used to analyse changes in gene expression in Chardonnay berries as they softened during ripening [19]. A cDNA array comprising 4,608 Shiraz berry clones was used to monitor changes in gene expression in the berry skin during ripening [20]. Finally, a cDNA array containing 4,224 clones from a subtraction library was used to investigate differences in gene expression between a pool of 11 varieties producing high levels of resveratrol and 10 varieties producing low levels of the compound. The varieties were compared at veraison, ripening and post-ripening phases [21].

In 2004, the first commercial high-density oligonucleotide arrays became available for the grapevine in the form of the Affymerix GeneChip® *Vitis vinifera* Genome Array. This was developed by selecting sequences from GenBank®, dbEST and RefSeq, and using sequence clusters created from the UniGene database (build 7, Oct 2003). With this platform, it was possible to monitor the expression of 14,000 *V. vinifera* transcripts and 1,700 transcripts from other *Vitis* species. Each transcript was represented by a probe set of 16 overlapping 25-mers including perfect match and mismatch sequences. The use of this platform was supported by the development of the Affymetrix Mas 5.0 algorithm and particularly by the development of bioconductor packages for RMA [22] and GCRMA methods for R languages [23] as useful tools for processing raw data.

Several reports were published with transcriptome data obtained using the Gene Chip® *Vitis vinifera* Genome Array (Table **1**). These included studies of tissue-specific expression in mature Cabernet Sauvignon berries [24] and developmental profiles of Cabernet Sauvignon [25] and Pinot Noir berries [26]. The array was also used to correlate gene expression and anthocyanin accumulation in response to heat in Cabernet Sauvignon berries [27], to compare developmental profiles in different Cabernet Sauvignon berries sampled at the same time [28], to profile Chardonnay and Cabernet Sauvignon berry development under water stress [29], and to investigate transcriptional changes in ripening Cabernet Sauvignon berry skin in response to abscisic acid treatment [30].

Another oligonucleotide array, the *Vitis vinifera* AROSE V1.0, was developed by Operon Biotechnology using sequence information from the Institute for Genomic Research (TIGR) Grape Gene Index Release 3.0 (August 13, 2003). The array contained 70-mer probes, allowing 14,562 transcripts to be monitored. This device was used to identify genes that were differentially expressed in Cabernet Sauvignon berries in response to ethylene treatment [31] and to analyse the gene expression profiles of Raboso Piave berry skin at different stages of weight loss in response to various postharvest dehydration rates [32].

When the complete grapevine genome sequence was published [1, 2] it finally became possible to profile grape berry developmental gene expression at the whole-genome level. The Centre for Plant Functional Genomics at the University of Verona used the CombiMatrix electrochemical oligonucleotide array assembly platform to create GrapeArray 1.2, consisting of 35–40-mer probes representing 25,471 transcripts based on TIGR Grape Gene Index (Release 5.0) and the 8.4-fold draft grapevine genome sequence [1]. GrapeArray 1.2 was initially used to profile gene expression during Corvina berry development and postharvest withering [33] but this study uniquely combined the transcriptome data set with proteome and metabolome data sets from the same berry samples. The combined data were used to identify putative stage-specific biomarkers and to describe development and withering using both hypothesis-free and hypothesis-driven systems biology approaches. A more advanced arrays based on Combimatrix (GrapeArray3.0) and NimbleGen (090918_Vitus_exp_HX12) technologies have since been developed using the 12.0-fold assembly gene prediction V.1 (http://ddlab.sci.univr.it/FunctionalGenomics/). The NimbleGen 12x135K array enables the hybridisation of up to 12 independent samples on a single slide, sensibly reducing the costs of the experiment. Each of the 12 sub-arrays contains 135,000 60-mer oligonucleotide probes and allows to detect the expression of 29,549 grapevine transcripts.

The recent technological advances in next-generation sequencing methods [34, 35] have been applied to transcriptomics, an approach known as RNA-seq [36]. In the case of the grapevine, more than 59 million cDNA sequence reads, each 36–44 bp in length, have been mapped onto the 8.4-fold draft sequence of the Pinot Noir 40024 reference genome [1], allowing gene expression to be analysed not only in terms of quantitative levels of mRNA, but also alternative splicing and expressed single nucleotide polymorphisms. A genome-scale transcriptional map has also been generated for Corvina berry development [37].

As well as identifying transcripts corresponding to protein-encoding genes, bioinformatic analysis of the grapevine genome sequence predicted the existence of 164 conserved micro-RNAs (miRNAs), grouped into several families. Micro-RNAs are small non-coding RNAs with an increasingly recognised role in gene regulation. The first attempt to validate the candidate grapevine miRNAs experimentally and to isolate those associated with grapevine berry development and ripening was reported by Carra *et al.* [38]. Other investigators prepared a library of small RNAs from three berry developmental stages, and generated a 12 K CombiMatrix custom array to confirm their authenticity and profile their expression [39].

The rapid increase in grapevine microarray data has provoked the development of customised tools for data analysis, visualisation and mining. For example, MapMan is a tool that displays large datasets in the context of metabolic or signalling pathways. Originally developed for Arabidopsis, MapMan has been extended to the grapevine using a previous annotation of TIGR *Vitis vinifera* Gene Index Release 5.0 [40]. The grapevine MapMan tool helps to visualise data obtained using Affymetrix and Operon microarrays. VitisNet is another tool for the visualisation of large grapevine data sets [41]. This can integrate transcript, protein and metabolite profiles into a combined molecular map. It is based on the manual annotation of 39,424 unique sequences obtained by pairing *V. vinifera* genome sequences [1] with *Vitis* ESTs, and by assigning 13,145 genes to 219 networks.

The metabolome, defined by analogy to the transcriptome, is the complete set of metabolites (small molecules, $M_r \leq 1{,}000$ D) produced in a single cell, tissue or organism [42]. Like the transcriptome, the metabolome is complex, dynamic and varies by cell type, developmental stage and in response to internal and external cues. Although a complete metabolomic analysis of grape berry development has yet to be reported, several studies have reported the profiles of selected metabolites [43-48] and others have comprehensively examined particular classes of compounds. For example, Parker *et al.* [49] performed a comprehensive untargeted analysis of volatile compounds in Shiraz berries by gas chromatography mass spectrometry (GC-MS) in order to predict the levels of pepperines (volatiles responsible for the peppery characteristics of the wine) using multivariate techniques such as principal component analysis (PCA) and projection to latent squares (PLS). Liquid chromatography followed by electrospray ionisation mass spectrometry (LC-ESI-MS) was used to simultaneously profile 40 phenolic compounds in the skins and seeds of one white and two red cultivars [50]. Finally, the metabolic profile of Corvina berries during development and post-harvest withering were characterised using untargeted LC-ESI-MS analysis, and stage-specific metabolites were identified using multivariate techniques [33, 51].

A comprehensive description of the metabolome can only be achieved using multiple experimental and analytical platforms to cover the diversity of plant metabolites, but no such analysis has been reported thus far for the grapevine [43].

TRANSCRIPTOMIC ANALYSIS REVEALS THE DYNAMIC NATURE OF GRAPE BERRY DEVELOPMENT

The earliest studies of gene expression during berry development sought to identify strongly modulated genes, but transcriptomics is a holistic approach and seeks a comprehensive picture of gene expression dynamics. Once microarrays had been developed containing several thousand probes, it became possible to obtain an extensive catalogue of gene expression profiles potentially covering all the predicted genes in the *V. vinifera* genome.

Table 1: Transcriptional analyses of grapevine berry development.

Reference	Techniques	Cultivar	Tissue	Sequence information and chip
Davies and Robinson [9]	Differential screening	Shiraz	deseeded berry	Shiraz post-veraison berry cDNA library
Ablet *et al.* [10]	EST analysis	Chardonnay	berry	2,479 Chardonnay berry ESTs
Terrier *et al.* [11]	EST analysis	Shiraz	berry	275 berry (3 stages) ESTs
Venter *et al.* [13]	cDNA-AFLP	Chardonnay	deseeded berry	
Burgher and Botha [14]	cDNA-AFLP	Cabernet Sauvignon; Clairette Blanche	berry	
Fei *et al.* [16]	Digital expression analysis			TIGR Grape Gene Index 2.0
Moser *et al.* [12]	EST analysis	Pinot Noir	berry	1,743 Pinot Noir berry (at veraison) EST
Terrier *et al.* [18]	cDNA microarray analysis	Chardonnay; Cabertnet Sauvignon; Shiraz	berry	3,175 Shiraz berry (9 developmental stages) unigenes
da Silva *et al.* [17]	Digital expression analysis	Chardonnay	berry	104,075 *Vitis* sequences deposited into GeneBank (NCBI) as of September 30, 2003
Waters *et al.* [20]	cDNA microarray	Shiraz	skin	4,608 Shiraz ripening berry (different developmental stages) cDNA clones
Mori *et al.* [27]	microarray analysis	Cabernet Sauvignon; Clairette Blanche	skin	Affymterix GeneChip® *Vitis* genome array ver. 1.0
Grimplet *et al.* [24]	microarray analysis	Cabernet Sauvignon	skin, pulp and seed	Affymterix GeneChip® *Vitis* genome array ver. 1.0
Deluc *et al.* [25]	microarray analysis	Cabernet Sauvignon	berry	Affymterix GeneChip® *Vitis* genome array ver. 1.0
Pilati *et al.* [26]	microarray analysis	Pinot Noir	berry	Affymterix GeneChip® *Vitis* genome array ver. 1.0
Chervin *et al.* [31]	microarray analysis	Cabernet Sauvignon	berry	Grape AROS V1.0
Gatto *et al.* [21]	cDNA microarray analysis	21 cultivars[a]	berry	4,224 cDNA clones (Pinot Noir; Moscato Bianco; Teroldego;Merzling) SSH clones
Glissant *et al.* [19]	cDNA microarray analysis	Chardonnay	berry	3,175 Shiraz berry unigenes
Iandolino *et al.*	EST analysis and MPSS	Cabernet	berry	30,737 and 26,878 berry (green, hard

[70]	signature analysis	Sauvignon		stage) distinct sequences
Lund *et al.* [28]	microarray analysis	Cabernet Sauvignon	berry	Affymterix GeneChip® *Vitis* genome array ver. 1.0
Zamboni *et al.* [15]	AFLP-TP	Corvina	deseeded berry	
Carra *et al.* [38]	miRNA analysis	Nebbiolo	berry	small berry (3 developmental stages) RNA library
Deluc *et al.* [29]	microarray analysis	Chardonnay; Cabertnet Sauvignon	berry	Affymterix GeneChip® *Vitis* genome array ver. 1.0
Mica *et al.* [39]	miRNA analysis; 454[b]	Pinot Noir	berry	1,974 miRNA-specific probes
Rizzini *et al.* [32]	microarray analysis	Raboso Piave	skin	Grape AROS V1.0
Zenoni *et al.* [37]	RNA-seq	Corvina	berry	*Vitis* genome 8.4X prediction
Zamboni *et al.* [33]	microarray analysis	Corvina	berry	GrapeArray 1.2
Koyama *et al.* [30]	microarray analysis	Cabernet Sauvignon	skin	Affymterix GeneChip® *Vitis* genome array ver. 1.0

[a] A pool of 11 high and a pool of 10 low resveratrol producer cultivars.

[b] miRNA analysis using an oligoarray; 454 sequencing.

The wealth of information generated by grapevine transcriptomic experiments in the last decade Table **2** is summarised in the following sections according to the principal metabolic pathways and biological process of berry development.

Primary Metabolism

The transcriptomic analysis of primary metabolism during berry development has focussed mainly on photosynthesis, and the metabolism of sugars, organic acids, amino acids and lipids.

Photosynthesis

Berry is a photosynthetic active organ in the first phase of berry formation until veraison [52]. Gene expression analysis supports this hypothesis because transcripts with a putative role in photosynthesis are more abundant in early development. These include transcripts encoding proteins belonging to photosystems I and II, subunits of ribulose-1,5-biphosphate carboxylase (RuBisCO), and light-harvesting chlorophyll *a/b* binding proteins. The down regulation of photosynthetic genes after veraison regardless of cultivar was first observed by EST profiling [11, 17] and confirmed by microarray analysis of berries [18, 25, 26, 33] or specifically berry skins [20], wherein photosynthesis-related transcripts are more abundant [24].

Water deficit stress appears to inhibit photosynthesis, although this is more pronounced in Chardonnay berries compared to red Cabernet Sauvignon [29]. Photosynthesis transcripts are also down regulated in response to abscisic acid (ABA) treatment [30]. The pre-veraison specificity of photosynthesis was also confirmed using a hypothesis-free evaluation of integrated datasets in developing Corvina berries, which identified a network of photosynthesis-related transcripts and proteins [33].

Sugar Metabolism

Sugars produced by photosynthesis in leaves (source organs) are transported to berries (sink organs) through the phloem. After veraison there is a shift from symplastic to apoplastic phloem unloading [53].

This correlates with the downregulation of *GIN1* and *GIN2*, encoding two vacuolar invertases [20, 25, 26], suggesting a putative role in symplastic unloading during the first berry growth phase.

The expression of proteins involved in cellulose depolymerisation, hexose transport and sucrose re-synthesis in the cytosol is supported by the observed upregulation of genes encoding sucrose-phosphate synthase, sucrose synthase, sucrose-6-phosphate phosphatase and UDP-sugar pyrophosphorylase [18, 25, 26, 28], while in the berry skin the levels of sucrose synthase mRNA decrease before veraison [20]. Different genes encoding hexose and sugar transporters are induced during ripening [18, 25, 26, 28] and they also play a role in apoplastic unloading. A recent phylogenetic analysis combined with macroarray-based profiling of hexose and sucrose transporter genes in vegetative tissues and berries showed that the regulation of sugar transport is more complex than previously suggested [54].

Deluc *et al.* [25] described the upregulation of transcripts involved in starch degradation during the development of Cabernet Sauvignon berries, including α- and β-amylase and water dikinase. This suggested that plastidial starch is mobilised for the accumulation of hexoses during the last phase of berry development. These hexoses could be used to produce osmolytes such as trehalose and sorbitol during ripening, as supported by the induction of genes encoding sorbitol dehydrogenase and trehalose-6-phosphate synthase/phosphatase during the last phase of development [25]. The induction of genes controlling sorbitol and trehalose synthesis has been observed in berries in response to water deficit stress [24], ABA treatment [30], ethylene treatment [31] and during withering [15].

Organic Acid Metabolism

Malic and tartaric acids reach their maximum levels during the first phase of berry development, and then fall as berries ripen. In Cabernet Sauvignon berries, the maximum tartrate levels were recorded at E-L stage 32 followed by a steady decline until harvest [25]. A similar profile was observed for three transcripts encoding L-idonate dehydrogenase, an enzyme required for tartrate synthesis [25]. Higher levels of tartrate synthase transcripts were detected in the seeds of ripe berries suggesting that tartrate is synthesised in seeds, but there is no biochemical evidence to support this hypothesis [24].

Malate concentrations also peak at E-L stage 32 in Cabernet Sauvignon berries, but they decline more rapidly than tartrate [25]. Malate is the only organic acid that is metabolised during berry ripening [52], and its pre-veraison biosynthesis is driven by the sugars deposited into berries and possibly also by sugars produced in berries by photosynthesis. After veraison, malic acid is released from the vacuole and acts as a carbon source for berry metabolism [52]. The synthesis of malate during the first berry growth phase can be driven by pyruvate produced from phosphoenolpyruvate (PEP) by pyruvate kinase (PK) or by the direct conversion of PEP into malate by the sequential activity of phosphoenolpyruvate carboxylase (PEPC) and cytosolic malate dehydrogenase [52]. This is supported by the induction of several transcripts related to malate synthesis in the first berry growth phase, *e.g.* glyoxysomal malate synthase [18], PK [15, 25] and PEPC [18, 25]. These transcripts are more abundant in the berry skin, which correlates with the higher levels of malate found in this tissue [24].

During ripening, malate is released from the vacuole into the cytosol and metabolised, which may reflect an increase in the proportion of the less efficient V-PPase in relation to the more efficient V-ATPase during ripening [52]. There is some evidence to support this suggestion from microarray experiments [25, 26]. The breakdown of malic acid during ripening correlates with the induction of transcripts encoding catabolic enzymes such as malic enzyme and phosphoenolpyruvate carboxykinase, as well as enzymes involved in the metabolism of pyruvate, such as pyruvate orthophosphate dikinase, pyruvate decarboxylase and pyruvate dehydrogenase [16, 18, 25, 26].

It has also been suggested that the aerobic fermentative pathway is activated during ripening and post-ripening [26, 33, 55]. This is supported by the up regulation of some transcripts encoding alcohol dehydrogenase and aldehyde dehydrogenase [18, 25, 26], which also occurs during and after veraison in response to ABA treatment [30].

Lipid Metabolism

The *in silico* analysis of EST collections by da Silva *et al.* [17] revealed the possible up regulation of an acyl-CoA-binding protein during berry ripening, putatively involved in the accumulation of epicuticular wax. A ω-6-fatty acid desaturase also appeared to be up regulated, suggesting a role in the synthesis of polyunsaturated fatty acids in seeds.

A transcript encoding lipoxygenase (LOX), which catalyses the peroxidation of lipids, was up regulated during the pre-bloom and bloom phases [17] but down regulated during berry development [25]. This class of enzyme helps to synthesise plant volatiles, and the corresponding transcripts were found to be more abundant in the berry skin than in the pulp and seeds [24]. LOX transcripts are also induced by water deficit stress [29] and withering [15].

ABA treatment also affected several genes involved in lipid metabolism, *e.g.* acyl-CoA synthase, ATP-citrate lyase β-subunit and fatty acid elongase-like protein, suggesting this hormone modifies the properties of cell membranes [30].

Amino Acid Biosynthesis

Berry ripening involves an increase in the free amino acid pool, especially the levels of proline [56]. Accordingly, many studies have revealed the induction of genes controlling amino acid metabolism during ripening [25, 26], particularly in the berry skin and seeds [24]. Transcripts corresponding to enzymes involved in proline synthesis showed only a small peak at veraison (*e.g.* pyrroline-5-carboxylate synthetase) or increased only at the end of ripening (*e.g.* proline dehydrogenase) [25].

Water deficit stress induced proline accumulation in Chardonnay and (more robustly) in Cabernet Sauvignon berries, but the up regulation of pyrroline-5-carboxylate synthetase transcripts was still restricted to veraison in both cultivars [29]. The stress-induced increase in proline levels reflects its osmoregulatory activity, and this can also be linked to the glutamate metabolism as suggested by the induction of genes encoding aspartate aminotransferase, glutamine synthetase (GS), asparagine synthetase and glutamate decarboxylase [29].

Early berry development also features the strong induction of a gene encoding a cytoplasmatic aspartate aminotransferase, which catalyses the reversible conversion of glutamate to aspartate [17]. Furthermore, the strong pre-veraison expression of genes encoding asparaginase, glutamate dehydrogenase and GS [25] suggests that the supply of ammonia for *de novo* protein synthesis is achieved by asparagine hydrolysis and subsequent assimilation *via* glutamate metabolism.

Secondary Metabolism

In the grapevine, secondary metabolism includes many pathways that are important determinants of berry quality traits, such as the synthesis of aromatic and volatile compounds, mainly terpenes, benzenoids and phenylpropanoids.

Initial attempts to analyse these pathways by transcriptional profiling were able to identify only a few genes clearly, although these tended to be modulated dramatically during berry development and ripening. Genes encoding cinnamyl-alcohol dehydrogenase (CAD), caffeic acid O-methyltransferase (COMT), chalcone-flavanone isomerase (CHI), chalcone reductase, isoflavone reductase (IFR), cytochrome b5 (DIF-F), and leucoanthocyanidin dioxygenase (LDOX) were the only ones related to secondary metabolism that were expressed during early berry development and/or at veraison and thereafter [11, 17].

The first microarray experiments proved to be more informative about berry secondary metabolism than the analysis of EST sequences [18, 20]. Remarkably, these studies showed that almost all developmentally modulated transcripts associated with secondary metabolism represented enzymes in the phenylpropanoid pathway, indicating that phenylpropanoid metabolism is under strict transcriptional control. Waters *et al.*

[20] found 11 secondary metabolism cDNAs (one phenylalanine ammonia-lyase, PAL; one 4-coumarate-CoA ligase, 4CL; two chalcone synthases, CHS; one CHI; two flavanone 3 hydroxylases, F3H; one DIF-F; one anthocyanin reductases, ANR; one CAD; one limonoid UDP-glucosyltransferase) that were up regulated during the herbaceous phase of Shiraz berry development but none from the berry skin that were up regulated during ripening. Terrier *et al.* [18] identified approximately 20 developmentally modulated phenylpropanoid pathway genes in Shiraz berries, but only five of these were involved in ripening and abruptly induced at the very beginning of berry colouration (IFR; COMT; flavonol synthase, FLS) and two genes involved in anthocyanin synthesis, *i.e.* UDP-glucose:flavonoid 3-O-glucosyltransferase (UFGT) and glutathione S-transferase 4 (GST4). The remaining transcripts (PAL; CHS; CHI; F3H; dihydroflavonol-4-reductase, DFR; flavonoid 3'5'-hydroxylase, F3'5'H; LDOX) were expressed predominantly during early development and later at ripening, confirming a bimodal pattern of transcript accumulation as initially described by Boss *et al.* [57] for the flavonoid pathway genes. This profile provides an explanation for the asynchronous accumulation of proanthocyanidins and anthocyanins in the berries of red cultivars.

In subsequent experiments using oligonucleotide chips representing up to 50% of the *V. vinifera* genome, many candidate genes with potential roles in secondary metabolism were shown to be induced or repressed during berry development. The combined results of the time course analysis carried out by Deluc *et al.* [25] and the spatial profiling carried out by Grimplet *et al.* [24] on Cabernet Sauvignon berries provided a great deal of information about the potential roles of differentially expressed genes, and were consistent with the known developmental profiles of phenylpropanoid and terpenoid compounds in the berry skin, pulp and seeds from fruit set to ripening.

Deluc *et al.* [25] reported that most of the differentially expressed transcripts involved in flavonoid metabolism declined in expression from the early development stage, whereas the small collection of genes directly involved in anthocyanin and flavonol synthesis were induced at the onset of ripening. Similarly, most of the transcripts related to phenolic acid metabolism declined during berry development. Consistent with the high levels of phenolic compounds in berry skins and seeds, genes related to phenylpropanoid synthesis were generally expressed at higher levels in these tissues compared to the berry pulp [24]. Pilati *et al.* [26] obtained similar data when investigating veraison in Pinot Noir berries. Taken together, these data provided a distinction between phase-specific and bimodal transcripts related to phenylpropanoid synthesis. Furthermore, many genes encoding homologous but differentially expressed enzymes were identified, *e.g.* three DFR and two UFGT transcripts differentially expressed in the skin, pulp and seed during ripening [24]. The proposed developmental role of these genes based on their differential expression profiles awaits confirmation by functional analysis.

In addition to genes encoding enzymes involved in secondary metabolism, transcriptional profiling has also confirmed the differential expression of transcription factors such as MYBA1, MYBA2, MYBPA1 and MYCA1, which control different branches of phenylpropanoid pathways. Many transcripts encoding other regulators, with possible although as yet unproven roles in the control of secondary metabolism, have been listed among the transcripts that are differentially expressed during berry development [18, 24, 26, 33, 37].

The analysis of post-ripening berries has revealed that transcripts representing general phenylpropanoid metabolism (*e.g.* PAL, 4CL) are co-expressed with stilbene synthase transcripts, correlating with the observed increase in stilbenes in late ripening and post-ripening [15, 32, 33].

Several genes putatively involved in terpene, benzenoid and monolignol synthesis are also modulated during development and differentially expressed in the skin, pulp and seeds [24, 25]. However, the extremely complex pathways that produce these metabolites, the diversity of products and the lack of knowledge about the precise enzymatic steps makes further interpretation of the transcriptomic data very challenging, particularly in terms of supporting the biochemical changes that affect the production of volatiles and their precursors during berry development.

Transcriptomic analysis has also revealed a number of secondary metabolic genes that are induced by water deficit stress in Cabernet Sauvignon and Chardonnay berries [24, 29]. The up regulation of anthocyanin-related genes such as PAL, cinnamate-4-hydroxylase (C4H), 4CL, UFGT, MYBA1 and/or MYBA2 is consistent with the higher anthocyanin levels in Cabernet Sauvignon berries from water-stressed plants. Genes involved in monolignol synthesis, such as COMT and cinnamoyl-CoA reductase, were also induced by water stress, but lignin content was not investigated. In Chardonnay berries, water stress induced the expression of FLS and increased the flavonol content.

Anthocyanin levels in berry skin are reduced under heat stress [27], but transcriptomic analysis has not indicated the significant inhibition of genes involved in anthocyanin synthesis. DFR and F3H are repressed, but only marginally. The decline in anthocyanin levels may therefore reflect an increase in catabolism, a hypothesis supported by the up regulation of a peroxidase under heat stress.

The stimulative effect of ABA on the onset of ripening was studied by transcriptomic approach to identify changes in gene expression related to this hormonal treatment. As expected, the increase of expression of anthocyanin-related genes such as PAL, F3H, CHS, MYBA1 and/or MYBA2 was observed in ABA-treated berries [30].

Hormone Biosynthesis and Regulation

Abscisic Acid

The accumulation of ABA correlates with the initiation of ripening and treatments that inhibit the accumulation of ABA also hold up ripening [58, 59]. Transcriptomics has identified several ABA-related transcripts that confirm the involvement of this hormone in the initiation and progress of ripening.

ABA is synthesised by 9-cis-epoxycarotenoid dioxygenases (NCEDs) such as VvNCED1 VvNCED2 and VvNCED4, which are induced after veraison [25, 26, 28]. Transcripts encoding ABA-responsive proteins are also induced after veraison and during ripening. Examples include transcripts encoding dehydrin and RD22 dehydration-responsive protein. ABA-induced transcription factors such as ABI3/VP1 are also up regulated at veraison [25, 26]. A putative ABA receptor, VvGCR2, is induced in ripening pericarp tissues, suggesting a possible increase in ABA perception as well as synthesis during berry ripening [28].

ABA is also produced in response to stresses such as water deficit. Microarray experiments investigating the effects of long-term seasonal water deficit on Cabernet Sauvignon and Chardonnay berries revealed that the ABA metabolic pathway was induced in both cultivars [29].

Auxin

Indole acetic acid (IAA) is required at the onset of fruit development and its levels are therefore highest at early developmental stages [60]. Transcriptomic analysis has confirmed that the expression of genes related to auxin synthesis, perception and response occurs in concert with the observed increase in auxin levels, supporting the prevalent role of this phytohormome in the regulation of cell division in the young berry. After veraison, there is a decline in the expression of a gene homologous to amidase *AtAM1*, which is thought to be required for IAA synthesis in *Arabidopsis* [26].

Many auxin carriers, which mediate auxin influx and efflux, appear either to be expressed only before veraison [25, 26] or induced at the onset of veraison [25]. At least nine transcripts encoding auxin responsive factors (ARFs) are expressed pre-veraison but repressed during ripening [25, 26], and a transcript encoding an auxin response repression factor (Aux22), which forms heterodimers with ARFs, increased after veraison [25]. Numerous auxin-regulated genes are induced after veraison, including those encoding Aux/IAA, small Auxin-UP RNA and GH3 [25, 26]. The expression profiles of two IAA amino acid hydrolase genes involved in IAA homeostasis *via* amino acid conjugation showed opposing expression profiles [25]. The expression of one gene increased steadily throughout berry development while the other declined. Tissue-specific microarray analysis has shown that auxin signalling-related transcripts are still present in mature berries, and are more abundant in the pericarp than in the seed [24]. The expression of miR160, miR164 and miR167, three miRNAs that regulate auxin perception genes, was shown to decline during berry ripening [38]. As auxin content remains low during ripening [61], the expression of these miRNAs in berries may be auxin-dependent, with a feedback mechanism operating only in the presence of high phytohormone concentrations.

Taken together these data suggest that IAA could have an active role not only during the first growth phase of berry development but also in the ripening phase.

Ethylene

The role of ethylene in grape berry development is indicated by the small transient increase in endogenous ethylene production that occurs just before veraison, corresponding to an increase in 1-aminocyclopropane-1-carboxylic acid (ACC) oxidase activity [62]. Transcripts encoding two isoforms of ACC oxidase were found to be expressed strongly in young berries until veraison, after which the levels declined steadily [25]. Pilati *et al.* [26] observed a peak of ACC oxidase expression at veraison and found that a different ACC oxidase and an ACC synthase were expressed until veraison and then repressed.

At least five transcripts encoding ethylene response factors (ERFs) were identified by Pilati *et al.* [26], whereas ten were identified by Deluc *et al.* [25]. Most of the ERF transcripts appeared to accumulate during early development or at veraison, but others were induced during ripening and reached peak expression in mature berries. Transcripts encoding two ethylene receptors declined until veraison but then increased during ripening, whereas a putative ethylene co-activator transcript showed biphasic accumulation during development [25]. At least six additional ethylene-induced genes showed very different expression patterns [26]. These intriguing profiles suggest that different ethylene signalling pathways may be induced at different times during berry development, possibly also in response to stress. Indeed Pilati *et al.* [26] found that PR1 and PR4, two PR (Pathogen Related) proteins involved in biotic stress responses (see below) and also known to be regulated by ERFs, were induced during berry ripening.

Zamboni *et al.* [15] used cDNA-AFLP analysis to show that ethylene is probably involved in post-harvest withering, as supported by the up regulation of S-adenosyl methionine (SAMe) synthase after ripening. Two SAMe synthases, two ACC oxidases and two ERFs were also induced in berries under water stress, suggesting that ethylene plays a key role in berry development under stress conditions [24].

Microarray analyses of berries treated with ethylene at veraison revealed the accumulation of several transcripts relating to cell wall metabolism. This was consistent with the observed physical changes induced by ethylene (an increase in diameter reflecting cell wall modifications that allowed for cell enlargement). The most strongly induced transcripts encoded xyloglucan endotransglucosylases (XET) and aquaporins, which appear to be good candidates to explain the mechanism of ethylene-induced berry expansion [31].

Cytokinin

The concentration of cytokinins is high during early berry development but declines towards veraison. Local cytokinin biosynthesis may help to determine the sink properties of the grape berry at the onset of ripening [61].

Transcriptomics has shed little light on the regulation of berry development by cytokinins, although a transcript encoding cytokinin oxidase was shown to decline during development by both Pilati *et al.* [26] and Deluc *et al.* [25]. Several of the small non-coding RNAs identified by Carra *et al.* [38] were also relevant in this context, since two of them (id4 and id65, matching the grapevine cytokinin synthase gene) appear to be developmentally regulated. Although neither of the transcripts fit the typical profile of a miRNA, id65 specifically accumulates in mature grape berries, in which expression of the cytokinin synthase gene is dramatically reduced. It is therefore possible that the gene is post-transcriptionally repressed in ripening berries by a mechanism involving double-stranded RNA. The post-transcriptional down-regulation of cytokinin synthase may lower the cytokinin content to allow for the progression of the senescence-like ripening programmes, which are thought to be driven mainly by ABA and brassinosteroids.

However several cytokinin response genes are also induced in ripening berries, *e.g.* those encoding a type-A response regulator [26], a cytokinin induced message [18, 25] and a cytokinin-repressed protein [25], suggesting a more complex role for cytokinins in berry ripening.

Gibberellins

Several pieces of evidence support the role for gibberellins (GAs) in the early berry development stages. Large-scale transcriptional analyses revealed the down regulation during early berry development of a gibberellin-2 oxidase [26] and several additional GA-responsive transcripts [25, 26], indicating the active metabolism of GAs before veraison. However, transcripts encoding a number of GA receptors and GA-responsive proteins appear to be induced after veraison [18, 25, 26].

Brassinosteroids

The onset of berry ripening corresponds to a dramatic increase in the levels of brassinosteroids (BRs) [61]. This matches the detection of transcripts relating to BR synthesis and signal transduction in early development, peaking at veraison [26]. Deluc *et al.* [25] additionally showed that the BR-responsive XET gene was up regulated at veraison, suggesting a role for BRs in fruit softening. Zamboni *et al.* [33] found that several BR-related transcripts, including one encoding 11-β-hydroxysteroid dehydrogenase, were identified as markers of the pre-veraison and veraison developmental phases.

Polyamines

The genes for ornithine decarboxylase and arginine decarboxylase, each of which is involved in the synthesis of polyamines from the amino acid arginine, are both strongly repressed prior to veraison (E-L stage 32). This suggests that polyamine precursors are required during the early stages of berry development [25].

Jasmonic Acid and Phytosulfokine

Transcripts related to jasmonate signalling were strongly induced during ripening [18], whereas a gene involved in methyl jasmonate (MeJA) synthesis was shown to be down regulated [25]. It has been suggested that jasmonate and MeJA may have overlapping roles in fruit ripening and pathogen defence [63].

Genes encoding a phytosulfokine peptide precursor, a phytosulfokine receptor, and other sulfokine-related proteins have been shown to be repressed in developing berries [18, 25]. This supports the proposed role for this growth factor in cell division during the herbaceous berry growth phase.

Table 2: Comparative sampling time-point map of berry development used in published transcriptional analyses, tentatively linked to growth stages according to the E-L systems. Days before (-) and after (+) veraison were reported for studies in which they are specified; in the other cases, sampling time-points are indicated as *.

Reference	Modified E-L system																			
	27	28	29	30	31	32		33		34		35		36		37		38		39
Ablet *et al.* [10]									*											
Davies and Robinson [9]									-14				+14							
Terrier *et al.* [11]							-14					*		+16						
Venter *et al.* [13] [a]					*													*		
Burgher and Botha [14]													+7					+35		
Moser *et al.* [12]												*								
Terrier *et al.* [18]					-38			-25			-10	*	4	10		29		45		
Water s*et al.* [20]			-42					-35		-28		*		14	21					
Deluc *et al.* [25]					-42	-21		-14		-7		*		21				49		

Grimplet *et al.* [24]																		3		
Mori *et al.* [27]												*		14		28		42		
Pilati *et al.* [26]								-14		-7				21						
Chervin *et al.* [31]										-7										
Gatto *et al.* [21]												*						*		*
Glissant *et al.* [19]					-54			-41		-20		*		11	16			26		36
Iandolino *et al.* [70]								*												
Lund *et al.* [28]								*		*		*						*		
Zamboni *et al.* [15]					-47			-24				*						41		* [b]
Deluc *et al.* [29] [c]					-28					-7		*	0.5			35			56	
Deluc *et al.* [29] [d]					-42			-21		-14	-7	*			21			42		
Carra *et al.* [38]								-28				*						56		
Mica *et al.* [39]							-28					*						42		
Rizzini *et al.* [32]																				* [b]
Koyama *et al.* [30]												*		14		28		37		
Zamboni *et al.* [33]								-12				*		27				41		* [b]
Zenoni *et al.* [37]						-35						*					35			

[a] Only the first and the latter sampling time-points were reported for Venter *et al.* [13].

[b] In these studies more than one post-ripening sampling time points were analysed.

[c] Cabernet Sauvignon.

[d] Chardonnay.

Cell Wall Metabolism

During berry development the skin and pulp cell walls undergo several structural modifications that allow for cell expansion and softening. This involves two distinct periods of rapid growth separated by a slower growth phase. The first phase of berry growth involves an increase in cell number and volume, whereas the second phase involves an increase in volume only and is also characterised by softening. Cell expansion requires the primary cell wall to loosen and incorporate new material, whereas softening involves the hydrolysis of cell wall polymers.

Several enzyme families contribute to cell wall metabolism and remodelling, *i.e.* pectin methylesterase (PME), polygalacturonidase (PG), pectate lyase (PL), β-galactosidase, xyloglucan endotransglycosylase transferase (XET), endoglucanase, cellulose synthase and expansin. The transcriptomic analysis of ripening berries has shown that members of the same gene family can be expressed specifically in the skin or pulp [24], suggesting that common polysaccharides are hydrolysed specifically within each tissue, or that some of the polysaccharides may themselves be tissue specific [24].

PMEs catalyse the demethylesterification of the homogalacturonan components of pectins. A grape ripening induced protein (GRIP 28) homologous to PME was shown to be upregulated after veraison in an early differential screening approach [9]. Subsequent microarray experiments have shown that different PME isoforms are expressed at different levels and at different times during berry development [18, 19, 25, 26]. This confirms that softening involves the degradation of pectic polymers, but pectin degradation is also associated with cell growth and modification of the mechanical proprieties of the cell wall before and after veraison [64, 65]. Moreover, EST screening and microarray analyses showed that transcripts for putative invertase/PME inhibitor proteins, which inhibit the activity of PMEs, are strongly expressed in pre-veraison berries but not detected during later stages of berry development [17, 25].

Microarrays have also shown that genes encoding PG and PL are abruptly induced at veraison, consistent with the onset of berry softening [18, 19, 25, 26, 28]. However, the expression of some isoforms of both enzymes has been shown to peak early in berry development, declining thereafter [19, 25]. PG activity could be fine-tuned by the expression of a PG-inhibiting protein, as reported by Deluc *et al.* [25] and Glissant *et al.* [19].

Non-reducing terminal galactosyl residues are removed from pectin side chains by β-galactosidase. Microarray analysis has revealed two β-galactosidase transcripts whose expression profiles mirror the two periods of berry enlargement [19], and others whose expression levels decline throughout development until the end of ripening [19, 25]. Other significant hydrolases such as α-arabinosidase, α-galactosidase, α-mannosidase, β-xylosidase and β-mannan endohydrolase are also expressed in developing berries. Arabinosidase transcripts were present at low levels during the later stages of ripening and three α-mannosidase transcripts were expressed before and after veraison [19]. Deluc *et al.* [25] showed that six β-xylosidase isoforms were down regulated during development, one β-mannan endohydrolase was up regulated during ripening and another was simultaneously repressed. One β-mannan endohydrolase was down regulated specifically during withering in Corvina berries [15]. Overall, these data show that hydrolases have important and complex overlapping roles in the control of fruit softening, reflecting the key role of pectin reduction and modification in this process [64].

XET and xyloglucan endoglucanase (XEG) promote xyloglucan reassembly, which is predominantly associated with fruit softening [64, 66]. However, transcriptomic analysis has shown that XET and XEG expression correlates with the two periods of berry enlargement, and therefore also appears to be related to cell wall elongation. One XET transcript was highly abundant in the pre-veraison phase [25], whereas at least four others were strongly up regulated after veraison [18, 26]. Transcripts encoding XEGs were shown to be expressed at their highest levels after veraison and at the mid-ripe stage [19] or declined steadily over the course of berry development [25, 26].

Cellulose synthases (CSs) add glucose residues to an existing (1,4)-linked glucan chain producing an inelastic fibre that is extruded from the cell surface and incorporated into matrix polysaccharides. Glissant *et al.* [19] reported that two members of the CS family were expressed during the first stages of berry development. In agreement, Deluc *et al.* [25] showed the steady decline of at least six CS transcripts during berry development, but four others displayed a more complex expression profile. Zamboni *et al.* [15] identified a CS transcript that was consistently suppressed in withering berries.

Expansins play an important role in cell wall loosening and disassembly. Most expansin transcripts increase or decrease steadily during berry development [25, 26], but some show characteristic expression profiles matching the second phase of fruit enlargement and softening [17, 19, 25]. Interestingly, one expansin transcript identified by Zamboni *et al.* [33] was shown to be a potential negative marker (*i.e.* conspicuous by its absence) of the pre-ripening/ripening phases.

The structure of the berry cell wall is influenced also by structural proteins that provide strength and attachment points for cross-linking. These include hydroxyproline-rich extensins, proline-rich proteins, leucine-rich repeat proteins, glycine-rich proteins and arabinogalactan proteins. Differential screening revealed that two extensin-like proteins (GRIP 3 and GRIP 4) and two proline-rich proteins (GRIP 13 and GRIP 15) were up regulated early in ripening [9]. In subsequent studies, proline-rich proteins and extensins were also shown to be expressed in early development [18, 19, 20, 25]. Deluc *et al.* [25] reported the down regulation of several transcripts encoding arabinogalactan and proline-rich proteins in early development, whereas two glycine-rich proteins were up regulated during ripening. Two proline-rich proteins were defined as potential molecular markers representing the pre-verasion/verasion stages [33]. These data support the hypothesis that certain structural cell wall proteins facilitate cell growth, while others might be involved in strengthening cell walls as a response to pathogen infection.

Cell wall metabolism can also be affected by water stress. This significantly reduces berry size, and is coincident with the down regulation of many transcripts encoding cell wall enzymes including PME, PG and expansin [24]. Ethylene and ABA induce the genes that modify the cell wall during ripening, *e.g.* PG, PME and XET [30, 31]. ABA also induces the expression of several proline-rich cell wall proteins. These results suggest that ethylene and ABA directly regulate the expression of genes encoding cell wall modifying enzymes, actively stimulating the expansion and modification of cell walls during berry ripening.

Response to Stress

Stress-response genes are not only differentially expressed during berry development and ripening, but also in the presence or absence of different forms of stress. In general, transcriptomic analysis has indicated that stress-related genes tend to be induced during berry ripening when the berry is more appealing and susceptible. The up regulation of stress-related genes during ripening was revealed in the very first EST-based experiments [11, 16, 17]. Many of the genes encoding grape ripening-induced proteins (GRIPs) are apparently involved in stress responses, including *Grip21*, *Grip22*, *Grip24*, *Grip32*, *Grip51*, *Grip55*, *Grip58*, *Grip61*, *VvTL2* and *gfh2* [9]. The largely post-veraison expression of these stress-related genes has been confirmed in several microarray experiments involving different cultivars [11, 18, 26, 37]. Interestingly, cultivars producing larger amounts of resveratrol expressed these GRIP genes at lower levels during berry development compared to cultivars with lower resveratrol levels [21].

Abiotic Stress

Several heat-shock protein (HSP) genes are induced predominantly during ripening [16, 17] or at other specific developmental stages [20]. The expression of HSP genes at the beginning of ripening reflects the large number of metabolic and developmental changes occurring at this time, all of which require new protein synthesis [17]. Ethylene and ABA treatment increased the expression of some HSP genes [31] as does the dehydration stress that occurs during withering [15]. HSP transcripts and HSPs belong to a transcript-protein-metabolite network that characterises the pre-veraison and veraison phases of Corvina berry development, as determined using a hypothesis-free data integration approach [33].

Several genes encoding enzymes involved in the reduction of H_2O_2 to water (ascorbate peroxidases, gluthatione peroxidases and peroxiredoxins) and enzymes controlling the balance between oxidised and reduced forms of ascorbic acid and glutathione (thioredoxins, glutathioredoxins, GST and metallothioneins) were shown to be up regulated starting from veraison [11, 17, 18, 26]. Indeed, an oxidative burst was observed in Pinot Noir berries at veraison, causing H_2O_2 accumulation that reached a peak after one or two weeks and then declined. This correlated with the induction of genes encoding enzymes responsible for the detoxification of H_2O_2 and other reactive oxygen species [26]. Transcriptomic analysis has not revealed a relationship between the oxidative burst and transcripts encoding catalase (CAT) or superoxide dismutase (SOD) [20, 26].

The expression of genes that respond to oxidative stress was shown to be characteristic of the pre-veraison and veraison phases of Corvina berry development [33]. By a hypothesis-free data integration, CAT and GST transcripts were assigned to a transcript, protein and metabolite network abundant during the pre-veraison and veraison phases. They could be involved in the scavenging of reactive oxygen species generated as a by-product of photosynthesis. This is justified by the presence in the same pre-veraison and veraison networks of transcripts and proteins involved in photosynthesis. A hypothesis-driven approach was also used to describe ripening and post-harvest withering [33]. This identified another set of detoxifying genes (encoding CAT, GST and glutaredoxins) that may respond to stress conditions during ripening and withering. Transcripts encoding proteins of the nudix (nucleoside diphosphate linked to X) family of phosphohydrolases appear to provide a response to oxidative stress during these phases in a similar way to the nudix transcripts putatively involved in the oxidative burst observed in Pinot Noir berries at veraison [26].

Dehydration response transcripts encoding DR22-like proteins, BURP domain proteins and dehydrins are induced in ripening berries [17, 25, 26]. These genes may also play a role in water deficit stress, in response to the loss of water during withering. Indeed, the up regulation of genes encoding dehydrins, other dehydration-induced proteins, ascorbate peroxidase, GST and enzymes required for trehalose synthesis was confirmed using both cDNA-AFLP transcriptional profiling [15] and hypothesis-free data integration [33]. The up regulation of dehydrin genes was also observed after Cabernet Sauvignon berries were treated with ABA [30].

Biotic Stress

Pathogen resistance mechanisms usually involve a signal transduction cascade triggered by infection, which induces a spectrum of resistance responses. These include strengthening the cell walls, the synthesis of PR proteins and antimicrobial compounds such as phytoalexins, and the hypersensitive response (HR), in which cells undergo programmed cell death in the infected region.

PR proteins are the most abundant proteins in wine. In early transcriptomic studies based on the differential screening of a ripening berry cDNA library, some transcripts homologous to thaumatin-like proteins (PR5) were identified and recognised as grape ripening-induced proteins. Grip51 (VvTL1) and VvTL2 cDNA sequences were shown to be strongly expressed in post-veraison berries [9]. In subsequent microarray experiments, the accumulation of several thaumatin and thaumatin-like transcripts was confirmed in ripening but also in green berries [18, 26]. Deluc *et al.* [25] identified other PR5 transcripts, showing expression patterns that span all stages of berry development.

The chitinase (PR3) family is also expressed strongly in berries, hydrolysing chitin in the cell walls of fungi and in the exoskeleton of insect pests. EST analysis showed that a class IV chitinase transcript was expressed at high levels during post-veraison development [17]. Several chitinase transcripts with distinct expression patterns spanning berry development and ripening were also identified in subsequent microarray experiments [18, 25].

Pathogen-related protein 1 (PR1) is one of the main downstream responses to salicylic acid signalling, and it plays an important role in the establishment of systemic acquired resistance. Microarray analysis revealed that four PR1-related transcripts were strongly expressed during early berry development, an expression profile that correlated with the accumulation of salicylic acid just before veraison [25]. AP2/EREBP-type transcription factor genes can also induce PR1 as part of the resistance mechanism against bacterial pathogens [67].

The induction of PR2 genes, encoding β-1,3-glucanases, is well documented in grape berries as a response to pathogens and their elicitors [68]. A putative β-1,3-glucanase transcript was shown to be differentially expressed during the first stages of berry development [17]. Five further β -1,3-glucanase transcripts were transiently expressed at different periods during berry development [25], and at least six putative β-1,3-glucanase transcripts, three active before veraison and three after, were identified by Pilati *et al.* [26].

PR4 transcripts (encoding endochitinase) were shown to be induced at veraison and during softening [17]. The up regulation of an endochitinase gene during ripening was also confirmed by microarray analysis [18, 26]. The expression of PR4 transcripts identified by microarray analysis was shown to peak at veraison, suggesting a role in pathogen responses during berry ripening [25].

The expression of transcripts encoding the antimicrobial proteins defensin (PR12) and thionin (PR13) was revealed in ripening Cabernet Sauvignon berries, but in pre-veraison, veraison and ripening Chardonnay berries, suggesting that the gene expression profiles differed between the two genotypes [17]. The up regulation of a defensin-like gene after veraison was confirmed by Pilati *et al.* [26].

PR14 proteins are nonspecific lipid transfer proteins (nsLTPs) whose role in disease resistance involves cuticle formation. Four PR14 proteins with different expression profiles were identified in the berry EST library constructed by da Silva *et al.* [17], one showing the second highest expression level of any differentially expressed EST at the onset of sugar accumulation. Terrier *et al.* [18] reported a PR14 gene with a biphasic expression profile, three PR14 genes expressed specifically in the berry skin were detected by Grimplet *et al.* [24], and PR14 transcripts that accumulated in ripening berries were detected by Pilati *et al.* [26].

Overall, most PR genes showed tissue-specific expression in the berry and many were also induced by water deficit stress [24]. This is consistent with cross-talk between the biotic and abiotic stress response pathways, suggesting that water deficit stress may improve resistance to microbial pathogens. Altogether, the expression patterns revealed by transcriptomics have confirmed previous results and revealed that defence-related proteins are expressed across all stages of berry development. This suggests the presence of a systemic acquired resistance strategy that could help to prevent pathogens from penetrating berries.

In addition to PR proteins, a large number of additional resistance proteins are induced during berry development. These include syringolide-induced proteins, cyanogenic glucosides and an alpha hydroxynitrile lyase that were induced by water deficit treatment [24] and, in some cases, by withering [15]. Microarray analysis showed that two transcripts encoding MLO-like proteins involved in barley mildew resistance, were induced in Cabernet Sauvignon berry skin cells and up regulated by water stress [24], and also in withering Corvina berries [15]. Microarray analysis also showed that ten genes involved in the regulation of programmed cell death were modulated during berry development, including suppressors of cell death peaking at veraison and promoters of cell death peaking after veraison [26].

Despite the prevalence of defence-related transcripts in developing and ripening berries, little is known of the signalling pathways and regulatory circuits that induce pathogen responses in the grapevine. Pathogens are recognised by receptors such as nucleotide-binding site leucine-rich repeat domain (NBS-LRR) proteins and receptor-like kinases (RLK). The specific expression of three NBS-LRR proteins in the pericarp of mature berries was reported by Grimplet *et al.* [24], as well as five members of the Avr9/Cf9 gene family, encoding LRR glycoproteins, which were expressed in all berry tissues [24]. Pilati *et al.* [26] described an Avr9/Cf9-related transcript that increased throughout development, and tags for a rapidly elicited Avr9/Cf9 protein were also detected by AFLP transcriptional profiling during withering [15].

The WRKY family of transcription factors is thought to be involved in the induction of pathogen responses [69]. Several WRKY transcripts are up regulated during berry development, ripening and withering [18, 26, 25, 15].

Transcriptomic analysis has also shown that six resveratrol synthase transcripts are up regulated after veraison in Pinot Noir berries [26]. This has been confirmed by RNA-seq in Corvina berries [37], where stilbene synthase genes were previously shown to be induced during withering [15]. Stilbene synthase is also induced by water stress in Cabernet Sauvignon berries [24].

Gatto *et al.* [21] found that genes involved in the response to biotic stress were induced more rapidly and strongly in cultivars producing high levels of resveratrol than in cultivars producing less resveratrol. These results suggest that cultivars producing greater amounts of resveratrol share a wider set of constitutive protection mechanisms against pathogens.

METABOLOMICS

Although grape berry transcriptomics remains a complex and challenging field, procedures for the preparation of samples and the extraction of RNA are now relatively straightforward and reliable. In contrast there remain significant hurdles for the comprehensive extraction of metabolites from berry tissues. The metabolome can also be influenced by genotype, which has a profound effect on secondary metabolism. Therefore, as stated earlier, few metabolomic studies of grape berry development have been carried out, and the small number of reported studies is discussed below.

The first untargeted multiplexed analysis of grape berry metabolites was carried out to identify marker compounds corresponding to the peppery aroma and flavour of Shiraz berries representing two vintages from different vineyards. This study combined GS-MS and chemometrics and used a new method for headspace analysis to capture all volatile compounds [49]. More than 13,000 signals were analysed using multivariate techniques to identify markers of the peppery/spicy aroma and to develop a model for the prediction of pepper flavour intensity based on GS-MS data. Many monoterpenes and sesquiterpenes were identified but only α-ylangene, the most abundant sesquiterpene, was identified as a putative marker of the peppery aroma in Shiraz grapes even if it has a weak aroma feature.

Cavaliere *et al.* [50] made the next step towards a comprehensive metabolomics analysis by using LC-MS for the simultaneous detection of phenolic compounds. The study compared table grape cultivars at ripening and after 6 weeks of storage in a refrigerated stockroom, showing that the phenolic content was unaffected by preservation treatment. Qualitative and quantitative differences in flavan-3-ols, flavonols and stilbenes were identified between cultivars although only flavan-3-ols were considered when comparing seeds [50].

The most recent and most ambitious study so far was a large-scale untargeted metabolic analysis of development, ripening and withering in Corvina berries by LC-MS to identify both qualitative and quantitative metabolic changes [33, 51]. The resulting data were analysed using multivariate statistical methods such as PCA for pattern recognition and bidirectional orthogonal projections to latent structures discriminant analysis (O2PLS-DA) to characterise the different stages of development and withering at the metabolic level. The investigation confirmed what was already known about the accumulation of different classes of compounds during berry development, but it also provided new insights into the accumulation of certain compounds, many linked to withering, that could be characteristic features of the Corvina berry. Pre-veraison and veraison were characterised by the accumulation of organic acids, flavan-3-ols and proanthocyanidins, while ripening and withering involved the accumulation of sugars, anthocyanins, stilbenes, viniferins, flavones and flavanones [51]. A slight pre-veraison reduction in tannin concentration was also observed but not the increase of high molecular weight tannins described for other cultivars. The levels of acylated-anthocyanin and stilbenes increased during ripening and mainly thereafter during withering. The results confirmed that changes in the concentrations of some metabolites during the withering process were not solely due to dehydration but also reflected new synthesis [51].

To promote the metabolomic approach as a tool to understand changes in grape berry composition, it will be necessary to maximise the number of metabolites that can be extracted, analysed and identified. Knowledge exchange and the development of common protocols and databases with detailed descriptions of experiments are also important issues that must be promoted in order to collect the greatest amount of information and allow for the integration of large datasets representing berry metabolites, transcripts and proteins.

ABBREVIATIONS

4CL	=	4-coumarate-CoA ligase
ABA	=	Abscisic acid; GIN, vacuolar invertase
ACC	=	1-aminocyclopropane-1-carboxylic acid
ANR	=	Anthocyanin reductase
ARF	=	Auxin responsive factor
ATP	=	Adenosine triphosphate
BR	=	Brassinosteroid
C4H	=	Cinnamate-4-hydroxylase
CAD	=	Cinnamyl-alcohol dehydrogenase
CAT	=	Catalase
cDNA-AFLP	=	Amplified fragment length polymorphism
CHI	=	Chalcone-flavanone isomerase
CHS	=	Chalcone synthase

COMT	=	Caffeic acid O-methyltransferase
CS	=	Cellulose synthase
dbEST	=	Expressed sequence tag database
DIF-F	=	Cytochrome b5 DIF-F
DRF	=	Dihydroflavonol-4-reductase
ERF	=	Ethylene response factor
EST	=	Expressed sequence tag
F3'5'H	=	Flavonoid 3'5'-hydroxylase
F3H	=	Flavanone 3 hydroxylase
FLS	=	Flavonol synthase
Gas	=	Gibberellins
GC-MS	=	Gas chromatography-mass spectrometry
GCRMA	=	GC robust multi-array average
GRIP	=	Grape ripening induced protein
GS	=	Glutamine synthetase
GST	=	Glutathione S-transferase
HR	=	Hypersensitive response
HSP	=	Heat-shock protein
IAA	=	Indole acetic acid
IRF	=	Isoflavone reductase
LC-ESI-MS	=	Liquid chromatography-electrospray ionisation-mass spectrometry
LC-MS	=	Liquid chromatography-mass spectrometry
LOX	=	Lipoxygenase
LDOX	=	Leucoanthocyanidin dioxygenase
miRNA	=	MicroRibonucleic acid
MeJA	=	Methyl jasmonate
MPSS	=	Massively parallel signature sequencing

NBS-LRR	=	Nucleotide-binding site leucine-rich repeat domain
NCED	=	9-cis-epoxycarotenoid dioxygenase
nsLTP	=	Nonspecific lipid transfer protein
NUDIX	=	Nucleoside diphosphate linked to X
O2PLS-DA	=	Bidirectional orthogonal projections to latent structures discriminant analysis
PAL	=	Phenylalanine ammonia-lyase
PCA	=	Principal component analysis
PEP	=	Phosphoenolpyruvate
PEPC	=	Phosphoenolpyruvate carboxylase
PG	=	Polygalacturonidase
PK	=	Pyruvate kinase
PL	=	Pectate lyase
PLS	=	Projection to latent squares
PME	=	Pectin methylesterase
PR	=	Pathogen related
RefSeq	=	The Reference Sequence collection
RLK	=	Receptor-like kinases
RMA	=	Robust multi-array average
RuBisCO	=	Ribulose-1,5-biphosphate carboxylase
SAGE	=	Serial analysis of gene expression
SAMe	=	S-adenosyl methionine
SOD	=	Superoxide dismutase
UFGT	=	UDP-glucose:flavonoid 3-O-glucosyltransferase
V-ATPase	=	Vacuolar-type H^+-ATPase
V-PPase	=	Vacuolar $H(^+)$-translocating pyrophosphatase
XEG	=	Xyloglucan endoglucanase
XET	=	Xyloglucan endotransglucosylase

ACKNOWLEDGEMENTS

This work was supported by: Fondazione Cariverona (Completamento del Centro di Genomica Funzionale Vegetale); MIPAF (VIGNA Project: Structural and functional characterization of the grapevine genome); ORVIT Consortium (BACCA Project).

REFERENCES

[1] Jaillon O, Aury JM, Noel B, *et al.* French-Italian Public Consortium for Grapevine Genome Characterization. The grapevine genome sequence suggests ancestral hexaploidization in major angiosperm phyla. Nature 2007; 449(7161): 463-7.

[2] Velasco R, Zharkikh A, Troggio M, *et al.* Plos One 2007; 2(12): e136.

[3] Rezaian MA, Krake LR. Nucleic acid extraction and virus detection in grapevine. J Vir Methods 1987; 17(3-4): 277-85.

[4] Boss PK, Davies C, Robinson SP. Analysis of the expression of anthocyanin pathway genes in developing Vitis vinifera L. cv Shiraz grape berries and the implications for pathway regulation. Plant Physiol 1996; 111(4): 1059-66.

[5] Gambino G, Gribaudo I, Leopold S, Schartl A, Laimer M. Molecular characterization of grapevine plants transformed with GFLV resistance genes: I. Plant Cell Report 2005; 24(11): 665-2.

[6] Gambino G, Bondaz J, Gribaudo I. Detection and elimination of viruses in callus, somatic embryos and regenerated plantlets of grapevine. Eur J Plant Pathol 2006; 114(4): 397-404.

[7] Moser C, Gatto P, Moser M, Pindo M, Velasco. Isolation of functional RNA from small amounts of different grape and apple tissues. Mol Biotechnol 2004; 26(2): 95–100.

[8] Wang Z, Gerstain M, Snyder M. RNA-seq: a revolutionary tool for transcriptomics. Nat Rev Genet 2009; 10(1): 57-63.

[9] Davies C, Robinson SP. Differential screening indicates a dramatic change in mRNA profiles during grape berry ripening. Cloning and characterization of cDNAs encoding putative cell wall and stress response proteins. Plant Physiol 2000; 122(3): 803-12.

[10] Ablett E, Seaton G, Scott K, *et al.* Analysis of grape ESTs: global gene expression patterns in leaf and berry. Plant Sci 2000; 159(1): 87-95.

[11] Terrier N, Ageorges A, Abbal P, Romieu C. Generation of ESTs from grape berry at various developmental stages. J Plant Physiol 2001; 158(12): 1575-83.

[12] Moser C, Segala C, Fontana P, *et al.* Comparative analysis of expressed sequence tags from different organs of *Vitis vinifera* L. Funct Integr Genomics 2005; 5(4): 208-17.

[13] Venter M, Burger AL, Botha FC. Molecular analysis of fruit ripening: the identification of differentially expressed sequences in *Vitis vinifera* using cDNA-AFLP technology. Vitis 2001; 40(4): 191-6.

[14] Burger AL, Botha FC. Ripening-related gene expression during fruit ripening in *Vitis vinifera* L. cv. Cabernet Sauvignon and Clairette blanche. Vitis 2004; 43(2): 59-63.

[15] Zamboni A, Minoia L, Ferrarini A, *et al.* Molecular analysis of post-harvest withering in grape by AFLP transcriptional profiling. J Exp Bot 2008; 59(15): 4145-59.

[16] Fei Z, Tang X, Alba RM, White JA, *et al.* Comprehensive EST analysis of tomato and comparative genomics of fruit ripening. Plant J 2004; 40(1): 49-57.

[17] da Silva FG, Iandolino A, Al-Kayal F, *et al.* Characterizing the grape transcriptome. Analysis of expressed sequence tags from multiple *Vitis* species and development of a compendium of gene expression during berry development. Plant Physiol 2005; 139(2): 574-97.

[18] Terrier N, Glissant D, Grimplet J, *et al.* Isogene specific oligo arrays reveal multifaceted changes in gene expression during grape berry (*Vitis vinifera* L.) development. Planta 2005; 222(5): 832-47.

[19] Glissant D, Dedaldechamp F, Delrot S. Transcriptomic analysis of grape berry softening during ripening. J Int Sci Vigne Vin 2008; 42(1): 1-13.

[20] Waters DLE, Holton TA, Ablet EM, Slade Lee L, Henry RJ. cDNA microarray analysis of developing grape (*Vitis vinifera* cv. Shiraz) berry skin. Funct Integr Genomics 2005; 5(1): 40-58.

[21] Gatto P, Vrhovsek U, Muth J, *et al.* Ripening and genotype control stilbene accumulation in healthy grapes. J Agric Food Chem 2008; 56(24): 11773-85.

[22] Irizarry RA, Bolstad BM, Collin F, Cope LM, Hobbs B, Speed TP. Summaries of Affymetrix GeneChip probe level data. Nucleic Acids Res 2003; 31: e15.

[23] Wu Z, Irizarry RA. Processing oligonucleotide array data. Nat Biotechnol 2004; 22(6): 656-8.

[24] Grimplet J, Deluc LG, Tillet LR, *et al.* Tissue-specific mRNA expression profiling in grape berry tissues. BMC Genomics 2007; 8: 187.

[25] Deluc LG, Grimplet J, Wheatley MD, *et al.* Transcriptomic and metabololite analyses of Cabernet Sauvignon grape berry development. BMC Genomics 2007; 8: 429.

[26] Pilati S, Perazzolli M, Malossini A, *et al.* Genome-wide transcriptional analysis of grapevine berry ripening reveals a set of genes similarly modulated during three seasons and the occurrence of an oxidative burst at vèraison. BMC Genomics 2007; 8: 428.

[27] Mori K, Goto-Yamamoto N, Kitayama M, Hashizume K. Loss of anthocyanins in red-wine grape under high temperature. J Exp Bot 2007; 58(8): 1935-45.

[28] Lund ST, Peng FY, Nayar T, Reid KE, Schlosser J. Gene expression analysis in individual grape (*Vitis vinifera* L.) berries during ripening initiation reveal that berry pigmentation intensity is a valid indicator of developmental staging within the cluster. Plant Mol Biol 2008; 68(3): 301-15.

[29] Deluc LG, Quilici DR, Decendit A, *et al.* Water deficit alters differentially metabolic pathways important flavor and quality traits in grape berries of Cabernet Sauvignon and Chardonnay. BMC Genomics 2009; 10: 212.

[30] Koyama K, Sadamatsu K, Goto-Yamamoto N. Abscisic acid stimulated ripening and gene expression in berry skin of Cabernet Sauvignon grape. Funct Integr Genomics 2010; 10(3): 367-81.

[31] Chervin C, Tira-Umphon A, Terrier N, Zouine M, Severac D, Roustan JP. Stimulation of the grape berry expansion by ethylene and effects on related-gene transcripts, over the ripening phase. Physiol Plant 2008; 134(3): 534-46.

[32] Rizzini FM, Bonghi C, Tonutti P. Postharvest water loss induced marked changes in transcriptional profiling in skin of wine grape berries. Postharvest Biol Technol 2009; 52(3): 247-53.

[33] Zamboni A, Di Carli M, Guzzo F, *et al.* Identification of putative stage-specific berry biomarkers and omics data integration into networks. Plant Physiol 2010; 154(3): 1439-59.

[34] Margulies M, Egholm M, Altman WE, *et al.* Genome sequencing in microfabricated high-density picolitre reactors. Nature 2005; 437(7057): 376-80.

[35] Bennet S. Solexa Ldt. Pharmacogenomics 2004; 5(4): 433-8.

[36] Mortazavi A, Williams BA, McCue K, Schaeffer L, Wold B. Mapping and quantifying mammalian transcriptomes by RNA-seq. Nat Methods 2008; 5(7): 621-8.

[37] Zenoni S, Ferrarini A, Giacomelli E, *et al.* Characterization of transcriptional complexity during berry development of Vitis vinifera using RNA-seq. Plant Physiol 2010; 152(4): 1787-95.

[38] Carra A, Mica E, Gambino G, *et al.* Cloning and characterization of small non-coding RNAs from grape. *Plant J* 2009; *59*(5): 750-63.

[39] Mica E, Piccolo V, Delledonne M, *et al.* High throughput approaches reveal splicing of primary microRNA transcripts and tissue specific expression of mature microRNAs in *Vitis vinifera.* BMC Genomics 2009; 10: 258.

[40] Rotter A, Camps C, Lohse M, *et al.* Gene expression profiling in susceptible interaction of grapevine with its fungal pathogen Eutypa lata: extending MapMan ontology for grapevine. BMC Plant Biol 2009; 9: 104.

[41] Grimplet J, Cramer GR, Dickerson JA, Mathiason K, Van Hemert J, Fennell AY. VitisNet: "Omics" integration through grapevine molecular network. PLoS One 2009; 4(12): 8365.

[42] Bais P, Moon SM, He K, *et al.* PlantMetabolomics.org: a web portal for plant metabolomics experiments. Plant Physiol 2010; 152(4): 1806-16.

[43] Mattivi F, Guzzon R, Vrhovsek U, Stefanini M, Velasco R. Metabolite profiling of grape: Flavonols and anthocyanins. J Agric Food Chem 2006; 54(20): 7692-702.

[44] Palomo ES, Diaz-Maroto MC, Vinas MAG, Soriano-Perez A, Perez-Coello MS. Aroma profile of wines from Albillo and Muscat grape varieties at different stages of ripening. Food Control 2007; 18(5): 398-403.

[45] Torregosa L, Pradal M, Souquet JM, Rambert M, Gunata Z, Tesniere C. Manipulation of Vv*Adh* to investigate its function in grape berry development. Plant Sci 2008; 174(2): 149-55.

[46] Guerrero RF, Liazid A, Palma M, *et al.* Phenolic characterisation of red grapes autochthonous to Andalusia. Food Chem 2009; 122(4): 949-55.

[47] Kalua CM, Boss PK. Evolution of volatile compounds during the development of cabernet sauvignon grapes (*Vitis vinifera* L.). J Agric Food Chem 2009; 57(9): 3818-30.

[48] Kalua CM, Boss PK. Comparison of major volatile compounds from Riesling and Cabernet Sauvignon grapes (*Vitis vinifera* L.) from fruitset to harvest. Aust J Grape Wine Res 2010; 16(2): 337-48.

[49] Parker M, Pollnitz AP, Cozzolino D, Francis IL, Herderich MJ. Identification and quantification of a marker compound for 'pepper' aroma and flavor in shiraz grape berries by combination of chemometrics and gas chromatography-mass spectrometry. J Agric Food Chem 2007; 55(15): 5948-55.

[50] Cavaliere C, Foglia P, Gubbiotti R, Sacchetti P, Sampieri R, Lagnanà A. Rapid-resolution liquid chromatography/mass spectrometry for determination and quantification of ployphenols in grape berries. Rapid Commun Mass Spectrom 2008; 22(20): 3089-99.

[51] Toffali K, Zamboni A, Anesi A, *et al.* Novel aspects of grape berry ripening and post-harvest withering revealed by untargeted LC-ESI-MS metabolomics analysis. Metabolomics in press, DOI: 10.1007/s11306-010-0259-y.

[52] Sweetman C, Deluc LG, Cramer GR, Ford CM, Soole KL. Regulation of malate metabolism in grape berry and others developing fruits. Phytochemistry 2009 70 (11-12): 1329-44.

[53] Zhang XY, Wang XL, Wang XF, *et al.* A shift of phloem unloading from symplastic to apoplastic pathway is involved in developmental onset of ripening in grape berry. Plant Physiol 2006; 142(1): 220-32.

[54] Afoufa-Bastien D, Medici A, Jeauffre J, *et al.* The Vitis vinifera sugar transporter gene family: phylogenetic overview and macroarray expression profiling. BMC Plant Biol 2010; 10: 245.

[55] Di Carli M, Zamboni A, Pè ME, *et al.* Two-Dimensional Differential in Gel Electrophoresis (2D-DIGE) Analysis of Grape Berry Proteome during Postharvest. J Proteome Res 2011; 10(2): 429-46.

[56] Kanellis AK, Roubelakis-Angelakis KA. Grape. In: Seymour G, Taylor J, Tucker G, Eds. Biochemistry of fruit ripening. Chapman and Hall, London 1993; pp. 189-234.

[57] Boss PK, Davies C, Robinson SP. Analysis of the Expression of Anthocyanin Pathway Genes in Developing Vitis vinifera L. cv Shiraz Grape Berries and the Implications for Pathway Regulation. Plant Physiol 1996; 111(4): 1059-66.

[58] Davies C, Boss PK, Robinson SP. Treatment of grape berries, a nonclimacteric fruit with a synthetic auxin, retards ripening and alters the expression of developmentally regulated genes. Plant Physiol 1997; 115(3): 1155-61.

[59] Gagné S, Saucier C, Gény L. Composition and cellular localization of tannins in Cabernet Sauvignon skins during growth. J Agric Food Chem 2006; 54(25): 9465-71.

[60] Aharoni A, Keizer LC, Van Den Broeck HC, *et al.* Novel insight into vascular, stress, and auxin-dependent and –independent gene expression programs in strawberry, a non-climacteric fruit. Plant Physiol 2002; 129: 1019-31.

[61] Symons GM, Davies C, Shavrukov Y, Dry IB, Reid JB, Thomas MR. Grapes on steroids. Brassinosteroids are involved in grape berry ripening. Plant Physiol 2006; 140(1): 150-8.

[62] Chervin C, El-Kereamy A, Roustan JP, Latche, Lamon J, Bouzayen M. Ethylene seems required for the berry development and ripening in grape, a non-climacteric fruit. Plant Sci 2004; 167(6): 1301-5.

[63] Polesani M, Bortesi L, Ferrarini A, *et al.* General and species-specific transcriptional responses to downy mildew infection in a susceptible (Vitis vinifera) and a resistant (V. riparia) grapevine species. BMC Genomics 2010; 18: 11:117.

[64] Nunan KJ, Sims IM, Cacic A, Robinson SP, Fincher GB. Changes in cell wall composition during ripening of grape berries. Plant Physiol 1998; 118(3): 783–92

[65] Nunan KJ, Davies C, Robinson SP, Fincher GB. Expression patterns of cell wall-modifying enzymes during grape berry development. Planta 2001; 214(2): 257-64.

[66] Yakushiji H, Sakurai N, Morinaga K. Changes in cell-wall polysaccharides from the mesocarp of grape berries during veraison. Physiol Plant 2001; 111(2): 188–195.

[67] Park JM, Park CJ, Lee SB, Ham BK, Shin R, Paek KH. Overexpression of the tobacco Tsi1 gene encoding an EREBP/AP2-type transcription factor enhances resistance against pathogen attack and osmotic stress in tobacco. Plant Cell 2001; 13(5): 1035-46.

[68] Jacobs AK, Dry IB, Robinson SP. Induction of different pathogenesis- related cDNAs in grapevine infected with powdery mildew and treated with ethephon. Plant Pathol 1999; 48(3): 325–336

[69] Kim CY, Zhang S. Activation of a mitogen-activated protein kinase cascade induces WRKY family of transcription factors and defense genes in tobacco. Plant J 2004; 38(1): 142-51.

[70] Iandolino A, Nobuta K, da Silva FG, Cook DR, Meyers BC. Comparative expression profiling in grape (Vitis vinifera) berries derived from frequency analysis of ESTs and MPSS signatures. BMC Plant Biol 2008; 8: 53.

CHAPTER 12

The Microbial Community of Grape Berry

V. Loureiro[1], M. M. Ferreira[1,*], S. Monteiro[1] and R. B. Ferreira[1,2]

[1]Centro de Botânica Aplicada à Agricultura, Instituto Superior de Agronomia, Technical University of Lisbon, 1349-017 Lisbon, Portugal and [2]Instituto de Tecnologia Química e Biológica, New University of Lisbon, 2781-901 Oeiras, Portugal

Abstract: The microbial community of grape berry is composed of an array of species exhibiting differential physiological characteristics and relevance to vine growing and winemaking. The most important phytopathogens responsible for grapevine diseases worldwide are the oomycete *Plasmopara viticola* (downy mildew) and the ascomycete *Erysiphe necator* (powdery mildew). The causal agent of grey rot is the saprophytic mould *Botrytis cinerea*. A wide diversity of yeast species are also common contaminants of berry surfaces, but the key agent of wine fermentation, *Saccharomyces cerevisiae*, is rarely recovered from the grapes. Bacterial groups include the spoiling acetic acid bacteria and lactic acid bacteria responsible for the malolactic fermentation. These microorganisms colonise grape surfaces from berry set to ripening, following a repeatedly cyclic pattern year after year. Highly complex interactions and chemical signalling take place among grapevines themselves and with the intervening biota, which also include insects, birds and mammals. The fundamental role played by the nonmicrobial biota in grape berry microbiota ranges from their role (especially in the case of insects) as microbial vectors to the damage directly inflicted on the grapes, which pave the way to the entrance of the saprophytes. The precise biota and the resulting interactions depend fundamentally on the berry development stage, on the intactness of the grape skin and on the prevailing environmental conditions, and exert a profound effect on the fruit quality. Given the great ecological, technological and economical importance of studying the grape microbiota, it is somewhat surprising to find scarce and fragmented information available on these topics.

Here we provide a balanced, highly multidisciplinary overview of the most relevant components of grape berry microbiota. Our proposal establishes four distinct groups of microorganisms - residents, adventitious, invaders and opportunists - which are defined on the basis of grape biochemical evolution, nutrient availability and ability to proliferate on berry surface. Their natural proliferation is particularly dependent on two main events: veraison and berry damage. The origin and the colonisation sequence on berry surface by those several groups of microorganisms will be tentatively settled.

Keywords: Bacteria, Berry damage, *Botrytis cinerea*, *Drosophila spp.*, *Erysiphe necator*, Grape berry microbiota, *Oenococcus oeni*, Ontogenic resistance, *Plasmopara viticola*, *Saccharomyces cerevisiae*, Skin integrity, Vectors, Volatile organic compounds, Yeast.

INTRODUCTION

Globally, grape berry microbiota may be grouped according to their ecological significance and technological and economic importance. The wine microbial consortium includes yeasts responsible for wine fermentation and spoilage, lactic acid bacteria responsible for metabolizing malic acid and for some wine disorders, and acetic acid bacteria, capable of converting wine into vinegar. The fungi relevant to viticulture include obligatory parasites and saprophytes, responsible for serious economical losses resulting from both grape quality and yield, and fungi producing mycotoxins present in grapes, raisins or wines, which affect public health. The microorganisms capable of growth on the intact berry surface, probably without economical importance but certainly with ecological relevance, include filamentous fungi and obligatory aerobic yeasts, mostly basidiomycetes. Finally, there is a wide diversity of microbial cells present on the berry surface which are unable to grow and, consequently are lacking any known economical or ecological significance.

***Address correspondence to M. M. Ferreira:** Centro de Botânica Aplicada à Agricultura, Instituto Superior de Agronomia, Technical University of Lisbon, 1349-017 Lisbon, Portugal; E-mail: mmalfeito@isa.utl.pt

Hernâni Gerós, M. Manuela Chaves and Serge Delrot (Eds)

To understand the evolution of grape microbiota during berry development, it is essential to describe the surface of grape berry, the potential nutrient sources available, the origin of chemical signals captured by plants to arm their defenses and by vectors carrying microorganisms to the berries, and the source of signals activating responses in invading fungi or triggering vine responses to the presence of these pathogens.

GRAPE BERRY MICROBIOTA

The different methodologies used to recover and identify grape microbiota influence dramatically both microbial counts and the number of species identified. Their major limitations are discussed to better understand the ecological significance of the available data.

Microbial Characterisation Approaches

Characterisation, identification and interpretation of grape berry microbiota depend heavily on the methods employed. A major distinction may be established between classical, cultivation-dependent and molecular biology methods, the latter based on the extraction and analysis of microbial DNA or RNA. Both exhibit advantages and disadvantages, possibly leading to conflicting and non-comparable results, of difficult interpretation, and that do not provide a real picture of grape microbiota.

The sampling scheme and grape selection are the first factors to consider. First, samples must account for the non-uniform microbial distribution in both space and time [1]. This limitation may be minimised through multiple sampling, using several berry samples collected from different vineyard locations and from two or more harvest years. Then, berries must be separated from bunches according to their health status to allow for sound ecological interpretations.

The classical methods are further influenced by 1) the microbial dislodgement technique from the berry surface, 2) the culture media and 3) the incubation conditions. Martini *et al.* [2] showed that a sequence of aggressive and disruptive actions on berry surfaces together with a previous enrichment culture on sterile grape must have allowed the recovery of much higher counts and number of species than did traditional isolation procedures. However, when the berries are subjected to "auto-enrichment", factors such as sugar concentration and anaerobiosis enact strong selective pressure in favour of fermenting yeasts, limiting the ecological value of the results [3]. Such enrichment techniques may be essential to recover minority populations of fermentative yeasts, like those of *Saccharomyces cerevisiae* or *Dekkera bruxellensis*. Generic culture media enable the recovery of dominant and faster growing populations while selective media must be used to have a wider view on microbial diversity [4, 5]. An advisable strategy may be to use a series of general purpose and selective media, directed to yeasts, bacteria and moulds [6-10]. In addition, the obligatory mould and oomycete parasites are not cultivable in culture media, requiring a totally different approach for detection and identification.

After strain isolation, molecular biology techniques have replaced the classical methods for species identification. They are particularly appropriate to identify species within heterogeneous genera like *Candida* or *Pichia*, for which the classical methods often do not go deeper than the genus level, or produce erroneous identifications. For these reasons, several yeast species have been recently associated with grapes for the first time, most probably because classical identification in former surveys did not proceed beyond the genus level. On the contrary, the major benefit of molecular methods for well defined and technologically relevant species of *Saccharomyces*, *Zygosaccharomyces*, *Torulaspora* or *Dekkera* is highlighted by the possibility of typing at the strain level. There are, nevertheless, a few exceptions regarding certain biochemical reactions or incubation temperatures that are required to identify species with similar molecular restriction profiles [9, 10].

Culture-independent methods have uncovered a wider range of species over the last decade. Direct extraction of DNA may establish how many live or dead species are present, whilst RNA analysis identifies those that are metabolically active. These analyses may be carried out by hybridisation with specific probes, using species-specific polymerase chain reaction (PCR), or by universal PCR and identification of the amplified products. The use of universal primers possibly allows the identification of all species from large microbial groups. Two important disadvantages have been noted for the former technique [11]: 1) there is a limit to the number of species detected/identified in a given sample; 2) previous knowledge on which microorganisms to look for in a sample is required. Another limitation of the molecular techniques is the detection threshold, being less sensitive than generic culture for determining the yeast ecology of grapes, as they do not allow reliable species detection at populations less than 10^4 $cfu.g^{-1}$ [12], although lower detection limits have been recently achieved [11].

The combined use of enrichment cultures and molecular biology is a particularly efficient procedure to detect minority species. In fact, the most recent hypothesis on *S. cerevisiae* natural dissemination was based on strain recovery using synthetic culture media with high ethanol content [13, 14]. The utilisation of selective broths or suspension media also enabled the easy recovery of the rare spoilage yeast *D. bruxellensis* and the common malolactic fermentation agent *Oenococcus oeni* from grape surfaces [6], which otherwise would not have been detected.

Systematisation of Grape Berry Microbiota

The overall characterisation of grape berry microbiota must take into account all microbial groups, commonly studied under different scientific approaches, gathering those concerned with wine fermentation and spoilage, those relevant to vine and grape pathology and those related to ecological issues. The model proposed in this chapter is an attempt to systematize grape berry microbiota, and is mostly based on the different abilities of microorganisms to utilise the nutrients available on berry surfaces, especially those occurring after berry damage.

The Influence of Nutrient Availability

Intact grape surfaces are part of the phylloplane, considered in the sense proposed by Fonseca and Inácio [1]. From a microbe perspective, the phylloplane is a continuously fluctuating physical environment, both spatially and temporally, which must not be considered as an extreme environment, but as a harsh environment due to the frequent, repeated and rapid changes in the microclimate factors which occur in different time scales, ranging from a few minutes up to a few months. The nature and size of epiphytic microbial populations are shaped by the availability of nutrients [1]. The oligotrophic nature of the phylloplane [15] derives from the multiple origins of its nutrients, which may be exogenous, such as compounds contained in debris, pollen and honeydew, as well as those derived from other organisms (microbes or insects), or endogenous. In addition, grape berries have been reported to exudate a variety of compounds through the cuticle and the epicuticular wax layer onto their surface, including phenols, sugars, lipids, malic acid, potassium and sodium [16]. Interestingly, grape berries exude onto their surface both microbial inhibitory [17] and stimulatory compounds [18-20], the former inhibiting the growth of *Botrytis cinerea* during the berry formation stage and the latter promoting it during the berry ripening stage [17, 18]. Some studies showed that the contents in sugars, organic acids and amino acids removed from sound berry surfaces by washing with water, increase during berry development [16, 21].

Microbial Groups

In microbial ecology, Sergei Winogradsky proposed the concepts of "autochthonous" and "allochthonous" respectively, to distinguish between microorganisms that dominate a particular ecosystem because they always find an adequate nutrient supply in it and organisms which are present essentially as contaminants [22]. Likewise, Davenport [23] considered two types of yeasts in each habitat – the residents and the transients – to distinguish those capable of growth in an environment from those transported to an environment where they are unable to grow. Our proposal to systematize grape berry microbiota goes a step further, taking also into account the differential technological significance of the berry microbial components.

The model proposed in this work groups grape microorganisms into residents (oligotrophs), adventitious, invaders and opportunists (copiotrophs), according to the access and origin of the nutrients consumed (Table **1**). The residents are microorganisms in the undamaged grape, from berry set to harvest, using the nutrients available on the surface of immature or mature berries. The adventitious are those detected on berry surfaces, at any phase of berry development, but without the ability to grow and whose presence derives from mere contamination by both biotic and abiotic vectors. Invaders are those which manage to penetrate through the intact skin tissue by their own means to gain access to grape pulp (or leaf) nutrients. They thrive from berry set to veraison and do not require the existence of skin lesions. The opportunists, who may also be coined as copiotrophic residents, are those able to grow, typically after veraison, at the expense of grape pulp nutrients as a consequence of microfractures or wounds in the berry skin. As in any classification, the borders between any two groups are not always clear-cut, as described below.

The group of microorganisms that dominate at any one time in a particular grape berry depends obviously on environmental factors, but is determined by the precise nutrient source under exploitation. In intact berries, the residents will override. However, depending on the berry stage of development (see below), when access is achieved to the nutrient-rich pulp, the residents are rapidly superseded by the invaders or the opportunists. Such access must be painstakingly conquered by the microorganisms themselves, which need to win a difficult chemical war resulting in skin perforation and the establishment of pathogenesis, or may be provided by a wound, either of abiotic (*i.e.*, a weathering factor such as hail) or biotic (*i.e.*, an insect or a larger animal) origin. In either case, the wound largely influences the microbiota which subsequently colonizes the damaged grape since it will induce a rapid change in the blend of grapevine volatile organic compounds (VOCs) emitted, which in turn will trigger an exchange of chemical signals or "chemical talk" among organism populations, including at minimum plants, microorganisms and their vectors.

Table 1: Grouping the main microbial species associated with grape development and disease.

Group	Microorganisms	Genera/Species	Significance
Residents	Ascomycetous moulds	*Aureobasidium pullulans*	Innocuous contaminants
	Basidiomycetous yeasts	*Cryptococcus spp., Rhodotorula spp., Rhodosporidium spp., Sporodiobolus spp.*	Innocuous contaminants
Adventitious	Several bacterial species	*Acinetobacter spp., Curtobacterium spp., Pseudomonas spp., Serratia spp., Enterobacter spp., Enterococcus spp., Bacillus spp., Staphylococcus spp.*	Innocuous contaminants
Invaders	Phytopathogens	*Plamospara viticola*	Downy mildew
		Erysiphe necator	Powdery mildew
		Guignardia bidwelli	Black rot
		Elsinoë ampelina	Anthracnose
		Pseudopezicula tracheiphila	Rotbrenner
Opportunists	Saprophytic moulds	*Botrytis cinerea*	Grey rot, Noble rot
		Aspergillus alliaceus, A. carbonarius, A. niger aggregate, A. ochraceus	*Aspergillus* rot, ochratoxin A producers
		Penicillium expansum	Green mold, patulin producer
		Cladosporium herbarum	Cladosporium rot
		Coniella petrakii	White rot
		Alternaria alternata	Alternaria rot
		Trichothecium roseum	Pink rot, thricothecen producer
	Ascomycetous yeasts	*Candida spp, Debaryomyces spp., Hanseniaspora spp., Issatchenkia spp., Kloeckera, spp., Metschnikowia, spp., Pichia spp.*	Must and wine contaminants

		Saccharomyces cerevisiae	Fermenting yeasts
		Zygosaccharomyces spp.	Wine spoilage yeasts
	Acetic acid bacteria	*Gluconobacter spp., Acetobacter spp., Gluconoacetobacter spp.,*	Wine spoilage, vinegar production
	Lactic acid bacteria	*Oenoccocus spp., Lactobacillus spp., Pediococcus spp., Weissella spp.*	Malolactic fermentation or wine spoilage

Residents

The oligotrophic residents use the nutrients present on the berry cuticle, including those fortuitously carried by rain or wind and those carried in the paws, bodies or mouthpart systems of visiting insects. In the common absence of the ephemeral, exogenous nutrient sources, intact grape berry surfaces are still usually colonised by microbial populations, probably at the expense of the endogenous nutrient sources. These may include the hydrophobic VOCs released by the berry which undergo dissolution by cutin or compounds which diffuse through the intact cuticle and undergo partial dissolution by condensed water. Consequently, residents are adapted to a quantitatively poor environment. Typical examples are the ascomycete yeast-like fungi *Aureobasidium pullulans* and some basidiomycetous non-fermentative yeasts (*Cryptococcus* spp., *Rhodotorula* spp., *Sporobolomyces* spp., *Sporidiobolus* spp. and *Rhodosporidium* spp.). Their populations constitute the dominant zymoflora of undamaged grapes, where they increase from berry set to harvest, and are similar to those recovered from the phylloplane of aerial plant parts [1]. They are also detected in damaged berries, where they suffer from the strong competition from the opportunists. Their significance is related to berry microecology and possible microbial interactions, since they do not survive for long during fermentation and are regarded as useless in winemaking.

The bacterial component of residents is poorly studied, and includes Gram – or Gram + bacteria [8]. It is reasonable to assume that they do not grow on cuticle nutrients, justifying their inclusion in the opportunist or adventitious groups. Concerning the lactic acid bacteria, Renouf *et al.* [6] detected, for the first time, the presence of *O. oeni* and *Pediococcus parvulus* before veraison. If regular isolation of these nutrient demanding species from immature berries is confirmed, then they should be considered as oligotrophic residents. However, at present they should be considered as opportunists because only they reach high counts in damaged berries. A similar justification applies to acetic acid bacteria.

Adventitious

The adventitious include those microorganisms unable to attain high populations in either sound or damaged berries, including mainly common environmental contaminants such as *Stenotrophomonas*, *Enterobacter*, *Serratia*, *Acinetobacter*, *Curtobacterium*, *Staphylococcus, Pseudomonas, Leifsonia, Enterococcus* and *Bacillus* species. Their recent detection on grapes is explained by the utilisation of direct molecular techniques [6, 8, 24]. Their inability to grow on sound berries should be related to their reduced capacity to use cuticle nutrients, while the low pH of grape juice should explain the absence of growth on damaged berries. The species *Burkholderia vietnamiensis*, recently reported by Renouf *et al.* [6], is an endophytic bacterium capable of entering the plant through the roots and of reaching the berries through the sap [25]. Therefore, it should not be included within the epiphytic microbiota.

Invaders

Fungal obligate parasites are able to penetrate through the intact grape skin by their own biochemical and mechanical activities and are responsible for high economic losses in the vineyards all over the world. The main species are the oomycete *Plasmopara viticola*, responsible for downy mildew, and the ascomycetes *Erysiphe necator* (powdery mildew), *Elsinoë ampelina* (anthracnose), *Guignardia bidwellii* (black rot) and *Pseudopezicula tracheiphila* (rotbrenner). Their biology and epidemiology, well-known by phytopathologists, are strongly dependent on weather conditions (mainly temperature and humidity). Berry susceptibility to these diseases decreases from berry set until veraison, after which development of ontogenic resistance explains the absence of parasite attacks, even in the absence of phytochemical treatments.

Opportunists

The opportunist group gathers the majority of species included in the wine microbial consortium. They are usually detected from veraison until harvest, particularly when damage is inflicted upon the berry skin. In favour of this hypothesis is the work of Belin [26] demonstrating that yeasts are not evenly distributed on the berry surface but concentrate in microcolonies around the pedicel insertion, peristomatic aureoles and near the stylar remnants, where it is more likely to find nutrients leaking from berry interior. Among yeasts, the genera *Candida, Metschnikowia* and *Pichia,* namely the species *C. stellata, M. pulcherrima* and *P. guilliermondii,* exhibit almost a universal distribution. They are characterised by a weak fermentative ability and by a high nutritional versatility. In addition, the genus *Hanseniaspora* (anamorph *Kloeckera*), particularly the species *H. uvarum* and *H. guilliermondii*, are probably the most frequent yeasts detected after veraison, ranging from 10^2 to 10^3 cells/berry, explaining their supremacy during the first phase of spontaneous wine fermentations.

Recently, Gaudory *et al.* [27] demonstrated that berries apparently sound to the naked eye, but affected by diffuse powdery mildew (DPM), changed from a predominance of *Aureobasidium* spp. to one of *Hanseniaspora* spp. Similar alterations were found with *C. stellata, M. pulcherrima* and the mould *B. cinerea,* but no significant changes were observed with sound and DPM berries for other species of *Candida, Sporidiobolus, Cryptococcus* and *Debaryomyces.* Therefore, while the former species behave typically as opportunists, these *Candida* and *Debaryomyces* species are better included in the oligotrophic residents.

S. cerevisiae is a classical example of the opportunist group because its occurrence is extremely rare on sound berries (see below). All other fermentative yeasts often detected in damaged berries and in fermenting juices are also included in this group, namely those belonging to the genera *Zygosaccharomyces, Torulaspora, Saccharomycodes, Schizosaccharomyces, Lachancea/Kluyveromyces, Issatchenkia* and *Dekkera/Brettanomyces.*

The bacterial fraction includes mainly species resistant to low pH, such as acetic acid and lactic acid bacteria, which outcompete the other bacteria when there is juice leaking from the berry pulp. The studies available are scarce, but in sound berries *Gluconobacter oxydans* predominate, while in rotten berries the supremacy goes for *Acetobacter aceti* and *A. pasteurianus* (see below). In what lactic acid bacteria are concerned, the dominance may be related to their ability to accumulate intracellular Mn(II) as a defense mechanism against endogenous O_2^- [28]. Their nutritional demand is the main argument to consider them within the opportunist group.

A wide diversity of saprophytic filamentous fungi is also universally present from berry set to harvest (Table **2**). In the case of sound berries, fungi are present as spores, predominantly asexual, vectored by wind and insects, and behaving as adventitious microorganisms. When the berry skin is damaged or shows microfractures, the spores may germinate and multiply in a limited way, not visible to the naked eye but clearly seen by scanning electron microscopy. Their presence is especially visible in damaged berries, as described below.

Table 2: Incidence of spores from major fungal genera during development and ripening of sound berries (expressed as a percentage of colonised berries after plate growth) [7].

Genera	Pea berry	Early Veraison	Harvest	Average
Cladosporium	25	26	24	25
Alternaria	32	25	16	24
Botrytis	12	17	16	15
Penicillium	5	8	14	9
Aspergillus	3	4	15	8
Epicoccum	8	6	3	6
Aureobasidium	4	4	5	4
Total number of colonised berries	1450	1645	1600	4695

GRAPE BERRY CHARACTERISATION

Anatomically, the grape berry is composed of three main parts, which vary considerably in composition: pulp, skin and seed, with the sheer bulk of must being derived from the pulp. Characterisation of the grape skin assumes a far greater importance than would be expected at first glance, an observation that may explain the relatively few studies especially dedicated to this grape component. Berry skins are metabolically active during development and ripening, exhibit endocrinal function [48] and constitute a physical barrier between the external environment and the inner tissues, meaning that their integrity is a key factor in preventing pathogen infections.

The aerial organs of all plants are covered by a thin, conspicuous lipophilic layer, the cuticle, at the interface between plant tissue and environment, which defines their boundaries and provides protection against biotic and abiotic stresses [49, 50]. Increasing evidence highlights that the cuticle cannot be regarded as a multifunctional passive physical barrier on the plant surface. Its function as a source of signals for invading pathogens and for the induction of innate immune responses in plants must also be taken into account [51]. The cuticle may also act as a source of signals for the host plant. Cuticle monomers and components of the plant cell wall released by pathogens during the early stages of infection can act as elicitors of plant defense responses [51]. Last but not least, the hydrophobic nature of the cuticle may dissolve, and thus retain some isoprenoid VOCs released by grapes – a potential nutrient carbon source for the berry microbiota, particularly the resident community.

Although the chemical composition of plant cuticles varies between species, and even between organs of a single plant, they are all broadly composed of two major components [51-53]: (i) the insoluble cutin, a chloroform-insoluble complex polymer consisting mostly of C_{16} or C_{18} ω-hydroxylated esterified fatty acids, the major structural component matrix; (ii) a mixture of waxes: the intracuticular waxes, which infiltrate the cutin framework, and a thin amorphous epicuticular wax layer which accumulates on the surface and is composed of a very complex mixture of long-chain lipids. In the specific case of the grape berry cuticle, the major components of cutin are C_{16} and C_{18} fatty acid esters [54]. The capacity of certain pathogens to metabolize cutin and cuticular waxes as well as pectic and hemicellulosic components may explain why certain species of fungi and aerobic yeasts, with high metabolic diversity, are so frequent on the sound berry surface. Also, the composition of the cuticular waxes changes during the period from flowering to maturity, revealing an increase in waxy deposits and significant modifications on the wax surface morphology. The content in cutin per unit surface decreases more than 2.5-folds between berry set (16 d after anthesis) and veraison, which might predispose the grape berry to fungal infection. Differentiation of the cuticle layers and a decrease in the thickness of the primary cuticle is further detected at harvest, partly justifying the well-known augmented susceptibility of ripe grapes to *B. cinerea* infection, either through an apparently intact or damaged berry surface. At the final stage of growth, berry has a smooth, continuous and homogenous, 3 μm thick cuticle. This homogeneity is no longer observed in tightly closed bunches, where abrasion between skins weakens cuticle thickness, creating conditions to the loss of pellicle integrity [55].

The morphology of berry surface presents several natural or artificial points of entry. In berries, as in leaves, stomata are natural pores bordered by two guard cells which regulate transpiration and gas exchanges. Stomata also constitute natural openings providing direct access to leaf tissue for numerous microbes. With time, functional stomata turn into wax-occluded lenticels, occasionally surrounded by crevices which may be perforated by hyphae [56].

The Changing Grape Berry, from Anthesis to Maturity

Grape berry development consists of two distinct but successive sigmoidal growth periods separated by a lag phase (Fig. **1**). The first, the berry formation stage, also termed herbaceous or green phase, is the first period of growth, lasting from bloom to approximately 60 days afterwards. During this stage, it is expected, at least from a theoretical point of view, for the grape skin to behave much like a leaf. The berry is formed, the seed embryos are produced, and rapid cell division occurs, so that by the end of this period, the total number of cells within the berry has been established. The berry expands in volume as solutes accumulate

[58]. From an ecological point of view and taking into account the interests of the plant, the first period of growth fulfills the first priority of the grapevine, which is to develop a viable seed and to produce compounds (organic acids, tannins, pyrazines) which combine to make foraging by birds and mammals a downright unpleasant experience, thus protecting it while doing so [59]. However, such compounds do not confer full protection or immunity to berries against pathogen attack. As a result, one or more chemical wars, each comprising many battles, may be initiated at any one time with pathogens attempting infection. The weaponry used by both sides consists of proteins and secondary metabolites. The outcome of each war determines whether pathogenesis develops or not [60].

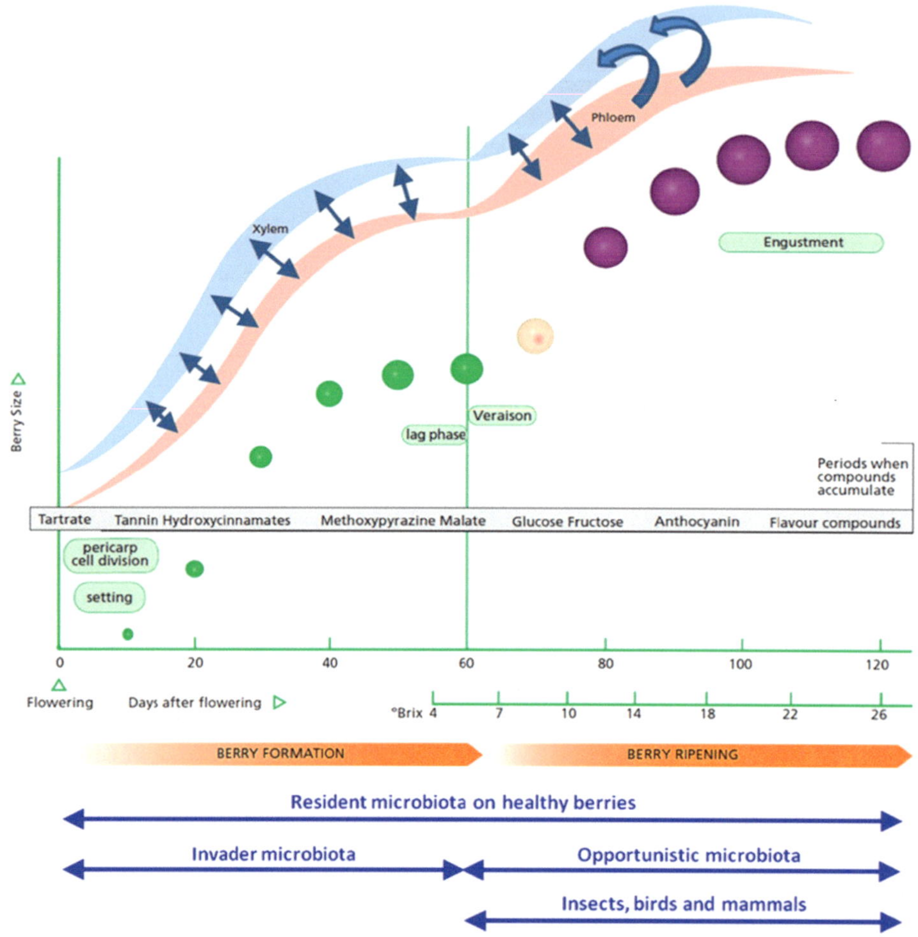

Figure 1: Diagram showing relative size, colour and composition of berries at 10-days intervals after flowering, passing through major developmental events (rounded boxes). Also shown are the periods when compounds accumulate, the levels of juice brix, an indication of the role of flow of xylem and phloem vascular saps into and/or out of the berry and the major biota present (adapted and modified from [57]).

The transition from first to second stage, or the inception of the ripening stage, occurring between 8 and 14 weeks after flowering, is termed turning or veraison [61]. Veraison is marked by a turning point in the grape skin phenylpropanoid pathway at the level of coumaroyl-CoA, which results in a shift from stilbene synthesis before veraison to anthocyanin synthesis following veraison (see below). After veraison berries typically accumulate sugar, soften, and undergo skin coloration which is most notorious in the case of red varieties due to the accumulation of reddish anthocyanins (restricted to skin tissue in most cultivars; [61]). Overall, the berry approximately doubles in size between the beginning of the second growth period and harvest [57]. Again from an ecological point of view and taking into account the interests of the plant, the goal during the second period of growth is to make the berry as appealing as possible to birds and mammals so that seed dispersal may occur [57].

Two main events during berry ripening have direct influence on the dynamics of microbial populations. The first is of chemical nature and reflects the grape metabolic shift occurring during veraison. The second, of physical nature, is related to the smooth release of VOCs from inner tissues culminating with full pulp nutrient availability at berry surface when skin rupture occurs.

In addition to insects, birds and mammals, the two developmental stages may also be distinguished by the microbiota populations which thrive on/in the grapes. While resident microorganisms are present on healthy grape surfaces at all times during their development, invader infections occur essentially during the berry formation stage, and opportunist infections are concentrated during the berry ripening phase. Grapes are known to develop ontogenic or age-related resistance to several pathogens as they age (*e.g. Eryshiphe necator* and *Plasmopara viticola*, and the invaders in general), but the opposite seems to occur in what the opportunists are concerned. Thus, *B. cinerea*, a necrotrophic opportunistic pathogen, assumes a totally different strategy when compared to the invaders. As an opportunist, *B. cinerea* degrades plant cells and absorbs the nutrients from dead tissue, whereas the invaders typically associate with living host cells without directly causing host cell death but redirecting their metabolism to obtain all the nutrients they require [62].

Berry Damage

The access to the nutrient-rich inner cells, through fissures or wounds on the berry skin changes completely the berry microbiota. The content in sugars, organic acids and amino acids is about 100 times higher in visibly damaged berries than in sound berries [21]. Grape damage is strongly dependent on grape skin integrity, as a function of grape variety, and biotic and abiotic stress factors. There is a positive correlation between skin thickness and pellicle elasticity on one hand and berry resistance to rot on the other. Abiotic factors capable of damaging the skin of such grapes assume a particular importance under these circumstances, as in the case of hail, rain or strong winds, late frosts, wide thermal amplitudes and intense radiation. In our experience, the most frequent situations arise from early rain in late summer which, occurring after a long drought period, may originate an abrupt grape volume increment (which may reach 30% of the berry weight). Such conditions lead to pellicle rupture, particularly in the case of fragile skin varieties. Biotic factors may also assume an important role in grape skin damaging. Thus, for example, *Lobesia botrana* and *Eupoecilia ambiguella*, whose larvae penetrate the intact berry skin, explain the positive correlation between their presence and grey mould or sour rot infections. Wasps, which unlike bees possess mouthparts capable of piercing the peel, also play an active role in this process, as do many bird species.

All factors, commonly described as influencing grape microbiota (rainfall, wind, temperature, diseases, pests, viticultural practices, *etc.*; [45]), affect primarily skin integrity, and so their impact must be analysed, taking into account all the changes induced by berry damage. Loss of skin integrity paves the way to massive growth of opportunists adapted to the low pH (3.0 to 3.8) and high sugar concentration (*ca.* 200 $g.L^{-1}$) of grape juice. These opportunists are all microorganisms of the wine microbial consortium, except for filamentous fungi, which may affect markedly the quality of grapes and wines but are unable to grow under wine conditions.

The number of yeast species recovered from grape berries provides an estimate of its wide diversity (Table **3**). However, as already mentioned, much of that variability may be explained by grape sampling when sound grapes are not separated from the damaged ones in apparently sound bunches. Damaged grapes possess much higher cell counts (up to 5 log cycles) and display wider species diversity than sound grapes (Table **4**). The exception is related to damage made by honeydew produced by mealybugs, where counts are similar in sound and affected berries. This may be explained by the presence of antimicrobial compounds and high sugar concentration, as in honey produced by bees [10]. Nevertheless, damage increases the amount and proportion of certain opportunist yeasts in detriment of the residents' *A. pullulans* and basidiomycetes. In particular, with reference to sour rot and honeydew, Barata *et al.* [9, 10] proposed that *Zygoascus hellenicus* and *Issatchenkia* spp. may be regarded as zymological indicators of these types of damage.

Concerning *S. cerevisiae*, the occurrence is about 0.05% to 0.1% in sound berries and 25% in damaged berries, usually with numbers of about 10^6 – 10^7/berry [63-65]. These observations explain why spontaneous fermentations are dominated, after grape pressing and a short period of time, by *S. cerevisiae*. The genus *Hanseniaspora*, particularly the species *H. uvarum*, seems to be the dominant fermentative yeast of sound and damaged berries at the early stages of spontaneous fermentations. There is no reported explanation for this dominance, which may be related to: 1) the role played by *Drosophila* spp., the so-called "fruit flies" or "yeast flies" [66], in berry microbial colonisation; 2) genetic factors determining specific growth rates and Monod constants in grape juice. The genus *Zygosaccharomyces* contains other relevant species in oenology, which are detected at higher frequencies in grapes affected by noble rot, sour rot and honeydew, suggesting their adaptation to conditions of reduced water activity and presence of weak organic acids (*e.g.* acetic acid; [9, 10]).

Table 3: Yeast and yeast-like species isolated from "sound" grapes at harvest (data collected from selected surveys). The symbol + indicates relative proportion of the detected species.

Species	France	Italy	Arkansas, USA	Japan	Spain	Canada	Australia	Argentina	Slovenia	Greece	Portugal	India	China	Spain	Brazil
Basidiomycetes [a]	+++	++	+++	+++	+++	+++	++	+	+++		+++		+++		+
Aureobasidium pullulans	+	+++			+++	+++	+++		++	+				+	+
Hanseniaspora spp.			+++	+						++		+++	+	+++	
H. uvarum	+++	++	+++	+++	+	++		+++	++	+++	++	+	+++	+++	+++
Metschnikowia spp.	+	++	+					+++	+		+		+	+	
Candida spp.		+	++	++	++			+			+	++	++		
C. stellata/ C. zemplinina								+		++	+		+		
Debaryomyces spp.									+			+			
Issatchenkia spp.								+		+		++	+	++	+++
Kluyveromyces spp./Lachancea spp.											+			++	
Pichia spp.		+	++					++	+		+	++	++	+	
Brettanomyces spp.		+													
Saccharomyces spp.		+	+												
S. cerevisiae												++		++	
Saccharomycopsis spp.		++	++												
Saccharomycodes ludwigii								+							
Torulaspora spp.											+			+	
Zygosaccharomyces spp.		+	+										+		
Z. bailii													+	+	
References	[29]	[30]	[31]	[32]	[33]	[34]	[12]	[35]	[36]	[37]	[9,10]	[38]	[39]	[40]	[41]

[a] Cryptococcus spp., Bulleromyces spp. , Sporidiobolus spp., Sporobolomyces spp., Rhodotorula spp., Trichosporon spp.

Table 4: Incidence of the most frequent yeast species, identified by molecular methods, in sound and damaged grapes at harvest time.

Type of damage	Sound berries	% [a]	Damaged berries	% [a]	Reference [b]
Mixed damage	*A. pullulans*	100/95	*Metschnikowia spp.*	75/64	[12]
	Cr. victoriae	-/4	*A. pullulans*	25/<1	
	R. laryngis	-/1	*Hanseniaspora spp.*	-/36	
Diffuse powdery mildew	*Aureoubasidium spp.*	65/66	*Hanseniaspora spp.*	67/80	[27]
	Hanseniaspora spp.	23/22	*Metschinikowia spp.*	19/3	
			Aureobasidium spp.	5/2	
			C. stellata	4/10	
Noble rot	*H. uvarum*	77	*H. uvarum*	75	[37]
	H. guilliermondii	9	*M. pulcherrima*	7	
	H. opuntiae	9	*H. opuntiae*	5.5	
	I. terricola	5	*H. guilliermondii*	3.5	
			Z. bailii	3.5	
			C. zemplinina	3.5	
			I. terricola	2	
Grey rot	*H. uvarum*	67	*H. uvarum*	76	[37]
	C. zemplinina	22	*C. zemplinina*	12	
	H. opuntiae	6.5	*I. occidentalis*	5	
	A. pullulans	4.5	*I. terricola*	3.5	
			H. opuntiae	3.5	
Sour rot	*Basidiomycetes*	89/100	*C. vanderwaltii*	59/76	[10]
	C. vanderwaltii	11/0	*H. guilliermondi*	23/3	
			P. membranifaciens	16/0.3	
			L. thermotolerans	2/6	
			Basidiomycetes	1/0.1	
			H. uvarum	-/12	

[a] Results of two samples separated by a slash.

[b] Frequency expressed as a percentage of specific colonies relative to total colony number on plate media.

Concerning bacteria, acetic acid and lactic acid bacteria predominate in both sound and injured grapes, but damage to grapes increases their numbers and influences the dominant species present. A good example is provided by *G. Oxydans*, which predominates in sound berries, but is superseded by *Gluconoacetobacter* and *Acetobacter*, namely *A. aceti* and *A. pasteurianus*, in rotten berries. Lactic acid bacteria, essentially represented by the genus *Lactobacillus,* are less numerous. Sound berries harbour less than 10^3 cfu.g^{-1} and damaged ones harbour up to 10^5 cfu.g^{-1}. The bacterial species diversity is higher in rotten berries but, surprisingly, *O. oeni*, the agent responsible for malolactic fermentation in most wines, is very rare and has not been isolated from either type of grapes by most studies (Table **5**).

Table 5: Bacterial species detected in both sound and damaged grapes.

Bacterial groups	Species	References [a, b]
Lactic acid	*Lactobacillus plantarum*	[6,8,42,43]
	L. hilgardii	[42]
	L. casei, L. sanfranciscansis	[6,43]
	L. lindneri, L. kunkeei	[8]
	L. kefiri [c], *L. mali* [c], *L. plantarum* [c]	[8]
	L. brevis	[24]
	Lactococcus lactis	[8,44]
	Leuconostoc fallax	[44]
	Lc. mesenteroides	[6,43]
	Oenococcus oeni [a]	[6,43]
	Pediococcus parvulus, P. damnosus, P. acidilactici	[6,43]
	Weissella paramesenteroides	[6,8,43]
Acetic acid	*Gluconobacter oxydans*	[6,24,43,45-47]
	Acetobacter aceti	[45-47]
	A. pasteurianus	[45,46]
	A. cerevisiae	[24]
	Gluconoacetobacter hansenii	[47]
Other species	*Burkholderia vietnamiensis, Pseudomonas jessenii, Serratia rubidae, Enterobacter gergoviae, Leifsonia xyli, Enterococcus faecium, Bacillus mycoides*	[6,43]
	Enterococus durans, E. faecium, E. avium	[8]
	E. hermaniensis [c]	[8]
	Acinetobacter spp., Curtobacterium spp., Enterobacter spp., Stenotrophomonas maltophila *Serratia spp., Staphylococcus spp.*	[24]

[a] Microbial detection after culture enrichment [6, 8, 24, 43].

[b] Microbial detection after plating [6,42-47].

[c] Recovery from damaged grapes only.

The saprophytic fungi which do not usually grow on healthy berry surfaces come into scene after skin rupture, leading to fast rotting under favourable environmental conditions. The classical example is given by *B. cinerea*, the causal agent of grey rot, responsible for intense damage when rainfall is abundant, especially towards the end of grape ripening, as a result of berry splitting. Under particular conditions of temperature and low humidity, *B. cinerea* infects white grape varieties of a rather thin cuticle (*e.g.* Furmint, Sémillon and Riesling), originating the highly valued noble rot. Grape quality is often enhanced by the increased concentration of berry constituents due to water evaporated through the mould-made holes. This species is a typical opportunist, but it remains unclear whether it can penetrate by its own means through the intact berry skin and/or if it requires the existence of micro-fissures known to be present under field conditions (see below). However, this behaviour is not typical of an invader, because it occurs after veraison and not before, as observed with obligatory parasites.

Other fungal species, classified as typical opportunists, may also assume some economical relevance in the final ripening phase as agents of several types of detrimental rot, such as *Penicillium expansum* (green mold), *Coniella petrakii* (white rot), *Alternaria alternata* (Alternaria rot), *Cladosporium herbarum* (Cladosporium rot), *Trichothecium roseum* (pink rot) and *Aspergillus* spp., namely *A. alliaceus, A. carbonarius, A. niger* aggregate and *A. ochraceus* (Aspergillus rot), possible ochratoxyn A (OTA) producers [67]. Their ability to colonize damaged berries depends strongly on temperature and humidity, as well as on the inoculum level. Thus, some OTA producing strains may be relevant in warm and humid regions [68], as in the Mediterranean basin. However, the presence of OTA in wine is mostly not of great concern because it may be easily controlled or eliminated [69].

Bunch *Versus* Berry

When studying individual berries, it is commonly assumed as straightforward to judge soundness by simple visual inspection. However, the presence of micro-fractures, micro-fissures or DPM, invisible to the naked eye, requires the use of a microscope. The situation is further complicated when the whole bunch is used, because the presence of a single hidden berry may alter dramatically, both qualitatively and quantitatively, the conclusions drawn on the berry microbiota. Unfortunately, numerous studies use bunches instead of single berries, whereas others do not specify the precise biological material, making it difficult, if not impossible, to establish comparisons and to interpret the published results.

Under field conditions, the tendency to rot is essentially genetically determined, varying with the grape variety according to skin thickness, bunch shape and density. High bunch density, as in the Portuguese varieties Baga and Trincadeira, augments the probability of berry burst due to the pressure exerted by the growing neighbouring grapes. This is the reason why most table grape varieties possess low density bunches. On the contrary, there is a general feeling that the higher the bunch density, the higher the propensity of each variety for wine production.

The strength of the bond between berry and pedicel is another factor which greatly influences the susceptibility of each variety to rot: a loose pedicel is indicative of a higher tendency for leakage. This is the reason why grapes from many varieties (*e.g.* the Portuguese table grape D. Maria) start rotting in the pedicel area, in a manner similar to strawberries and raspberries. In conclusion, grape bunch morphology and berry skin resistance are major determinants underlying the number and diversity of grape berry microbiota and, consequently, are prerequisites to take into account when studying grape microbial ecology.

THE MAIN MICROBIAL SPECIES

Three major fungal/oomycete diseases affect grape cultivation throughout the world: *viz* powdery mildew, downy mildew and *Botrytis* bunch rot or grey mould, whereas one yeast and one bacterium assume a leading role at the winery: *Saccharomyces cerevisiae and Oenococcus oeni.* Their ecological behaviour has many contact points, but their full comprehension continues to defy the scientific community.

Plasmopara viticola, an Invader

Downy mildew caused by the oomycete, obligate biotroph *Plasmoparaviticola* (Berk. & M. A. Curt.) Berl. & de Toni, is one of the most widespread and destructive diseases of grapevine and occurs in most of the grape growing areas of the world where warm and wet weather conditions prevail during the vegetative growth of the vine [70, 71]. Most cultivars of *V. vinifera* are susceptible to downy mildew [71]. All green parts with stomata of the host grapevine may be infected with *P. viticola*, and grapes are no exception. *P. viticola* infects only through stomata. The occurrence, development and spread of disease are mainly dependent on rainfall and temperature [71]. The pathogen overwinters mainly as oospores in fallen leaves but can also survive as mycelium in buds and in persistent leaves. The oospore, the primary source of infection, germinates on moist soil or in the presence of free water when temperatures rise above 11 ^{o}C, producing sporangia. The zoospores produced by sporangia swim with the aid of flagella and are projected onto vine tissues near the soil by rain splashes. In the presence of water, zoospores swim for a few minutes with whirling motion and near stomata loose their flagella, assume a spherical shape, are enclosed in a membrane called encyst and under moisture conditions produce a single, nonseptate, flexuous, germ tube which penetrates the stomatal aperture. When the nutrient substrate is exhausted the pathogen generates sporangiophores exterior to underside of leaves through ostioles of stomata [71].

During the *P. viticola* life cycle, stomata play an important role in the infection and sporulation stages [72], since they represent the natural openings which provide direct access to the inside (infection stage) or the outside (sporulation stage). In the course of grape berry development, stomata become non-functional as they are gradually converted to lenticels when berries are "pea-sized" [19, 27]. *P. viticola* is capable of infecting older, lenticel-containing berries by growing through the pedicels into the berries [70] (Fig. **2**). Resistance to *P. viticola* is associated with the loss of the infection court as stomata are converted to lenticels, but the time of onset, cultivar variation and seasonal variation in ontogenic resistance remain to be elucidated [70].

Figure 2: Symptoms and signs of grapevine downy mildew. (A) Sporulation on pedicel of berry prior to appearance of berry symptoms. (B) Discoloration on berries with necrosis of pedicel and/or sporulation upon pedicels. (C) Necrotic pedicels of berries prior to appearance of berry symptoms. (D) Emergence of sporangiospores through incisions in the epidermis of symptomatic postveraison berries. Conversion of stomata to lenticels following infection prevented emergence of sporangia prior to creating the incisions (adapted and modified from [70]).

Erysiphe necator, an Invader

Powdery mildew of grapevine incited by the obligate biotrophic ascomycetous fungus *Erysiphenecator* (synonym *Uncinula necator* [Schw.] Burr.) is a grapevine disease of wide occurrence. It is the most serious fungal disease of the cultivated grapevine and therefore, of great economical importance [73-75]. Most cultivars are highly susceptible to *E. necator* because *V. vinifera* was not exposed to this pathogen during its evolution [76]. In contrast, other grapevine species, such as *V. labrusca*, *V. rupestris*, *V. aestivalis* and *M. rotundifolia* co-evolved with *E. necator* on the North American continent and acquired various levels of resistance to the pathogen [76]. As with downy mildew, symptoms of the disease appear on all the green parts of vine, leaves, inflorescence and berries, and greatly affect their cuticle.

For the powdery mildew fungus, penetration of the plant surface corresponds to penetration of the epidermic cell wall [73] (Fig. **3**). Upon contact with an epidermal cell, an *E. necator* conidiospore germinates almost immediately, producing one or two germ tubes. A lobed structure called appressorium then forms at the tip of a germ tube. An infection peg then forces penetration through the surface cuticle and the cell wall of the epidermal cell below. After cuticle and cell wall penetration the fungus begins to invade into the cell but does not pierce the plasma membrane of the epidermal cell; instead the plasma membrane invaginates and expands to surround the fungal body. At this point the peg differentiates into a bulbous and more elaborate structure named the haustorium [74]. This fungal feeding structure takes up nutrients for further mycelial growth, which occurs only on aerial epidermal tissue, allowing continuation of the infection cycle: extension of secondary hyphae, infection of additional cells and production of conidiophores bearing new conidia. After 5–8 days, the asexual life cycle is completed by the formation and release of conidiospores [62]. The process of penetration and haustorium formation in *V. vinifera* may be as short as 14 h under optimal conditions [74].

The pathogen survives or overwinters from one growing season to the next as mycelia and conidia in dormant buds and as ascospores produced in cleistothecia, meaning that there are two sources of primary inoculum for *E. necator*. The primary infection leads to the production of conidia, which then serve as means of secondary infection.

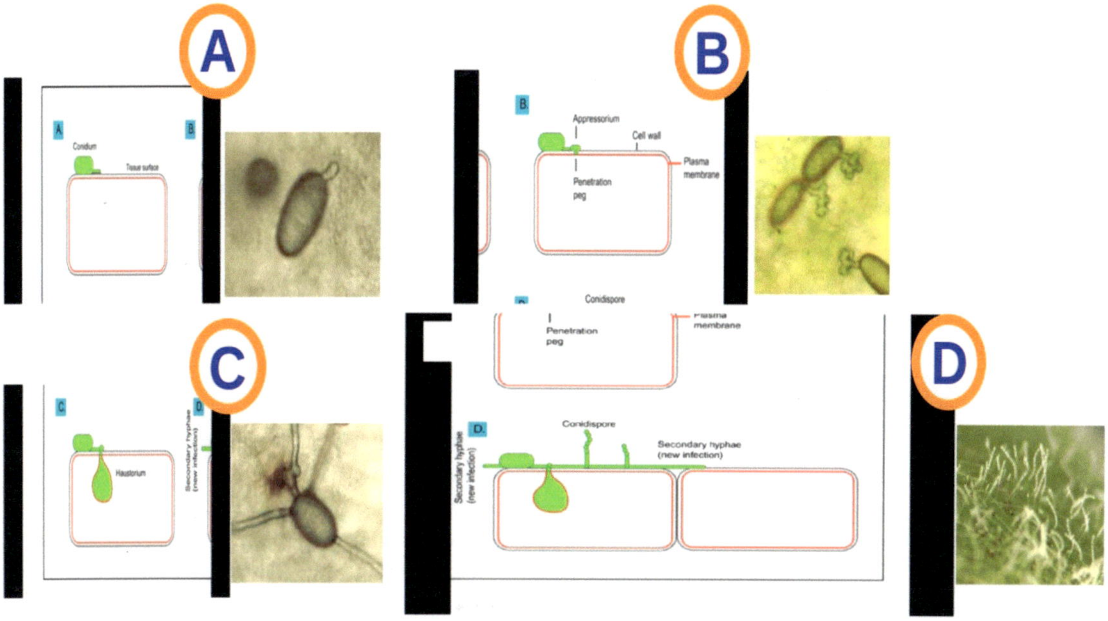

Figure 3: Stylised diagrams (left) and microscopic images (right) of the infection structures of *Erysiphe necator* growing on *Vitis vinifera* leaves cv. Chardonnay. (A) Germination and hyphae extension. (B) Appressorium formation. (C) Secondary hyphae extension. (D) Conidiophores on leaf surface. A, B, C: 40x magnification, D: 10x magnification (adapted and modified from [73]).

Under appropriate conditions, the overwintering mycelia present in buds or leaf primordia are activated at bud break or with the onset of vegetative growth and cover the emerging shoots with abundant conidia containing mycelia [71]. Ascospores are the other source of primary infection, which germinate under suitable conditions to form normal powdery mildew colonies [71]. Conidia are dispersed by wind or physical contact between infected and healthy plant parts [73].

Grape berries are generally considered to be susceptible to infection by *E. necator* during the berry formation stage but resistant after veraison, *i.e.*, during the berry ripening stage [77]. The ontogenic resistance of mature grapes to *E. necator* infection has been attributed to a thicker grape skin, as well as to the presence of PR proteins and phenolic compounds.

A brief mention should be made to the so-called diffuse powdery mildew (DPM, [27]), very likely mistaken in field surveys for powdery mildew even by experienced observers, which represents a colony-level reaction by the dying pathogen to the gradual development of ontogenic resistance. The end result is a wounded but macroscopically unblemished epidermis. An interesting study was reported by Gadoury and colleagues [27]: grape berries, nearly immune to infection by *E. necator* due to the development of ontogenic resistance, which appear to be healthy and free of powdery mildew, in late-season vineyard assessments with the naked eye may still support diffuse and inconspicuous mildew colonies. The authors showed that the presence of these colonies on berries was associated with (i) elevated populations of spoilage microorganisms; (ii) increased evolution of volatile ethyl acetate, acetic acid and ethanol; (iii) increased infestation by a number of fruit-feeding insects known to be attracted to the aforementioned volatiles, which act as sensory cues to locate sources of food; (iv) and increased rotting by *B. cinerea*.

Botrytis cinerea, an Opportunist

Botryotinia fuckeliana [teleom.] (de Bary) Whetzel, *Botrytis cinerea* [anam.] Pers.: Fr, is a cosmopolitan filamentous fungus and a necrotrophic plant pathogen which is able to infect more than 200 dicotyledonous plant species and which causes one of the most serious diseases in grapevine, *i.e.*, *Botrytis* bunch rot or grey mould [78].

Botrytis bunch rot occurs under cool, wet and humid conditions, which favour sporulation and infection. The severity of grey mould is closely related to environmental conditions and is especially dependent on temperature and relative humidity. The infection of grapes by the fungus requires at least 25 h at 15-20 °C and saturated relative humidity.

There is a considerable degree of uncertainty/controversy concerning the entry port of *B. cinerea* into grape berry cells. *B. cinerea* does not usually penetrate through stomata, with infection taking place *via* wounds. It has been well established that wounds on the grape surface, either of abiotic or biotic origin, constitute a preferential way for *B. cinerea* entry. Wounds and cracks of unknown origin are frequently observed on the surface of apparently healthy grape berries (Fig. **4**). In addition, the presence of wounds of exogenous origins has also been shown to promote infection by the pathogen, *e.g.* injuries caused by insects which can also act as vectors of conidia [78].

Botrytis infection is dependent on the evolution of berry resistance during fruit development, *i.e.*, ontogenic resistance. The susceptibility of grapes to this infection has been shown to greatly increase at the onset of grape ripening (veraison) [78-80]. In addition, high free water level, suitable temperature and the presence of nutrients on the skin surface are considered as necessary to induce spore germination and fungal growth. Spore germination is greatly stimulated by the presence of low concentrations of simple sugars, such as glucose and fructose, as well as by amino acids on the grape berry surface. Nitrogen nutrition is also important, as demonstrated by field observations showing that nitrogen fertiliser increases bunch rot probably due to nitrogen causing increased canopy density, which in turn causes a microclimate more conducive to the development of botrytis bunch rot [78].

The disease cycle of *B. cinerea* includes inoculation, penetration, invasion, growth and reproduction, dissemination, and the survival stage of the pathogen [81]. The fungus utilizes three sources of primary inoculum that play a role in the epidemiology of the disease, namely mycelia, conidia and sclerotia [82]. A major determinant factor in the occurrence of secondary infections is the infection of floral tissues and/or fruit pedicel, as occurs in late spring, followed by a period of latency until veraison [78, 82]. Soon after the fungus becomes active, it infects both the mature and immature grapes in veraison as the secondary infection [82].

DPM is likely to play an important role in *Botrytis* infection, since it apparently provides a court for direct infection by *B. cinerea*. Sporophores of *B. cinerea* were observed emerging occasionally not only from patches of dead epidermal cells but also from lenticels and the calyx and stem end of the fruit (Fig. **4**).

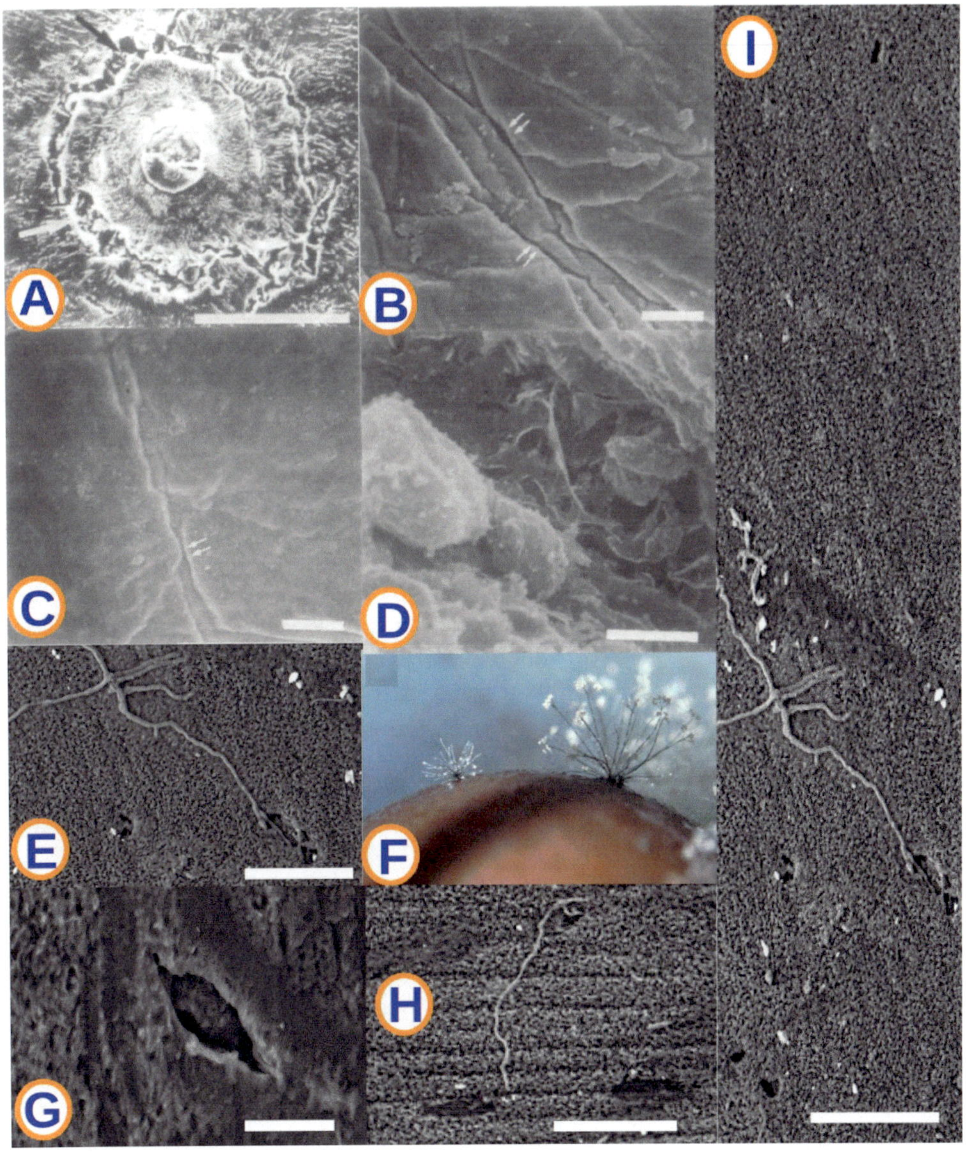

Figure 4: Scanning electron micrographs of a lenticel and the appearance of cracks in the skin of grapes at maturity. (A) Lenticel with microfissures in the peristomatic area (arrow). (B, C) Cracks in a disease-susceptible and disease-tolerant clone skin grapes, respectively (double arrow). (D) Germination of a *Botrytis cinerea* conidium, near a crack. (E) Mycelium from an unknown fungal species growing onto the surface of an apparently healthy ripen grape berry. (F) *Botrytis cinerea* emerging from lenticel (left) and calyx end (right) of Chardonnay berry bearing diffuse powdery mildew colonies. (G, H, I) Cracks in the skin of apparently healthy ripen grape berries. (D) Bar = 2 µm; (G) bar = 25 µm; (H) bar = 50 µm; (A, B, C, E, I) bar = 100 µm. A to D were adapted and modified from [79] and F from [27]. E, G, H and I were obtained from mature and apparently sound berries of Portuguese cv. Touriga Franca by Dr. Olfa Zarruk (sample preparation) and Dr. Victoria Fernández (scanning electron microphotography).

Saccharomyces cerevisiae, an Opportunist

S. cerevisiae should probably be considered as an irrelevant object of study given its scarcity in nature. However, it becomes a main protagonist due to its major role in fermentation, widely recognised after the pioneering work of Louis Pasteur in late 19th century. Since then it has been overwhelmingly used as a model microorganism in yeast research. Its common recovery from grapes using enrichment procedures and predominance in alcoholic fermentations may explain why it is popularly known as a widespread microorganism. In fact, outside fermentative environments, the isolation of *S. cerevisiae* is rare, even in vineyards. This has led to the hypothesis that *S. cerevisiae* in nature is a migrant from human-associated fermentations [3, 83], but more recent evidence indicates that fermenting strains have derived from natural populations unassociated with alcoholic beverage production [84].

Davenport [85] was probably the first to make a wide survey on vineyard environments and realised that this species was absent from grapes. This author isolated the yeast only from vine flowers, soil, compost and mummified pears on the ground [85], which is in accordance with former references associating it to decaying plant tissues, mushrooms and soil litter, independent from the proximity of vineyards [86]. More recently, ecological studies outside the vineyard have shown that this species is found all year long in the bark of oak trees and soil [13], as well as in fruit tree leaves during vegetative growth [87]. The predominance and total numbers are always low in these sources. When present, *S. cerevisiae* occurs in high numbers in damaged grapes. This species becomes dominant only in fermenting juices in the winery, being expectable that its recovery increases in the winery neighbourhood or near water running off the winery [88]. Without a vector, these yeasts are not found far from the winery. Commercial starters do not dominate over natural grape contaminants [89], although they may suffer genetic modifications reflecting adaptive mechanisms to the grape environment [88].

Having in mind that grapes, and particularly damaged grapes, occur naturally for only about 2 months, from veraison to harvest or decay, the understanding of *S. cerevisiae* ecology depends on the enlightenment of the dissemination from the permanent sources (mainly tree bark and soil) to the grape surface. Sampaio and Gonçalves [13] proposed a new model for the ecological niche of *S. cerevisiae* and its close relatives, *S. kudriavzevii*, *S. paradoxus*, and *S. uvarum*, for which tree bark constitutes the primary reservoir. Sugar-rich environments related to the maturation of different kinds of flowers and fruits typically have a seasonal occurrence and consequently would constitute the secondary habitat. This suggestion implies that besides the adaptation to thrive in sugar-rich environments, *Saccharomyces* yeasts have also evolved mechanisms to cope with quite distinct environmental conditions, such as those prevailing in tree bark. Accordingly, Goddard *et al.* [14] showed that *S. cerevisiae* strains residing in soil, vine bark and surrounding flowers are dispersed by bees and are present in local winery ferments. The authors suggested that *S. cerevisiae* exists primarily in such niches as spores, which are dispersed from the diffuse reservoir surrounding ephemeral fruits (bark and soil) by insects to fruits as they come into season.

The question of *S. cerevisiae* origin and possible strain relation with a particular habitat has been debated under the concept of "terroir" used in wine production. This concept justifies wine distinctive quality mainly by the climatic and soil characteristics of its region of production. The role of yeasts in wine distinctiveness has also been claimed on a regional basis, but this issue is not supported by a solid scientific background. The scientific evidence required to validate this assumption would need to prove (i) whether strains isolated from grapes can be responsible for wine fermentations; and (ii) if specific *S. cerevisiae* strains can be spatially coincident with wine production regions. The answers to these two questions have not yet been definitely established. Some reports state that grape *S. cerevisiae* strains are not responsible for wine fermentations [90, 91] and that they have scarce persistence on sound or damaged grapes [92]. The winery resident *Sacharomyces* microflora has been reported to be independent from the strains present in grapes [91], but Budroni *et al.* [93] and Le Jeune *et al.* [94] demonstrated that part of the fermentative *S. cerevisiae* derived from the winery and part from the vineyard. Then, if a "terroir"-related population may be found it is not certain that it conducts or directs wine fermentation. Even when considering exclusively the winery resident populations, the available evidences are controversial. Although some reports refer the consistent isolation of specific strains over the years in single wineries [95], it seems that strain variability, within the same winery and over the years, is more likely to occur [96, 97]. The possibility of frequent crossings between strains and genome interchanging during vegetative propagation may also contribute to the different predominant strains collected over repeated vintages [14, 98]. This high-level diversity forms the basis for *S. cerevisiae* evolution and has been explained by the model of "genome renewal" [83]. Therefore, until now there is no evidence for the existence of a perennial *S. cerevisiae* population resident in a winery or originating from grapes collected in any particular wine region.

The second question, concerning spatial coincidence between strain distribution and wine region, is perhaps more difficult to prove. As mentioned above, yeast dissemination takes place all year round whereas grapes, considered as a favourable substrate for yeast propagation, only occur during 2 months. Thus, assuming the absence of vine resident strains, the existence of "terroir" strains should be first sought after in vineyard environs and then checked for similarity with the grape isolates. Although insects exhibit short dissemination ranges, the role of vectors (*i.e.*, migrating birds) should also be evaluated since association of yeast strains to specific regions would require proving that yeasts are not coming from outside regional mesoclimatic boundaries.

According to our winemaking experience, wine distinctiveness is mainly given by the quality and characteristics of grapes, independently of the strain of the fermenting yeast. Despite the absence of evidence indicating the existence of "terroir yeasts", the studies on this subject will continue, in our opinion, as long as there are commercial interests behind the association of yeast consortia with the uniqueness of particular wines.

Oenococcus oeni, an Opportunist (?)

As in the case of *S. cerevisiae*, the lactic acid bacterium *O. oeni* has a rare occurrence in nature despite the essential role it plays in winemaking: it is responsible for the malolactic fermentation, as it occurs in most low-pH wines throughout the world, and it is a potential producer of biogenic amines, which justifies the long held interest of researchers [99, 100]. Nevertheless, studies on its ecology are virtually nonexistent and it remains to be established whether the absence of reports on its isolation from vineyards or from healthy, damaged or rotten grapes is due to its presence in very low numbers or to the inefficacy of the isolation procedures utilised. Most studies failed to detect it in grapes or vineyards [8, 42, 44], despite its frequent isolation from winery environments after fermentation [42, 101-103]. With the use of molecular methodology, independent of microbial growth, it has been recently detected at the berry surface, both in cv Cabernet Sauvignon after berry set and in cv Merlot after veraison [6]. Such observations make it hard to understand *O. oeni* behavior and the size of its populations. Therefore, *O. oeni* ecological origin remains a mystery, leading to the assumption that its role in the malolactic fermentation derives from two main aspects: (i) the presence in the wineries as a resident bacterium [104]; and (ii) the high capacity to adapt to wine conditions (low pH, alcohol and tannins in particular), allowing its dominance over the other lactic acid bacteria.

For the reasons expressed above, it remains uncertain whether *O. oeni* should be considered as an opportunist (copiotrophic) or a resident (oligotrophic) – further studies are required to clarify its nature.

THE INTERACTIONS AMONG GRAPEVINE, VECTORS AND MICROORGANISMS

Each year the grape berry surface is colonised by microorganisms, being a repetition of the previous cycle which initiates with flower fecundation and finishes with harvest or decay in the soil. To understand the cycle renewal it is necessary to consider the vectors conveying the microbes to the berry, its primary source of microorganisms, and the physical and chemical stimuli that attract vectors to the berry. Unfortunately, such studies are still sparse, making it difficult to establish an explanatory model adequately supported by scientific evidence. Therefore, in the next section we will formulate a hypothetic model based on the scarce information available, our own experience and plausible speculations.

Natural Sources of Grape Berry Microbiota

In the case of moulds, their life cycles are relatively well known, mainly those from parasites, as already described, given the need to optimize plant protection schemes and an appropriate timing of fungicide treatments. The spores are produced massively and so are easily detected, mainly when fruit bodies are visible, being spread in the vineyard surroundings. When humidity, temperature conditions and nutrient availability are adequate, spores germinate, and soon after a new generation of spores is massively spread by wind and insects. On the contrary, lactic acid bacteria, acetic acid bacteria and yeasts do not form spores massively, nor do they have specialised structures to resist the winter stress. It is then expectable that microbial numbers are reduced to their lowest values during the 8 months in which grapes are not available. This explains their difficult and often unsuccessful detection in nature. Although most studies have been directed to study the microbiota of wine fermentations, many authors succeeded in isolating wine consortium microorganisms, namely yeasts, from different habitats. Yeast cells are generally found in large amounts, up to several millions per gram, in the following substrata: sweet, juicy fruits, especially in a decomposing state; agarics, especially decaying and decomposing fruit bodies; tree exudates; flower-visiting insects. However, in most flower samples no yeast cells or only a few were detected, *i.e.*, they seem to occur in similar small amounts as might be found on leaves and bark, although their numbers increase considerably during summer months [86].

Specific studies on vineyard soils have been reported [85, 105-107]. Recent reports describe the consistent and high frequency isolation of *Saccharomyces sensu stricto* species from natural samples of oak tree bark in the United States [108], England [109] and Portugal [13], strongly suggesting that tree barks, and particularly those of certain oaks, are natural habitats for *Saccharomyces* yeasts [13] and probably for other opportunistic microbiota.

Acetic acid bacteria have generally been regarded as environmental and ubiquitous bacteria. However, recent research on microbe-insect symbiosis has shown that bacteria of the genera *Acetobacter, Gluconoacetobacter, Gluconobacter, Asaia, Saccharibacter* and the novel genus *Commensalibacter* establish symbiotic relationships with several insects of the orders Diptera, Hymenoptera, Hemiptera and Homoptera, all relying on sugar-based diets, such as nectars, fruit sugars or phloem sap [110]. The close association between acetic acid bacteria and their insect hosts has been confirmed by demonstrating the multiple modes of bacteria transmission among individuals and to their progeny, including vertical and horizontal transmission routes and comprising a venereal one [110], suggesting that the insects may well be considered both as natural sources and as vectors of acetic acid bacteria.

Various species of lactic acid bacteria (*Bifidobacterium*, *Lactococcus*, *Lactobacillus* and *Leuconostoc*) can be isolated from the bodies and intestinal contents of insects that visit ripe grapes, such as honey bees and wasps [111-114]. However, most of the bacterial species isolated are different from those commonly encountered on grape berries, implying that insects cannot be regarded as their natural sources. Other bacterial species associated with honey bees were reported by Gilliam [112]: Gram variable pleomorphic bacteria, *Bacillus* sp. and *Enterobacteriaceae* were consistently found, although the intestinal microflora of mature, worker bees may vary to a certain extent with the age of the bee, season and geographical location [115].

VOCs Emission by Grapevine

The change in skin colour and the increase in berry aromatic intensity after veraison are the main stimuli that attract insects and other living organisms to sound grapes in their final stage of ripening. There are a few studies concerning insect attraction by grape berries, which take into consideration their importance in the process of berry microbial colonisation. However, the principle of insect attraction to their plant hosts by volatile compounds is well established [116, 117].

Grapevine provides a particularly fascinating example for the multiple functions of VOCs, which are exploited by insect herbivores and predators (*e.g. Lobesia botrana*) as long-range signals for host location [118-122]. Another well known example is the attraction of the fruit fly (*Drosophila* spp.) by the grape berry, mainly when it is damaged. The natural environment for the fruit fly consists of fermenting plant material (with yeasts and acetic acid bacteria), rich in ethanol and other alcohols, acids, acetone, acetoin, fruit esters, and highly volatile esters such as ethyl acetate [123]. The invasion of wineries by fruit flies at the time of fermentation is a well known visible consequence of the attraction of *Drosophila* by acetic acid and ethyl acetate, metabolites typically produced by yeast and acetic acid bacteria. However, the chemical signalling reason explaining why *Drosophila* is attracted to recently damaged berries without signals of fermentation remains to be elucidated.

Vectors

A number of vectors, both of abiotic and biotic nature, are actively involved in the transportation of microbiota to and from grape berries, playing a decisive role in berry colonisation by microorganisms. However, they have been poorly studied, especially in the case of animals, which managed to establish chemical communication with plants, using volatile compounds that are typically present in the atmosphere in minute amounts.

Wind and Water

The role of wind in fungal spore dispersion is well known. Spore structure and the massive amounts in which they are produced evolved to allow their long distance transport by the wind. The history of European viticulture evidences clearly the role of wind in the dispersion of those fungi which were brought to Europe in mid-19th century from the American continent, namely the powdery and downy mildews that devastated most European vineyards in just a few years. However, the transatlantic crossing was only possible by sea transport, when commercial trade between both sides became intense. Such observations indicate that there is a yet unknown range limit for wind to disseminate fungal spores.

Water, mainly from rainfall and fog, is another vector with the ability to drag microbes suspended in aerosols, namely fungal spores and bacteria resistant to UV radiation. Water drops, splashed when heavy rain hits the ground or grapevine stems and trunks, are known to be part of the life cycle of major grapevine fungal pathogens. Conversely, under heavy rain, microbes may be expelled and washed away from berries.

Grapevine, like other plants, emits VOCs under normal, healthy conditions. However, assuming that between berry set and veraison there are no motives on either side to attract insects and birds, it is reasonable to accept that most microorganisms, until veraison, reach the berry surface being carried by water and/or wind.

Birds

Vineyard avifauna and its role in berry microbiota have also been little studied. Recently it was shown that blackbirds may carry yeasts of the wine consortium, and that grapes pecked by birds harboured $7.94 \times 10^4 \pm 7.94 \times 10^3$ $CFU.g^{-1}$ [44]. Much of these birds are local, but migrant birds, such as the classic example of starlings, may have devastating effects in that they destroy berries and pave the way for a subsequent invasion by opportunists. If these migrant birds are shown to act as microbial vectors, the hypothesis of "terroir" microbiota becomes seriously compromised.

Insects

Insects are recognised as essential vectors in microbial dispersion through paws and wing contact and, mainly, *via* their mouthparts. The insect mouthparts are decisive in the pollination of many plant species and in microbial colonisation of many fruits, including grapes. Thus, it seems highly relevant to know precisely the vine entomofauna, particularly those species which participate actively after veraison. Among the few studies reported on the subject, Stan *et al.* [124] describe the vineyard entomofauna structure in the Craiova region, Romania. In a survey performed in South Africa the arthropod community composition was found to be influenced by a combination of management practices, the surrounding landscape and geographic locality, highlighting the interdependence of the cultivated area with its surroundings [125].

In addition to the insects mentioned above, there is a diversity of ants, moths, weevils, cicadas, spiders and grasshoppers that visit and attack grapevine leaves and berries at various stages of development [126]. It is not uncommon to observe such insects crawling over brunches of berries as they grow and mature in vines (Fig. **5**). It is highly likely that all these visitors will contribute generally and specifically to the microbiota of grapes, although there is little information available on this topic.

COLONIZATION SEQUENCE OF GRAPE BERRY MICROBIOTA

Based on the annual lifecycle of grapevine, on the nature and habitats of microorganisms and their vectors, and on the complex chemical signaling which takes place between plants on one hand and insects and birds on the other, it is possible to formulate a sequence hypothesis in an attempt to explain the way berry surfaces become colonised, mainly as far as the wine microbial consortium is concerned.

Figure 5: Insects differentially attracted to grape berries bearing diffuse powdery mildew colonies. (A) Ants (*Leptothorax* spp.) congregating around cavity excavated during feeding by yellowjackets (*Vesper* spp.). (B) Sap beetle (*Glischrochilus quadrisignatus*) on Pinot Noir berry. (C and D) Yellowjackets preparing to feed upon berries with unbroken skin. (E) Damage to Pinot Noir berry inflicted by single yellowjacket during 30 s of feeding. Wound cavity (wc) is ≈ 1.0 mm in diameter and extends through the epidermal layer into the berry pulp. (F) Exudation of sap from wound cavity in Pinot Noir berry after 30 s of feeding by a single yellowjacket (adapted from [27]).

Around harvest time, the vine plant environs should have the maximum load of wine microbial consortium, resulting from grape damage and fall to the ground, as well as from juice dropping. The microbial cell concentration should increase as a function of grape damage. Conditions become favourable to a fast and massive multiplication of moulds, yeasts, acetic acid bacteria and possibly lactic acid bacteria. Intense activities of the entomofauna, namely *Drosophila*, honey bees and wasps, and avifauna, are expected, dispersing the microorganisms under a radius of a few kilometers. With the arrival of winter and the onset of plant dormancy, microbial populations undergo a gradual reduction over the entire ecosystem, reaching a minimum before the beginning of bud burst, when a new cycle of growth begins. When conditions become favourable, most life forms – vine plants, other plants, insects, birds and microorganisms – experience increasing activity. The grapevine phylloplane will host the residents, adventitious and the invaders, until veraison, depending on the prevailing environmental conditions. The adequate use of phytochemicals will prevent the proliferation of the invaders. At this stage, vine tissue colonisation should be made by wind and rainfall, although some insects that hatch in this season, as some mite species, might assume a relevant role. Meanwhile, precocious plants might have already been visited by pollinating insects that initiate the dispersion of wine microbial consortium species. Recently, the main grape yeast and yeast-like species have been consistently isolated from the leaf surface of fruit (apple, plum, cherry, apricot and peach; [127]) and forest trees (spruce, pine, willow, oak, beech, maple, hornbeam, linden, acacia and ash; [87]), suggesting that they may share a similar colonisation process. Before vine flowering, strawberries, loquats and cherries attain maturity, attracting several insect species, particularly to the damaged fruits. Chemical signaling between plants and biotic vectors increases dramatically. The appearance of fermentative yeasts in damaged fruits allows the simultaneous or sequential development of acetic acid and lactic acid bacteria and a high concentration of insect vectors. This process will proceed in other fruits – peaches, pears, precocious figs, plums, apples, late figs and quinces, until the time for grapes arrives. When berries are sound, they should be visited by fewer insects to suffer less contamination by the opportunists of the wine microbial consortium. They can, however, be visited by insects or birds that feed on intact grapes, particularly after veraison, with devastating effects, opening wounds and boosting chemical signaling, which attracts other microbial-conveying insects and enhance an almost out-of-control intense microbial multiplication. If not inhibited by fungicides and under appropriate climatic conditions, *B. cinerea* may dominate berry microbiota and induce grey rot. Local birds will carry local microorganisms, but migrant birds, like starlings, may introduce microorganisms from distant regions. However, insects should have a more important role in microbial dissemination than birds. Results from our laboratory (under publication) have shown that wounds mechanically provoked on berries protected by nets, and thus not visited by *Drosophila*, are healed after some time, while unprotected berries rotten quickly. These results support the view that during the final stages of grape ripening, insects, rather than wind or rain, play a predominant role as microbial vectors.

A comparative analysis of the microbiota sharing different fruits, grapes and insect vectors in the same region should provide an adequate indicator to validate the proposed sequence of events.

ABBREVIATIONS

DPM = Diffuse powdery mildew

FISH = Fluorescence *in situ* hybridisation

PCR = Polymerase chain reaction

VOCs = Volatile organic compounds

ACKNOWLEDGEMENTS

The authors thank Dr. Cecília Rego (Laboratório Veríssimo de Almeida, Instituto Superior de Agronomia) and Dr. José Carlos Franco (Instituto Superior de Agronomia) for the information provided on grapevine phytopathology and enthomology, respectively. Dr. Olfa Zarruk (Molecular Ecophysiology Laboratory, Instituto de Tecnologia Química e Biológica, Oeiras, Portugal) and Dr. Victoria Fernández (Plant Nutrition Department, Estación Experimental de Aula Dei, Zaragoza, Spain) were responsible for the Touriga Franca grape scanning electron micrographs.

REFERENCES

[1] Fonseca A, Inácio J. Phylloplane yeasts. In: Biodiversity and Ecophysiology of Yeasts. Rosa CA, Péter G, Eds. Springer-Verlag, Berlin, Heidelberg 2006; pp. 263-301.

[2] Martini A, Federici F, Rosini G. A new approach to the study of yeast ecology of natural substrates. Can J Microbiol 1980; 26: 856-9.

[3] Martini A, Ciani M, Scorzeti G. Direct enumeration and isolation of wine yeasts from grape surfaces. Am J Enol Vitic 1996; 47(4): 435-40.

[4] Loureiro V, Querol A. The prevalence and control of spoilage yeasts in foods and beverages. Trends Food Sci Technol 1999; 10(11): 356-65.

[5] Loureiro V, Malfeito-Ferreira M. Yeasts in Spoilage. In: Caballero B, Trugo L, Finglas P, Eds. Encyclopedia of Food Sciences and Nutrition, 2nd ed. Academic Press, London 2003; pp 5530-6.

[6] Renouf V, Claisse O, Lonvaud-Funel A. Understanding the microbial ecosystem on the grape berry surface through numeration and identification of yeast and bacteria. Aust J Grape Wine Res 2005; 11: 316-27.

[7] Serra R, Braga A, Venâncio A. Mycotoxin-producing and other fungi isolated from grapes for wine production, with particular emphasis on ochratoxin A. Res Microbiol 2005; 156: 515-21.

[8] Bae S, Fleet G, Heard G. Lactic acid bacteria associated with wine grapes from several Australian vineyards. J Appl Microbiol 2006; 100: 712-27.

[9] Barata A, Seborro F, Belloch C, Malfeito-Ferreira M, Loureiro V. Ascomycetous yeast species recovered from grapes damaged by honeydew and sour rot. J Appl Microbiol 2008; 104: 1182-91.

[10] Barata A, González S, Malfeito-Ferreira M, Querol A, Loureiro V. Sour rot-damaged grapes are sources of wine spoilage yeasts. FEMS Yeast Res 2008; 8: 1008-17.

[11] Cocolin L, Campolongo S, Alessandria V, Dolci P, Rantsiou K. Culture independent analyses and wine fermentation: an overview of achievements 10 years after first application. Ann Microbiol 2010; 61(1): 17-23.

[12] Prakitchaiwattana C, Fleet G, Heard G. Application and evaluation of denaturing gradient gel electrophoresis to analyse the yeast ecology of wine grapes. FEMS Yeast Res 2004; 4: 865-77.

[13] Sampaio JP, Gonçalves P. Natural populations of *Saccharomyces kudriavzevii* in Portugal are associated with oak bark and are sympatric with *S. cerevisiae* and *S. paradoxus*. Appl Environ Microbiol 2008; *74:* 2144-52.

[14] Goddard M, Anfang N, Tang R, Gardner R, Jun C. A distinct population of Saccharomyces cerevisiae in New Zealand: evidence for local dispersal by insects and human-aided global dispersal in oak barrels. Environ Microbiol 2010; 12(1): 63-73.

[15] Andrews J, Harris R. The ecology and biogeography of microrganisms on plant surfaces. Annu Rev Phytopathol 2000; 38: 145-80.

[16] Padgett M, Morrison J. Changes in grape exudates during fruit development on their effect on mycelial growth of *Botrytis cinerea*. J Am Soc Hort Sci 1990; 115: 269-73.

[17] Pezet R, Pont V. *Botrytis cinerea*: activité antifongique das les jeunes grappes de Vitis vinifera*,* varieté Gamay. Phytopathol Z 1984; 111: 73-81.

[18] Kosuge T, Hewitt W. Exudates of grape berries and their effect on germination of conidia of *Botrytis cinerea*. Phytopathol 1964; 54: 167-72.

[19] Bessis R. Caractères et development des exsudats au cours de la maturation du raisin. Proc. Symp. IIF et OIV, Science et Technique Froid, Paris 1977.

[20] Doneche B. La nature des exsudats de raisins e leur rôle dans la germination de *Botrytis cinerea*. Agronomie 1986; *6:* 67-73.

[21] Prakitchaiwattana C. Investigation of yeasts associated with Australian wine grapes using cultural and molecular methods. PhD thesis, University of South Wales, Australia 2004.

[22] Lachance MA. Yeast selection in nature. In: Panchal J, Ed. Yeast Strain Selection. M. Dekker: New York 1990; vol. 8: pp. 2-41.

[23] Davenport R. An introduction to yeasts and yeast-like organisms. In: Skinner FA, Passmore SM, Davenport RR, Eds. Biology of Yeasts. Academic Press, London 1980; pp 1-27.

[24] Prieto C, Jara C, Mas A, Romero J. Application of molecular methods for analysing the distribution and diversity of acetic acid bacteria in Chilean vineyards. Int J Food Microbiol 2007; 115(3): 348-55.

[25] Compant S, Kaplan H, Sessitsch A, Nowak J, Ait Barka E, Clément C. Endophytic colonization of Vitis vinifera L. by Burkholderia phytofirmans strain PsJN: From the rhizosphere to inflorescence tissues. FEMS Microbiol Ecol 2008; 63(1): 84-93.

[26] Belin JM. Recherches sur la répartition des levures à la surface de la grappe de raisin. Vitis 1972; 11: 135-45.

[27] Gadoury D, Seem R, Wilcox WF, *et al.* Effects of Diffuse Colonization of Grape Berries by Uncinula necator on Bunch Rots, Berry Microflora, and Juice and Wine Quality. Phytopathol 2007; 97(10): 1356-65.

[28] Daeschel M, Andersson R, Fleming H. Microbial ecology of fermenting plant materials. FEMS Microbiol Rev 1987; 46: 357-67.

[29] Barnett J, Delaney M, Jones E, Magson A, Winch B. The numbers of yeasts associated with wine grapes of Bordeaux. Arch Mikrobiol 1972; 83: 52-5.

[30] Guerzoni E, Marchetti R. Analysis of yeast flora associated with grape sour rot and of the chemical disease markers. Appl Environ Microbiol 1987; 53:571-6.

[31] Moore K, Johnson M, Morris J. Indigenous Yeast Microflora on Arkansas White Riesling (*Vitis vinifera*) Grapes and in Model Must Systems. J Food Sci 1988; 53(6): 1725-8.

[32] Yanagida F, Ichinose F, Shinohara T, Goto S. Distribution of wild yeasts in the white grape varieties at Central Japan J Gen Appl Microbiol 1992; 38: 501-4.

[33] Sabate J, Cano J, Esteve-Zarzoso B, Guillamón J. Isolation and identification of yeasts associated with vineyard and winery by RFLP analysis of ribosomal genes and mitochondrial DNA. Microbiol Res 2002; 157: 267-74.

[34] Subden R, Husnik JI, van Twest J, van der Merwe G, van Vuuren H. Autochthonous microbial population in a Niagara Peninsula icewine must Food Res Int 2003; 36(7): 747-51.

[35] Combina M, Mercado L, Borgo P, *et al.* Yeasts associated to Malbec grape berries from Mendoza, Argentina J Appl Microbiol 2005; 98(5): 1055-61.

[36] Raspor P, Milek D, Polanc J, Mozina S, Cadez N. Yeasts isolated from three varieties of grapes cultivated in different locations of the Dolenjska vine-growing region, Slovenia Int J Food Microbiol 2006; 109: 97–102.

[37] Nisiotou A, Nychas G-J. Yeast Populations Residing on Healthy or Botrytis-Infected Grapes from a Vineyard in Attica, Greece. Appl Environ Microbiol 2007; 73(8): 2765-8.

[38] Chavan P, Mane S, Kulkarni G, *et al.* Natural yeast flora of different varieties of grapes used for wine making in India Food Microbiol 2009; 26(8): 801-8.

[39] Li S, Cheng C, Li Z, Chen J, Yan B, Han B, Reeves M. Yeast species associated with wine grapes in China. Int J Food Microbiol 2010; 138(1-2): 85-90.

[40] Clavijo A, Calderón I, Paneque P. Diversity of Saccharomyces and non-Saccharomyces yeasts in three red grape varieties cultured in the Serranía de Ronda (Spain) vine-growing region. Int J Food Microbiol 2010; 143(3): 241-5.

[41] Baffi M, Bezerra C, Arévalo-Villena M, Briones-Pérez A, Gomes E, Silva R. Isolation and molecular identification of wine yeasts from a Brazilian vineyard. Ann Microbiol 2011; 61(1): 75-8.

[42] Lafon-Lafourcade S, Carre E, Ribéreau-Gayon P. Occurrence of lactic acid bacteria during the different stages of vinification and conservation of wines. Appl Environ Microbiol 1983; 46(4): 874-80.

[43] Renouf V, Claisse O, Lonvaud-Funel A. Inventory and monitoring of wine microbial consortia Appl Microbiol Biotechnol 2007; 75(1): 149-64.

[44] Francesca N, Chiurazzi M, Romano R, Aponte M, Settanni L, Moschetti G. Indigenous yeast communities in the environment of ''Rovello bianco'' grape variety and their use in commercial white wine fermentation World J Microbiol Biotechnol 2010; 26(2): 337-51.

[45] Joyeux A, Lafon-Lafourcade S, Ribéreau-Gayon P. Evolution of acetic acid bacteria during fermentation and storage of wine. Appl Environ Microbiol 1984 48(1): 153-6.

[46] Barbe J, Revel G, Joyeux A, Bertrand A, Lonvaud-Funel A. Role of botrytized grape microorganism in SO2 binding phenomena. J Appl Microbiol 2001; 90: 34-42.

[47] González A, Hierro N, Poblet M, Mas A, Guillamón J. Application of molecular methods to demonstrate species and strain evolution of acetic acid bacteria population during wine production. Int J Food Microbiol 2005; 102: 295-304.

[48] Kanellis AK, Roubelakis-Angelakis KA. Grape Biochemistry of fruit ripening. In: Seymour G, Taylor J, Tucker G, Eds. Chapman and Hall, London 1993; pp. 189-234.

[49] Jeffree CE. The fine structure of the plant cuticle. In: Riederer M, Muller C, Eds. Biology of the Plant Cuticle. Blackwell Publishing, Oxford, UK 2006; pp. 11-125.

[50] Nawrath C. Unraveling the complex network of cuticular structure and function. Curr Opinion Plant Biol 2006; 9(3): 281-7.

[51] Chassot C, Métraux JP. The cuticle as source of signals for plant defense. Plant Biosyst 2005; 139(1): 28-31.

[52] Comménil P, Brunet L, Audran JC. The development of the grape berry cuticle in relation to susceptibility to bunch rot disease. J Exp Botany 1997; 48(313): 1599-607.

[53] Isaacson T, Kosma DK, Matas AJ, *et al.* Cutin deficiency in the tomato fruit cuticle consistently affects resistance to microbial infection and biomechanical properties, but not transpirational water loss. Plant J 2009; 60: 363-77.

[54] Walton TJ, Kolattukudy PE. Determination of the structures of cutin monomers by a novel depolymerization procedure and combined gas chromatography and mass spectrometry. Biochem 1972; 11(10): 1885-97.

[55] Casado C, Heredia A. Ultrastructure of the cuticle during growth of the grape berry (Vitis vinifera). Physiol Plant 2001; 111: 220-4.

[56] Rogiers S, Hatfield J, Jaudzems V, White R, Keller M. Grape berry cv. Shiraz epicuticular wax and transpiration during ripening and preharvest weight loss. Am J Enol Vitic. 2005; 55(2): 121-7.

[57] Kennedy J. Understanding grape berry development. Winery and Vineyard. July/August, 2002 (assessed at www.practicalwinery.com on the 28th October 2010).

[58] Deytieux C, Geny L, Lapaillerie D, Claverol S, Bonneu M, Donèche B. Proteome analysis of grape skins during ripening. J Exp Bot 2007; 58 (7): 1851-62.

[59] Ferreira RB, Monteiro S, Freitas R, *et al.* Fungal pathogens: The battle for plant infection. Crit Rev Plant Sci 2006; 25: 505-24.

[60] Ferreira RB, Monteiro S, Freitas R, *et al.* The role of plant defence proteins in fungal pathogenesis. Molec Plant Pathol 2007; 8: 677-700.

[61] Sarry JE, Grimplet J, Sommerer N, *et al.* Combined Mass Mapping and Biochemical Characterization of Grape β-Glycosidade-enriched. Extract Protein J 2008; 27: 258-66.
[62] Schulze-Lefert P, Vogel J. Closing the ranks to attack by powdery mildew. Trends Plant Sci 2000; 5(8): 343-8.
[63] Pretorius I, van der Westhuizen T, Augustyn O. Yeast biodiversity in vineyards and wineries and its importance to the South African wine industry. A review. S Afr J Enol Vitic 1999; 20(2): 61-74.
[64] Rosini G, Federichi F, Martini A. Yeast flora of grape berries during ripening. Microb Ecol 1982; 8: 83–9.
[65] Mortimer R, Polsinelli M. On the origin of wine yeast. Res Microbiol 1999; 150: 199-204.
[66] Lachance MA. The Phaff school of yeast ecology. Int Microbiol 2003; 6: 163-7.
[67] Kassemeyer H-H, Berkelmann-Löhnertz B. Fungi of Grapes. In: König H, Unden G, Fröhlich J, Eds. Biology of Microorganisms on Grapes, in Must and in Wine. Springer-Verlag, Berlin, Heidelberg 2009.
[68] Leong LS, Hocking AD, Pitt JI, Kazi BA, Emmett RW, Scott ES. Black *Aspergillus* Species in Australian Vineyards: From Soil to Ochratoxin A in Wine. In: Hocking AD, Pitt JI, Samson RA, Thrane U, Eds. Adv Exp Med Biol 2006; 571: 153-71.
[69] Malfeito-Ferreira M, Barata A, Loureiro V. Wine spoilage by fungal metabolites. In: Polo C, Moreno-Arribas MV, Ed. Wine Chemistry and Biochemistry. Springer, New York 2009; pp. 615-45.
[70] Kennelly MM, Gadoury DM, Wilcox WF, Magarey PA, Seem RC. Seasonal Development of Ontogenic resistance to Downy Mildew in Grape Berries and Rachises. Phytopathol 2005; 95: 1445-52.
[71] Thind TS, Arora JK, Mohan C, Raj P. Epidemiology of Powdery Mildew, Downy Mildew and Anthracnose Diseases of Grapevine. Kluwer Academic Publishers (I) 2004; pp. 621-38.
[72] Díez-Navajas AM, Merdinoglu SW, Greif C, Merdinoglu D. Nonhost Versus Host resistance to the Grapevine Downy Mildew, Plasmopara viticola, Studied at the Tissue Level. Phytopathol 2008; 98: 776-80.
[73] Dry IB. Mechanism of powdery mildew infection in grapevines. Cooperative Research Centre for Viticulture: Australia 2006.
[74] Fung RWM, Gonzalo M, Fekete C, *et al.* Powdery Mildew Induces Defense-Oriented Reprogramming of the Transcriptome in a Susceptible But Not in a resistant Grapevine. Plant Physiol 2008; 146: 236-49.
[75] Marsh E, Alvarez S, Hicks LM, Barbazuk WB, Qiu W, Kovacs L, Schachtman D. Changes in protein abundance during powdery mildew infection of leaf tissues of Cabernet Sauvignon grapevine (*Vitis vinifera* L.). Proteom 2010; 10: 2057-64.
[76] Mullins M, Bouquet A, Williams L. Biology of the Grapevine. Cambridge University Press: New York 1992.
[77] Gadoury DM, Seem RC, Ficke A, Wilcox WF. Ontogenic resistance to Powdery Mildew in Grape Berries. Phytopathol 2003; 93: 547-55.
[78] Deytieux-Belleau C, Geny L, Roudet J, Mayet V, Donéche B, Fermaud M. Grape berry skin features related to ontogenic resistance to *Botrytis cinerea.* Eur J Plant Pathol 2009; 125: 551-63.
[79] Comménil P, Brunet L, Audran J. The development of the grape berry cuticle in relation to bunch rot disease. J Exp Bot 1997; 48: 1599-607.
[80] Mlikota-Gabler F, Smilanick JL, Mansour M, Ramming DW, Mackey BE. Correlations of Morphological, Anatomical, and Chemical Features of Grape Berries with resistance to *Botrytis cinerea.* Phytopathol 2003; 93(10): 1263-73.
[81] Agrios GN. Plant Pathology. Academic Press: New York 1997.
[82] Nair NG, Hill GK. Bunch rot of grapes caused by *Botrytis cinerea.* In: Kumar J, Chaube HS, Singh US, Mukhopadhyay AN, Eds. Plant diseases of international importance. v. III: Diseases of fruit crops. Prentice-Hall, Englewood Cliffs, NJ, USA 1992; pp. 147-69.
[83] Mortimer R. Evolution and Variation of the Yeast (*Saccharomyces*) Genome. Genome Res 2000; 10: 403-79.
[84] Fay J, Benavides J. Evidence for Domesticated and Wild Populations of *Saccharomyces cerevisiae.* PLoS Genet 2005; 1(1): 66-71.
[85] Davenport RR. Microecology of yeasts and yeast-like organisms associated with an Englishn vineyard. Vitis 1974; 13: 123-30.
[86] Lund A. Studies on the Ecology of Yeasts. Munksgaard: Copenhagen 1954.
[87] Sláviková E, Vadkertiová R, Vránová D. Yeasts colonizing the leaf surfaces. J Basic Microbiol 2007; 47: 344-50.
[92] Comitini F, Ciani M. Survival of inoculated Saccharomyces cerevisiae strain on wine grapes during two vintages. Lett Appl Microbiol 2005; 42(3): 248-53.
[88] Schuller D, Casal M. The genetic structure of fermentative vineyard-associated *Saccharomyces cerevisiae* populations revealed by microsatellite analysis. Ant Leeuwen 2007; 91: 137-50.

[89] Valero E, Cambon B, Schuller D, Casal M, Dequin S. Biodiversity of *Saccharomyces* yeast strains from grape berries of wine-producing areas using starter commercial yeasts. FEMS Yeast Res 2007; 7: 317–29.

[90] Ciani M, Mannazzu I, Marinangeli P, Clementi F, Martini A. Contribution of winery-resident *Saccharomyces cerevisiae* strains to spontaneous grape must fermentation. Ant Leeuwen 2004; 85: 159-64.

[91] Mercado L, Dalcero A, Masuelli R, Combina M. Diversity of *Saccharomyces* strains on grapes and winery surfaces: analysis of their contributions to fermentative flora of Malbec wine from Mendoza (Argentina) during two consecutive years. Food Microbiol 2007; 24: 403-12.

[93] Budroni M, Farris GA, Vodret T. On the origin of wine yeast. Yeasts of spontaneous fermentations (2nd note). Riv Vitic Enol 1995; 3.

[94] Le Jeune C, Erny C, Demuyter C, Lollier M. Evolution of the population of Saccharomyces cerevisiae from grape to wine in a spontaneous fermentation. Food Microbiol 2006; 23(8): 709-16.

[95] Renouf V, Perello MC, Strehaiano P, Lonvaud-Funel A. Global Survey of the Microbiol Ecosystem During Alcoholic Fermentation in Winemaking. J Int Sci Vigne Vin 2006; 40 (2): 101-16.

[96] Santamaría P, Garijo P, López R, Tenorio C, Gutiérrez AR. Analysis of yeast population during spontaneous alcoholic fermentation: Effect of the age of the cellar and the practice of inoculation. Int J Food Microbiol 2005; 103(1): 49-56.

[97] Lopes CA, Rodríguez ME, Sangorrín M, Querol A, Caballero AC. Patagonian wines: implantation of an indigenous strain of *Saccharomyces cerevisiae* in fermentations conducted in traditional and modern cellars. J Ind Microbiol Biotechnol. 2007; 34(2): 139-49.

[98] Csoma H, Zakany N, Capece A, Romano P, Sipiczki M. Biological diversity of *Saccharomyces* yeasts of spontaneously fermenting wines in four wine regions: Comparative genotypic and phenotypic analysis. Int J Food Microbiol 2010; 140(2-3): 239-48

[99] Wibowo D, Eschenbruch R, Davis C, Fleet G, Lee T. Occurrence and growth of lactic acid bacteria in wine: a review. Am J Enol Vitic 1985; 36: 302-13.

[100] Van Vuuren H, Dicks L. *Leuconostoc oenos*: a review. Am J Enol Vitic 1993; 44: 99-112.

[101] Davis C, Wibowo D, Eschenbruch R, Lee T, Fleet G. Practical implications of malolactic fermentation. A review. Am J Enol Vitic 1985; 36: 290-301.

[102] Edwards C, Jensen K, Spayd S, Seymour B. Isolation and characterization of native strains of Leuconostoc oenos from Washington State wines. Am J Enol Vitic 1991; 42: 219-26.

[103] Garijo P, López R, Santamaría P, *et al.* Presence of lactic bacteria in the air of a winery during the vinification period. Int J Food Microbiol 2009; 136: 142-6.

[104] Fleet G, Prakitchaiwattana C, Beh A, Heard G. The yeast ecology of wine grapes. In: Ciani M, Ed. Biodiversity and biotechnology of wine yeasts. Research Signpost, Kerala, India 2002.

[105] Capriotti A. Yeasts from U.S.A. soils. Arch Mikrobiol 1967; 57(4): 406-13.

[106] Benda I. The yeast flora of the district of Franconia. Weinberg Keller 1964; 11: 67–74.

[107] Parle J, Di Menna M. The source of yeast in New Zealand wines. J Agr Res 1966; 9: 98-107.

[108] Sniegowski PD, Dombrowski PG, Fingerman E. *Saccharomyces cerevisiae* and *Saccharomyces paradoxus* coexist in a natural woodland site in North America and display different levels of reproductive isolation from European conspecifics. FEMS Yeast Res 2002; 1(4): 299-306.

[109] Johnson L, Koufopanou V, Goddard M, Hetherington R, Schafer S, Burt A. Population genetics of the wild yeast *Saccharomyces paradoxus*. Genetics 2004; 166: 43-52.

[110] Crotti E, Rizzi A, Chouaia B, *et al.* Acetic acid bacteria, newly emerging symbionts of insects. Appl Environm Microbiol 2010; 76 (21): 6963-70.

[111] Ruiz-Arguesso T, Navarro-Rodriguez A. Gluconic acid- producing bacteria from honey bees and ripening honey. J Gen Microbiol 1973; 76: 211-6.

[112] Gilliam M. Identification and roles of non-pathogenic microflora associated with honey bees. FEMS Microbiol Lett 1997; 155: 1-10.

[113] Reeson AF, Jankovic T, Kasper ML, Rogers S, Austin AD. Application of 16S rDNA-DGGE to examine the microbial ecology associated with a social wasp Vespula germanica. Insect Molec Biol 2003; 12(1): 85-91.

[114] Audisio M, Torres M, Sabaté D, Ibarguren C, Apella M. Properties of different lactic acid bacteria isolated from Apis mellifera L. bee-gut. Microbiol Res 2011; 166: 1-13.

[115] Gilliam M, Valentine D. Enterobactericeae isolating from foraging worker honey bees. J Inverteb Pathol 1974; 23: 38-41.

[116] Engelberth J, Alborn HT, Schmelz EA, Tumlinson JH. Airborne signals prime plants against insect herbivore attack. Proc Natl Acad Sci USA 2004; 101: 1781-5.

[117] Goff SA, Klee HJ. Plant volatile compounds: sensory cues for health and nutritional value? Science 2006; 311: 815-19.

[118] Vancanneyt G, Sanz C, Farmaki T, *et al.* Hydroperoxide lyase depletion in transgenic potato plants leads to an increase in aphid performance. Proc Natl Acad Sci USA 2001; 98: 8139-44.

[119] Kulakiotu EK, Thanassoulopoulos CC, Sfakiotakis EM. Biological control of Botrytis cinerea by volatiles of 'Isabella' grapes. Phytopathol. 2004; 94(9): 924-31.

[120] Lund ST, Bohlmann J. The molecular basis for wine grape quality – a volatile subject. Science 2006; 311: 804-5.

[121] Tasin M, Backman A-C, Bengtsson M, Varela N, Ioriatti C, Witzgall P. Wind tunnel attraction of grapevine moth females, Lobesia botrana, to natural and artificial grape odour. Chemoecol 2006; 16: 87-92.

[122] Tasin M, Anfora G, Ioriatti C, *et al.* Antennal and behavioral responses of grapevine moth Lobesia botrana females to volatiles from grapevine. J Chem Ecol 2005; 31: 77-87.

[123] Xiao ZJ, Xu P. Acetoin metabolism in bacteria. Crit Rev Microbiol 2007; 33: 127-140.

[124] Stan C, Tuca O, Mitrea I. The spectrum of harmful arthropod fauna from the vineyard SD. Banu Maracine – Craiova. Proceedings of the 44th Croatian & 4th International Symposium on Agriculture, 2009; 876-9.

[125] Ghaiger R. The effect of different management systems on the epigaeic arthropod assemblages in the Cape Floristic Region, South Africa. M. Sc. Thesis, Faculty of AgriSciences, University of Stellenbosch, 2008.

[126] Buchanam G, Amos T. Grape Pests. In: Coombe BG, Dry PR, Eds. Viticulture Practices, vol. 2. Winetitles, Adelaide 1992; pp. 209-31.

[127] Sláviková E, Vadkertiová R, Vránová D. Yeasts colonizing the leaves of fruit trees. Ann Microbiol 2009; 59(3): 419-24.

CHAPTER 13

Wine and Health

P. -L. Teissedre*

Institut des Sciences de la Vigne et du Vin, Université Bordeaux II, UMR 1219 INRA Œnologie, Faculté d'œnologie, Laboratoire de Chimie Appliquée; 210, chemin de Leysotte, CS 50008; 33882 Villenave d'Ornon Cedex, France

Abstract: The results of epidemiological cohort studies show the separate influence of a moderate consumption for different types of alcoholic beverages (beer, wine, spirits) and their effects on coronary heart diseases, with a superior protective effect for wine. Wine possesses specificity due to the phenolic antioxidants (flavonoids and non-flavonoids). Several ways for metabolism and polyphenol excretion of wine are eligible. Moderate wine consumption can lead to decreased platelet aggregation and vasodilatation. Physiological effects derived from the nutritional consumption of wine polyphenol extracts, ethanol or their possible synergistic combination may lead to a prevention of atherosclerosis, diabetes or hypertension. This synergy permits:

- In the case of atherosclerosis, a decrease in the fatty streak area and cholesterolemia, and an increased level in specific antioxidant enzymes active against free radicals,
- To restore the antioxidant capacity of plasma, thus allowing for an improvement of the defences against oxidative stress of diabetes and a favourable action on insulin secretion, to normalise the systolic pressure, and a high reduction of cardiac hypertrophy, as well as free radical generation to the thoracic aorta and heart tissues in the case of hypertension. Wine can play a role in preventive nutrition, when its regular and daily consumption with moderation is added to the diet.

Keywords: "French Paradox", Alcohol, Atherosclerosis, Cardiovascular diseases, Diabetes, Oxidative stress, Phenolic antioxidants, Preventive nutrition, Vasodilatation, Wine composition.

INTRODUCTION

Several international teams have studied the topic "wine and health", providing us with numerous epidemiological and scientific data. Various epidemiological inquiries realised during the last 40 years in industrial nations have confirmed that wine-drinking populations presented low rates of mortality due to cardiovascular diseases. In particular, a protective activity of wine against cardiovascular diseases is widely described and certain studies suggest that wine could decrease by 40% the risk of myocardial infarction, as well as the risk of cerebral vascular thrombosis by 25%. One of the most famous echoes is the broadcast "60 minutes", presented in November, 1991 on the American TV channel CBS where Dr. Serge Renaud [1] made the "French Paradox" known to tens of millions of Americans.

This is the designation of an epidemiological report showing that if in most countries a high consumption of saturated fats is strongly correlated with mortality due to cardiovascular diseases, this is not the case in France, and more particularly in the region of Toulouse, where the mortality of coronary origin is low in spite of the consumption of saturated fats. Among the hypotheses advanced to explain this paradox is the regular and moderate consumption of wine.

Which could be the processes responsible for this beneficial effect of wine-consumption on health? What is it in the wine that produces this favourable effect on health: is it the alcohol ingested in moderate amounts and on a regular basis at every meal?

***Address correspondence to P. -L. Teissedre:** Institut des Sciences de la Vigne et du Vin, Université Bordeaux II, UMR 1219 INRA Œnologie, Faculté d'œnologie, Laboratoire de Chimie Appliquée ; 210, chemin de Leysotte, CS 50008 ; 33882 Villenave d'Ornon Cedex, France; E-mail: p.teissedre@u-bordeaux2.fr

Hernâni Gerós, M. Manuela Chaves and Serge Delrot (Eds)

As suggested in other studies, the answer lies in certain substances found specifically in the non-alcoholic fraction of the wine: is it the antioxidant phenolic compounds or these same substances in synergy with alcohol?

Moreover, we must not forget that the composition of wine is extremely complex. The various constituents of wine are represented rather unevenly and some, in largest number, can only be found in the state of traces. Wine contains water (first constituent), which represents 80 to 85% of all compounds. Ethylic alcohol (the second constituent) produced during the yeast fermentation is an indispensable element for the solubilisation of fundamental constituents such as phenolic compounds or may act as an enhancer of given aroma compounds. But wine also contains organic acids (tartaric acid, malic acid, citric acid) as well as major minerals (potassium, calcium, magnesium, sodium, iron, sulfates, phosphor, chlorides), all of which contribute to cover a part of the daily intake needs of humans. Wine further contains polyols (such as glycerin), which contribute to the sensory perception of softness and present diverse sugars in low amounts, close to 2 g/L in the case of dry wines, and in higher amounts able to exceed 100 g/L in the case of sweet wines. Wine contains nitrogenous substances in small quantities and does not contain liposoluble vitamins (vitamins A-D-E-K-F do not exist in wine). On the other hand, it contains hydrosoluble vitamins of the B group and reduced quantities of vitamin C. It is also composed of elements that can play the role of oxidisers, reducers and metallic catalysts (selenium, chromium, zinc, copper, manganese, fluorine, iodine, arsenic, *etc.*), and enzymes, which are essential for the realisation of chemical reactions necessary to life and cellular reproduction.

Finally wine contains more specific constituents (still poorly known), which confer it personality as aromatic compounds or numerous phenolic compounds (Fig. **1**) subdivided into several groups (acids phenols, cinnamates, stilbenes, tannins, anthocyanines, flavonols) and which play a fundamental role in the sensory perception of the product.

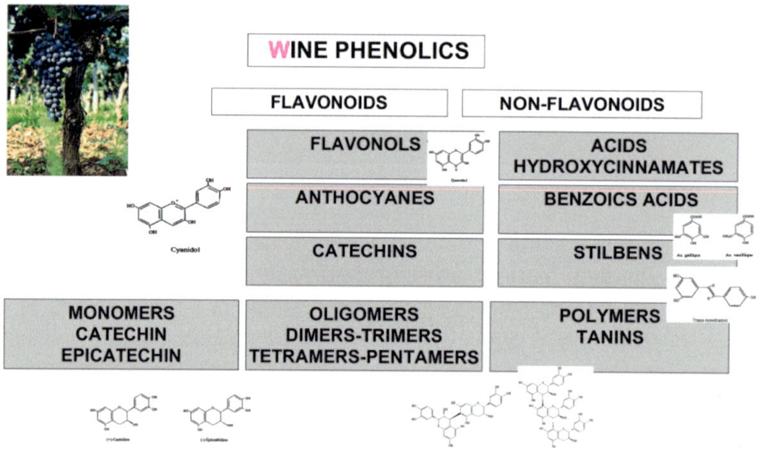

Figure 1: Classification of phenolics found in grapes and wines.

Alcoholic Drinks, Wines and Coronary Diseases

As far as the consumption of alcoholic drinks is concerned, it was in 1970 that research on alcohol and cardiovascular diseases was undertaken by Professor Arthur Klatsky [2], a cardiologist at the Hospital Centre of Oakland in California. He began a study on long-term health care with more than 100,000 persons to try to discover the factors which significantly affected the risks of cardiovascular diseases. The results were published in 1974 and appeared as pioneers in the context of American research by indicating that moderate consumers of alcoholic drinks (those who drink from 1 to 3 glasses a day) have a lower risk of death by cardiovascular diseases than those who abstain or those who drink excessively. These results have since been known as a U or J curve. This relation has been confirmed by researchers in several different countries working on diverse populations. Alcohol in moderation is indeed known to increase HDL (High Density Lipoproteins) levels and reduce atherogenic lipoproteins as well as platelet aggregation.

However, there is great variability of composition and consumption amongst different alcoholic drinks. For example, wine, which contains approximately 12.5% vol. ethanol, is mainly consumed during the meals, while beer and spirits which contain approximately 4.5% vol. and 40% vol. of ethanol, respectively, are mainly consumed outside the meals. These parameters are to be considered, although at present their importance is unknown. These peculiarities engender the existence of a very heterogeneous group of consumers of alcoholic drinks. Besides, numerous works suggest that wine, and more particularly red wine, could induce an antioxidant effect due to the polyphenols that it contains in significant amounts, thus leading to an additional adavantage in the mechanisms of coronary protection [3].

If we consider the inquiries separately according to the consumption of wine, beer and spirit there are five which concern important populations and reveal interesting data [4]. Fig. **2** shows a reduction of the relative risk of coronary mortality for regular consumers (in regard to abstinents) and moderate consumers (1 to 2 glasses a day, which represents amounts from 15 to 30 g of alcohol/day) of an alcoholic drink regardless of its nature, with an exception concerning a group of spirits. Among these works [2, 3, 5-7], it is important to observe that in four of the studies the consumer groups of wine benefit from a reduction of the relative risk of coronary mortality when compared to the consumer groups of beer and spirits. In the other study, although there was no reduction in the risk of coronary mortality, moderate wine consumption was not related to an increased risk. Furthermore it should be noted that during his study of 1995, Dr. Groenback [3] also showed that moderate consumers of wine present the lowest risks of mortality of any causes in regard to the consumers of beers and spirits. Certain scientists argue, however, that epidemiological studies merely show associations and cannot prove their real causes and effects. Nonetheless, some correlations between moderate wine consumption and, for instance, the decreased risk of cardiac disease are simply too strong to be scientifically discarded. Indeed, experimental science has provided some biological explanations for this.

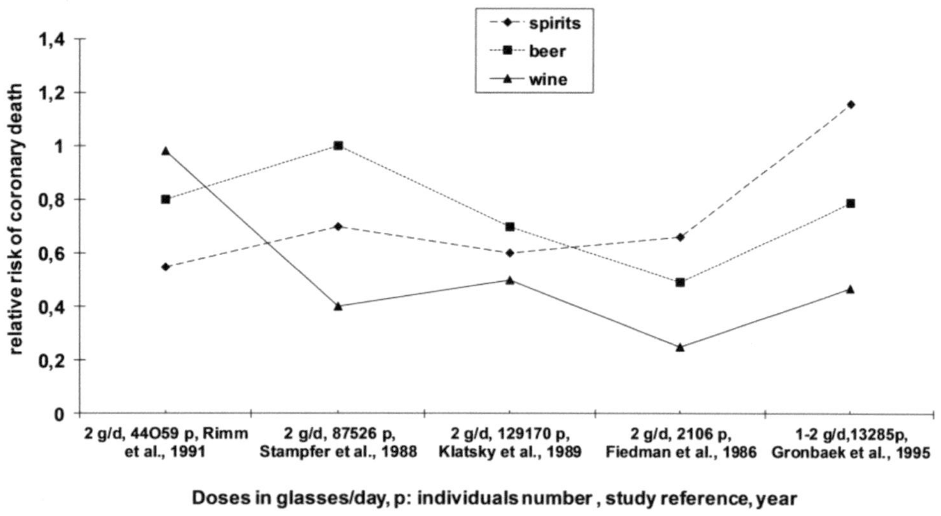

Figure 2: Risk of mortality from coronary artery disease: epidemiological studies on the consumption of wine, beer and spirits [4].

Potential Role of the Phenolic Antioxidants

The hypocholesteremic power as well as the stabilisation of the collagen fibres of the blood vessels bound to the polyphenols of wine were very early described by Professor Masquelier, in 1961 [8]. One of the essential properties of the polyphenols which could explain their beneficial activities for health is their capacity to get free radicals. Produced permanently, the free radicals are for the greater part destroyed by the diverse antioxidising defences of the body (enzymatic or non-enzymatic), a large part of which comes with food (Fig. **3**). However, with age, this balance changes and the oxydative stress increases with a change of the biological molecules, thus facilitating the emergence of chronic pathologies.

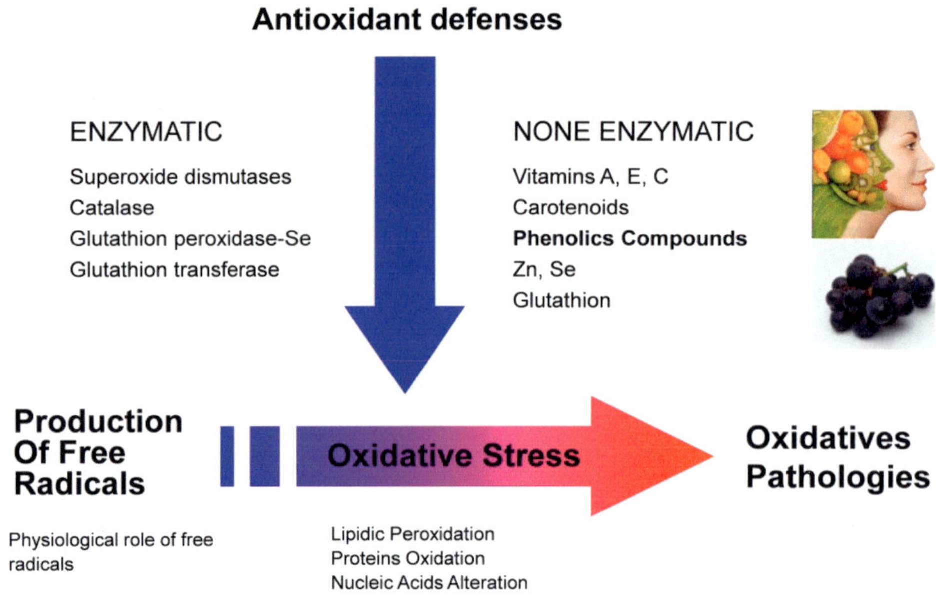

Figure 3: Balance between free radicals and antioxidants.

More recent research on the phenolic compounds of the non-alcoholic fraction of wine further demonstrated their antioxidant properties which can protect lipids from oxidative damage and could be responsible for noticeable cardioprotective effects. Possible mechanisms on the protective role of antioxidants in the cardiovascular diseases and cancers were described. Due to the fact that oxygen is a reactive and plentiful molecule in the environment, lipids of the body are constantly subject to oxidation through reactions of free radicals. If these reactions are not controled, oxidised lipids lead to extensive tissues damage, which can cause diseases including formation of atheromas as in atherosclerosis (Fig. **4**) and, in extreme cases, death.

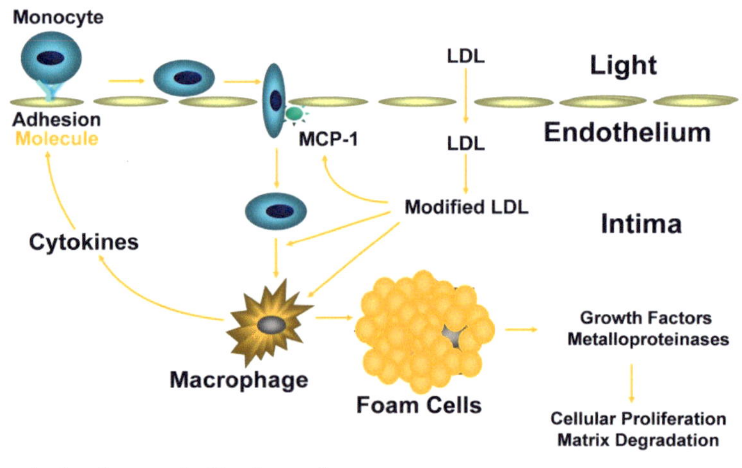

Figure 4: Atherosclerosis development in blood vessels.

Fortunately, the body is protected from the oxidative damage of the lipids by antioxidants. Lipids in the particles of LDL (Low Density Lipoproteins) are typically protected from oxidative damage by compounds such as vitamin E, ß-carotene, ascorbic acid and enzymes containing selenium. The absence of the aforementioned antioxidants in the food leads to damage in the blood vessels. The wine phenolic compounds could provide significant antioxidant protection through several mechanisms. The first one would be the direct capture of free radicals before they react with the lipid LDL as well as the inhibition of free radical formation before causing any damage. The second mechanism would be the reduction of the activity of oxidative enzymes, and the third mechanism would be the decrease in the concentration of peroxidised lipids in the plasma.

As regards the molecular mechanism of the protective effect of wine compounds, [9-12] it was demonstrated that the phenolic compounds inhibited the oxidation of LDL (one of the first phenomena leading to the constitution of the fatty streak atheroma) *in vitro*, and in a more important way than vitamin E, which is considered as a reference antioxidant (Fig. **5**). These results were confirmed *in vivo* by [13], who demonstrated that the susceptibility of the plasma and the LDL in lipid peroxidation was reduced by 20% in eight volunteers (25 - 45 years), having absorbed a quantity of 400 mL of red wine per day during 2 weeks. These results illustrate the fundamental role of the anti-oxidising properties of the phenolic compounds of wine in their protective effect against cardiovascular diseases.

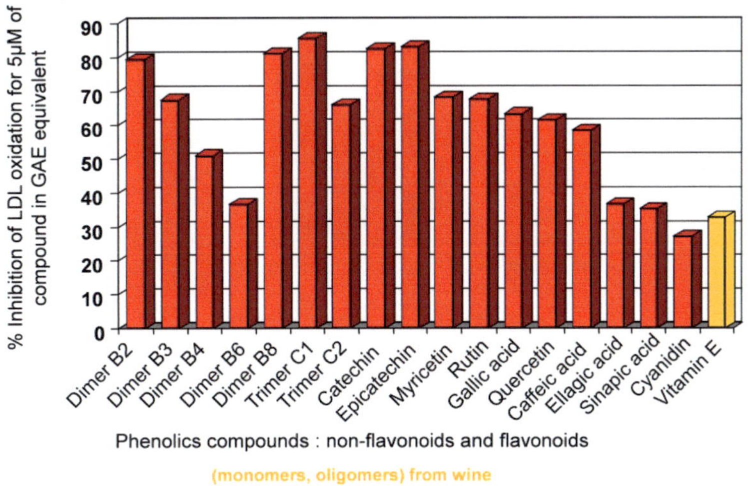

Figure 5: Inhibition of LDL oxidation by phenolic compounds in wine [11].

Red wines, which are particularly rich in catechins, are important contributors of polyphenols to the human diet. Therefore, moderate consumption of red wine (180 mL a day) corresponds to an average consumption of 400.2 mg of total phenolic compounds per day and per person and to a daily consumption of 83.2 mg of catechines, *i.e.* 21% of the total polyphenols under forms of monomers (8.4%) and of dimers (12.6%). In the case of white wines the contribution in phenolic compounds corresponds only to an average consumption of 44.1 mg per day and per person or nearly 10 times less than in the case of red wine. The contribution in catechins made by white wine represents only about 3% (1.17 mg) of the phenolic contents during this consumption [14].

More specifically, an epidemiological study [15] was carried out to establish the preventive role of antioxidant molecules (catechin) of blood serum in cardiovascular accidents. The research and quantification of active antioxidant molecules, and first of all catechin, in serums of regular and moderate consumers of wine in comparison to non-wine consumer groups (control) shows, after some adjustments, that the concentration of (+)-catechin in the plasma is 3 times as that of the consumers of fruits and vegetables and 4 times as that of the moderate consumers of wine. The concentration of (+)-catechin seems to be at its highest (4.7 times more) when the consumption of fruits, vegetables and wine are combined. The contribution of this antioxidant and antiaggregant molecule in the food, and particularlyin wine, could partially explain the protective role against cardiovascular diseases. The data concerning the effective bioavailability of the potentially involved phenolic compounds remains fragmented. The absorption and the metabolism depend on the nature and on the shape of the phenolic compounds (aglycone, polymer, glycosylated) as well as of the intestinal bacterial flora. Several metabolites and by-products can thus be produced and absorbed (free or conjugated phenolic acids and lactones). Consequently, the bioavailability of flavane-3-ols, anthocyanins and phenolic compounds of the products of the vine is still poorly known

[16, 17]. Only limited data exist concerning wine flavanols [18, 19] and the grape seed [20] as well as wine anthocyanins [21]. In the case of flavanols, the authors characterised well the presence of catechin [22, 15] and of epicatechin [23] in the plasma, with important proportions of conjugated (glucuronides, sulphates) and methylated metabolites. In fact, having undergone diverse transformations in the digestive tract, the phenolic compounds are going to be absorbed by the intestinal cells. In the body, they will undergo diverse conjugations: methylation, glucuronidation and\or sulfatation. These transformations are under the dependence of enzymes in the endoplasmic reticulum. The conjugation of glucuronic acid with polyphenols brings in the action of the UDP glucuronosyl transferase (UGT; EC 2.4.1.17). The liver is the organ which presents the strongest capacity to produce glucuronide polyphenols [24]. The catechol-O-methyl transferase (COMT; EC 2.1.1.6) is responsible for the methylation of polyphenols and is present in numerous tissues. The position of the replacements by a methyl group will be determined by the specificity of the enzyme towards every phenolic compound. Sulfo-tranferases (SULT; EC 2.8.2.1) are cytosolic enzymes which catalyse the conjugation of a sulfate group on polyphenols. All the reactions are in fact reactions of phase II which aim at facilitating the transportation of the phenolic compounds by their solubilisation in the aqueous mediums represented by the bile, the urine and the plasma (Fig. **6**).

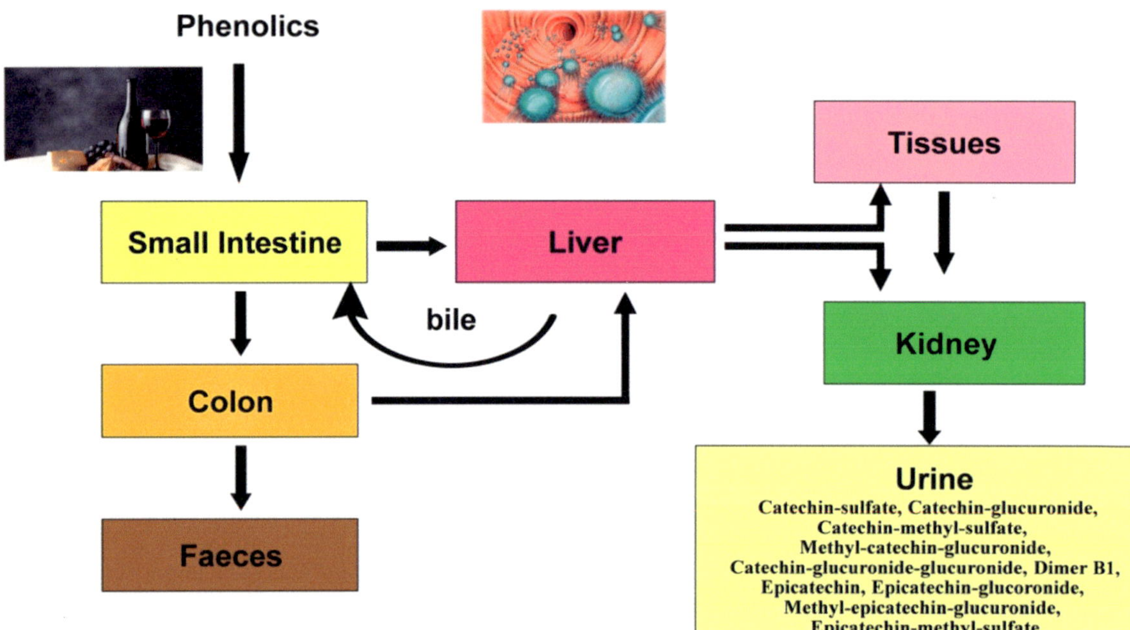

Figure 6: Possible routes for consumed wine phenolics in humans (absorption, metabolism, excretion).

Few studies allow for a complete vision in the domain and the absorption of the procyanidin oligomers. Indeed, a study of Donovan *et al.* shows that procyanidin dimers are not found in the blood or the urine of rats after ingestion [25], while Baba *et al.* identified dimer B2 in the blood and the urine of rats after oral administration of the compound [26]. Furthermore, the unabsorbed compounds could be hydrolysed by the flora in molecules of smaller size, such as phenylvaleric or phenylacetic acids, but also in monomers of catechins, which could then be absorbed [27]. The health effects related to grape and wine consumption may well be due to these poorly understood phenolic acid metabolites [28]. It was not until recently (Fig. **7**) that the identification of procyanidins dimers was able to be obtained in the plasma as well as a quantification in the urine of rats after absorption of tannins of grape seeds [29]. On the other hand, anthocyanins were essentially found under free forms. These data, although partial, suggest that the nature of the studied molecules operates in their bioavailability. They also suggest that the monomeric forms (catechin and epicatechin) effectively cross the intestinal epithelium to lead to active metabolites. Furthermore, it was demonstrated that oligomeric forms in grape seeds possessed *in vivo* antioxidant effects to the rat [30]. These results thus make it necessary to develop follow-up studies on the bioavailability of the phenolic compounds.

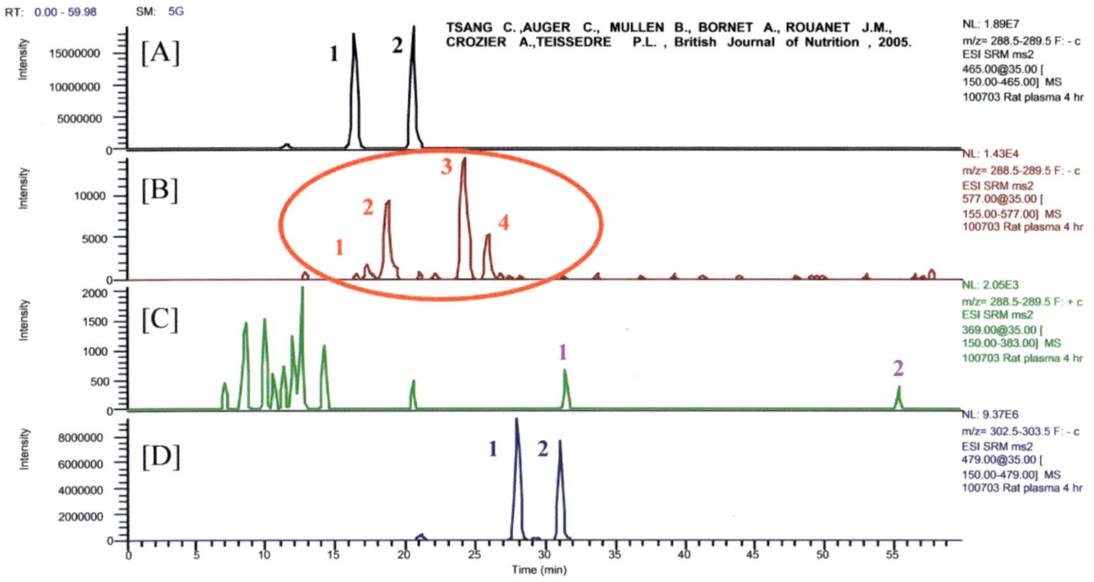

Trace [A]: [M-H]- 465; Peak A1: (+)-catechin glucuronide (R_t min 16.4); Peak A2: (+)-epicatechin glucuronide (R_t min 20.5); Trace [B]: [M-H]- 577;Peak B1: dimer B_3 (R_t min 17.3); Peak B2: dimer B_1 (R_t min 18.7); Peak B3: dimer B_4 (R_t min 24.1); Peak B4: dimer B_2 (R_t min 26); Trace [C] : [M-H]- 369; Peak C1: (+)-catechin sulfate (R_t min 31.3); Peak C2: (-)- epicatechin sulfate (R_t min 55.4); Trace [D] : [M-H]- 479; Peak D1: methyl catechin glucuronide (R_t min 27.9); Peak D2: methyl epicatechin glucuronide (R_t min 31).

Figure 7: Identification of flavanols and metabolites in rat plasma 4 h after ingestion of a grape seed extract with detection by selective reaction monitoring (SRM).

Besides antiplatelet aggregation and vasodilation, anti-inflammatory activities of wine flavonoids were reported. The flavonoids of the wine could decrease the thrombotic tendency by reducing the rate of hydroperoxided lipids and by inhibiting cyclooxygenases and the formation of the precursors of the Thromboxane TXA2. This action could be similar to the beneficial effect of the aspirin which, in low doses, inhibits the formation of the TXA2 and reduces the thrombotic tendency.

Generally, the protective effect of wine consumption on the inhibition of platelet aggregation is similar to that following the ingestion of an equivalent amount of ethanol. However, the absorption of red wine after a period of weaning of eighteen hours is not associated with an aggregation platelet rebound as it is observed with alcohol alone [31]. The inhibitive effect of the red wine on the aggregation of platlets can be completely reproduced by adding some alcohol to tannins extracted from grapes. The authors suggest that tannins would be absorbed only in the presence of alcohol.

A vasodilatation action was also described for red wine [32]. Besides the observed reduction of TXA2 (a vasoconstrictor) formation by some flavonoids, wine may stimulate the synthesis of the radical nitric oxide (NO) in the endothelial cells. One of the *in vivo* functions of this radical is to relax the smooth muscular cells of the blood vessel walls. The wine could thus decrease the vasoconstriction of arteries and reduce the thrombotic tendency.Vasodilator effects of phenolic extracts of wine and anthocynanin compounds, possibly implying the synthesis of NO, were observed *in vitro* [33]. In a test concerning 16 red wines, the total contents in phenols are very strongly correlated with antioxidant activity and vasodilatation capacity, but only the total of anthocyanins is correlated with vasodilatation capacity [34]. There are numerous pharmacological studies demonstrating a preventive effect of wine phenolic compounds on vascular infringements in experimental models of cardiovascular diseases [35-39]. Besides, the quasi-totality of the *in vivo* studies was carried out on total phenolic extracts, which makes it difficult to determine the relative role of the various families of compounds. The isolated phenolic compounds of grape seeds which show the greatest efficiency for the endothelium-dependent vasorelaxation are proanthocyanidins trimers, tetramers, pentamers and polymers as well as their gallates and dimer gallates [40]. An effect of wine on the synthesis of Endotheline 1 was studied. Endotheline-1 (ET-1) is a vasoconstrictor peptide, whose production is a key factor in the development of vascular diseases and atherosclerosis. In a recent work [41] it was shown that due to their polyphenols, read wines decrease the synthesis of ET-1 by deleting the transcription of the gene ET-1 (inhibition until 75 %). On the other hand, white wines have no effect on the synthesis of ET-1 (inhibition < to 5 %). Finally, it should be noted that an anti-inflammatory activity through the flavonoids of the wine would be possible by inhibiting the activity of lipooxygenases in platelets, leukotriens and macrophages.

The antioxidant phenolic compounds contained in wines could also play a role in the prevention of cancers and degenerative diseases through several mechanisms. As an example, a study [42] was conducted using carrier transgenic mice of a transactivating gene of an inductive human virus of cancerous tumours. These mice, predisposed to develop girdles of tumours (at the nerve level) similar to those that occur in the case of human neurofibromatosis, were subjected to two types of daily food rations with aminoacids, one of which was supplemented with extracts of red wine. The first tumour appeared in the transgenic mice fed with the normal diet after 55 days, but not until 74 days for those fed with the diet supplemented with red wine extracts. These results suggest that the phenolic compounds of wine could play a protective role in carcinogenesis by delaying the appearance of cancerous tumors. Extracts of white wines have since been tested on lineages of HCT-15 cells derived from human colon cancer as well as AGS cells derived from human gastric cancer. The extracts of white and red wines allow for the abolition of cell growth. However, it was noted that the anthocyanin fractions of red wine show the highest rates of deletion [43].

At the epidemiological level, there are two studies [44, 45] indicating that a moderate consumption of wine is associated with a reduction of the risk of cancer appearance by 15% for 2 glasses/day. The administration of some stilbenes such as resveratrol to rats developing hepatic tumours has been studied. A significantdecrease in cancer cells by 25% has been observed. Resveratrol would induce apoptosis of the tumoural cells [46]. This molecule therefore possesses an interesting potential in the treatment of certain tumours [47]. Resveratrol, *in vitro*, causes the phosphorylation of enzymes (extracellular signal-regulated kinases) ERK1 and ERK2 of the family of the MAP kinases, which have a role in the regulation of gene expression in cellular transduction but also in the synaptic plasticity of the neurones [48]. Resveratrol possesses an interesting potential in the treatment of neurological disorders, but the mode of action is still not totally clarified [49]. It is worth noting that polydatin, a glycoside of resveratrol which is present in grape juice and wine, has been shown to be rapidly metabolised to resveratrol in the small intestines and livers of rats and then further metabolised to the glucuronidated resveratrol [50].

PAQUID is an epidemiological study, whose general objective is to analyse the intellectual and functional ageing after 65 in order to distinguish its normal and pathological modalities and to identify the subjects with high risk of physical or intellectual deterioration, where preventive action would be possible. This objective was fulfiled by means of the implementation of a cohort of 4,134 senior citizens in Gironde and in Dordogne that were followed since 1988 until at least 2003. The results obtained with alcohol consumption are quite surprising. Alcohol is considered as a major toxin at the intellectual level, responsible for chronic encephalopathies caused by direct toxic effect or vitamin deficiency. In the region of Bordeaux, wine is the only alcoholic drink consumed regularly by the subjects of more than 65 years of age. Other drinks (beer, apéritifs, liquors, *etc.*) are consumed by less than 2.5% of the subjects. Initially there was evidence of a significant association between consumption of wine and cognitive performances, with moderate consumers (between 0.25 and 0.5 litres of wine a day) having the best performances. However, this association was no longer significant after correction of the results according to the social and occupational group [51]. This relation also included the the mentally ill subjects of the cohort. A moderate initial consumption of wine is associated with a 5-time lower risk of incidental dementia within three years, which remains practically unaltered after adjustment of the social and occupational group, as well as the initial cognitive performances or the social activities. The same relation is observed for Alzheimer's disease [52, 53]. These unexpected results appeared rather convincing to justify follow-up studies to look for the mechanism of action of a possible protective effect of wine (antioxidant effect or action on the metabolism of lipids). During the three-year follow-up, the subjects of Gironde answered a questionnaire of frequency of consumption of the main classes of food, while in Dordogne a nutritional detailed survey (PAQUINUT) was conducted by a dietitian with 169 subjects. A study of the relation between consumption of flavonoids contained in the food and the risk of dementia was performed on 1,300 subjects that were followed over a period of 5 years. The quantity of ingested flavonoids was estimated on the basis of a nutritional questionnaire and the subjects were classified in 3 groups according to their consumption of flavonoids. The risk of incidental dementia was significantly decreased (RR=0.55, $p < 0.05$) for the subjects with the highest consumption of flavonoids compared with the two other groups. The relation remains stable after the consideration of gender, level of studies, weight and consumption of vitamin C [54]. All in all, this study showed that the contribution of flavonoids was conversely proportional to the risk of dementia. It consolidates the hypothesis of a connection between oxidative stress and the risk of dementia. On the other hand, it strengthens the idea that the protective effect of wine could partially be due to the flavonoids it contains.

More recently, it has been demonstrated that resveratrol activates sirtuin, a protein deacetylase promoting cell survival in *Saccharomyces cerevisiae*, increasing its lifespan by about 80% [55]. In humans, resveratrol is metabolised by the liver to produce metabolites of glucuronide and sulfate of resveratrol. Today it is thus necessary to estimate the antioxidant and antiproliferative effects of these metabolites.

EFFECTS OF WINE POLYPHENOLS, ALCOHOL OR THEIR COMBINATION IN CHRONIC PATHOLOGIES

Atherosclerosis

We studied [56] the effect of phenolic compounds on the prevention of atherosclerosis of nutritional origin on the Golden Syrian Hamster model. This model is fed with a lipid diet rich in cholesterol and saturated fatty acids. The chronic consumption (8 weeks of treatment) of nutritional doses of phenolic compounds is based on the consumption of two glasses of red wine per meal for a 70 kg man, *i.e.* 500 mL a day providing 1g of total phenolic compounds. A phenolic extract of red wine was tested in aqueous solution and in hydro-alcoholic solution, using ethanol alone as comparison. At the end of the treatments, the cholesterol level was significantly reduced for animals receiving the wine extract dissolved in water or in 12 % ethanol, when compared to the controls. The envelope of lipoproteins is established by phospholipids and by apolipoproteins A-1 for the HDL and B for the LDL (when the ratio Apo A-1/Apo B increases it is favourable to a decrease in the risk of progression of atherosclerosis). The extract decreases the rate of apolipoprotein B by 40% and 32%, respectively, in solution in water or 12% ethanol. The ratio Apo A-1/Apo B increases by 45% with ethanol alone, 61% with the extract dissolved in ethanol, and by 80% with extract dissolved in water. There is no effect of the extract on the plasmatic rates of Apo A-1, triglycerides, iron, copper or cholesterol bound to the HDL. Concerning the hepatic antioxidant enzymes, the extract dissolved in water significantly increases the specific activities of SeGSHPx or the glutathion peroxidase enzyme (67%) and catalase (73%), and decreases the activity of Superoxide dismutase (32%). The extract dissolved in ethanol and the 12% ethanol only significantly increase the activity of SeGSHPx by 27% and 28%, respectively. But the result which we deem most interesting is the significant decrease of the surface of the aortic fatty streak area occupied by lipid deposits in animals receiving the extract dissolved in water (29%), ethanol alone (60%) and the extract dissolved in ethanol 12% (73%), suggesting an additive preventive effect of both fractions (ethanol and polyphenols) of some red wines (Fig. **8**). Also, the antioxidant capacity of plasma is significantly increased by the extract dissolved in water (18%), ethanol alone (24 %), and the extract in ethanol (35%).

Décordé *et al.* [57] were able to demonstrate that the phenolic compounds of red grapes prevented the development of atherosclerosis early induced by food atherogenic model in the Syrian golden hamsters. In this work, consumption of grapes (cv. Hambourg Muscat) at a dose equivalent to 600 grams/day for a man of 70kg, or their juice at a dose equivalent to 500mL/jour for a man of 70 kg, led to a reduction of the surface of fat deposition in the aortic arch by 78% and 93%, respectively, compared to control animals (atherogenic food only). Likewise, with the consumption of grapes or grape juice, total cholesterol decreased by 30.4% and 34.6% and non-HDL cholesterol (High Density Lipoproteins) decreased by 64% and 58.9%. In parallel, levels of antioxidant capacity of blood plasma of animals receiving the grapes or grape juice increased by 41% and 61% and hypolipidemic activity was found. Recently, red wines prepared from different types of grapes (Grenache, Syrah) and processing were tested in hamsters receiving an atherogenic diet for 12 weeks and a daily gavage with 12% ethanol, red wines from vinification by flash release, by tanisage, by traditional vinification, or water as control. Consumption of wines lowered plasma total and LDL-cholesterol. Aortic lipid deposits were reduced by ethanol (30%), wines (54% on average) or wines obtained by tanisage (65%). Cardiac production of superoxide anion decreased from 20% to 33% (flash release wines). The expression of NAD(P)H oxidase decreased by 44% for flash release wine and tanisage wine, 26% for traditional wine. Wines increased uricemia by 15% on average. These findings indicate that moderate and chronic consumption of red wine has potential beneficial effects to prevent the development of atherosclerosis. Prevention of NAD(P)H oxidase induction and preservation of aortic lipids oxidation likely contribute to this effect [58].

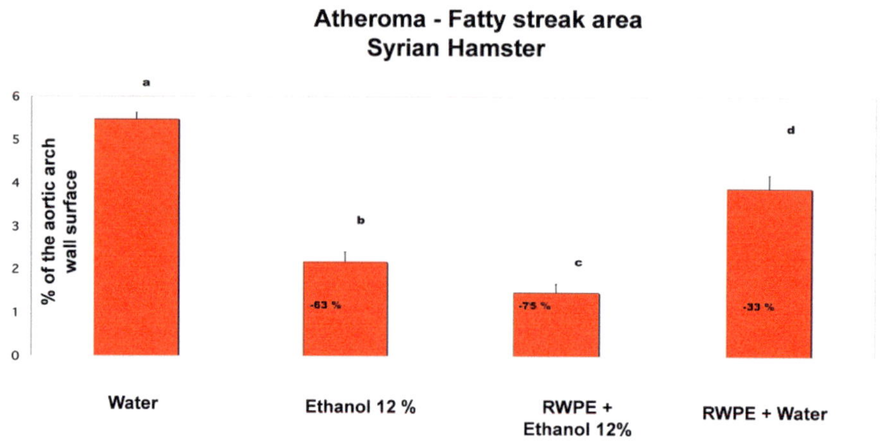

Figure 8: Atherosclerosis prevention by polyphenols of red wine and ethanol [56].

Moreover, the effects of acute, food-induced moderate increase of plasma uric acid (UA) on arterial stiffness and markers of oxidative damage in plasma have been explored in healthy males exposed to 100% normobaric oxygen in a cross-over design over the period of 4 weeks for 10 subjects. Acute elevation of plasma UA was induced by the consumption of red wine, as well as the combination of ethanol and glycerol. Before exposure to hyperoxia, plasma antioxidant capacity and plasma UA increased. Significant increase of lipid hydroperoxides which occurred during 30 min of hyperoxia in the water group was largely prevented in the groups that consumed red wine and glycerol + ethanol. In contrast to chronic hyperuricemia, which is generally considered as a risk factor for cardiovascular diseases and metabolic syndrome, acute increase of UA acts protectively against hyperoxia-induced oxidative stress and related increase of arterial stiffness in large peripheral arteries [59]. Although chronic hyperuricemia is associated with gout [60] and ethanol intake is a well-established risk factor [61], recent epidemiological studies have shown that moderate consumption of wine is not associated with higher incidence of gout, in contrast to the consumption of beer and spirits [62, 63]. Taken together, it can be concluded that acute plasma urate increase after wine consumption is not likely to cause detrimental effects to human health associated with chronic hyperuricemia. Quite the opposite, if wine is consumed with meals, the timed elevation of plasma uric acid may significantly contribute to the wine's protective effects against postprandial oxidative stress [13]. Indeed, in the most recent review by Covas *et al.* [64], who analysed the studies on moderate wine consumption and oxidative damage in humans, protective effects of wine were most convincing during the meal-related postprandial oxidative stress. In line with this is the finding of an epidemiological study that showed that cardioprotective effects of wine intake were observed only in participants who consumed wine with meals, but not in those who consumed wine separately from the food intake [65].

Diabetes

There are two types of diabetes: Type 1 diabetes (insulin-dependent), which represents 20% of the diabetic population; and Type 2 diabetes (non-insulin dependent), representing 80% of diabetics. Type 1 diabetes often appears in childhood, more rarely in adulthood, and Its cause is probably genetic. Type 1 diabetes is characterised by the incapacity of the pancreas to secrete insulin. Consequently, even after a meal, there is little or no insulin in the blood, thus causing an increase in glycemia. Injections of insulin must be made periodically to maintain normal levels of glycemia. Contrary to type 1 diabetes, in type 2 diabetes it is not the absence of insulin that is responsible for the increase in blood glucose but rather the resistance of cells to insulin. Consequently, glucose stays in the blood, increasing the levels of glycemia. Type 2 diabetes can be more or less serious. It does not generally require the injection of insulin but rather control of the food, weight loss as well as exercise, so that the levels of glycemia return to normal.

This disease affects more particularly the adults of more than 35 years of age. Furthermore, it is now known that obesity plays a role in the development of type 2 diabetes, because it creates the resistance of cells to insulin. In agreement with the criteria of the American Diabetes Association of 1997, Type 2 diabetes is diagnosed when the fasting blood glucose level is superior to 1.26 g/L. An excellent diabetic model is the one of the streptozotocine rat (model of diabetes with and without insulin dependence). This antibiotic was injected intravenously (i.v.) in a dose of 60 mg/kg. It destroyed selectively most of the ß cells (90 %) of langheran islands of the pancreas. Thus, the state of chronic diabetes is justified by a constant hyperglycemia in the order of 3-4 g/liter of blood (or 16.5-22 mmol/L), with insulin-resistance and glucose intolerance appearing with a glycosuria without ketonuria. Animals present the usual complications associated with diabetes as well as a major oxidative stress. Both Type 1 and 2 diabetes are associated with the existence of a major oxidative stress partially due to the state of chronic hyperglycemia, which is in connection with the neurological, retinal and renal complications following the structural changes and hemodynamics of the small blood vessels. Oxidative stress is also a factor for insulin resistance. Indeed, oxidative stress inhibits the transduction of the insulinic signal of cell lines by micromolar concentrations of free radicals (as hydrogen peroxide): real mediators of the insulin resistant effect.

We were able to elaborate a white Chardonnay wine "with a red winemaking process", which allowed us to obtain an important increase in the rate of total phenolic compounds (4.26 times: 1,346 mg of total polyphenols in gallic acid equivalent) and of catechins in regard to traditional winemaking, and to appreciate its antioxidant activity: (13.8 mmol/L) comparable to that of certain red wines [66]. The effect of the consumption of white Chardonnay wines enriched in phenolic compounds, complete or whithout alcohol on diabetic rats (streptozotocine model) deficient in insulin was estimated. Treatments were made by force-feeding during 6 weeks with 4.3 mL/kg administration, twice a day and allowing for the restoration of the antioxidant capacities of the plasma at a level not significantly different from that of the non-diabetic animals [67]. In particular, we shall note that the antioxidant capacity of the plasma of diabetic animals treated with Chardonnay wine (complete or without alcohol) enriched with polyphenols increased by 20.7% (Fig. **9**).

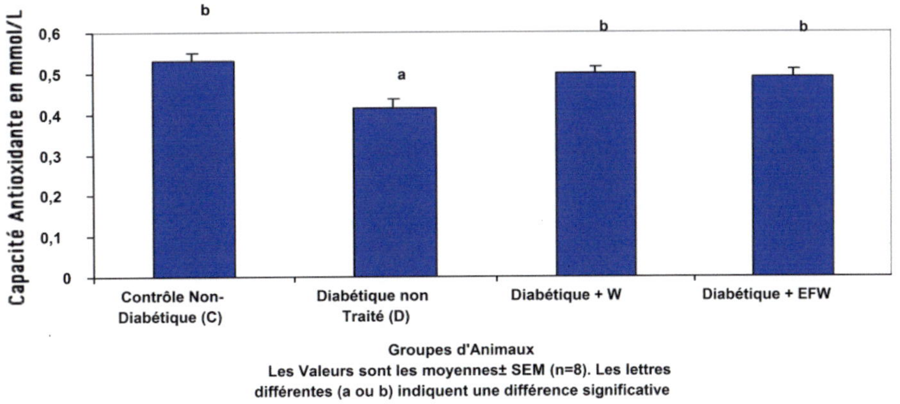

Figure 9: Effects of treatment with white wine enriched in phenolics (W) or de-alcoholised white wine enriched in phenolics (EFW) on Plasma Antioxidant Capacity of diabetics rats [67].

The natural enrichment in phenolic compounds of a white wine allows for the restoration of the antioxidant capacity of plasma when it is administered *in vivo* and suggests an *in vivo* improvement of the defences against the oxidative stress of diabetes.

These results led us to pursue our work on the streptozotocine model rat with an extract of polyphenols of red wine and ethanol, which we used alone and in combination with moderate doses [68]. Diabetes was induced by a unique injection of streptozotocine (60 mg/kg, I.V.) and the treated animals received a daily administration for 4 weeks of red wine polyphenol extract (200 mg/kg), of ethanol with a nutritional dose (1mL/kg: corresponding in 500 mL of wine to 13% vol. for an adult of 70 kg), an administration of their association, or some water for controls. In parallel, the same administration doses were performed on non-diabetic animals. The results show that the extract of polyphenols of red wine, ethanol, or their association, possessed anti-diabetic properties. The administration of the polyphenol extract was associated with a progressive decrease in glycemia, which is probably linked to a decrease in the absorption of carbohydrates or taking of food and that. The anti-diabetic activity of polyphenols appears to be directly linked to this activity in this model. The administration of ethanol in moderate doses also allows for the correction of the diabetic condition. It is likely that this effect of ethanol, which becomes noticeable only in moderate doses, is bound to its insulin-sensitising.

In every case, glycemia in animals was reduced by approximately 50% (Fig. **10**). The rates of insulin normalised with the ethanol treatment or in particular when polyphenols and ethanol are associated. It is likely that this effect of ethanol, which becomes visible only in moderate doses, generates an increase in glucose captation with a reduction of the blood rates in glucose and insulin following ingestion. Recently, the effects of a phenolic grape seed extract (GSE) on obesity and oxidative stress in hamsters receiving a high-fat diet (HF) has been investigated. Hamsters received a HF diet plus a daily gavage with water (Control, HF) or a solution of GSE (HF+GSE) for 12 weeks. HF diet increased abdominal fat. GSE avoided this feature. HF diet led to higher plasma glucose, triglycerides, insulin and insulin resistance values. GSE partially prevented these effects, reducing insulinemia and leptinemia by 16.5% and 45% respectively, whereas adiponectin level increased by 61% compared with obese controls. GSE lowered glycemia and insulin resistance and strongly prevented cardiac NAD(P)H oxidase activity (74%) and expression (30%). For the first time the chronic consumption of grape seed tannins was found to have potential effects against the development of obesity [69].

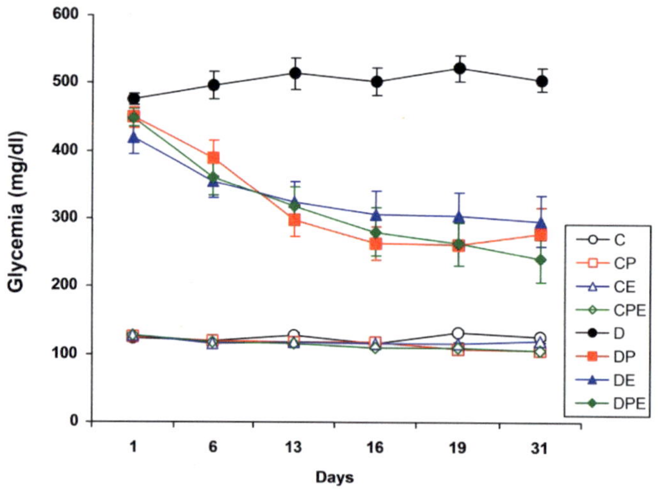

Figure 10: Glycemia evolution for animal models (diabetic and non-diabetic rats) treated with a polyphenol red wine extract (RWPE), ethanol in moderate doses, or their association [68].

High Arterial Blood Pressure

The World Health Organisation has established that a subject is hypertensive when, at rest, his/her diastolic pressure is superior to 95 mm Hg and his/her systolic pressure is superior to 160 mm Hg. There are two other types of high blood pressure: essential high blood pressure, which represents 90% of the cases and has no detectable causes; and secondary high blood pressure, which represents 10% of the cases and has a precise and recognisable cause, mostly arteriosclerosis. 90% of the causes of high blood pressure are unknown. The rat fed with a diet enriched in fructose is an experimental model of insulin resistance reproducing most of the characteristics of the syndrome X or metabolic syndrome: glucose intolerance, visceral obesity, high blood pressure, insulin resistance, hyperinsulinemia and dyslipidemia. Besides, this model is associated with the existence of a significant oxidative stress as well as a cardiac hypertrophy, in particular of the left ventricle. Our last works on the fructose-fed rat, which becomes hypertensive at the end of three weeks, show that after six weeks of red wine polyphenol extract consumption (P) used in a "nutritional" dose of 100mg/kg, ethanol in 10% vol. (E) with a moderate consumption dose (corresponding to 500 mL of wine in 10% vol.), or their association, the systolic pressure reaches normal levels. We evaluated the weight of cardiac mass which corresponds to the cardiac hypertrophy provoked by the fructose-rich diet. The cardiac hyperthrophy decreases appreciably with the consumption of polyphenols or ethanol. However, the best results were obtained when polyphenols of red wine and the ethanol are associated (FPE), in which case the cardiac hypertrophy no longer exists.

Additionally (Fig. **11**), the oxidative stress was measured (production of anion superoxide by the heart at the level of the left ventricle and of the thoracic aorta) in association with the model of resistance in the insulin induced by fructose. If a hyperproduction of anion superoxide tissular appears in the case of animals subjected to the fructose diet (F), we can observe that the polyphenol extract administered significantly decreases the production of reactive oxygen species by the heart and the aorta. The best effects appear with the association of ethanol and the polyphenols (FPE) in moderate doses with a total normalisation of the heart in particular, in comparison to the control group (C) [70]. These effects are associated with a decrease in gene expression of the NADPH oxidase enzyme membrane (generating the production of the anion superoxide) with the treatments of animals using wine polyphenols [71].

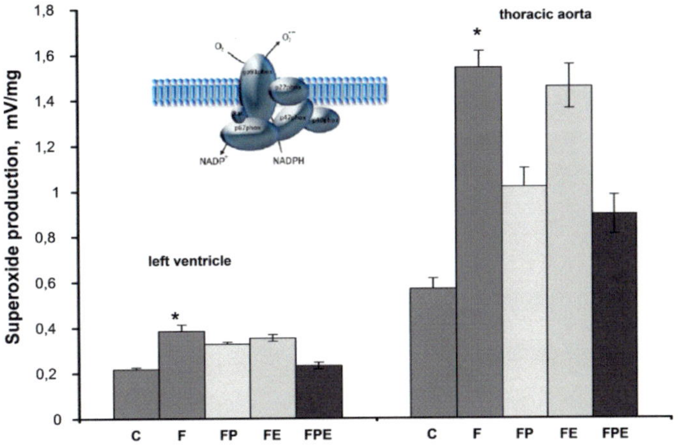

Figure 11: Hypertension prevention by wine polyphenols, ethanol or their combination, anion superoxide production in function of treatment [70].

CONCLUSIONS

If moderate and regular wine consumers possess more chances to live longer that excessive drinkers or those who never drink and to see decreased the risks of appearance of certain chronic pathologies such as atherosclerosis, diabetes or high blood pressure, then it is important to define the optimised conditions for wine consumption in particular when the wine is integrated in the diet. The crucial consideration to avoid any abuse is that the effects of the alcohol depend on the dose, type and time of administration. The controlled use of wine in nutrition is not the abuse. The key for a moderate consumption of wine, beneficial and without risks must be determined in every individual case, and must also take into consideration the context, the environment and the circumstances of this consumption.

Wine can thus play a role in preventive nutrition when it is consumed regularly, with moderation and integrated into the diet. The phenolic compounds of the grape and wine possess unmistakably therapeutic properties, in particular for certain chronic pathologies such as atherosclerosis, diabetes, high blood pressure and certain cancers. Among the mechanisms of action of the phenolic compounds implied in the prevention of chronic pathologies, we have to retain the following possibilities:

- A direct trap effect on free radicals,
- An economy effect of endogenous antioxidants (vitamin E, vitamin C, ß-carotene),
- An economy effect of antioxidant enzymes (SOD: Superoxide Dismutase, Catalase, SeGSHPx: Glutathion peroxydase),
- A decreased effect on the cholesterol levels and equilibrium of the blood lipids (HDL / LDL),
- A chelation effect on oxidation cofactors of fatty acids as certain metals (Fe^{2+}, Cu^{2+}),
- An inhibition effect on oxidative enzymes such as cyclooxygenases and lipooxygenases,
- An effect on the endothelial NO synthesis: at the cellular level of the arterial wall conducting to a vasorelaxation and an hyperpolarisation of the membrane by extracellular potassium release,
- An inhibition effect on the genesis of the NAPH oxidase production at the level of the cells of the vascular wall (thoracic aorta and heart) leading to a decrease in free radical production.

These antioxidant phenolic compounds give a real specificity to the wine, and research into the effects and mechanisms of action of wine compounds on the chronic pathologies have to continue. It is thus advisable to hold a balanced speech towards the consumption of wine. If we have to warn the population against the risks bound to abuse, then all individuals legally have the right to be informed about the potentially beneficial effects to health of a moderate, controled and regular wine consumption, thus allowing them to make an informed decision on whether or not they wish to consume it.

ABBREVIATIONS

HDL = High Density Lipoproteins

LDL = Low Density Lipoproteins

NO = Nitric oxide

ET-1 = Endotheline-1

ERK = Extracellular signal-regulated kinase

UA = Uric acid

GSE = Grape seed extract

HF = High-fat (diet)

REFERENCES

[1] Renaud S, Lorgeril M De. Wine, Alcohol, plateted and french paradox for coronary heart disease. Lancet 1992; 339: 1523-6.

[2] Klatsky AL, Armstrong MA. Alcoholic beverage choice and risk of coronary artery disease mortality: do red wine drinkers fare best? Am J Epidemiol 1993; 71: 467-9.

[3] Groenbaek M, Deis A, Sorensen Tia, Becker U, Schnohr P, Jensen G. Mortality associated with moderate intakes of wine, beer, or spirits. Br Med J 1995; 310: 1165-9.

[4] Carando S, Teissedre P-L, Ferriere M, Descomps B, Cabanis JC. Cahiers de Nutrition et de Diététiques. 1998; 33(3): 182-7.

[5] Rimm EB, Giovannucci EL, Willet WC, *et al.* Prospective study of alcohol consumption and risk of coronary disease in men. Lancet 1991; 338: 464-8.

[6] Stampfer MJ, Colditz GA, Willett WC, Hennekens CH. A prospective study of moderate alcohol consumption and the risk of coronary heart disease and stroke in women. N Engl J Med 1988; 319: 267-73.

[7] Friedman LA, Kimball AW. Coronary heart disease and alcohol consumption in Framingham. Am J Epidemiol 1986; 124(3): 481-9.

[8] Masquelier J. Les constituants du vin présentant une action hypocholestérolémiante. C. R. Congrès Médical International pour l'Étude Scientifique du Vin et du Raisin, Bordeaux 1961.

[9] Frankel EN, Kanner J, German JB, Parks E, Kinsella JE. Inhibition of oxidation of human low-density lipoprotein by phenolic substances in red wine. Lancet 1993; 341: 454-7.

[10] Frankel EN, Waterhouse AL, Teissedre PL. Principal phenolic phytochemicals in selected California wines and their antioxidant activity in inhibiting oxidation of human low-density-lipoproteins. Journal of Agricultural and Food Chemistry 1995; 43: 890-4.

[11] Teissedre PL, Frankel EN, Waterhouse A, Peleg H, German JB. Inhibition of *in vitro* human LDL oxidation by phenolic antioxidants from grapes and wines. Journal of the Science of Food and Agriculture 1996; 70: 55-61.

[12] Waterhouse AL, German JB, Frankel EN, Walzem RL, Teissedre PL, Folts J. The phenolic phytochemicals in wine fruit and tea: Potential nutritional effects and dietary levels. In: Finley JW, Armstrong DJ, Nagy S, Robinson S, Eds. Hypernutritious Foods. Agscience, Auburndale FL, USA 1996; pp. 219-38.

[13] Fuhrman B, Lavy A, Aviram M. Consumption of red wine with meals reduces the susceptibility to human plasma and low density lipoprotein to lipid peroxydation. Am J Clin Nutr 1995; 61: 549-54.

[14] Teissedre PL, Landrault N. Wine phenolics: contribution to dietary intake and bioavailability. Food Research International 2000; 33: 461-7.

[15] Ruidavets JB, Teissedre PL, Carando S, Ferrieres M, Cabanis JC. Catechin in the mediteranean diet : vegetable, fruit or wine? Atherosclerosis 2000; 153(1): 107-17.

[16] Hollman PCH, Katan MB. Absorption, metabolism, and bioavailability of flavonoids. In: C.A. Rice Evans CA, Packer L, Eds. Flavonoids in Health and Disease. Marcel Dekker Inc., New York, Basel, Hong-Kong 1998; pp. 483-522.

[17] Hollman PCH, Katan MB. Dietary flavonoids: intake, health effects and bioavailability. Food Chem Toxicol 1999; 37: 937-42.

[18] Donovan LJ, Bell JR, Kasim-Karakas SE, *et al.* Catechin is present as metabolites in human plasma after consumption of red wine. J Nutr 1999; 129: 1662-8.

[19] Bell JR, Donovan LJ, Wong R, Waterhouse A, *et al.* (+)-Catechin in human plasma after ingestion of a single serving of reconstitued red wine. Am J Clin Nutr 2000; 71: 103-8.

[20] Koga T, Moro K, Nakamori K, *et al.* Increase of antioxidative potential of rat plasma by oral administration of proanthocyanidin-rich extract from grape seeds. J Agric Food Chem 1999; 47: 1892-7.

[21] Tsuda T, Horio F, Osawa T. Absorption and metabolism of cyanidin 3-*O*-ß-D-glucoside in rats. FEBS Letters 1999; 449: 179-82.

[22] Donovan JL, Luthria DL, Stremple P, Waterhouse AL. Analysis of (+)-catechin, (-)-epicatechin, and their 3'- and 4'-O-methylated analogs. A comparison of sensitive methods. J Chrom B Biomed Sci Appl 1999; 726 (1-2): 277-83.

[23] Piskula MK, Terao J. Accumulation of (-)-Epicatechin metabolites in rat plasma after oral administration and distribution of conjugation enzymes in rat tissues. J Nutr 1998; 128: 1172-8.

[24] Mojarrabi B, Mackenzi PI. Characterization of two UDP glucuronosyltransferases that are predominantly expressed in human colon. Biochem Biophys Res Commun 1998; 247: 704-9.

[25] Donovan JL, Manach C, Rios L, Morand C, Scalbert A, Remesy C. Procyanidins are not bioavailable in rats fed a single meal containing a grape seed extract or the procyanidin dimer B3. Br J Nutr 2002; 87: 299-306.

[26] Baba S, Osakabe N, Natsume M, Terao J. Absorption and urinary excretion of procyanidin B2 [epicatechin-(4ß-8)-epicatechin] in rats. Free Rad Biol Med 2002; 33: 142-8.

[27] Gonthier M-P, Donovan JL, Texier O, Felgines C, Remezy C, Sclabert A. Metabolism of dietary procyanidins in rats. Free Rad Biol Med 2003; 35(8): 837-44.

[28] Sarah C. Forester and Andrew L. Waterhouse, (2009) Metabolites Are Key to Understanding Health Effects of Wine Polyphenolics, The Journal of Nutrition, Supplement: Grapes and Health, 138: 1824S–31S.

[29] Tsang C, Auger C, Mullen B, *et al.* The absorption, metabolism and excretion of flavanols and procyanidins, following the ingestion of a grape seed extract by rats. British Journal of Nutrition 2005; 94: 170-81.

[30] Tebib K, Rouanet JM, Besançon P. Antioxidant effects of dietary polymeric grape seed tannins in tissues of rats fed a high cholesterol-vitamin E-deficient diet. Food Chem 1997; 59: 135-41.

[31] Ruf JC, Berger JL, Renaud S. Platelet rebound effect of alcohol withdrawal and wine drinking in rats. Arteriosclerosis, Trombosis and Vascular Biology 1995; 15(1): 140-4.

[32] Fitzpatrick DF, Hirschfield SL, Coffey RG. Endothelium-dependent vasorelaxing activity of wine and other grape products. Am J Physiol 1993; 265: H774-8.

[33] Stoclet JC, Kleschyov A, Andriambeloson E, Diebolt M, Andriantsitohaina R. Endothelial NO release caused by red wine polyphenols. J Physiol Pharmacol 1999 Déc; 50(4): 535-40.

[34] Burns J, Gardner PT, O'neil J, *et al.* Relationship among antioxidant activity, vasodilation capacity, and phenolic content of red wines. J Agric Food Chem 2000; 48: 220-30.

[35] Xu R, Yokohama WH, Irving D, Rein D, Walzem RL, German JB. Effect of dietary catechin and vitamin E on aortic fatty streak accumulation in hypercholesterolemic hamsters. Atherosclerosis 1998; 137: 29-36.

[36] Da Luz Protasio L, Serrano CV Jr, Charca AP, Monteiro HP, Yoshida VM, Furtado M. The effect of red wine on experimental atherosclerosis: lipid-independant protection. Experimental and Molecular Pathology 1999; 65(3): 150-9.

[37] Munday JS, Thompson KG, James KAC, Mankelow BW. The effect of moderate alcohol consumption as either red or white wine in the C57BL/6 mouse atherosclerosis model. Coronary Artery Disease 1999; 10: 97-102.

[38] Wollny T, Aiello L, Di Tommaso D, Bellavia V, Rotilio D, Donati MB. Modulation of haemostatic function and prevention of experimental thrombosis by red wine in rats : a role for increased nitric oxide production. British Journal of Pharmacology 1999; 127: 747-55.

[39] Mizutani K, Ikeda K, Kawai Y, Yamori Y. Extract of wine phenolics improves aortic biomechanical properties in stroke-prone spontaneously hypertensive rats (SHRSP). Journal of Nutritional Science and Vitaminology 1999; 45: 95-106.

[40] Fitzpatrick DF, Fleming RC, Bing B, Maggi DA, O'malley RM. Isolation and characterization of endothelium-dependent vasorelaxing compounds from grape seeds. J Agric Food Chem 2000; 48: 6384-90.

[41] Corder R, Douthwaite JA, Lees D, *et al.* Endothelin-1 synthesis reduced by red wine. Nature 2001; 414: 863-4.

[42] Clifford AJ, Powers TJ, Ebeler SE, *et al.* American Journal Clinical Nutrition, Delayed tumors onset in transgenic mice fed an amino acid diet supplemented with red wines solids 1996; 64: 748-56.

[43] Kamei H, Hashimoto K, Koïde T, Kojima T, Hasegawa M. Anti-tumor effect of methanol extracts from red and white wines. Cancer Biother Radiopharm 1998; 13(6): 447-52.

[44] Renaud SC, Guegen R, Schenker J, D'houtaud A. Epidemiology, Alcohol and mortality in middle-aged men from eastern France. Mar 1998; 9(2): 184-8.

[45] Gronbaek M, Becker U, Johansen D, Tonnesen H, Jensen G, Sorensen TI. Population based cohort study of the association between alcohol intake and cancer of the upper digestive tract. British Medical Journal 1998 Sep 26; 317 (7162): 844.

[46] Weng CJ, Yang YT, Ho CT, Yen GC. Mechanisms of Apoptotic Effects Induced by Resveratrol, Dibenzoylmethane, and Their Analogues on Human Lung Carcinoma Cells. J Agric Food Chem. 2009; 57 (12): 5235-43.

[47] Carbó N, Costelli P, Baccino FM, Lopez-Soriano FJ, Argilés JM. Resveratrol, a natural product present in wine, decreases tumour growth in a rat tumour model. Biochem. Biophys Res Comm 1998; 254: 739-43.

[48] Miloso M, Bertelli AA, Nicolini G, Tredici G. Resveratrol-induced activation of the mitogen-activated protein kinases, ERK1 and ERK2, in human neuroblastoma SH-SY5Y cells. Neurosci Lett 1999 Apr 2; 264(1-3): 141-4.

[49] Bishayee A, Politis T, Darvesh AS. Resveratrol in the chemoprevention and treatment of hepatocellular carcinoma. Cancer Treatment Reviews 2010; 36(1): 43-53.

[50] Zhou S, Yang R, Teng Z, *et al.* Dose-Dependent Absorption and Metabolism of trans-Polydatin in rats. J Agri Food Chem 2009; 57(11): 4572-9.

[51] Letenneur L, Dartigues JF, Orgogozo JM. Wine consumption in the elderly. Ann Intern Med 1993 Feb 15; 118(4): 317-8.

[52] Lemeshow S, Letenneur L, Dartigues JF, Lafont S, Orgogozo JM, Commenges D. Illustration of analysis taking into account complex survey considerations: the association between wine consumption and dementia in the PAQUID study. Personnes Ages Quid Am J Epidemiol 1998 Aug 1; 148(3): 298-306.

[53] Orgogozo JM, Dartigues JF, Lafont S, *et al.* Wine consumption and dementia in the elderly: a prospective community study in the Bordeaux area. Rev Neurol (Paris) 1997 Apr; 153(3): 185-92.

[54] Commenges D, Scotet V, Renaud S, Jacqmin-Gadda H, Barberger-Gateau P, Dartigues JF. Intake of flavonoids and risk of dementia. Eur J Epidemiol 2000 Apr; 16(4): 357-63.

[55] Howitz KT, Bitterman KJ, Cohen HY, *et al.* Small molecule activators of sirtuins extend *Saccharomyces cerevisiae* lifespan. Nature 2003 Sep 11; 425(6954): 191-6.

[56] Auger C, Caporiccio B, Landrault, *et al.* Red wine phenolic compounds reduce plasma lipids and apolipoprotein B, and prevent early aortic atherosclerosis in hypercholesterolemic Golden Syrian Hamsters (Mesocricetus auratus). Journal of Nutrition 2002; 132: 1207-13.

[57] Décordé K, Teissèdre P-L, Auger C, Cristol J-P, Rouanet J-M. Phenolics from purple grape, apple, purple grape juice and apple juice prevent early atherosclerosis induced by an atherogenic diet in hamsters. Mol Nutr Food Res 2008; 52(4): 400-7.

[58] Qian Y, Agne A, Chira K, *et al.* A moderate consumption of Côtes du Rhône red wines affects the progression of aortic lesions, and reduces oxidative stress and p22 phoxNADPH oxidase activation in an experimental model of diet-induced atherosclerosis, according to the vinification process. European Food Research and Technology 2009; 229(3): 485-93.

[59] Vukovic J, Modun D, Budimir D, *et al.* Acute, food-induced moderate elevation of plasma uric acid protects against hyperoxia-induced oxidative stress and increase in arterial stiffness in healthy humans. Atherosclerosis 2009; 207(1): 255-60.

[60] Choi HK, Mount DB, Reginato AM. American College of Physicians. American Physiological Society. Pathogenesis of gout. Ann Intern Med. 2005; 143: 499-516.

[61] Yamamoto T, Moriwaki Y, Takahashi S. Effect of ethanol on metabolism of purine bases (hypoxanthine, xanthine, and uric acid). Clin Chim Acta 2005; 356:35-57.

[62] Choi HK, Atkinson K, Karlson EW, Willett W, Curhan G. Alcohol intake and risk of incident gout in men: a prospective study. Lancet 2004; 363: 1277-81.

[63] Choi HK, Curhan G. Gout: epidemiology and lifestyle choices. Curr Opin Rheumatol. 2005; 17: 341-5.

[64] Covas MI, Gambert P, Fitó M, de la Torre R. Wine and oxidative stress: up-to-date evidence of the effects of moderate wine consumption on oxidative damage in humans. Atherosclerosis 2010; 208: 297-304.

[65] Trevisan M, Schisterman E, Mennotti A, Farchi G, Conti S. Risk Factor And Life Expectancy Research Group. Drinking pattern and mortality: the Italian Risk Factor and Life Expectancy pooling project. Ann Epidemiol 2001; 11: 312-9.

[66] Landrault N, Poucheret P, Ravel P, Gasc F, Cros G, Teissedre PL. Antioxidant capacities and phenolics levels of french wines from different varieties and vintages. Journal of Agricultural and Food Chemistry 2001; 49(7): 3341-8.

[67] Landrault N, Poucheret P, Azay J, *et al.* Effect of a polyphenols-enriched chardonnay white wine in diabetic rats. Journal of Agricultural and Food Chemistry 2003; 51(1): 311-8.

[68] Al-Awwadi N, Azay J, Poucheret P, *et al.* Antidiabetic activity of red wine polyphenols, ethanol or both, in streptozotocin-treated rats. J Agric Food Chem 2004; 52: 1008-16.

[69] Décordé K, Teissèdre P-L, Sutra T, Ventura E, Cristol J-P, Rouanet J-M. Chardonnay grape seed procyanidin extract supplementation prevents high-fat diet-induced obesity in hamsters by improving adipokine imbalance and oxidative stress markers. Molecular Nutrition and Food Research 2009; 53(5): 659-66.

[70] Al-Awwadi NA, Bornet A, Azay J, *et al.* Red Wine Polyphenols Alone or in Association with Ethanol Prevent Hypertension, Cardiac Hypertrophy, and Production of Reactive Oxygen Species in the Insulin-Resistant Fructose-Fed Rat. J Agric Food Chem 2004; 52: 5593-7.

[71] Al-Awwadi NA, Araiz C, Bornet A, *et al.* Extracts enriched in different polyphenolic families normalize increased cardiac NADPH oxidase expression while having differential effects on insulin resistance, hypertension and cardiac hypertrophy in high fructose-fed rats. J Agric Food Chem 2005; 53(1): 151-7.

INDEX

A

B

C

D

E

Hernâni Gerós, M. Manuela Chaves and Serge Delrot (Eds)

Made in the USA
Las Vegas, NV
04 November 2026

57438644R00183